Sounds in the Sea
From Ocean Acoustics to Acoustical Oceanography

The oceans are a vast, complex, mostly dark, optically opaque but acoustically transparent world that has been only thinly sampled by today's limited technology and science. Underwater acousticians and acoustical oceanographers use sound as the premier tool to determine the detailed characteristics of physical and biological bodies and processes at sea. Myriad components of the ocean world are being discovered, identified, characterized, and imaged by their interactions with sound.

Sounds in the Sea is a comprehensive and accessible textbook on ocean acoustics and acoustical oceanography. "Ocean acoustics" describes the traditional way in which our knowledge of ocean temperature and salinity allows us to use sound to find fish, submarines, icebergs, and the depth of the ocean. "Acoustical oceanography" interprets the distinctive details of time-varying, sound amplitudes, and phase, over acoustical paths to deduce the physical and biological parameters of the specific ocean through which the sound has traveled.

This is an invaluable textbook for any course in ocean acoustics in the physical and biological ocean sciences, engineering, and physics. It will also serve as a reference for researchers and professionals in ocean acoustics, and an excellent introduction to the topic for scientists from related fields.

Chapters 1 to 9 provide the basic tools of ocean acoustics. The following 15 chapters are written by many of the world's most successful ocean researchers, who use sound in innovative ways to learn about the sea and its contents. These chapters describe modern developments, and are divided into four parts: Studies of the near-surface ocean; Bioacoustical studies; Studies of ocean dynamics; Studies of the ocean bottom.

HERMAN MEDWIN is Emeritus Professor at the Naval Postgraduate School, Monterey, California. He is a Fellow and Past President of the Acoustical Society of America, and has won both the Silver and Gold Medals in Acoustical Oceanography from the Society. He is co-author, with C. S. Clay, of the influential textbooks *Acoustical Oceanography* (1977) and *Fundamentals of Acoustical Oceanography* (1998). He has authored over 100 professional articles in the *Journal of the Acoustical Society of America* and *Journal of Geophysical Research*, and others.

Sounds in the Sea
From Ocean Acoustics to
Acoustical Oceanography

Herman Medwin
Naval Postgraduate School,
Monterey, California

With contributions from

Joseph E. Blue, Leviathan Legacy Inc., Florida
Mike Buckingham, University of California, San Diego
Douglas H. Cato, DSTO, Australia
Ching-Sang Chiu, Naval Postgraduate School, Monterey
Daniela Di Iorio, University of Georgia
Orest Diachok, Naval Research Laboratory
David R. Dowling, University of Michigan
David M. Farmer, University of Rhode Island
Ann E. Gargett, Old Dominion University, Virginia
Edmund R. Gerstein, Leviathan Legacy Inc., Florida
Gary J. Heald, DSTL, UK.
D. Vance Holliday, BAE Systems, San Diego
John K. Horne, University of Washington, Seattle
Josef M. Jech, National Oceanic and Atmospheric Administration
T. G. Leighton, University of Southampton, UK
Rob McCauley, Curtin University, Australia
Nicholas Makris, Massachusetts Institute of Technology
Christopher Miller, Naval Postgraduate School, Monterey
Michael J. Noad, University of Queensland, Australia
Jeffrey A. Nystuen, University of Washington, Seattle
David Palmer, National Oceanic and Atmospheric Administration
Peter Rona, Rutgers University, New Jersey
Heechun Song, University of California, San Diego
Robert Spindel, University of Washington, Seattle
Timothy K. Stanton, Woods Hole Oceanographic Institution

CAMBRIDGE
UNIVERSITY PRESS

University Printing House, Cambridge CB2 8BS, United Kingdom

One Liberty Plaza, 20th Floor, New York, NY 10006, USA

477 Williamstown Road, Port Melbourne, VIC 3207, Australia

4843/24, 2nd Floor, Ansari Road, Daryaganj, Delhi - 110002, India

79 Anson Road, #06-04/06, Singapore 079906

Cambridge University Press is part of the University of Cambridge.

It furthers the University's mission by disseminating knowledge in the pursuit of education, learning and research at the highest international levels of excellence.

www.cambridge.org
Information on this title: www.cambridge.org/9781108448147

First published 2005
First paperback edition 2017

A catalogue record for this publication is available from the British Library

Library of Congress Cataloging in Publication data
Medwin, Herman, 1920–
    Sounds in the Sea : From Ocean Acoustics to Acoustical
Oceanography/Herman Medwin, with contributions from
Joseph E. Blue... [*et al.*].
    p. cm.
    Includes bibliographical references and index.
    ISBN 0 521 82950 X (hardback)
    1. Underwater acoustics.   2. Seawater – Acoustic properties.
3. Oceanography. I. Title.
    QC242.2.M44 2005
    551.46'54 – dc22   2004051867

ISBN 978-0-521-82950-2 Hardback
ISBN 978-1-108-44814-7 Paperback

This book is dedicated to my dear wife, Eileen

Contents

Notes on contributors

Herman Medwin (Chapters 1–9) is Emeritus Professor, Naval Postgraduate School, Monterey, California; Fellow, Silver Medalist in Acoustical Oceanography, Gold Medalist, and Past President of the Acoustical Society of America. He is co-author, with C. S. Clay, of graduate textbooks published in 1977 (Russian translation 1980) and 1998, which defined the field of acoustical oceanography. He plays classical string quartets weekly with a geophysicist-musician, a violin-making organic chemist, and a well-known marine biologist. E-mail <oceanac@mbay.net>

David M. Farmer (Chapter 10) is Dean of the Graduate School of Oceanography, University of Rhode Island, Narragansett, Fellow of the Royal Society of Canada, Fellow of the American Geophysical Union, Fellow of the Acoustical Society of America and recipient of the Rosenstiel Award in Marine Sciences, The US Navy and Oceanography Society's Walter Munk Award, and the Canadian Meteorological and Oceanography Society President's Prize. E-mail <TheDean@gso.uri.edu>

Jeffrey A. Nystuen (Chapter 11) is Principal Oceanographer, Applied Physics Laboratory, College of Ocean and Fisheries Sciences, University of Washington, Seattle, where he was recognized in 2000 for the Development and Disclosure of Innovative Technology. In 2003 he received the Medwin Prize in Acoustical Oceanography from the Acoustical Society of America and became a Fellow of ASA. He has authored or co-authored 27 refereed scientific publications. He travels widely, enjoying birdwatching, dancing and exploring foreign cultures. E-mail <nystuen@apl.washington.edu>

D. Vance Holliday (Chapter 12) is Director of Analysis and Applied Research, Applied Technologies, Electronic Systems Division, BAE Systems. He is a Fellow and Silver Medalist in Acoustical Oceanography of the Acoustical Society of America and a senior member of the US delegation to the International Council for the Exploration of the Sea (ICES). He is a member of American Society of Limnology and Oceanography (ASLO) and serves on the Editorial Board of the new ASLO Journal on Methods. A charter member of The Oceanography Society (TOS), he is currently the representative for technology on the TOS Executive Council. With David Farmer, Dr. Holliday co-chairs SCOR Working Group 118 on the introduction of new technology for the detection and study of marine life. In the summer of 2002, for his work in bioacoustics and its impact on naval systems, the Chief of Naval Research presented him the US Navy's Meritorious Public Service Award. He continues to search for more time to play with his two granddaughters and one grandson. E-mail <van.holliday@baesystems.com>

Tim K. Stanton is Senior Scientist and former Chair of the Department of Applied Ocean Physics and Engineering, Woods Hole Oceanographic Institution (WHOI). He is also on the teaching staff in the Massachusetts Institute of Technology/WHOI Joint Graduate Education Program. Trained in physics, Dr. Stanton has conducted research on a wide range of topics in underwater acoustics. He is a Fellow of the Acoustical Society of America (ASA) and member of The Oceanography Society (TOS) where he has chaired or co-chaired several international meetings. He has served as Associate Editor of JASA, and Guest Associate Editor of IEEE Journal of Oceanic Engineering and Deep Sea Research. He has been awarded the A. B. Wood Medal for his contributions to research in underwater acoustics. Outside of his professional activities, Stanton performs trombone in various groups, including the Metropolitan Wind Symphony in Boston, and plays piano accompaniment for his church choir. E-mail <tstanton@whoi.edu>

John K. Horne (Chapter 13) splits his time between Research Assistant Professor at the University of Washington, School of Aquatic and Fishery Sciences, and member of the Fisheries Acoustics Group, Alaska Fisheries Science Center in Seattle, Washington. He is a member of the Acoustical Society of America, the American Society of Limnology and Oceanography, and the Fisheries Society of the British Isles. When not working, traveling for work, or out on a research cruise, he enjoys exploring the northwest by following the dog up a mountain. E-mail <john.horne@noaa.gov>

J. Michael Jech is a Research Fisheries Biologist, Northeast Fisheries Science Center, Woods Hole, Massachusetts. He has had the opportunity to work on a variety of aquatic environments and sail on an assortment of vessels in the Laurentian Great Lakes, Chesapeake Bay, Southern and Northern Atlantic Ocean, and in estuaries along the US east coast. He is a member of the Acoustical Society of America, American Fisheries Society, and Sigma Xi. When not out on a research cruise or in front of a computer, he enjoys fishing, mountain biking, and going to the beach. E-mail <michael.jech@noaa.gov>

Orest Diachok (Chapter 14) is a research physicist with the Naval Research Laboratory, Washington, DC. He is a Fellow of the Acoustical Society of America and a member of the American Fisheries Society. He is the author of numerous papers on acoustical sensing of the biological and physical properties of the ocean. The Naval Research Laboratory recently recognized his work on applications of matched field processing to acoustic remote sensing as one of NRL's top 75 achievements. Between 1980 and 1992 he served as Head of the Applied Ocean Acoustics Branch at NRL, and between 1992 and 1996 held the post of Chief Scientist of the NATO Undersea Research Centre in Italy. He enjoys billiards, archeology, art history, Puccini arias, B. B. King laments, Ukrainian folk songs, pale ale and Italian wines. He and his wife Olha, are proud parents of three sons, one of whom, Mateo (10), wants to become an acoustical oceanographer. E-mail <OrestDia@aol.com>

Douglas Cato (Chapter 15) is Head of the Shallow Water Environment Group of the Defence Science and Technology Organisation in Sydney, Australia. He is a Fellow of the Acoustical Society of America and has honorary positions at the Universities of Sydney and Queensland, and Curtin University in Perth. His main research contributions have been in ambient noise in the ocean and marine bioacoustics. When he has time, he enjoys playing classical guitar or returning to his earlier interest as a visual artist. E-mail <Doug.Cato@bigpond.net.au>

Michael Noad is Postdoctoral Research Fellow at the University of Queensland, Australia, who recently completed his Ph.D. in Marine Biology at the University of Sydney. His postdoctoral work is a continuation of his doctoral work, concerning the use of sound by humpback whales as well as the way their songs change as a cultural phenomenon. He continues to work in close collaboration with Doug Cato who supervised his Ph.D. During his spare time he surfs, sails, skis, and travels. E-mail <mnoad@uq.edu.au>

Robert McCauley is Senior Research Fellow with the Centre for Marine Science and Technology at Curtin University in Western Australia. As a biologist, his main research focus is on the use of passive acoustics to study marine animals and on the impacts of noise on marine animals. He enjoys coaching hockey and surfing with his son. E-mail <r.mccauley@cmst.curtin.edu.au>

Joseph E. Blue (Chapter 16) retired from US Navy's Senior Executive Service in 1996, prior to which he served as Superintendent of the Underwater Sound Reference Division (USRD) for the Naval Research Laboratory in Orlando, FL. He was a Fellow of the Acoustical Society of America, is listed in *Who's Who in Science in America*, and in *1000 Great Americans*, International Biographical Centre, Cambridge,

England. When his grandson was the manatee spokesman for his elementary school class and asked him to help save the manatees, Dr. Blue volunteered his services and began to work with Edmund Gerstein. From that beginning, his avocation was centered on acoustical causes of collision between watercraft and marine mammals. Dr. Blue spent much time assisting students in solving marine mammal acoustical measurement problems. Joe Blue passed away on January 7, shortly after writing his contribution to Chapter 16.

Edmund R. Gerstein (Chapter 16) has a Ph.D. in Psychobiology and Neuroscience. He holds faculty appointments in both the Departments of Psychology and Biology at Florida Atlantic University, where he is Director of Marine Mammal Research and Behavior. His research is focused on marine mammal hearing and underwater bioacoustics. Working with Joseph E. Blue in their research corporation, Leviathan Legacy Inc., he has studied marine mammal behavioral ecology and has revealed the underlying sensory and acoustical causes for vessel collisions on manatees and whales. When not studying marine mammals, he relaxes by watching and raising birds in his backyard aviary in Boca Raton, where his neighbors affectionately refer to him as the Birdman of Boca. E-mail <gerstein2@aol.com>

Ching-Sang Chiu (Chapter 17) is Professor of Oceanography, Naval Postgraduate School, Monterey, California, Fellow of the Acoustical Society of America, and Editor-in-Chief, Journal of Computational Acoustics. He has authored or co-authored over 40 refereed publications in the areas of ocean acoustics and acoustical oceanography. E-mail <chiu@nps.edu>

Christopher W. Miller is an electrical engineer, working as Research Associate at the NPS Ocean Acoustics Laboratory, and is Manager of the NPS Ocean Acoustic Observatory at Point Sur, CA. He has also been a Docent at the Monterey Bay Aquarium for the past 16 years. His 10 years' experience working with sea otter rescue and rehabilitation has provided his background and stimulated his interest in applying acoustics to marine biology. E-mail <cwmiller@nps.edu>

Robert C. Spindel (Chapter 18) is Director Emeritus of the Applied Physics Laboratory, University of Washington, Seattle, and Professor of Electrical Engineering. He is the recipient of the A. B. Wood Medal, British Institute of Acoustics, and the Walter Munk Award, US Navy and The Oceanography Society. He is a Fellow of the Acoustical Society of America and the Institute of Electrical and Electronics Engineers, and Fellow and Past President of the Marine Technology Society. In his spare time he enjoys rebuilding cars; old Porsches are a specialty. E-mail <spindel@apl.washington.edu>

David R. Dowling (Chapter 19) is Associate Professor of Mechanical Engineering and Applied Mechanics at the University of Michigan, Ann Arbor where he received an Outstanding Accomplishment Award in 2001 for contributions to research and education. He is a Fellow of the Acoustical Society of America and has authored or co-authored 36 refereed publications, about half of which cover topics in acoustics. He lives with his wife and their six children, and enjoys recreational swimming and other water sports. E-mail <drd@engin.umich.edu>

Heechun Song is Research Scientist, Marine Physical Laboratory, Scripps Institution of Oceanography, La Jolla, California. He has authored or co-authored 20 refereed scientific publications. He is a Fellow of the Acoustical Society of America. E-mail <hcsong@mpl.ucsd.edu>

Daniela Di Iorio (Chapter 20) is Assistant Professor, Department of Marine Sciences, University of Georgia, Athens, Georgia. She enjoys teaching acoustical and physical oceanography at all student levels and is committed to be a soccer-, bike- and music-mom for her two little energetic boys. E-mail <daniela@arches.uga.edu>

Ann Gargett is Professor of Physical Oceanography at the Center for Coastal Physical Oceanography, Old Dominion University, Norfolk, Virginia, She is a Fellow of the Royal Society of Canada and Emeritus Senior Scientist at the Institute of Ocean Sciences, Canada. Mother of a grown daughter, she now has more time for rowing.

Tim Leighton (Chapter 21) is Professor of Ultrasonics and Underwater Acoustics at the Institute of Sound and Vibration Research, University of Southampton, England. His interest in acoustical oceanography began when, as an undergraduate at Cambridge University, he began to research the sources of sound in babbling brooks. At age 28 he completed his monograph *The Acoustic Bubble* and moved to Southampton University, where he has researched physical, oceanographic and biomedical acoustics, publishing over 200 articles. He received the Inaugural Medwin Prize in Acoustical Oceanography of the Acoustical Society of America (ASA), and is the only person to have been awarded both the A. B. Wood and Tyndall Medals of the UK Institute of Acoustics. He is a Fellow of the ASA, UK Institute of Acoustics, and the UK Institute of Physics. In 2000 he was awarded a Leverhulme Trust Senior Research Fellowship from the Royal Society. E-mail <T.G.Leighton@soton.ac.uk>

Gary Heald is an Acoustics Technical Expert with the Defence Science and Technology Laboratory (DSTL), Winfrith, England. His Ph.D. was from the Physics Department, University of Bath, for research on sediment classification using acoustic scattering. He has worked on underwater acoustics and sonar for over 20 years, with a special interest in high frequency underwater acoustics and scattering. He is a Fellow of the Institute of Acoustics (UK) and in 2000 was the recipient of the A. B. Wood medal. He is a Fellow of the Acoustical Society of America and has served on the ASA Technical Committee on Acoustical Oceanography since 1998. In April 2003 he was elected Chair of the Underwater Acoustics Group of the Institute of Acoustics. In 2001 he was given an achievement award from the Non-Atomic Research and Development Technical Cooperation Panel for his international collaboration on environmental reconnaissance. Gary is a visiting senior lecturer at the Institute of Sound and Vibration Research, University of Southampton. In his spare time Gary is interested in cabinet making, woodturning, photography, and ham radio. E-mail <gjheald@mail.dstl.gov.uk>

David R. Palmer (Chapter 22) is Research Physicist, Atlantic Oceanographic and Meteorological Laboratory in Miami. He has over 25 years experience conducting experimental and theoretical research in acoustical oceanography and related fields and has published over 150 scientific papers. He is a member of several honor and professional societies including Phi Beta Kappa and the Acoustical Society of America. When time permits he teaches physics at Florida International University. He grew up in Colorado and spent much of his summers hiking the mountains searching for ghost towns from the mining era. E-mail <David.R.Palmer@noaa.gov>

Peter Rona is Professor of Marine Geology and Geophysics at Rutgers University. His interest in ocean acoustics began in the 1960s when he worked under his mentor C. S. Clay at the former Hudson Laboratories of Columbia University. He was Senior Research Geophysicist with NOAA before coming to Rutgers in 1994 to build marine programs. He is presently working with others in applying innovative methods to acoustically image and measure plumes that buoyantly rise from black smoker vents and diffuse hydrothermal flow on the sea floor. He is Fellow of Acoustical Society of America, Geological Society of America, Society of Economic Geologists, and the American Association for the Advancement of Science, and recipient of Francis Shepard Medal for Excellence in Marine Geology, Gold Medal of the US Department of Commerce for exceptional scientific contributions to the nation, and the Hans Pettersson Bronze Medal of the Royal Swedish Academy of Sciences. He enjoys exploration of the Earth wherever it leads. E-mail <rona@imcs.rutgers.edu>

Nicholas Makris (Chapter 23) is a Professor at the Massachusetts Institute of Technology. He is a Secretary of the Navy/Chief of Naval Operations Scholar of Oceanographic Sciences, a recipient of the A. B. Wood Medal, the Doherty Professorship of Ocean Utilization, the MIT

Edgerly Fellowship, the ONR Young Investigator Award, NRL's Alan Berman Award and is a Fellow of the Acoustical Society of America. He teaches acoustical sensing to both undergraduates and graduate students at MIT and is currently helping to prove that there are oceans of liquid water on some of Jupiter's moon, as part of NASA's Jupiter Icy Moons Orbiter (JIMO) Science Definition Team. He likes to play an old-fashioned Fender Stratocaster and brings one on oceanographic cruises to jam with his piano-obsessed friend from WHOI. He used to sail and windsurf a lot in his spare time but now finds traveling around the planet more fun. E-mail <makris@keel.mit.edu>

Michael J. Buckingham (Chapter 24) is Professor of Ocean Acoustics, Marine Physical Laboratory, Scripps Institution of Oceanography, University of California, San Diego and Visiting Professor, University of Southampton, UK. Previously an Individual Merit Senior Principal Scientific Officer, Royal Aerospace Establishment, UK, he has used a jet aircraft for acoustics research in ice-covered seas. Recipient of the A. B. Wood Medal (UK Institute of Acoustics), Clerk Maxwell Premium (UK Institution of Electronic & Radio Engineers). Fellow, Acoustical Society of America, UK Institute of Acoustics, UK Institution of Electrical Engineers, Explorers Club; member, New York Academy of Sciences. Has published some 200 scientific papers and a book on solid-state physics. Weekend photographer and flier, he holds a private pilot's license with instrument and glider ratings. E-mail <mjb@ucsd.edu>

Preface
The world of ocean sounds

This book is the reader's gateway to a science that spans physics, oceanography and marine biology. Wherever possible we perform the trick of Janus, the mythological Roman God, who simultaneously faces in opposite directions. One view, called "ocean acoustics," is the traditional direction in which the knowledge of (or assumptions about) the ocean temperature and salinity allows one to use sound to find fish, submarines, icebergs, and the depth of the ocean. The opposite view, "acoustical oceanography," interprets the distinctive details of time-varying, sound amplitudes and phases over acoustical paths to deduce the physical and biological parameters of the specific ocean through which the sound has traveled . . . It is best to look in both directions.

We will be considering the diverse potentialities of passive listening, as well as benign probing by unobtrusive sound: in rough seas and smooth seas; deep seas and shallow seas; clean seas and seas made locally dirty by dumping of man's garbage; seas of uniform temperature and those that are thermally layered; dead seas and seas noisily filled with abundant life ranging from the grand whales to microscopic zooplankton and phytoplankton. It is a vast, complex, mostly dark, optically opaque, but acoustically transparent world that has been only thinly sampled by today's limited technology and science.

Tragic beginnings

In retrospect, the impetus for the effective use of sound in the sea occurred in 1912 when the steamship TITANIC struck an iceberg. The subsequent loss of hundreds of lives triggered man's use of sound to sense scatterers in the oceans of the world. Within a month of the disaster, a patent application was filed by L. R. Richardson in the United Kingdom (10 May 1912) for "detecting the presence of large objects under water by means of the echo of compressional waves – directed in a beam – by a projector." The basic idea was that a precise knowledge of the speed of sound in water, and the travel time of the sound from source to scatterer and back to the source/receiver, permits the calculation of the distance to the scattering body. This was to be the beginning of the use of underwater sound projectors and receivers. They were to be called "SONARs" i.e. devices for **SO**und **NA**vigation and **R**anging.

In fact, the speed of sound in fresh water had already been measured very accurately almost a century before. Corrections for depth dependence and salinity dependence of the speed were soon determined. A grand variety of new commercial and military activities were immediately practical, including acoustical fish finding (patented in 1935), acoustical measurements of ocean depth, and acoustical detection of submarines.

Present status, future promise

Since those early days, underwater acousticians and acoustical oceanographers have used sound as the premier tool to determine the detailed characteristics of physical and biological bodies and processes at sea. Myriad components of the ocean world are being discovered, identified, characterized, and imaged by their interactions with sound. Chapters 1 through 9 of this textbook, "Fundamentals," provide the basic tools of ocean acoustics. The following 15 chapters, written by some of the world's most successful ocean researchers, who are using sound in innovative ways to learn about the sea and its contents, describe several modern developments. Their contributions are divided into the four sections, titled: "Studies of the near-surface ocean"; "Bioacoustical studies"; "Studies of ocean dynamics"; "Studies of the ocean bottom."

Acknowledgements

Inspiration for this book has come from my professors at the University of California at Los Angeles, my students and associates at the Naval Postgraduate School and colleagues in the Acoustical Society of America. I had the great fortune of being a student at UCLA during those vital, early post-WWII years when the Physics Department at UCLA was the world's finest academic environment for teaching and research in physical acoustics. The names of my professors read like a "who's who" of acoustics in the latter half of the twentieth century: Leo Delsasso, Carl Eckart, Vern O. Knudsen, Robert W. Leonard, and my outstanding thesis advisor, Isadore Rudnick.

Let me pay tribute to my Naval Officer students at the Naval Postgraduate School, who were some of the brightest, most energetic people I have ever had the pleasure of working with. They wrote theses, part of the requirement for their M.S. degrees in Acoustical Engineering, that would have satisfied the academic quality and significance of Ph.D. research at many of the world's colleges and universities. Most of these energetic students went on to become Admirals and Captains in the navies of the USA and Turkey and West Germany; their fine student research is appropriately identified and referenced throughout this book.

My years of book co-authorship with C. S. Clay (1977, 1998) remain a high point of my work in the field of acoustics. The first nine chapters of this book have been extracted from the graduate level textbook, Medwin and Clay, *Fundamentals of Acoustical Oceanography*, Academic Press (1998), (abbreviated *M&C*). Much of that material has been updated and rewritten here for undergraduates or beginning graduate students in physical and biological sciences. Many of the unidentified figures were drawn by C. S. Clay, for *M&C*, and I am grateful that he allowed them to be re-used here.

Some topics, based on the research of others, are identified simply by the author names and the date of publication. More complete references will be found in the References or Bibliography sections, or on the Internet. Descriptions of a special few publications are in "Further reading" at the ends of each of the first nine chapters.

Valuable extensive discussions have been held with Aubrey Anderson, Steven Baker, Mohsen Badiey, Jonathan Berkson, William Carey, Jacques Chamuel, N. Ross Chapman, Dezang Chu, Lawrence Crum, William Cummings, Peter Dahl, Grant Deane, Chris Feuillade, Fred Fisher, Charles Greenlaw, Mark Hamilton, Richard Keiffer, Saimu Li, Michael Longuet-Higgins, James Lynch, Kendall Melville, James H. Miller, Jorge Novarini, Wesley Nyborg, John Potter, Andrea Prosperetti, Jeffrey Simmens, Kevin Smith, Eric Thorsos, Alex Tolstoy, Ivan Tolstoy, and O. Bryan Wilson.

My co-authors acknowledge the significant contributions by their many students and their research colleagues: Karen G. Bemis, Kevin Conley Michael Czarnęcki, Henry S. Fleming, Eric Giddens, Charles Greenlaw, Thomas Hahn, Darrell R. Jackson, Christopher D. Jones, Grace Kamitakahara-King, Duncan McGehee, Kyohiko Mitsuzawa, Michael Richardson, Deborah Silver, Fernando Simonet, Ron Teichrob, Svein Vagle, Timothy Wen, Peter Wiebe, Norman Zabusky.

Several ocean scientists have looked at early versions of the marine biology sections of the book and have made useful suggestions. I am greatly indebted to Susanna Blackwell, Steven Haddock, and Redwood W. Nero who have helped me to bridge the gap between physical scientists and biological scientists.

I am most grateful to my oceanographer colleague at the Naval Postgraduate School, Professor Ching-Sang Chiu, who has been generously critical of my attempt to open our subject to undergraduates in the ocean sciences. Also special thanks to Dr. Steve Haddock, marine biologist of the Monterey Bay Aquarium Research Institute, who has gently led me into twenty-first-century word processing techniques.

I am particularly indebted to the 25 prominent acoustical oceanographers from Australia, Canada, England, and the United States of America who wrote Chapters 10 to 24 to describe their recent important research. My co-authors are: David M. Farmer; Jeffrey A. Nystuen; D. Vance Holliday and Timothy K. Stanton; John K. Horne and J. Michael Jech; Orest Diachok; Douglas Cato, Michael Noad, and Robert McCauley; Edmund Gerstein and Joseph Blue; Ching-Sang Chiu and Christopher W. Miller; Robert C. Spindel; David R. Dowling and Heechun Song; Daniela Di Iorio and Ann Gargett; T. G. Leighton and Gary Heald; David Palmer and Peter Rona; Nicholas Makris; Michael J. Buckingham.

A bow to the many anonymous professionals of the Acoustical Society of America, whose collegial conversations and e-mails I so thoroughly enjoy at, and between, the semi-annual meetings of the ASA.

Research support for the works reported here has come from several sources; most importantly from US Office of Naval Research (ONR), but also from US National Science Foundation (NSF), US Strategic

Environmental Research Development Program (SERDP), US National Oceanographic and Atmospheric Administration (NOAA)'s National Undersea Research Program, West Coast and Polar National Undersea Research Center, NASA, USA, Navy Submarine Development Group ONE, Deep Submergence Group of the Woods Hole Oceanographic Institution, Department of Fisheries and Oceans, Canada.

The encouragement and early offer of support from the US Office of Naval Research, was particularly appreciated; that branch of the US Navy was among the first to see a need for an undergraduate textbook on ocean acoustics.

Finally, my appreciation to my contacts at Cambridge University Press: Matt Lloyd, Publisher for Earth and Space Sciences, and Jo Bottrill, Production Editor, for their wise comments and challenging questions.

Herman Medwin

Prologue: Active and passive sensing in the sea

An innocent person, fully informed by the news media about the fantastic successes of radar and other electromagnetic frequencies in studies of the universe, can be excused for asking "Why should one use sound to probe or communicate in the sea? Why not use light or radar?"

In fact, sound has overwhelming advantages, compared to light or radar, as a tool for active or passive studies and for communication within the ocean. The great advantages of acoustics compared to electromagnetics at sea, depend on two crucial characteristics of all waves:

(*a*) the wavelength needed to "see" a body by its backscatter or radiation;
(*b*) the wave attenuation (decrease with propagation distance) at "usable" frequencies.

First, the effect of wavelength on wave backscatter.

In the nineteenth century Lord Rayleigh proved that, if a simple spherical body is to be sensed by an observer, the wavelength of the sound or light that is used as an active probe should be less than the circumference of the sphere. In fact, if the wavelength is larger than the circumference, the cross-section of the object appears to be very much *smaller* than it actually is. The reduction is inversely proportional to the fourth power of the wavelength. That is, if the wavelength is twice the circumference, the backscattered sound intensity (which one needs to identify an object) is approximately one-sixteenth as great as when the wavelength is equal to, or less than, the circumference!

A strong argument against the use of light or radar for detection or communication under the ocean surface is quite simple. Wavelengths of visible light range from 0.4 to 0.8 microns (slightly less in water). This means that ocean particles with circumference greater than about one micron will scatter light very effectively. In fact, it is well known that light beams are greatly attenuated by scatter from the large amount of minute suspended particles in the sea. In most oceans, one must get very close to take a photograph of an object. Skin divers talk about a typical maximum range of about 2 meters (6 feet) and point out that side lighting and strobe lighting is generally essential. It has been said that the opacity of the water in the English Channel is so great that, regardless of the illumination, a diver is unable to see his fingers at the end of his outstretched arm! Submersible vehicles that travel to great depths suffer from the same limitation. In conclusion, the extraordinarily large scatter of short optical and radar wavelengths, accompanied by the enormous absorption of all electromagnetic waves due to the electrical conductivity of salt water, combine to make light and radar almost unusable except for very short distances in the ocean.

Active sensing by light or sound

Let us summarize *active* detection of bodies at sea.

To see and describe an object *optically*, one can use the visual part of the electromagnetic spectrum from blue to red light (frequency ratio, two to one). Unfortunately, the typical ocean range for visible light is only a few meters.

Because of its very much smaller attenuation in the sea, a far greater span of *acoustic* frequencies can be employed productively. The more than 1000 to one frequency ratio for active uses of sounds includes *fractions of a kilohertz* (mid-keyboard of the piano) which are effective in the study of large scale ocean motions (eddies) kilometers in extent, *kilohertz* sounds (near the upper limit of human hearing) common for active sonar detection by submarines, *tens of kilohertz* frequencies for acoustical fish finders, *hundreds of kilohertz* sounds to measure ocean particle motions and search for buried objects, and *thousands of kilohertz* (megahertz) sonars for remote, acoustical, non-destructive, *in situ* "counting" of 0.1 mm diameter zooplankton!

These acoustical studies can be accomplished at acoustic intensities that are not harmful to marine life. In fact, almost all ocean acoustical research is performed at sound levels that are barely detectable above the ambient noise in the sea.

The superiority of active acoustical probing compared to optical photography can be comprehended in a practical case by considering the detail in the "acoustical image" of a 57 m long shipwreck, sunk in turbid water of depth 38 m. The photo below was produced by using *acoustical* backscatter of a 500 kHz side-scan sonar at a range of 75 meters. The *optical* visibility was only about one or two meters at the time of this test! Note the missing stern section, the three large hatches designed for convenient handling of timber, and the crater due to sediment erosion.

Photo courtesy of EDGE TECH, Milford, Massachusetts.

Passive sensing by light or sound

Finally, consider *passive* detection in the sea. Marine biologists have studied several species, such as some jellyfish, which make their presence known by emitting light. But one must be within a few meters of the animals to observe them.

On the other hand, some "sound" (defined here for the frequency range from a fraction of a hertz to several megahertz) is emitted by all whales, many species of fish and plankton, all oceanographic and meteorological phenomena, and all man-made vehicles. This makes it possible to detect, identify, and study these sources by simply listening. Recent passive underwater sound research includes: low frequency detection of secret underwater explosions and microseisms at distances over 10 000 kilometers; eavesdropping on voices of the great whales hundreds of kilometers away; and measuring the amount and detailed characteristics of mid-ocean rainfall by remote satellite transmission of the sounds of the raindrop impacts and the ringing microbubbles created by the splash.

Part I
Fundamentals

HERMAN MEDWIN

Chapter 1
Sound propagation in a simplified sea

Summary

Sound is a mechanical disturbance that travels through a fluid. The sound wave can be a short-duration pulse or a continuous wave oscillation (CW) that is usually, for simplicity, sinusoidal. Because most detectors of ocean sounds are pressure sensitive devices, the propagating disturbance is most often identified as a time-varying incremental pressure, i.e., an *acoustic pressure*. Sometimes the description is in terms of the incremental density, the incremental temperature, the material displacement from equilibrium, or the transient particle velocity of the sound.

In this chapter we assume that the medium is *homogeneous* (same physical properties at all points) and *isotropic* (same propagation properties in all directions). We also assume that there is no sound *absorption* (no sound energy conversion to heat) and no *dispersion* (no dependence of sound speed on sound frequency). And we assume that the acoustic pressure increment is very, very small compared to the ambient pressure (no finite amplitude, non-linear effects).

Several wave phenomena that occur in the sea will be discussed here. When sound encounters an obstacle, it is *scattered*; part of the scattered energy bends around the obstacle (this is called *diffraction*) and part is backscattered toward the source; when it is incident on a boundary surface where it meets a different density or different sound speed, some *reflection* occurs, accompanied by some *refraction* (i.e., transmission at an angle different from the incident angle). When a sound wave meets another sound wave, the two pressures may add constructively or destructively, and *interference* ensues.

We seek to exploit the capabilities of underwater sound to discover more about the world's oceans and its inhabitants. This is done by interpreting the behavior of several descriptors of sound such as its pressure amplitude, particle velocity, density, intensity, radiated power, and propagation speed, all of which are introduced in this chapter. Also in this chapter, we study the effect of the ocean on sounds, including the basic behaviors such as: "reflection" from ocean surfaces; "refraction" of sound rays (the bending caused mostly by the spatial variation of the water temperature and salinity); "interference," (the superposition of sounds from different sources, or sounds that have traveled different paths from the same source). These same phenomena may be known from one's study of optics, but the omnipresent effects are crucially important to understand and interpret sounds in the sea.

Various simplifying approximations are developed at this time to derive the sound propagation equations that describe the effects of the ocean surface, volume, and bottom. We discuss waves in plane, cylindrical, and spherical coordinate systems, which will be needed to express the propagation at sea simply and appropriately to the sources and the environment.

Contents

1.1 Sound speed in water

Knowing the sound speed in water is critical to ocean communication and much of biological and geophysical ocean research. The earliest measurement was by Colladon and Sturm (1827) in the fresh water of Lake Geneva, Switzerland (Fig. 1.1). A value of 1435 m/s was found, but it was soon realized that the speed in saline water is somewhat greater than this and that, in general, the temperature of the water is an even more important parameter than salinity.

* This section contains some advanced analytical material.

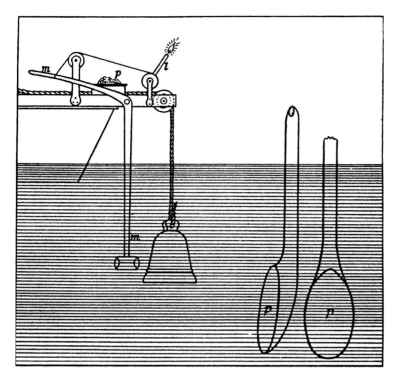

Fig. 1.1. Colladon and Sturm's apparatus for measuring the speed of sound in water. A bell suspended from a boat was struck under water by means of a lever *m*, which at the same moment caused the candle *l* to ignite powder *p* and set off a flash of light. An observer in a second boat used a listening tube to measure the time elapsed between the flash of light and the sound of the bell. The excellent results were published in both the French and German technical literature. (*Annales de Chimie et de la Physique* 36, [2], 236 [1827] and Poggendorff's *Annalen der Physik und Chemie* **12**, 171 [1828].)

Numerous laboratory and field measurements have now shown that the sound speed increases in a complicated way with increasing temperature, hydrostatic pressure, and the amount of dissolved salts in the water. A very simple formula for the speed in m/s, accurate to 0.1 m/s, but good only to 1 kilometer depth, was given by Medwin (1975),

$$c = 1449.2 + 4.6T - 0.055T^2 + 0.00029T^3$$
$$+ (1.34 - 0.010T)(S - 35) + 0.016z \qquad (1.1)$$

In (1.1), temperature T is in degrees centigrade, salinity S is in parts per thousand of dissolved weight of salts, and the depth z is in meters. A better, longer, but still simple expression (Mackenzie, 1981) is in Chapter 2. The best equation (Del Grosso, 1974) involves some 19 terms containing coefficients with 12 significant figures.

Portable sound "velocimeters," which measure the time of travel of a megahertz pulse, have an accuracy of 0.1 m/s in non-bubbly water. But everpresent microbubbles, which are not considered in any of the sound speed equations above, can cause the actual speed of propagation at frequencies below about 100 kHz to be different from the velocimeter readings by tens of meters/s, particularly near the ocean surface, see Chapter 6.

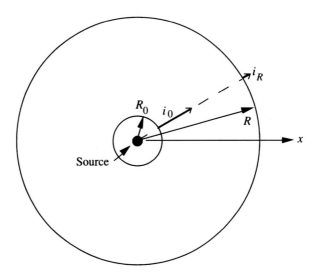

Fig. 1.2. Spherical spreading of a pulse wave front. The instantaneous intensity is i_0 at the radius R_0 and later is i_R at the radius R.

1.2 Pulse wave propagation

1.2.1 Intensity of a diverging compressional pulse

In a medium that is homogeneous and isotropic, a tiny sphere expands suddenly and uniformly and creates an adjacent region of slightly higher density and pressure. This higher density region is called a *condensation*. Assume that it has a thickness dr. The condensation "impulse" or "pulse," will move outward as a spherical wave shell and will pass a reference point during time δt. It is called a *longitudinal wave* because the displacements in the medium are along the direction of wave propagation. As it propagates, the energy of the impulse is spread over new spherical shells of ever larger radius, at ever lower acoustic pressure. By conservation of energy, the energy in the expanding wave front is constant in a lossless medium.

The *acoustic intensity is the fluctuating energy per unit time that passes through a unit area.* The total energy of the pulse is the integral of the intensity over time and over the spherical surface that it passes through. Figure 1.2 shows the expanding wave front at two radii. Applying the conservation of energy, the energy that passes through the sphere of radius R_0 is the same as the energy passing through the sphere of radius R. Conservation of energy gives the sound intensity relationship, where i_0 and i_R are the intensities at R_0 and R,

$$4\pi i_R R^2(\delta t) = 4\pi i_0 R_0^2(\delta t) \tag{1.2}$$

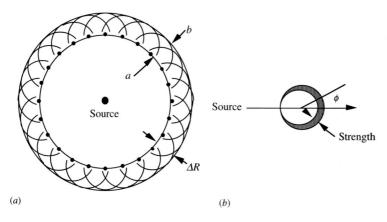

Solving for the intensity at R, one gets

$$i_R = \frac{i_0 R_0^2}{R^2} \qquad (1.3)$$

The sound intensity decreases as $1/R^2$ due to spherical spreading. Later, in Section 1.5.3 "Near field and far field approximations," we will show that the sound intensity is proportional to the square of the sound pressure. Therefore, sound pressure decreases as $1/R$ in a spherically diverging wave. We would have had the same result if the sphere at the origin had imploded instead of exploded. Then a *rarefaction* pulse, a propagating region of density less than the ambient value, would have been created.

1.3 Pulse wave reflection, refraction, and diffraction

A useful qualitative description of wave propagation was first given by Christian Huygens, Dutch physicist–astronomer (1629–1695). Huygens proposed that each point on an advancing wave front can be considered as a source of secondary waves which move outward as spherical wavelets in a homogeneous, isotropic medium. The outer surface that envelops all these wavelets constitutes the new wave front (Fig. 1.3).

The sources used in underwater sound measurements are sometimes condensation pulses, for example the shock wave from an explosion. The application of Huygens' Principle to an idealized pulse wave front is particularly simple and physically direct.

Baker and Copson (1950) provided a secure mathematical basis for Huygens' Principle. The concept is extensively used in optics, as well as acoustics. e.g., see A.D. Pierce (1981).

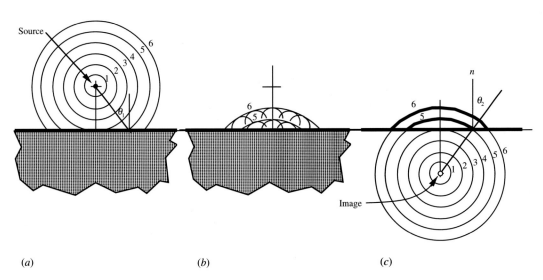

Source

Image

(a)   (b)   (c)

Fig. 1.4. A spherical pulse from a point source and its reflection at a rigid, plane reflector. The usual penetration of the pulse into the lower half space behind the plane face of the reflector is not shown. (a) Successive positions of the incident pulse wave over a half space. (b) Huygens constructions of successive positions of the reflected pulse wave fronts. (c) The reflected pulse wave fronts appear to come from an image source in the lower half space. A homogeneous, isotropic medium is assumed. The geometry shows that $\theta_2 = \theta_1$.

Stokes (1849) derived an obliquity factor which describes the pressure amplitude of the expanding wavelet with lesser side radiation and no back radiation, which agrees with observation. The shading in Fig. 1.3(b), the "strength," follows the law

$$\text{amplitude} \sim \cos\frac{\varphi}{2} = \sqrt{\frac{1 + \cos\varphi}{2}} \tag{1.4}$$

In the short time Δt the disturbance from each of the secondary sources on the wave front travels a distance ΔR (see Fig. 1.3). The outward surfaces of the wavelets coalesce to form the new wave front b. The strength of the wavelet is maximum in the direction away from the source and zero in the backward direction.

1.3.1 Reflection at a plane surface: law of reflection

The propagation of a pulse will be demonstrated graphically without the preceding details of the Huygens' construction. See Fig. 1.4(a) where each successive position of the pulse at equal time intervals Δt is indicated by 1, 2, 3, etc. In a homogeneous, isotropic medium the wave front travels the same distance ΔR during each interval. In the ray direction,

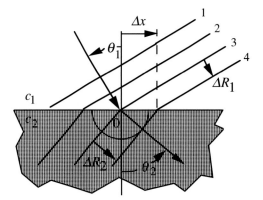

Fig. 1.5. Huygens construction for a compressional plane pulse at a sequence of times and positions in adjacent media showing Snell's Law of Refraction.

normal to the wave front for the isotropic medium, the distance of advance of the pulse is given by $R = c\Delta t$ where c is the sound speed.

The Huygens construction of the interaction of a spherical pulse wave at a plane boundary, Fig. 1.4(b), suggests that the wave front of the reflection is expanding as if it has come from a source beneath the reflecting surface. The apparent source after reflection is called the *image*.

A way to treat the image and the real source is shown in Fig. 1.4(c). The image wave of the proper strength is initiated at the same time as the source, and when it moves into the real space it becomes the reflected wave.

The simple geometry shows that the angle of reflection θ_2 of the *rays* (perpendicular to the wave front) is equal to the angle of incidence θ_1, and is in the same plane.

$$\theta_2 = \theta_1 \qquad (1.5)$$

This is called the "Law of Reflection." *Note:* Some people prefer to give the law of reflection in terms of its complement, the grazing angle that the ray makes with the surface, instead of the angle with the normal.

1.3.2 Pulse refraction at a plane interface: Snell's Law

Now we assume that the pulse wave front has come from a very distant point source, that the curvature of the spherical wave front is negligible, and that it is effectively a plane wave front in our region of interest. Figure 1.5 shows that the wave is incident on the plane boundary between two media which have sound speeds c_1 and c_2.

The figure is drawn for $c_2 > c_1$ (which is the case for sound going from air to water); the reader can easily sketch the figure for the other case, $c_1 > c_2$. Successive positions of a plane pulse are shown as it moves

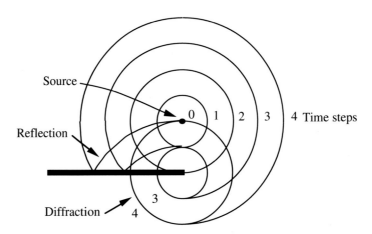

Fig. 1.6. Huygens construction for diffraction of a pulse at a reflecting half plane. Incident, reflected, and diffracted pulse positions are shown for a sequence of times. The transmitted wave is omitted to simplify the sketch.

across the interface. In general, there will also be a reflected pulse but it is omitted here for simplicity. In the time Δt the pulse has moved a distance ΔR_1 in medium 1 and ΔR_2 in medium 2. In the same time, the contact of the pulse at the interface has moved a distance Δx along the horizontal x axis. The angles are measured between the rays and the normal to the interface, or the grazing angle between the pulse front and the interface. The propagation distances in the two media are

$$\Delta R_1 = \Delta x \sin\theta_1 \quad \Delta R_2 = \Delta x \sin\theta_2$$

The speeds are $c_1 = \Delta R_1/\Delta t$ and $c_2 = \Delta R_2/\Delta t$. Therefore,

$$\sin\theta_1/c_1 = \sin\theta_2/c_2 \tag{1.6}$$

This is the well-known *Snell's Law of Refraction*. We use Snell's Law throughout the book.

Sometimes the law for refraction is given in terms of the grazing angle $\phi = 90 - \theta$ in which case the sine θ is replaced by the cosine ϕ.

1.3.3 Diffraction at the edge of a plane

Assume that the pulse source is above a semi-infinite plane that permits part of the pulse wave to be transmitted (not shown), part to be reflected, and part to be diffracted. The situation is in Fig. 1.6, where the diffracting plane (heavy line) extends from the boundary edge, infinitely to the left, and infinitely in front and behind the page.

The impulse wave front spreads from the source and interacts with the plane. The interactions at the plane become sources of Huygens wavelets. The envelopes of the wavelets coming only from the plane become the reflected waves. The outgoing wave beyond the edge of the plane continues, unaffected. The envelope of the Huygens wavelets originating at the edge form a wave front which appears to spread from the edge. That wave is called the *diffracted* wave. (The transmitted wave is omitted for simplicity.)

The diffracted wave is a separate arrival. It is strongest in the direction of propagation but it is more easily detected in any other direction because of its later arrival. The diffracted wave exists because there is a reflecting plane to the left of the edge and none to the right.

In general, "scattering" is a redirection of sound when it interacts with a body. Scattered sound in a fluid is made up of three components: the transmitted, reflected, and diffracted waves. Pulse sounds are helpful in analyzing scattering problems because they have distinctive arrival times that depend on their path lengths. Scattered and diffracted waves are particularly important because they are often used to identify the invisible bodies that cause them. They are discussed in quantitative detail in later chapters.

1.4 Sinusoidal, spherical waves in space and time

When a sinusoidally excited "point" source expands and contracts periodically it produces a continuous spherical wave. The resulting *condensations* (density and pressure above the ambient) and *rarefactions* (below ambient) in the medium move away from the source at the sound speed c, in the same manner as the disturbance from a pulse source. A representation of the sinusoidal fluctuations at some later instant would resemble the cartoon in Fig. 1.7(a). The distance between adjacent condensations (or adjacent rarefactions) along the direction of travel is the *wavelength*, λ.

The disturbances sketched in Fig. 1.7(a) radiate outward from a point source which is small compared to λ. As a condensation moves outward, the acoustic energy is spread over larger and larger spheres. Correspondingly, the *pressure amplitude* (the acoustic pressure at the peak of the sinusoid) decreases. Later we prove that, in spherical radiation, the pressure amplitude decreases as $1/R$, where R is the distance from the source. The distance between adjacent crests continues to be λ.

The simplest functions that repeat periodically are sines and cosines. The spatial dependence of the instantaneous sound pressure at large ranges may be written as, for example,

$$p = \frac{P_0 R_0}{R} \sin \frac{2\pi R}{\lambda} \quad \text{or} \quad p = \frac{P_0 R_0}{R} \cos \frac{2\pi R}{\lambda} \tag{1.7}$$

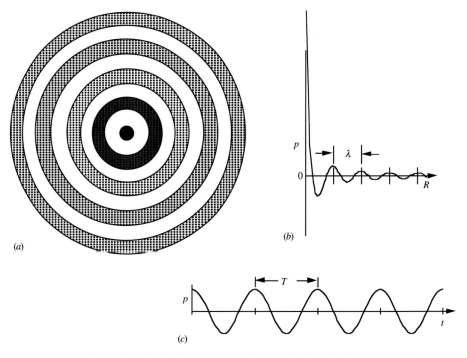

Fig. 1.7. Radiation from a very small periodically pulsating source. (*a*) Pressure field at an instant of time. The dark condensations are lightened at increasing range to show the decreasing acoustic pressure. (*b*) Graph of range-dependent pressure at an instant of time. (*c*) Time-dependent pressure signal at a point in space.

where P_0 is the amplitude of the pressure oscillation at reference range R_0. Equation (1.7) includes the decrease of pressure with increasing R. The amplitude at R is $P(R) = P_0 R_0/R$.

The time between adjacent crests passing any fixed point is the *period*, T (Fig. 1.7(*c*)). The temporal dependence of the pressure oscillation at R is, for example,

$$P = P(R)\sin 2\pi ft \quad \text{or} \quad P = P(R)\cos 2\pi ft \qquad (1.8)$$

where f is the *frequency* of the oscillation, measured in cycles per second or hertz (Hz).

The simplest functions that repeat periodically are sines and cosines which repeat themselves for every increment of 2π. For example, $\sin(\theta + 2n\pi) = \sin\theta$ where n is an integer. For two adjacent times, t_1 and $t_2 = t_1 + T$, the functions repeat so that $2\pi ft_2 = 2\pi ft_1 + 2\pi$. Therefore, $2\pi fT = 2\pi$, and the period, T, is

$$T = 1/f \qquad (1.9)$$

Consider Fig. 1.7. In the time T the disturbance has moved the distance λ. Therefore, the speed at which it travels, the *sound speed*, is

$$c = \lambda/T = f\lambda \qquad (1.10)$$

The units of λ are generally meters, so c is in meters per second.

In (1.8) the dimensionless product ft is in cycles and $2\pi ft$ is in radians. The latter is sometimes called the *temporal phase*. It is customary to absorb the 2π into the frequency and to define the *angular frequency*,

$$\omega = 2\pi f = 2\pi/T \quad \text{radians/s} \qquad (1.11)$$

The spatial dependence of pressure at any instant is described, for example, by $\sin(2\pi R/\lambda)/R$ where the dimensionless argument $(2\pi R/\lambda)$ is sometimes called the *spatial phase*.

The spatial *wave number*, k, is defined by

$$k = 2\pi/\lambda \quad \text{radians/meter} \qquad (1.12)$$

Note that the spatial *wave number, k* is analogous to the temporal *angular frequency, $\omega = 2\pi/T$*. The two quantities are related through the equation for the speed (1.10). The relation is

$$k = \omega/c \qquad (1.13)$$

As shown later, when spherical wave propagation is described in space and time these two concepts combine in forms such as

$$p = (P_0 R_0/R)\sin(\omega t - kR) \qquad (1.14)$$

or

$$p = (P_0 R_0/R)\sin[\omega(t - R/c)] \qquad (1.15)$$

or

$$p = (P_0 R_0/R)\sin[2\pi(t/T - R/\lambda)] \qquad (1.16)$$

and so forth, where P_0 is the pressure amplitude at the reference range $R = R_0$.

It is easy to prove that the foregoing equations describe a radially propagating wave having the speed c. For example, pick an arbitrary phase at time t, position R. At a later time, $t + \Delta t$, the same phase will exist at position $R + \Delta R$. Equating,

$$[t + \Delta t - (R + \Delta R)/c] = \omega[t - (R/c)] \qquad (1.17)$$
$$\text{Therefore}, c = \Delta R/\Delta t \qquad (1.18)$$

To summarize: combinations such as $(t - R/c)$ or $(R/c - t)$ or $(\omega t - kR)$ indicate a wave traveling in the positive, increasing, R direction.

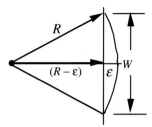

Fig. 1.8. Geometry for the local plane wave approximation.

Others, such as $(t + R/c)$ or $(R/c + t)$ or $(\omega t + kR)$ describe a wave traveling in the negative R direction.

1.5 Wave interference, effects and approximations

Wave interference phenomena are the result of sound pressures from more than one source being present at the same position and at the same time. In "linear acoustics" the resulting sound pressure is the algebraic sum of the contributions. The addition holds, regardless of the directions of travel, time dependencies, and amplitudes of the components. In other words, instantaneous acoustic pressures at a point are scalars and they add algebraically. Non-linear acoustics, which describes effects of very intense sounds, is considered in Chapter 4.

Approximations have a long and distinguished history in the physical and applied mathematical sciences. They are a guide to understanding, and they permit one to obtain useful solutions to otherwise intractable analytical problems. Approximations continue to be important, even in the era of high speed digital computers, because they assist the modeler and the experimentalist in evaluating whether the results of a complex computer calculation are realistic or not.

1.5.1 Local plane wave approximation

It is often convenient to assume that when the spherical wave is being studied at a large distance from the source it appears to be a plane wave within the region of interest. This "local plane wave approximation," which simplifies the problem, is extensively used in air acoustics, underwater sound, and geophysics. The usual condition for this approximation is given in terms of the sagitta of the arc, ε, compared to the wavelength over the restricted region of width W. As shown in Fig. 1.8, the geometry gives

$$R^2 = (R - \varepsilon)^2 + W^2/4 \tag{1.19}$$

$$\frac{W^2}{4} = (2R - \varepsilon)(\varepsilon) \approx 2R\varepsilon \tag{1.20}$$

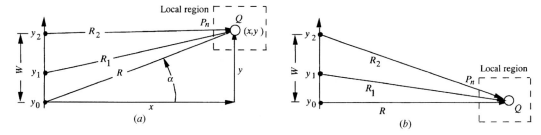

Fig. 1.9. (a) Geometry for several sources with receiver at Q. The local region for the plane wave approximation is in the dashed line rectangle. As a plane wave approximation in the region, the incident sound pressures have the amplitudes P_n. The sources are at y_0, y_1, y_2, etc. The distances from the sources to the listening point Q at R, α are R, R_1, R_2, etc. The acoustic pressure at Q is the sum of the pressures contributed by the several sources. (b) Redraw of (a) to have $\alpha = 0$.

A common assumption is that the plane wave approximation may be used over the restricted region W provided that

$$\varepsilon \leq \lambda/8 \qquad (1.21)$$

Therefore, the region for this plane wave approximation has the extent

$$W \leq (\lambda R)^{1/2} \qquad (1.22)$$

1.5.2 Fresnel and Fraunhofer approximations

To add the signals due to several sinusoidal point sources we need the distances to the observation point. The geometry is shown in Fig. 1.9.

The pressure waves incident at the point Q are spherical and, recalling (1.14), (1.15), (1.16) the amplitudes of the pressures depend on the ranges. At large ranges, we suppress that range dependence and let the *incident sound pressure* amplitudes *in the local region* have the values P_1, P_2, and so forth. The total sound pressure is

$$p = \sum_n P_n \sin(\omega t - kR_n) \qquad (1.23)$$

where the P_n are the pressure amplitudes at ranges R_n. The relative phases of the terms are most important because they determine whether the total instantaneous pressure will be greater or less than the individual pressures. To separate the time dependence from the summation we expand the $\sin(\dots)$ and get

$$p = \sin(\omega t) \sum_n P_n \cos(kR_n) - \cos(\omega t) \sum_n P_n \sin(kR_n) \qquad (1.24)$$

The sin (ωt) and cos (ωt) terms have a $\pi/2$ phase difference and are said to be in phase "quadrature."

One can make numerical evaluations of the summations in (1.24) by using the following expressions for R and R_n

$$R^2 = y^2 + x^2 \tag{1.25}$$

$$R_n^2 = (y - y_n)^2 + x^2 \tag{1.26}$$

$$R_n^2 = (R \sin\alpha - y_n)^2 + R^2 \cos^2(\alpha) \tag{1.27}$$

This can be rearranged to

$$R_n = R \left(1 - \frac{2y_n}{R}\sin\alpha + \frac{y_n^2}{R^2}\right)^{1/2} \tag{1.28}$$

Simpler forms are possible at very long ranges when $W/R \ll 1$. Then expansion as a binomial gives

$$R_n \approx R[1 - \frac{y_n}{R}\sin(\alpha) + \frac{y_n^2}{2R^2}(1 - \sin^2(\alpha)) + \cdots] \tag{1.29}$$

Depending on how small y_n/R is, one can neglect most of the higher-order terms. When only the first order term in y_n/R is kept, we get the *Fraunhofer Approximation* for very long ranges

$$R_n \approx R \left(1 - \frac{y_n}{R}\sin\alpha\right) \tag{1.30}$$

On the other hand, when both the first order term y_n/R and the second-order term $(y_n/R)^2$ are kept, we have the approximation for nearer ranges, which is called the *Fresnel Approximation*:

$$R_n \approx R \left[1 - \frac{y_n}{R}\sin(\alpha) + \frac{y_n^2}{2R^2}(1 - \sin^2(\alpha))\right] \tag{1.31}$$

1.5.3 Near field and far field approximations

Often the terminologies "near field" and "far field" are used to describe distinctive parts of the acoustic field due to a "large" source or an array of sources. The "near field" is the region where the differential distances to the elements of the source are large enough for the phase differences to cause constructive and destructive interferences. The "far field" is where the range is greater than this "critical range" and there are no peaks and troughs of interference. Actually, far field approximations are useful only for ranges that are *much* greater than a critical range. Often, a range at least four times the critical range is used to define the beginning of the "far field."

To derive the minimum critical range let $\alpha = 0$ for simplicity in Fig. 1.9(a), and redraw the geometry as in Fig. 1.9(b). The critical range R_c is a distance at which it is no longer possible for wavelets traveling the longest path, (from y_2 the source farthest away from the axis) to interfere destructively with those traveling the shortest path, (from y_0, on the axis). A pressure minimum cannot occur when these two distances differ by less than $\lambda/2$. From the geometry and using the binomial expansion,

$$R_2 = (R^2 + W^2)^{1/2} \approx R \left(1 + \frac{W^2}{2R^2}\right) \qquad (1.32)$$

The condition in this case is

$$\Delta R = R_2 - R \approx \frac{W^2}{2R} \leq \frac{\lambda}{2} \qquad (1.33)$$

The critical range on the axis of the array, beyond which there can be no minimum, is

$$R_c = W^2/\lambda \qquad (1.34)$$

and the "far field" is defined by $R > W^2/\lambda$.

In practice, one usually goes far beyond this critical range to be sure that the simplified calculations and measurements are securely in the far field. Often, a range four times the critical range is used to define the "far field." The complexity of the acoustic near field and the transition to the far field can be appreciated by calculating the acoustic pressure for various ranges and frequencies, as suggested in the problems at the end of the chapter.

1.5.4 Interference between distant sources: use of complex exponentials

We now introduce the powerful complex exponential description which greatly simplifies many mathematical operations in acoustics.

The relations between trigonometric functions and complex exponential functions are

$$\exp(i\Phi) = e^{i\Phi} = \cos\Phi + i\sin\Phi \qquad (1.35)$$

and

$$\cos\Phi = \frac{e^{i\Phi} + e^{-i\Phi}}{2}, \quad \sin\Phi = \frac{e^{i\Phi} - e^{-i\Phi}}{2i} \qquad (1.36)$$

In Section 1.4 we used $\sin(\omega t - kR_n)$ to describe the change of phase of a spherical traveling wave. But notice that the expression $\exp[i(\omega t - kR_n)]$ contains both the sine and cosine components. Using the complex exponential, and keeping only the imaginary part which is the sine, two pressure oscillations p_1 and p_2, of the same slowly-changing amplitude (at long range) P, can be expressed as

$$p_1 = P\exp[i(\omega t - kR_1)] \qquad (1.37)$$
$$p_2 = P\exp[i(\omega t - kR_2)] \qquad (1.38)$$

The exponential description is particularly useful in differentiation and integration, which we will need to do later. For example, differentiation of p with respect to time is equivalent to multiplication by $i\omega$; integration with respect to time is simply division by $i\omega$.

The sum of the pressures is

$$p = p_1 + p_2 = P \{\exp[i(\omega t - kR_1)] + \exp[i(\omega t - kR_2)]\} \quad (1.39)$$
$$p = P \exp(i\omega t) [\exp(-ikR_1) + \exp(-ikR_2)] \quad (1.40)$$

Now compute the square of the absolute value $|p|^2 = pp^*$, where $*$ denotes the complex conjugate. (The complex conjugate is obtained by changing the sign of the imaginary.) The effect of this operation is to eliminate the time dependence because the product $\exp(i\omega t) \exp(-i\omega t)$ is unity.

Then, (1.40) becomes

$$|p|^2 = P^2[\exp(-ikR_1) + \exp(-ikR_2)][\exp(ikR_1) + \exp(ikR_2)] \quad (1.41)$$

and with the aid of (1.35) we obtain

$$|p|^2 = 2P^2\{1 + \cos[k(R_1 - R_2)]\} \quad (1.42)$$

The maximum value is $4P^2$ and the minimum value is 0. The co-existence of the two pressures, p_1 and p_2, produces interferences with maxima at $k(R_2 - R_1) = 0, 2\pi, 4\pi, \ldots$ and minima at $\pi, 3\pi, 5\pi, \ldots$

The phase difference $k(R_2 - R_1)$ can cause constructive interference with pressure $2P$ at the maxima, destructive interference with zero pressure at the minima, and any amplitude between 0 and $2P$, depending on the phase difference.

1.5.5 Plane wave interference near an interface: standing waves

Sound pressures near an interface consist of the incident and reflected components of the signal. The signals may overlap and interfere significantly within a few wavelengths of the surface. The result for a continuous wave is that there are near-surface regions of high acoustic pressure (anti-nodes of pressure) and low pressure (nodes). The warning for experimentalists is that the signal that is received by a near-surface, or near-bottom, hydrophone depends on both the wavelength and the distance from the interface.

To illustrate this important effect we first assume that an incident plane wave is traveling vertically downward and that it reflects upward at a perfectly reflecting sediment. The sound pressure will be the sum of the two pressures. Let p_i and p_r be the incident and reflected pressures which we assume have the same amplitude, P. Define z as positive upward and write

$$p_i = P \exp[i(\omega t + kz)] \tag{1.43}$$
$$p_r = P \exp[i(\omega t - kz)]$$

The sum is

$$p = p_i + p_r = P \exp(i\omega t)[\exp(ikz) + \exp(-ikz)] \tag{1.44}$$
$$p = 2P \exp(i\omega t) \cos(kz)$$

The acoustic pressure has an envelope with magnitude $|2P \cos(kz)|$ and a time dependence $\exp(i\omega t)$. The envelope is stationary in time; therefore the result of the interference is called a "stationary wave" or a "standing wave." For example, the nulls at $kz = \pi/2$ are nulls at all times.

Reflection at a partially reflecting interface also produces a standing wave. In this case the standing wave consists of the sum of the incident wave and the reflected fraction of the incident wave. The remainder of the incident wave moves through the minima on the way to the second medium so, in this situation, the interference minima are not zero.

An *obliquely* incident plane wave at a *perfectly reflecting* interface produces a standing wave perpendicular to the interface and a traveling wave parallel to the interface. An *obliquely* incident wave at a *partially reflecting* surface creates a weaker standing wave perpendicular to the surface and a traveling wave perpendicular to the surface, plus a traveling wave parallel to the surface.

1.5.6 Point source interference near the ocean surface: Lloyd's Mirror effect

During World War II it was discovered that a sinusoidal point source near the ocean surface produced an acoustic field with major interferences between the spreading direct sound and the phase-shifted reflected sound. The latter appeared to have diverged from an above-surface image of the source (Fig. 1.10). In underwater acoustics the interfering sound field is sometimes called "the surface interference effect." The original observation of the phenomenon, described as "Lloyd's Mirror interference," had been discovered in mid-nineteenth century laboratory optics experiments.

The geometry and results of some World War II experiments, are shown in Fig. 1.10. For a point source, the direct sound at range R_1 is

$$p_1 = \frac{P_0 R_0}{R_1} e^{i(\omega t - k R_1)} \tag{1.45}$$

The smooth ocean surface reflected sound experiences a 180 degree

Fig. 1.10. Geometry and World War II experimental results showing Lloyd's Mirror effect at sea (Eckart, 1946). Source depth 14 ft, hydrophone depth 50 ft, ocean depth 2000 fathoms. (*Note*: kc, kilocycles, called kHz, kilohertz, today. One yard is slightly smaller than a meter. A foot is approximately 30 cm.)

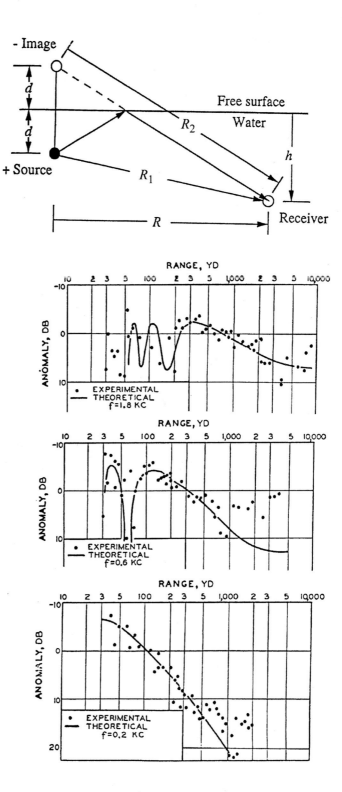

phase shift, and appears to have come from the image at range R_2.

$$p_2 = \frac{P_0 R_0}{R_2} e^{i(\omega t - k R_2)} \tag{1.46}$$

It is left as a student problem to show that when the source is at shallow depth d (much less than R) and the hydrophone is at shallow depth h (much less than R), the two sounds produce an interference pattern at range R with peaks and troughs. The usual amplitude decrease that goes as $1/R$ for a point source in an infinite medium is modified by pressure doubling in the near regions. Beyond the last peak, the interference causes the pressure amplitude to decrease as R^{-2}.

This is shown in Fig. 1.10, which is a copy of an original figure from a WWII experimental result. The negative transmission anomaly describes a sound pressure greater than expected. For a perfectly smooth surface the peaks would occur at a negative transmission anomaly 6 dB, instead of the 4 to 5 dB in the actual experimental results.

1.6 One-dimensional wave equation

Our discussion in the preceding sections dealt with several phenomena of wave propagation, following the Huygens' description. Now we develop the algebraic connections between the properties of the medium and the sound that propagates in it. These relations are based on Newton's Second Law of Mechanics, Conservation of Mass, and a form of the Equation of State that describes pressure and density in a fluid. All three equations are encompassed in a single second-order partial differential "*wave equation*," which describes the acoustic pressure in time and space. It is derived below.

1.6.1 Newton's Law, Conservation of Mass, and the Equation of State for Acoustics

Assume that a disturbance in the ocean is caused by the sudden expansion of a small spherical source. This causes the local density and pressure to increase because the rest of the medium does not instantaneously move to allow space for the expansion. Consider a small region at a very large distance from the source where the plane wave approximation holds (Section 1.5.1). In this region the variations of pressure, velocity, and acceleration of a fluid particle are approximately functions of the direction of propagation. The total pressure at any fixed point (i.e. Eulerian coordinates) is

$$p_{\mathrm{T}} = p_{\mathrm{A}} + p \tag{1.47}$$

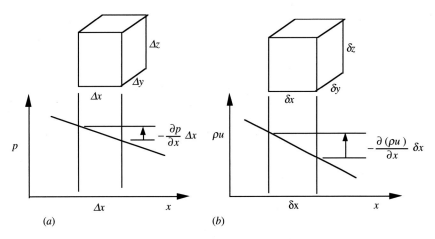

Fig. 1.11. Pressure differential across a small volume. (*a*) Lagrangian coordinates. The pressure differential causes the mass $\rho_A \Delta x \Delta y \Delta z$ to move to the right. (*b*) Eulerian coordinates. Mass flow is through the small volume $\delta x \delta y \delta z$ (*u* is the component of velocity of flow along the *x* axis; ρ_A is ambient density).

where p_A is the ambient, static pressure which does not change over our region, and where $p \ll p_A$ is the incremental, acoustic pressure.

Similarly, the total density at a point is

$$\rho_T = \rho_A + \rho \tag{1.48}$$

where the acoustic density ρ is very much less than the ambient density, $\rho \ll \rho_A$

Newton's Law for Acoustics

We consider the moving water particle in a description called the *Lagrangian* coordinate system as shown in Fig. 1.11(*a*). The net pressure in the $+x$ direction, acting on our fluid particle is written in terms of the *partial* rate of change (curved letter, ∂)

$$\text{net pressure} = -\left(\frac{\partial p}{\partial x}\right)\Delta x \tag{1.49}$$

The net force is $-(\partial p/\partial x)\Delta x \Delta y \Delta z$. The mass of the water particle is $\rho_A \Delta x \Delta y \Delta z$.

We define u as the particle velocity in the $+x$ direction and $\partial u/\partial t$ as the local acceleration. Then Newton's Law, $F = ma$, gives

$$-\left(\frac{\partial p}{\partial x}\right)\Delta x \, \Delta y \, \Delta z = \rho_A \frac{\partial u}{\partial t} \Delta x \, \Delta y \, \Delta z$$

$$-\frac{\partial p}{\partial x} = \rho_A \frac{\partial u}{\partial t} \tag{1.50}$$

We keep in mind that *Newton's Law for Acoustics* (1.50) is a point statement. It applies to the pressure and particle velocity at position x at time t. For the plane wave propagating in the $+x$ direction, $p = p(t - x/c)$ and $u = u(t - x/c)$. For a plane wave propagating in the $-x$ direction, $p = p(t + x/c)$ and $u = u(t + x/c)$. The speed, c, may be a function of position.

Conservation of Mass for Acoustics

Now consider a small cage that is fixed in space (Fig 1.11(b)). This is called an *Eulerian*, or fixed, coordinate system, Mass flows in one face and out the other. Since the fluid is compressible, more mass may flow in than out, and the density within the cage may increase. The net mass flowing per unit time into the cage is $-[\partial(\rho_T u)/\partial x]\delta x\,\delta y\,\delta z$. This causes a rate of density increase which may be written simply as $\delta\rho_T/\partial t = \partial\rho/\partial t$ because the ambient density ρ_A is constant in $\rho_T = \rho_A + \rho$. Equating the two rates of change of mass gives

$$-\left(\frac{\partial(\rho_T u)}{\partial x}\right)\delta x\,\delta y\,\delta z = \left(\frac{\partial\rho}{\partial t}\right)\delta x\,\delta y\,\delta z \qquad (1.51)$$

Since $\partial(\rho_T u)/\partial x$ is approximately equal to $\rho_A \partial u/\partial x$ for acoustic waves, we simplify to the *acoustical equation of conservation of mass,*

$$-\rho_A\frac{\partial u}{\partial x} = \frac{\partial\rho}{\partial t} \qquad (1.52)$$

Equation of State for Acoustics

Hooke's Law states that, for an elastic body, the stress is proportional to the strain. In the acoustical version of Hooke's Law the stress (force per unit area) is the acoustic pressure, p, and the strain (relative change of dimension) is the relative change of density, ρ/ρ_A. The proportionality constant is the ambient bulk modulus of elasticity, E (or E_A). Except for intense sounds (see Chapter 4), Hooke's Law holds for fluids. This relation between acoustic pressure and acoustic density is sometimes called the *acoustical equation of state,*

$$p = \left(\frac{E}{\rho_A}\right)\rho \qquad (1.53)$$

The equation above also assumes that an instantaneous applied pressure p causes an instantaneous proportional increase of density, ρ. Actually there is generally a time lag in the response of the fluid to applied pressure due to "molecular relaxation." This causes acoustic energy to be absorbed, as described later.

1.6.2 Acoustic pressure, acoustic density, particle velocity, Mach number, and the impedance of the medium

The wave equation

The one-dimensional "wave equation," which incorporates all three laws developed in the previous section, is obtained by taking the $\partial/\partial x$ of (1.50) and the $\partial/\partial t$ of (1.52) and eliminating the common second derivative $\partial^2 u/\partial x\, \partial t$. To put it in terms of the acoustic pressure, one uses the pressure density relation (1.53). The result is the one-dimensional linear wave equation

$$\frac{\partial^2 p}{\partial x^2} = \frac{\rho_A}{E} \frac{\partial^2 p}{\partial t^2} \qquad (1.54)$$

If we had eliminated p instead of ρ we would have obtained an equation identical to (1.54), but in terms of ρ. In fact, the wave equation (1.54) could also be derived in terms of the particle velocity component u, or the particle displacement, or the incremental temperature, or any parameter that is characteristic of the acoustic wave. We choose to work with p because virtually all hydrophones that are sensitive to underwater sound are *pressure* sensitive.

Equations (1.37) and (1.43) where the directions of propagation were called R or z, are all solutions of the plane wave equation (1.54), or are long range approximations to the plane wave solution. In fact, any linear combination of solutions is also a solution of (1.54), as may be easily proven by substitution.

When we substitute any one of these solutions into the wave equation we obtain

$$c^2 = \frac{E}{\rho_A} \qquad (1.55)$$

so that the **one dimensional wave equation** can be written in its more common form

$$\frac{\partial^2 p}{\partial x^2} = \frac{1}{c^2} \frac{\partial^2 p}{\partial t^2} \qquad (1.56)$$

We have assumed that the elasticity, E, and the speed, c, are not dependent on the direction of propagation. If this were not the case, (for example, in solids) we would write E_x and c_x for those quantities.

Impedance of the medium

There is an important relation between acoustic particle velocity and acoustic pressure in a plane wave. Recall that a wave traveling in the $+x$ direction has particle velocity component $u = u(x - ct)$; therefore

$$\frac{\partial u}{\partial t} = -c\frac{\partial u}{\partial x} \tag{1.57}$$

Substitution of (1.57) into (1.50) gives

$$\frac{\partial p}{\partial x} = \rho_A c\frac{\partial u}{\partial x}$$

Integration yields

$$p = +(\rho_A c)u \tag{1.58}$$

where the plus sign is for waves traveling in the positive x direction; a minus sign is used for waves traveling in the negative x direction.

Equation (1.58) resembles Ohm's Law, with the acoustic pressure taking the place of voltage, acoustic particle velocity replacing electric current, and $(\rho_A c)$ being the impedance. The analogy is used frequently, and the $(\rho_A c)$, "rho-c," which is called the "specific acoustic impedance" is a common acoustical characterization of the medium. Often, the subscript A is dropped for simplicity.

Acoustic Mach number of a sound wave

The ratio of the acoustic particle velocity to the speed of sound, u/c, may be calculated by starting with (1.52). Then, since $u = u(t - x/c)$ and $\frac{\partial u}{\partial x} = -\frac{1}{c}\frac{\partial u}{\partial t}$, we find

$$M = u/c = \rho/\rho_A \tag{1.59}$$

where M is sometimes called the acoustic Mach number because it relates the acoustic velocity to the sound speed. But it is much more than that. The acoustic Mach number is a measure of the strength of the sound wave, and thereby the linearity of the signal propagation. For intense sounds, when the ratio is large enough, extraordinary, very useful, non-linear propagation effects occur as described in Chapter 4.

Acoustic pressure/acoustic density relation

The preceding equations (1.58) and (1.59) allow us to go directly to the useful relation between the acoustic pressure and the acoustic density,

$$p = \rho c^2 \tag{1.60}$$

which can be used to calculate the speed of sound if the equation of state $p = p(\rho)$ is known. For sound in gases, there is a relatively simple equation of state; consequently, the accurate theoretical speed in air has been known for over 200 years. But for liquids the equation of state is so complicated that the inverse calculation is used. That is, the equation of state for water is calculated from accurate measurements of the speed of sound in water.

1.6.3 Acoustic intensity

The intensity of a wave is the power passing perpendicularly through a unit area; usually the area is 1 m^2. Intensity is a vector which has a direction normal to the unit area. Suppose a plane wave is traveling in the $+x$ direction and the unit area is in the yz plane. The instantaneous intensity is the product of the instantaneous pressure and the in-phase particle velocity along the x direction, u_x. The x-component of intensity is

$$i_x = pu_x \qquad (1.61)$$

We would need a subscript on c (e.g., c_x) if the speed were a function of direction of propagation.

Since the plane wave is traveling in the $+x$ direction, we apply (1.58) so that the intensity can be written in terms of pressure alone,

$$i_x = \frac{p^2}{\rho_A c} \qquad (1.62)$$

Similarly, using the long range plane wave approximation to a spherical wave, if the unit area is normal to the direction of R from the source, the intensity along R is

$$i_R = \frac{p^2}{\rho_A c} \qquad (1.63)$$

When the wave is a continuous sinusoid $p = P \sin(kx - \omega t)$, the value of i_x is

$$i_x = \frac{P^2}{\rho_A c} \sin^2(kx - \omega t) = P^2 \frac{1 - \cos[2(kx - \omega t)]}{2\rho_A c} \qquad (1.64)$$

The instantaneous intensity i_x oscillates between 0 and $P^2/(\rho_A c)$ and has the frequency 2ω.

For a sinusoidal wave it is useful to calculate the average intensity by integrating over time. The time average intensity at x is

$$I_x = i_x = (P^2)/(2\rho_A c) = P_{rms}^2/(\rho_A c) \qquad (1.65)$$

where P is the peak pressure and $P_{rms} = 0.707\, P$ at x. Again, the analogy

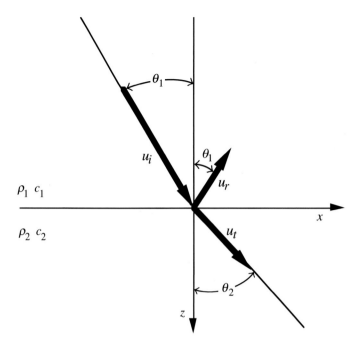

Fig. 1.12. Reflection and refraction geometry for vector particle velocities (heavy arrows) at an interface between two fluids.

of (1.65) to power in AC circuits provides a useful mnemonic for electrical engineers and physicists.

1.7 Plane wave reflection and refraction at a plane interface

While the derivations for reflections and transmissions at an interface are given here for infinite plane waves and a plane interface, the results provide a practical technique when one can use the local plane wave approximation to a spherical wave, Section 1.5.1 "Local plane wave approximation."

1.7.1 Reflection and transmission coefficients: critical angle for total internal reflection

A very powerful way to do interesting problems uses the physical boundary conditions at an interface between two fluids. The simplest example is the reflection and transmission of plane waves at a fluid interface. Figure 1.12 shows the ray directions and the components of the incident, reflected, and transmitted particle velocities at the interface, u_i, u_r, and u_t.

The first condition at the boundary is the equality of pressures on each side of the interface, evaluated at $z = 0$,

$$p_i(t) + p_r(t) = p_t(t) \tag{1.66}$$

where the subscripts designate the incident, reflected, and transmitted sound pressures at time t.

The second condition at the boundary is the equality of the normal components of particle velocity, evaluated at $z = 0$, at time t,

$$u_{zi}(t) + u_{zr}(t) = u_{zt}(t) \tag{1.67}$$

Suppose a ray is incident at angle θ_1 at the interface. The angle of refraction is θ_2, given by Snell's Law,

$$\theta_2 = \arcsin[(c_2/c_1)(\sin\theta_1)] \quad \text{for} \quad (c_2/c_1)\sin\theta_1 < 1 \tag{1.68}$$

The three vertical components of particle velocity at the fluid interface, $z = 0$, are

$$u_{zi}(t) = u_i(t)\cos\theta_1 \tag{1.69}$$
$$u_{zr}(t) = u_r(t)\cos\theta_1 \tag{1.70}$$
$$u_{zt}(t) = u_t(t)\cos\theta_2 \tag{1.71}$$

where u_i, u_r, and u_t are in the incident, reflected, and transmitted directions, respectively.

The computation of particle velocities in terms of pressure follows from (1.58) where the negative value of ρc is used for propagation in the $(-z)$ direction:

$$u_{zi}(t) = \frac{p_i(t)}{\rho_1 c_1}\cos\theta_1 \tag{1.72}$$

$$u_{zr}(t) = -\frac{p_r(t)}{\rho_1 c_1}\cos\theta_1$$

$$u_{zt}(t) = \frac{p_t(t)}{\rho_2 c_2}\cos\theta_2$$

At a simple plane fluid interface, the time dependencies of the incident, reflected, and transmitted waves are the same. The pressure reflection and transmission coefficients for a wave going from medium 1 to medium 2, evaluated at $z = 0$, are

$$\mathcal{R}_{12} \equiv \frac{p_r(t + z/c_1)}{p_i(t - z/c_1)} \quad \text{and} \quad \mathcal{T}_{12} \equiv \frac{p_t(t - z/c_2)}{p_i(t - z/c_1)} \tag{1.73}$$

In terms of the coefficients, the pressure condition can be written

$$1 + \mathcal{R}_{12} = \mathcal{T}_{12} \tag{1.74}$$

and the velocity condition takes the form

$$(\rho_2 c_2 - \rho_2 c_2 \, \mathcal{R}_{12}) \cos \theta_1 = \rho_1 c_1 \, \mathcal{T}_{12} \cos \theta_1 \qquad (1.75)$$

The equations at the boundary are now solved for the pressure reflection and transmission coefficients,

$$\mathcal{R}_{12} = \frac{\rho_2 c_2 \cos \theta_1 - \rho_1 c_1 \cos \theta_2}{\rho_2 c_2 \cos \theta_1 + \rho_1 c_1 \cos \theta_2} \qquad (1.76)$$

and

$$\mathcal{T}_{12} = \frac{2\rho_2 c_2 \cos \theta_1}{\rho_2 c_2 \cos \theta_1 + \rho_1 c_1 \cos \theta_2} \qquad (1.77)$$

Snell's Law gives the connection between θ_1 and θ_2,

$$\theta_2 = \arcsin\left(\frac{c_2}{c_1} \sin \theta_1\right) \qquad (1.78)$$

There are two particularly important applications for this theory of reflection and transmission between two fluids: the ocean surface and the ocean bottom. Consider the ocean surface under the greatly simplifying assumptions: the surface is a smooth plane between sea water (density 1000 kg/m^3, sound speed 1500 m/s) and air (density 1.03 kg/m^3 and sound speed 330 m/s).

Assume underwater sound is normally incident to the interface ($\cos \theta_1 = 1$ and $\cos \theta_2 \cong 1$). Since $\rho_1 c_1 \gg \rho_2 c_2$, (1.76) and (1.77) give $\mathcal{R}_{12} \cong 1$ and $\mathcal{T}_{12} = 4.5 \times 10^{-4}$. Further, from (1.66) we find $p_r \cong -p_i$ (i.e., a phase-reversed pressure) which thereby results in a *near zero total pressure at the surface*, and from (1.67) we find $u_r \cong 2u_i$, so there is a particle velocity *doubling* at the surface. The water/air interface is called a "pressure release" or "soft" surface for underwater sound.

But notice that if the direction of propagation had been reversed, sound going from air to the ocean would find a pressure doubling interface, with essentially zero particle velocity. Viewed from the air, the same surface would be called acoustically "hard."

The water-to-air propagation is an extreme case of $c_2 < c_1$ which always results in $(c_2/c_1) \sin \theta_1 < 1$ and $\theta_2 < 90°$ for all angles of incidence.

However, for sound going from the ocean to a sediment bottom, when $c_2 > c_1$ there is the possibility of "total reflection." Total reflection occurs at angle of incidence $\theta_1 \geq \theta_c$ where θ_c is the "critical angle" defined by

$$\theta_c = \arcsin(c_1/c_2) \qquad (1.79)$$

When the angle of incidence is greater than θ_c, Snell's Law can be written as

$$\cos\theta_2 = \left[1 - \left(\frac{c_2}{c_1}\right)^2 \sin^2\theta_1\right]^{1/2} \equiv \pm ig_2 \tag{1.80}$$

where the magnituide of the cosine is

$$g_2 \equiv \left[\left(\frac{c_2}{c_1}\right)^2 \sin^2\theta_1 - 1\right]^{1/2} \tag{1.81}$$

We choose the solution $(-ig_2)$ because it describes an acoustic pressure which becomes weaker with increasing depth of penetration into the second medium (in the z direction) while it propagates in the x direction. The incorrect choice of the plus sign would have led to the physically absurd description of a wave which had its origin in the first medium and which is stronger at greater distances from the interface in the passive second medium. Using (1.76)

$$\mathcal{R}_{12} = \frac{\rho_2 c_2 \cos\theta_1 + i\rho_1 c_1 g_2}{\rho_2 c_2 \cos\theta_1 - i\rho_1 c_1 g_2} \quad \text{for} \quad \theta > \theta_c \tag{1.82}$$

The numerator is the complex conjugate of the denominator. Therefore, the magnitude of the ratio is unity, $|\mathcal{R}_{12}| = 1$.

For angles of incidence greater than critical we write $\mathcal{R}_{12}$ to allow us to calculate the phase shift

$$\mathcal{R}_{12} = e^{+2i\Phi} \tag{1.83}$$

where

$$\Phi \equiv \arctan\frac{\rho_1 c_1 g_2}{\rho_2 c_2 \cos\theta_1} \tag{1.84}$$

1.7.2 Plane wave reflection at a sedimentary bottom

C. S. Clay and colleagues have made extensive measurements of the geophysical structure and acoustical propagation in the shallow water south of Long Island, New York. The sediment parameters are shown in Fig. 1.13. The values of the speed and density in the water, and in the uppermost sediment, are known. We assume that the sediment acts as a fluid and therefore we use the equations of the previous section to calculate the reflection coefficient and the phase shift, as a function of the angle of incidence. The fractional pressure reflection, $\mathcal{R}_{12}$, is sometimes called the "bottom loss" and it is usually expressed in decibels as the positive number, $BL = -20\log_{10}\mathcal{R}_{12}$. The values of $\mathcal{R}_{12}$ and the phases plotted in Fig. 1.14 should be verified by the student.

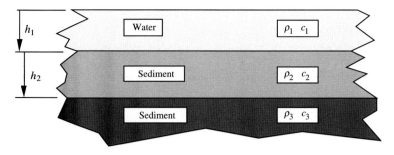

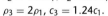

Fig. 1.13. Water and sediment structure south of Long Island, New York. Parameters are: $\rho_1 = 1033$ kg/m^3, $c_1 = 1508$ m/s, $h_1 = 22.6$ m; $\rho_2 = 2\rho_1$, $c_2 = 1.12c_1$, $h_2 = 0.9h_1$; $\rho_3 = 2\rho_1$, $c_3 = 1.24c_1$.

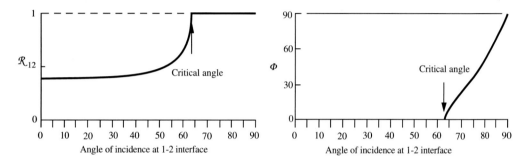

Fig. 1.14. Relative amplitude $\mathcal{R}_{12}$ and phase shift Φ of the reflection at the 1–2 interface shown in Fig. 1.13.

The sea floor is usually covered by layers of sediments. The layers are called "thin" when the reflection from a sequence of layers can be replaced by the reflection from a composite layer. The local plane wave assumption is used when the thickness of the sequence is very small relative to the distances to the source and receiver.

Thin layers are also used in the design of sonar windows. The sound passes from the transducer, through a thin protective window, into the water. The transmission coefficient through the window depends on the thickness of the window and the sound speed and ambient density of all three media. The thickness and physical parameters are chosen to maximize the sound transmission through the window at the required frequency.

1.7.3 Plane wave reflection beyond critical angle

In Fig. 1.14 it is demonstrated that for plane wave incidence beyond the critical angle there is perfect reflection with an accompanying phase shift. Weston (1960) and Buckingham (1987) have described the usefulness of introducing a virtual, displaced, pressure release surface (i.e., $\mathcal{R} = -1$) to represent that situation. Figure 1.15 shows the geometry of the virtual reflector concept.

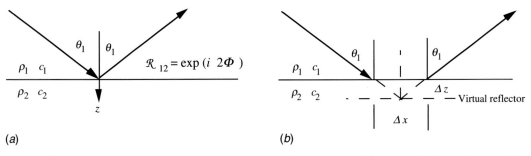

(a)   (b)

Fig. 1.15. Reflection beyond critical angle. (a) Actual reflection and phase shift. (b) Equivalent reflection at a virtual reflector at depth Δz.

Fig. 1.16. Geometry for head wave propaqation at bottom sediment. The impulse source is in a lower-speed medium (e.g., water over most sediments). The Huygens sources move along the interface at the speed in the bottom (higher speed) medium.

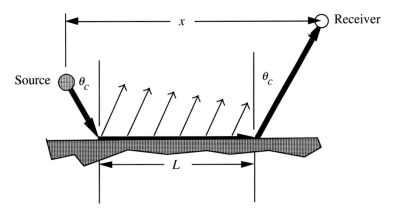

1.7.4 Spherical waves beyond critical angle: head waves

When there is a higher speed medium adjacent to the source medium, and the incident spherical wave front meets the interface at the critical angle, a *head wave* is produced. The head wave moves at the higher speed c_2 along the interface and radiates into the source medium, c_1. Because it travels at higher speed, at sufficient ranges the head wave arrives significantly ahead of the spherically diverging direct wave. The energy appears to be continually shed into the lower speed source medium at the critical angle, calculated from $\sin\theta_c = c_1/c_2$, as it propagates along the interface, Fig. 1.16.

Huygens wavelets and wave front constructions can explain the processes that produce a head wave that supplements the well-known reflected and refracted waves at the interface (see *M&C*, Section 2.6.5). Analytical developments of the head wave are in Cerveny and Ravindra

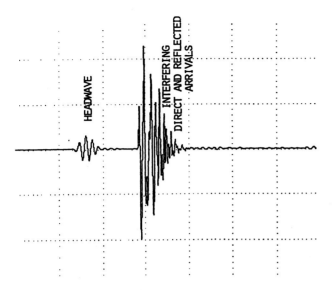

Fig. 1.17. Impulse response for a shallow point source of sound under a model of the Arctic ice canopy. The source was driven by 2 cycles of 62.5 kHz; $\lambda =$ 2.37 cm in water. Source and receiver depths were 0.4 wavelengths in water; source and receiver separation, $X = 27$ wavelengths; path length in plate, $L =$ 26 wavelengths. The figure is from Browne (1987) and Medwin *et al.* (1988).

(1971). One significant conclusion is that, for a point source, the amplitude of the head wave in medium 1 is given by

$$\text{amplitude of head wave} \sim x^{-1/2}L^{-3/2}k^{-1} \qquad (1.85)$$

where x and L are defined in Fig. 1.16. Here it is assumed that attenuation is negligible.

The conditions for a head wave also exist for a sound source under Arctic ice even when it is only a fraction of a wavelength in thickness. Arctic "pack ice" often consists of "plates" that are hundreds of meters in extent and 1 or 2 meters thick; the plates are bounded by broken sections called "ridges." A laboratory scale model study of Arctic ice reported by Medwin *et al.* (1988) verified that the predicted amplitude dependence on range and frequency, as given in (1.85) for seismic waves, occurs in this case as well.

To model Arctic propagation it is important to use laboratory materials that have the same physical parameters as the Arctic ice. Then the spatial dimensions of plate thickness, source range, and receiver range are scaled relative to the wavelength. For example, a 3.3 mm acrylic plate floating on water in the laboratory, when insonified at 62.5 kHz, can represent a 1 meter thick Arctic plate insonified by 200 Hz sound. In both cases, the point source under the "ice" canopy generates a compressional

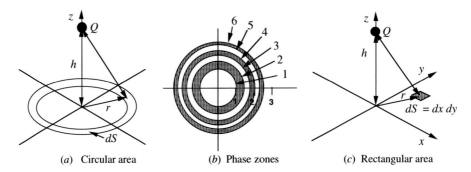

Fig. 1.18. Reflection from a finite circular plane area. (*a*) Circular area. (*b*) Phase zones. (*c*) Rectangular area.

wave at the critical angle for entry into the ice. As the compressional wave propagates in the plate, it "sheds" a head wave into the water at the critical angle. The hydrophone will receive a reflected wave, and a direct wave as well. When the range is large enough, the head wave precedes the other two arrivals by a significant time.

In geometrical–acoustical scale modeling it is also necessary to select the frequency, pulse length, source and receiver depths in order to isolate, and identify, the head wave. Since the compressional speed in the laboratory plate and laboratory water were essentially the same as in the Arctic, the critical angle for the compressional wave was approximately the same, $\theta_c = \sin^{-1}(c_1/c_2) \approx \sin^{-1}(1480/2353) = 39°$.

Because the thickness of Arctic ice is often a small fraction of a wavelength, early discussions of reflection from the ice cover *incorrectly* assumed that the "thin" ice covered by air was equivalent to a simple, water–air, pressure–release interface. Model experiments such as the above showed that the problem is far more complex than that. Taking advantage of this complexity, the existence of a head wave in the Arctic provides a technique for measuring the low frequency compressional wave speed in an ice plate. See also a series of ingenious laboratory experiments by Chamuel, e.g., Chamuel and Brooke (1988).

1.7.5 Spherical wave reflection from a finite reflector: Fresnel zones

Consider the reflection of a spherical wave from a circular plane surface. The geometry for a point source and receiver is sketched in Fig. 1.18. The problem is solved most easily by dividing the circular surface area into circular rings, called Fresnel zones. The radii of the rings are set to cause one half wavelength difference for adjacent rings. Summing the backscatter from the rings shows that, for **finite** reflecting planes, *the* *magnitude of the reflection depends on the wavelength of the sound, the distance to the surface, and the size of the plane surface as well as the reflection coefficient,* $\mathcal{R}_{12}$.

Different circular elements are different distances from the source/receiver; therefore, the phases of the ring backscatter contributions will differ and there is a possibility of cancellation or increased signal for a finite disk compared to an infinite radius reflector. For a fixed range, the sum depends on the radius of the disk.

For simplicity, we will not consider the small amount of diffraction at the disk edge. Referring to Fig. 1.18, considering the wavefront that travels from the source at Q to the typical scattering area element dS then back to the receiver, a total distance of $2R$, the phase is

$$\Phi = \omega t - 2kR \tag{1.86}$$

$$\text{where } R = (h^2 + r^2)^{1/2} \tag{1.87}$$

In summing the contributions from different elements, as r increases from zero, the quantity of interest is the *phase change*, $2kR$. When we let the smallest value, $2kh$, be the reference phase, the relative phase difference for the element dS is

$$\Delta\Phi = 2kR - 2kh \tag{1.88}$$

Solve for R and then for r as a function of $\Delta\Phi$,

$$R = (r^2 + h^2)^{1/2} = \frac{\lambda\Delta\Phi}{4\pi} + h \tag{1.89}$$

$$r^2 = \frac{h\lambda\Delta\Phi}{2\pi} + \left(\frac{\lambda\Delta\Phi}{4\pi}\right)^2 \tag{1.90}$$

We are interested in the situation where the reflector is many wavelengths from the source, $h \gg \lambda$, in which case (1.90) can be approximated by the first term only. Then

$$\Delta\Phi \cong \frac{2\pi r^2}{h\lambda} \tag{1.91}$$

The contributions are positive for elements at radii such that $\Delta\Phi$ is in the range 0 to π; this is the first phase zone, the central white circle in Fig. 1.18(b). The contributions are negative when Φ is in the range π to 2π; this is the first shaded ring in the figure.

In general, letting $\Delta\Phi = n\pi$, the phase changes sign at radii given by

$$r_n = \left(\frac{\lambda h}{2}\right)^{1/2} n^{1/2} \tag{1.92}$$

Each ring with the same phase sign is called a "Fresnel zone" or a "phase zone."

The central circle, with $n = 1$, is the first Fresnel zone. The radius of this important Fresnel zone, in which there is a minimum of destructively interfering phase change, is

$$r_1 = \left(\frac{\lambda h}{2}\right)^{1/2} \tag{1.93}$$

A complete calculation shows that a signal reflected at a disk having radius r_1 has a maximum acoustic pressure *almost twice as large as for an infinite plate*. On the other hand, when the radius of the disk is r_2 the first zone and oppositely phased second zone both contribute, and the reflected signal is nearly zero. Other maxima and minima alternate as the radius is increased, and as additional zones contribute. For a disk of infinite radius an analytical solution gives the reflected signal proportional to

$$\frac{1}{2h}e^{i(\omega t - 2kh)} \tag{1.94}$$

The interpretation of (1.94) is that the amplitude and phase of the reflected pressure from a very large plate are equivalent to those from a virtual image at distance h behind the reflector.

That is, using the image construction, the reflected pressure seems to come from an image below the interface. The factor $1/(2h)$ appears because, for spherical divergence, the acoustic pressure is inversely proportional to the range. If the plate is not perfectly reflecting, the radiation from the image is multiplied by the fractional reflection coefficient at the interface.

The exact solution of the reflection of a spherical wave at an infinite plane interface is a more difficult theoretical problem than we have implied above. The image approximation in this case is acceptable only when the distances R_1 and R_2 are many acoustic wavelengths, and the angle of incidence is not too near the critical angle.

1.8 Three-dimensional wave equation*

The simple use of an infinite plane wave, or of a point source whose spherical waves at long ranges can be approximated by a local plane wave, have served us well. However, in a realistic world and for many applications, we use wave fields that are more conveniently described in spherical or cylindrical coordinates. Examples are: (1) sound transmissions in shallow water is more effectively stated in cylindrical coordinates; (2) computations of the scattering of sound waves by fish are obviously closer to reality if one uses cylindrical coordinates to describe the bodies; (3) computations of the scattering of sound waves by spheres

* This section contains some advanced analytical material.

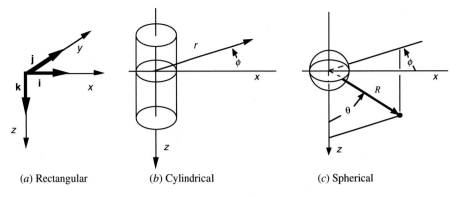

(*a*) Rectangular   (*b*) Cylindrical   (*c*) Spherical

Fig. 1.19. Three coordinates systems. The *z* axis is drawn as positive downward. The (longitude) angles Φ are measured in planes perpendicular to the *z* axis. The (latitude) angle θ is measured to the *z* axis. (*a*) Rectangular. (*b*) Cylindrical. (*c*) Spherical.

are described best in spherical coordinates although an important part of the solution is often accomplished by the expansion of spherical functions in terms of incident plane waves. The three most useful coordinate systems are shown in Fig. 1.19. We will consider the application of each of these geometries when we need them, later in the text.

To generate the wave equation in three dimensions we combine the results for each of the coordinate directions. It is easiest to start with rectangular coordinates, and use the unit vectors **i**, **j**, and **k**, Fig. 1.19.

The *three-dimensional wave equation* in terms of the acoustic pressure, is expressed succinctly as

$$\nabla^2 p = \frac{1}{c^2}\frac{\partial^2 p}{\partial t^2} \tag{1.95}$$

where ∇^2 is the *Laplacian* (the divergence of the gradient of p), which takes different forms in the different coordinate systems. The three dimensional wave equation encompasses the three laws of propagation described in Section 1.6 (see *M&C* Section 2.7.1 for derivations in various coordinate systems).

In rectangular coordinates:

$$\nabla^2 = \nabla \bullet \nabla = \frac{\partial^2}{\partial x^2} + \frac{\partial^2}{\partial y^2} + \frac{\partial^2}{\partial z^2} \tag{1.96}$$

which is the form convenient for plane waves and plane surface interactions.

One uses **cylindrical coordinates** for cylindrically propagating waves such as in shallow water, or waves interacting with cylindrical surfaces. The gradient and Laplacian in coordinates r, z, ϕ are shown in Fig. 1.19(*b*). Note that range in cylindrical coordinates is r, whereas range in spherical coordinates is R. The Laplacian in cylindrical coordinates is

$$\nabla^2 = \frac{\partial^2}{\partial r^2} + \frac{1}{r}\frac{\partial}{\partial r} + \frac{1}{r^2}\frac{\partial^2}{\partial \phi^2} + \frac{\partial^2}{\partial z^2} \tag{1.97}$$

We use the **spherical coordinate system,** as shown in Fig. 1.19(*c*) involving R, θ, ϕ, for spherical waves in free space, or interactions with spherical surfaces. The Laplacian in spherical coordinates:

$$\nabla^2 = \frac{\partial^2}{\partial R^2} + \frac{2}{R}\frac{\partial}{\partial R} + \frac{1}{R^2\sin^2\theta}\frac{\partial^2}{\partial \phi^2} + \frac{1}{R^2}\left(\frac{\partial^2}{\partial \theta^2} + \cot\theta\frac{\partial}{\partial \theta}\right) \tag{1.98}$$

Solutions of the cylindrical and spherical wave equations are considered later, when needed.

1.8.1 Continuous waves in rectangular coordinates

The solution of the wave equation in rectangular coordinates gives a simple illustration of a very powerful method from classical physics, *the separation of variables.* Rectangular coordinates are simple because all of the functions are sines or cosines or complex exponentials. The wave equation in rectangular coordinates is

$$\nabla^2 p = \frac{\partial^2 p}{\partial x^2} + \frac{\partial^2 p}{\partial y^2} + \frac{\partial^2 p}{\partial z^2} = \frac{1}{c^2}\frac{\partial^2 p}{\partial t^2} \tag{1.99}$$

We assume that the waves along the three coordinate directions are independent of one another. Therefore, the pressure can be written as a product of functions of the four variables

$$P = X(x)Y(y)Z(z)T(t). \tag{1.100}$$

Substitution into the wave equation in rectangular coordinates gives

$$X''YZT + Y''XZT + Z''XYT = \frac{T''}{c^2}XYZ \tag{1.101}$$

where X'' is $\partial^2 X/\partial x^2$ and so on. Now, rearrange to

$$c^2\frac{X''}{X} + c^2\frac{Y''}{Y} + c^2\frac{Z''}{Z} = \frac{T''}{T} \tag{1.102}$$

Notice that each term is a function of only one variable. That is, T''/T is only a function of time, and the first term, $c^2 X''/X$, is a function of x only, etc. Therefore, since the equation is true for all values of x, y, z, and t, *each of the terms must be a constant.*

Assume that the time dependence is harmonic and evaluate the constant. For example, try the solutions:

$$T = A_1 e^{i\omega t} \text{ or } T = A_2 e^{-i\omega t} \text{ or } T = A_1 e^{i\omega t} + A_2 e^{-i\omega t} \qquad (1.103)$$

where A_1 and A_2 are constants. They all satisfy

$$\frac{T''}{T} = -\omega^2 \qquad (1.104)$$

For the present illustration we assume that c is constant. Again we use the logical argument that if X, Y, and Z are independent of each other, then X''/X, Y''/Y, and Z''/Z must each equal constants. Using X as an example, we try the solution

$$X = B_1 \exp(ik_x x) + B_2 \exp(-ik_x x) \qquad (1.105)$$

where k_x is the wave number in the x direction.

Evaluation of X''/X gives

$$\frac{X''}{X} = -k_x^2 \qquad (1.106)$$

Similar expressions for Y and Z give

$$\frac{Y''}{Y} = -k_y^2 \qquad (1.107)$$

and

$$\frac{Z''}{Z} = -k_z^2 \qquad (1.108)$$

The substitution of these into (1.102) yields

$$c^2 \left(k_x^2 + k_y^2 + k_z^2 \right) = \omega^2 \qquad (1.109)$$

or

$$k^2 \equiv \frac{\omega^2}{c^2} \qquad (1.110)$$

where

$$k^2 = k_x^2 + k_y^2 + k_z^2 \qquad (1.111)$$

Our solution gives us flexibility to describe any set of plane waves propagating in any direction. For example, the equation of a plane wave traveling in the $+x$, $+y$, and $+z$ directions would be

$$p = P \exp[i(\omega t - k_x x - k_y y - k_z z)] \qquad (1.112)$$

whereas if propagation is in the negative z direction, the sign of the z term changes to $(+k_z z)$.

The vector wave number k which defines the direction of propagation of the plane wave can be resolved into its components in the x, y, and

z directions by knowing the cosines of the angles with the coordinate axes. Usually it is obvious from the context whether the bold letter k is the vector wave number, as here, or the unit vector k in the z direction. Of course, to make it really simple one can often realign the coordinate axes so that the direction of propagation is along one of them!

1.8.2 Omnidirectional continuous waves in spherical coordinates

We assume spherical symmetry. The wave equation is thereby greatly simplified and the solutions are familiar, and useful in many applications.

A very important form of the wave equation is used for the pressure wave caused by a pulsating point source that radiates the same in all directions in an ideal isotropic, homogeneous medium. The wave equation in spherical coordinates, with range and time dependence only, reduces to

$$\frac{\partial^2 p}{\partial R^2} + \frac{2}{R}\frac{\partial p}{\partial R} - \frac{1}{c^2}\frac{\partial^2 p}{\partial t^2} = 0 \qquad (1.113)$$

This can be rearranged to

$$\frac{1}{R}\frac{\partial^2 (Rp)}{\partial R^2} - \frac{1}{c^2}\frac{\partial^2 p}{\partial t^2} = 0 \qquad (1.114)$$

or

$$\frac{\partial^2 (Rp)}{\partial R^2} - \frac{1}{c^2}\frac{\partial^2 (Rp)}{\partial t^2} = 0 \qquad (1.115)$$

This equation is identical to the one dimensional version of the wave equation in rectangular coordinates, (1.56), but with p replaced by Rp and x replaced by R.

The solutions to (1.115) are therefore the same as for plane waves, but with the above replacements. In general, including the functional dependence on time and range,

$$pR = R_0 p_0(ct \pm R) \quad \text{or} \quad p = \frac{p_0(ct \pm R)R_0}{R} \qquad (1.116)$$

where $p_0(ct \pm R)$ describes the acoustic pressure as any function of $(ct \pm R)$. The $\pm$ sign in (1.116) is written to indicate that the pressure p may have either, or both, functional dependences. The choice $(ct - R)$ would designate an outward radiating wave such as the wave from a pulsating sphere. The choice $(ct + R)$ would give an inward radiating wave which might be, for example, an implosion or the inward reflection from a spherical reflector. In many situations, the waves are known to be traveling away from a source, therefore the inward traveling wave is dropped on physical grounds to keep it simple.

Normally, p_0 is the sound pressure referred to the unit distance R_0, usually 1 meter.

Acoustic pressure for sinusoidal, omnidirectional waves

In general, the solution of (1.115) is the product of a radial function and a temporal function

$$p = P\mathcal{R}(R)T(t) \tag{1.117}$$

For sinusoidal waves, the time-dependent function is a sine or a cosine, written

$$T = e^{i\omega t} \tag{1.118}$$

Substitution of (1.118) into the omnidirectional spherical wave equation (1.113) gives,

$$\frac{1}{R}\frac{\partial}{\partial R^2}(R\mathcal{R}) + k^2\mathcal{R} = 0 \tag{1.119}$$

where $k = \omega/c$.

The radial wave equation has two independent solutions

$$\mathcal{R} = \frac{\cos(kR)}{R} \quad \text{and} \quad \mathcal{R} = \frac{\sin(kR)}{R} \tag{1.120}$$

In complex notation the outward propagating sinusoidal wave is

$$p = \frac{P_0 R_0}{R} e^{i(\omega t - kR)} \tag{1.121}$$

where the amplitude is written in terms of the pressure P_0 at range R_0.

Particle velocity

Computations of the particle velocity start with the acoustic force equation. For the case of spherical symmetry there is only the radial component which is similar to (1.50) so we write simply

$$\frac{\partial p}{\partial R} = -\rho_A \frac{\partial u}{\partial t} \tag{1.122}$$

When we use u in complex exponential notation, $\frac{\partial u}{\partial t} = i\omega u$, so that (1.122) becomes

$$u = -\frac{1}{\rho_A}\frac{\partial p}{\partial R}\left(\frac{1}{i\omega}\right) \tag{1.123}$$

Form $\frac{\partial p}{\partial R}$ from (1.121), and get

$$\frac{\partial p}{\partial R} = -p\left(\frac{1}{R} + ik\right) \tag{1.124}$$

Therefore,

$$u = \frac{p}{\rho_A c}\left(1 - \frac{i}{kR}\right) \tag{1.125}$$

The important conclusion is that close to the source, small kR, the particle velocity has a "quadrature" component (the imaginary term) which lags the acoustic pressure by $90°$. A similar conclusion is found for an impulse sound such as an explosion where there is a great "whooshing" motion as the medium moves outward following the pressure wave. The non-propagating energy carried by out-of-phase components such as exist for small kR in (1.125) has been discussed by Stanzial *et al.* (1996).

At large kR, the particle velocity in the spherical wave is much simpler. It is proportional to the sound pressure, and abides by the same equation as we derived for a plane wave (1.58).

Far field intensity

In Section 1.2 we used the conservation of energy to show that the intensity of sound from an outwardly radiating omnidirectional source is proportional to i_0/R^2, (1.3). We now know that for pressure signals that satisfy $kR \gg 1$ the particle velocity along the direction of R is proportional to the pressure. Therefore, for outward propagating waves, now inserting the functional dependence $(ct - R)$, at many wavelengths range,

$$u_R \cong \frac{p_0(ct - R)R_0}{\rho_A c R} \quad \text{for} \quad kR \gg 1. \tag{1.126}$$

The instantaneous intensity i_R is $p u_R$ and, *when calculated at long ranges*, it is simply the product of p and u_R. Therefore,

$$i_R = p u_R = \frac{p_0^2(ct - R)R_0^2}{\rho_A c R^2} \tag{1.127}$$

We have come full circle and verified our assumed solutions in Section 1.2.1. Using our properly derived solutions to the wave equation we have proven that, *in the far field of a point source,* as the range increases both the sound pressure and particle velocity decrease as $1/R$, and the intensity decreases as $1/R^2$.

Problems and some answers

1.1 Use a sketch and Huygens wavelets to derive the law of reflection for plane waves incident at a plane interface.

1.2 Use a sketch and Huygens wavelets to derive Snell's Law of Refraction for spherical waves from a point source.

1.3 Drop a small object in a shallow pond, or a bathtub, to observe reflection of a circular wave from a nearby small plane reflector, or a small concave curved reflector, or a small convex reflector. Sketch what you observe and identify the reflected waves and diffracted waves.

1.4 Excite a ripple by briefly touching the surface of water. Observe and compare the wave with the waves from a harmonically disturbed surface when you wiggle your finger up and down in the water. Compare for different frequencies of the "wiggle." Repeat the experiment using two adjacent fingers of one hand; comment. Repeat using two separated fingers of one hand: comment.

1.5 Use your computer, or pencil and paper, to plot $p = P \sin \omega t$ over a range from $\omega t = -2\pi$ to $\omega t = +3\pi$. Do the same for $p = P \cos \omega t$. Compare with $p = P \cos (\omega t + \pi)$. Compare with $p = P \cos (\omega t + \pi/2)$.

1.6 Use your computer, or pencil and paper, to add $p = P \sin \omega t$ to $p = P \sin 2\omega t$. Now add $p = P \sin \omega t$ to $p = P \sin 3\omega t$. Comment on the effect of adding these harmonics.

1.7 Use your computer, or pencil and paper, to add $p = P \sin \omega t$ to $p = P \sin (\omega t + \phi)$ and observe the changing wave form when ϕ takes on values that range from $0°$ to $180°$.

1.8 Later we discuss the various sounds in the sea. Consider a 20 Hz sound of pressure amplitude P_0, which radiates omnidirectionally from a whale which is 2 meters below a smooth (perfectly reflecting) ocean surface. (*a*) What is the total sound pressure as a function of angle at range 1000 m? (*b*) What is it if the whale is 10 m deep? (*c*) If 100 m deep (see Fig. 1.10)?

1.9 Assume that the sound from a plane wave is normally incident at a smooth, hard, perfectly reflecting bottom. A hydrophone is fixed on a tripod two meters above the bottom. The incident sound has equal pressure components of magnitude P at frequencies 1000 Hz, 2000 Hz, 5000 Hz. What is the total (incident plus reflected) pressure at the hydrophone for each of these frequencies? Comment about the dependence of pressure measurements on the height of the hydrophone. What would be the effect if the bottom were only partially reflecting?

1.10 The speed of sound in gases can be calculated from the equation for adiabatic propagation, $p\rho^{-\gamma} = $ constant where γ is the ratio of specific heats of the gas.
(*a*) Form $dp/d\rho$ and show that the speed of sound in gases is given by $c = (\gamma P_A/\rho_A)^{1/2}$.
(*b*) For air $\gamma = 1.4$. Calculate the speed of sound in air at sea level where $P_A = 10^5$ pascals and $\rho_A = 1.29$ kg/m³.

1.11 The particle velocities in air and in the water at the air–water ocean surface interface are necessarily the same because the air is in contact with the water. (*a*) Compare the pressures in air and water, for the same particle velocities. Assume the ambient constants:

$\rho_A = 1.29$ kg/m^3, $c_A = 335$ m/s; $\rho_w = 1000$ kg/m^3, $c_w = 1500$ m/s.

(b) From the previous answer, comment on the effectiveness of an air sound source when used in water, or an underwater sound source when used in air. Assume that the source transducer maintains a constant velocity at its face, regardless of the medium.

1.12 Plot graphs of magnitude and phase of $\mathcal{R}_{12}$ and $\mathcal{T}_{12}$ versus θ_1 for plane wave propagation from water to sediment. Use values $\rho_1 = 1000$ kg/m^3, $c_1 = 1500$ m/s; $\rho_2 = 1400$ kg/m^3, $c_2 = 1480$ m/s.

1.13 Plot graphs of magnitude and phase of $\mathcal{R}_{12}$ and $\mathcal{T}_{12}$ versus θ_1 for plane wave propagation fom water to sediment. Use values $\rho_1 = 1000$ kg/m^3, $c_1 = 1500$ m/s; $\rho_2 = 2000$ kg/m^3, $c_2 = 2000$ m/s.

1.14 Plot graphs of magnitude and phase of $\mathcal{R}_{12}$ and $\mathcal{T}_{12}$ versus θ_1 for plane wave propagation from water to air. Use values $\rho_1 = 1000$ kg/m^3, $c_1 = 1500$ m/s; $\rho_2 = 1.3$ kg/m^3, $c_2 = 340$ m/s.

1.15 Plot graphs of magnitude and phase of $\mathcal{R}_{12}$ and $\mathcal{T}_{12}$ versus θ_1 for plane wave propagation fom air to water. Use values $\rho_1 = 1.3$ kg/m^3, $c_1 = 340$ m/s; $\rho_2 = 1000$ kg/m^3, $c_2 = 1500$ m/s.

1.16 Solve for the plane wave reflection and transmission coefficients for particle velocity rather than for pressure.

(a) Calculate these coefficients as a function of θ_1 for propagation from air to water.

(b) Calculate these coefficients as a function of θ_1 for propagation from water to air.

1.17 Solve for the plane wave reflection and transmission coefficients for sound intensity rather than for pressure, at an interface.

1.18 Solve for the plane wave reflection and transmission coefficients for sound power rather than for pressure at an interface. Comment on the difference between the coefficients for sound intensity and those for sound power at an interface.

1.19 The wave equation has been given in terms of the acoustic pressure. Re-derive the wave equation in terms of the acoustic density. Re-derive in terms of the particle velocity in the x direction. Re-derive in terms of the particle displacement in the x direction. Re-derive in terms of the excess temperature for a wave propagating in a gas describable by $p_A = \rho_A R_G T_A$ where p_A is the ambient pressure, ρ_A is the ambient density, R_G is a gas constant, T_A is the ambient temperature in Kelvin.

1.20 A pair of sources are on the z axis at $+5$ meters and -5 meters. The amplitudes are a and the frequency f is 1500 Hz. Assume the speed of sound is 1500 m/s. Calculate and graph the interference as a function of angle at range 2000 m.

1.21 Same geometry as in previous problem, but now there are sources, at -7.5 m, -2.5 m, $+2.5$ m, and $+7.5$ m. Calculate and graph the interference.

1.22 Assume that echo sounders of frequency 3.5 kHz, 12.5 kHz, and 45 kHz are available and that the sound speed is 1500 m/s. Estimate the radius of the first phase zone at the ocean bottom for these three instruments operating from the surface, at the following water depths: 100 m, 1000 m, 3200 m. **Answers:** At depth 100 m, for $f = 3.5$ kHz, 12.5 kHz, 45 kHz, radii 4.6 m, 14.6 m, 26 m.

1.23 Wave equation (1.95) is in terms of the acoustic pressure. (*a*) Re-derive the wave equation in terms of the acoustic density. (*b*) Re-derive in terms of the particle velocity in the x direction. (*c*) Re-derive in terms of the particle displacement in the x direction. (*d*) Re-derive in terms of the excess temperature for a wave propagating in a gas describable by $p_A = \rho_A R_G T_A$ where p_A is the ambient pressure, ρ_A is the ambient density, R_G is a gas constant, T_A is the ambient temperature in Kelvin.

Further reading

Colladon, J. D. and Sturm, J. K. F. (1827). The compression of liquids (in French), *Ann. Chim. Phys.* Series **2**(36), part IV, Speed of sound in liquids 236–57.
The first accurate measurements of the speed of sound in water, performed with two ships, an underwater bell, the light of a gunpowder explosive, and a Swiss stop watch. (See Fig. 1.1.)

Baker, B. B. and Copson, E. T. (1950). *The Mathematical Theory of Huygens' Principle*. Oxford: Oxford University Press.
Mathematical justification for the wave propagation concepts that were intuitively proposed by Christian Huygens in the seventeenth century, after he carefully observed the propagation of water waves on the surface of a lake.

Chapter 2
Transmission and attenuation along ray paths

Summary

The wave concepts based on Huygens' Principle, which were used in Chapter 1, have guided the field of sound propagation, including diffraction, reflection, refraction, and interference, for almost four centuries. In this chapter we first assume, for simplicity, that the sound energy in the ocean moves along frequency-independent ray paths from the source to the receiver. The development of ray methods is particularly appropriate to the use of impulse sources because the different travel times of multiple arrivals provide invaluable information that is not obtained in a simple continuous wave (CW) frequency description.

Generally, temperature, salinity, and depth vary along the ray path in the ocean. Therefore, the sound speed at sea is a function of position. The spatially varying sound speed can cause even greater ray divergence and a greater rate of decrease of sound intensity than simple spherical divergence. On the other hand, in some cases the rays may converge so that there may actually be an increase in sound intensity at certain increased ranges. A simple ray treatment of multi-path transmissions in typical oceans aids in the understanding of the time-varying sound intensity at distant points.

At large propagation ranges, the incremental divergence loss in dB/meter, which was significant near the source, becomes less than the loss in dB/meter due to energy absorption and scatter out of the ray direction. This spatial attenuation, which is described best in terms of its dependence on sound frequency, is due both to energy absorption and scatter out of the ray direction. Attenuation by energy absorption is a physical phenomenon caused principally by shear viscosity and molecular relaxation; the latter is particularly sensitive to the sound frequency. It is common practice to correct the ray predictions of sound intensity by simply using the separately calculated loss due to frequency-dependent attenuation. Those formulae, which give the rate of attenuation due to energy absorption by viscous and thermal losses as well as by molecular relaxation phenomena, were derived and tested toward the end of the twentieth century.

However, the frequency-dependent attenuation due to scatter is the more interesting component, because it provides an opportunity to determine the number and characteristics of the physical and biological scatterers along those ocean paths, as will be seen later.

Propagation is described by researchers in terms of frequencies or times, and waves or rays, depending on which formulation is more fruitful.

A little history of rays and waves

When differential wave equations were derived in the nineteenth century, they were, for the most part, used to understand the propagation of *harmonic* waves – that is, waves described in terms of frequency as parameter. Frequency was a natural parameter because of the human perception of musical sounds, in which there is a close connection between pitch, frequency, and the harmonic, sinusoidal components that determine the quality of a musical sound.

Although it failed to describe correctly wave phenomena such as interference and diffraction, Isaac Newton's seventeenth-century concept of corpuscular energy moving along "rays" radiating from a source proved to be fruitful in the description of reflection by mirrors and refraction by lenses. Ray-path techniques are popular today not only in optics but also in underwater acoustics and seismology; their great virtue is that they provide some insight to propagation in complex environments.

The first effective use of an underwater impulse source (a bell) was the measurement of the speed of sound in 1827 (Fig. 1.1). By the 1930s, pulsed ocean sources, "sonars," had been introduced to determine distances, by measuring the time for reflections of the ping off the sea floor or from schools of fish. In those early days, some underwater acousticians still assumed that the sound paths could be described by straight line rays.

Modern ocean acoustics stands on the shoulders of research done in World War II, 1939–45. For example, during the war, Ewing and

Worzel extended their experience in geophysics by using explosives as impulse sources and following ray paths. They discovered and exploited the deep sound SOFAR (SOund Fixing And Ranging) channel, where the sound from a few kilograms of explosive can be heard at very great ranges across the ocean. In 1960 they dropped 100 kg-depth charges in the Pacific Ocean off Perth, Australia, and found that the low-frequency sounds reached hydrophones half-way around the world, off the coast of Bermuda, in the Atlantic Ocean 3.5 hours later.

Our development of ray methods emphasizes the use of impulse sources because the different travel times of multiple arrivals provide valuable information not contained in the frequency descriptions.

As we shall see later, there is a duality in the two descriptions, and the two parameters are often advantageously interconverted.

Contents

2.1 Energy transmission in ocean acoustics*

2.1.1 Impulse sources

Chemical explosions, airguns, and sparks are commonly used as short duration "impulse" sources for sound transmissions. The underwater implosion of a collapsing light bulb also produces a simple impulse pressure. These sources initially radiate sound omnidirectionally.

Typically the sound pressure is very large at the instant of the explosion and then decays rapidly. Depending on the size of the explosive, the duration of the peak pressure pulse is usually a few milliseconds. For convenience, physicists and engineers sometimes replace the explosive exponential wave form by a simplified impulse wave with time dependence given by the ideal *delta function* $\delta(t)$ (Fig. 2.1). A definition of

---

* This section contains some advanced analytical material.

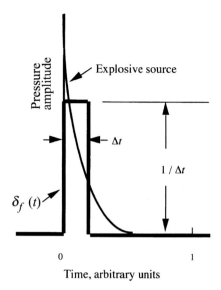

Fig. 2.1. Explosive pressure represented by a single discrete delta function $\delta_f(t)$ of duration Δt and amplitude $1/\Delta t$.

the finite/discrete delta function that can be used in digital calculations is

$$\delta_f(t) = 1/\Delta t \quad \text{for} \quad 0 \leq t \leq \Delta t \tag{2.1}$$

and $\delta_f(t) = 0$ for all other t. Also,

$$\int_0^{\Delta t} \frac{dt}{\Delta t} = 1 \tag{2.2}$$

Commonly Δt is one time step, the sampling interval. In the limit, one can let $\Delta t \to 0$ and still require that the integral be unity. This limiting form commonly appears in the literature. In its proper usage, the delta function is defined by an integration operation. For a function $g(t)$

$$\int_{-\infty}^{\infty} g(t)\,\delta\,(t - t_1)\,dt = g(t_1) \tag{2.3}$$

The delta function integration evaluates $g(t)$ at $t = t_1$. The concept was introduced by the physicist P. A. M. Dirac, and is known as the Dirac delta function.

2.1.2 Pressure, particle velocity, and intensity in a pulse

From the preceding discussion of energy conservation, the *far-field* sound pressure, particle velocity, and instantaneous intensity for waves spreading spherically from a point source in a lossless, iso-speed medium are a function of $(t - R/c)$

$$p(R, t) = p_0(t - R/c) \frac{R_0}{R} \tag{2.4}$$

$$u_R(R, t) \approx \frac{p}{\rho c} = \frac{p_0(t - R/c)}{\rho c} \frac{R_0}{R} \tag{2.5}$$

$$i_R(R, t) = p u_R = \frac{p_0^2(t - R/c)}{\rho c} \frac{R_0^2}{R^2} \tag{2.6}$$

where $p_0(t)$ is the source pressure at the range R_0, and $p(t)$ is the outward traveling pressure wave. Although there is a time delay and amplitude change, the waveforms of $p_0(t)$ and $p(t)$ are the same in such a medium.

A suitable transient signal $p_0(t)$ satisfies two conditions:

$$p_0(t) = 0 \quad \text{(for } t < 0) \tag{2.7}$$

and

$$\int_0^\infty |p_0(t)| dt \quad \text{is finite} \quad \text{(for } t \geq R/c) \tag{2.8}$$

Assume for simplicity that the far-field pressure after an explosion (actual details later) is approximated by the exponential decay

$$p(R, t) = \frac{p_0 R_0}{R} \exp\left[-\frac{1}{\tau_s}\left(t - \frac{R}{c}\right)\right] \tag{2.9}$$

where τ_s is the time for the pressure to decay to e^{-1} of its peak value p_0 at 1 m. For $t \geq R/c$,

$$\int_{\frac{R}{c}}^{\frac{R}{c}+t_g} \frac{p_0 R_0}{R} \exp\left[-\frac{1}{\tau_s}\left(t - \frac{R}{c}\right)\right] dt$$
$$= \frac{\tau_s p_0 R_0}{R}\left[1 - \exp\left(\frac{t_g}{\tau_s}\right)\right] \tag{2.10}$$

where t_g is the duration of the integral after the arrival of the pressure signal or the "gate time," for acceptance into the integration operation. As the gate time, t_g, goes to infinity, the integral tends to $\tau_s p_0 R_0/R$. The integral is within 98 percent of the limit for $t_g > 4\tau_s$.

2.1.3 Energy transmission in a pulse

At many wavelengths from a point source in the "far field," the radial component of particle velocity (units, m/s) and the instantaneous radial intensity (units, watts/m^2) are, respectively,

$$u_R = p/(\rho_A c) \quad \text{and} \quad i_R = p u_R \tag{2.11}$$

For acoustic transmission in a lossless medium, one can define the "message" energy, ΔE_m (units, joules), which passes through an element of surface ΔS at range R in gate open time t_g and is a function of $(t - R/c)$

$$\Delta E_m = \Delta S \int_{\frac{R}{c}}^{\frac{R}{c}+t_g} pu_R dt \approx \frac{R_0^2 \Delta S}{\rho_A c} \int_{\frac{R}{c}}^{\frac{R}{c}+t_g} \frac{[p_0(t-R/c)]^2}{R^2} dt \quad (2.12)$$

where the integration limits are over the effective duration of the transient pressure, t_g.

The element of surface area can be expressed as

$$\Delta S = R^2 \Delta \Omega \quad (2.13)$$

where $\Delta \Omega$ is an element of solid angle.

All energy flows outward through the spherical surface and along R. The radiation from a directional source can be expressed in terms of the solid angle subtended by ΔS. The entire solid angle viewed from the center of a sphere is 4π. Assuming the source radiation is spherically symmetric, the integral over 4π of solid angle, which gives the total energy, E_m, in the "message," is simply (2.12) with $\Delta S = 4\pi R^2$. (Sometimes we drop the subscript m for a general statement of radiated energy and call it simply E.)

Therefore, for isotropic radiation of a pulse,

$$E_m \approx \frac{4\pi R_0^2}{\rho_A c} \int_{\frac{R}{c}}^{\frac{R}{c}+t_g} [p_0(t-R/c)]^2 dt \quad (2.14)$$

The total energy transmitted by the source is given in joules, or watt seconds.

The *energy per unit area* (joules/m²), measured at range R, transmitted during the time interval t_g is

$$\varepsilon_R \equiv \int_{\frac{R}{c}}^{\frac{R}{c}+t_g} pu_R \, dt \approx \frac{[tips]}{R^2} \frac{R_0^2}{\rho_A c} \quad (2.15)$$

where (tips) is the time interval of the pressure squared (2.14).

Equations (2.14) and (2.15) are energy equations for spherically divergent transmission in an ideal, lossless, homogeneous medium. They give the total sound energy, and the sound energy per unit area, respectively, of a single transient arrival that has the travel time R/c.

2.1.4 Power radiated by a continuous wave signal

Assume a point source transmits the continuous wave sinusoid written in exponential form as $P_0 e^{i2\pi ft}$ with peak value P_0. The source power is the average energy transmitted per unit time over the period T of the sinusoidal wave. If there is no loss by scatter or absorption, the total power Π (watts) over all angles at a large range is the same as at the source:

$$\Pi \approx \frac{4\pi R_0^2}{\rho_A c} \frac{P_0^2}{2} \quad (2.16)$$

where ρ_A is the ambient density

$$\text{and} \quad P_0^2/2 = \frac{1}{T}\int_0^T |P_0\sin(2\pi ft)|^2\,dt \tag{2.17}$$

Sometimes $(P_0^2)/2$ is written P_{rms}^2 for a sinusoidal wave.

In addition to sinusoidal source transmissions (2.16) and (2.17) may be used also to define the *average* power of the transducer output. In that case, the time T is the duration of the transmission.

2.2 Ray paths and ray tubes

The solution of the spherical wave equation in Section 1.8 is the start of our discussion of ray paths. That solution, using the separation of variables, is the product of a spatial function and a temporal function. We cast our discussion of ray paths in that context. First, the exact solution is taken apart, and each term is identified as either space- or time-dependent.

In a homogeneous, isotropic, lossless medium, the exact spherical wave solution for the space–time variation in the acoustic pressure is a function of $(t - R/c)$,

$$p(t, R) = p_0(t - R/c)\frac{R_0}{R} \tag{2.18}$$

The sound input to the medium is the temporal function $P_0(t)$. The travel time for the message to arrive, R/c, and the dimensionless ratio, R_0/R, are components of the spatial function. The temporal output function at the receiver is $p(t)$.

2.2.1 Reflections along ray paths

The Hugyens construction of a reflected wave was shown in Fig. 1.4. In a ray-path construction, the reflected wave appears to come from an "image" source behind the interface, as shown in Fig. 2.2.

We assume that there is a point source, that the medium is homogeneous and lossless, that propagation is the same in all directions (isotropic) and that the interfaces are smooth. We make the approximation that the ray is reflected locally as if it were a plane wave. Then the plane-wave reflection coefficient is computed for the local angle of incidence using (1.76) for simple reflections and (1.80) through (1.84) for reflections beyond the critical angle. Figure 2.2 shows the ray paths for reflections at the bottom and top interfaces.

The pressure signal for the bottom reflected path, shown in Fig. 2.2, is

$$p(t) = p_0\left(t - \frac{R_a + R_b}{c}\right)\frac{R_0}{R_a + R_b}\mathcal{R}_{12} \tag{2.19}$$

where $\mathcal{R}_{12}$ is the pressure reflection coefficient at the water/bottom, 1, 2, interface.

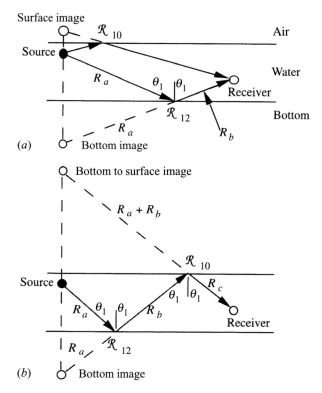

Fig. 2.2. Image constructions of (a) singly reflected rays, and (b) a doubly reflected ray path. The segments of the ray paths are labeled $R_a, R_b,$ and R_c. The direct ray path from the source to the receiver is not shown. It is assumed that the sound speed is the same at all points of the water medium.

The travel time is

$$t_{path} = (R_a + R_b)/c \tag{2.20}$$

The pressure signal for the bottom and surface reflected ray path, shown in Fig. 2.2, is

$$p(t) = p_0\left(t - \frac{R_a + R_b + R_c}{c}\right)\frac{R_0}{R_a + R_b + R_c}\mathcal{R}_{12}\mathcal{R}_{10} \tag{2.21}$$

where $\mathcal{R}_{10}$ is the pressure reflection coefficient at the water–air interface.
    The time delay is

$$t_{path} = (R_a + R_b + R_c)/c \tag{2.22}$$

2.2.2 Multiple ray paths

Transmissions from a source can travel by many paths, and for each path the signal has its own amplitude, phase, and time delay t_{path}. It is

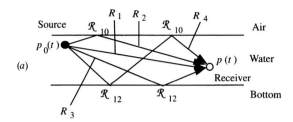

Fig. 2.3. (*a*) Four ray paths for a transmission from an explosion. The ray paths are labeled R_1, R_2, R_3, and R_4. (*b*) The top trace is the source signal $p_0(t)$. The bottom trace is the pressure signal at the receiver $p(t)$. The reflection coefficients are assumed to be $\mathcal{R}_{12} = 0.3$ and $\mathcal{R}_{10} = -1$, for all angles of incidence. Note the phase shifts of 180° for the surface reflected paths, 2 and 4 and the assumed zero degree phase shift for the bottom reflected paths, 3 and 4.

important to recognize that, in acoustic propagation, the signal pressure amplitudes for each path add vectorally and may therefore interfere at the receiver. Sketches of four paths are shown in Fig. 2.3(*a*). The sound speed is assumed to be the same at all points in the water.

Ray path R_1 is direct. The other ray paths have reflections at the surface and bottom.

A model source transmission, such as from an explosive, has the exponentially decaying $p_0(t)$ labeled 1 in the trace of Fig. 2.3(*b*). It is constructed by delaying $p_0(t)$ by the travel time R_1/c and multiplying $p_0(t)$ by the amplitude factor that depends on the range. Similarly, arrival 2 has the time delay R_2/c and a magnitude that depends on the range and reflection coefficient. Since arrivals 1 and 2 overlap, the signal pressure is the algebraic sum of the pressures. Arrivals 3 and 4 are calculated similarly.

Figure 2.3 is an example of sound transmissions to ranges of about three water depths. In this case straight ray paths are good approximations to the actual ray paths.

2.2.3 Conservation of energy in ray tubes

Ray tubes are constructed by drawing many rays from a source to an area of the wavefront; the rays form the surface of the ray tube, Fig. 2.4. For a constant speed (iso-velocity) medium, the rays are straight lines,

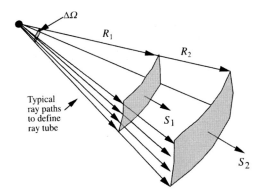

Fig. 2.4. Ray-tube construction for a homogeneous, non-absorbing, non-scattering, iso-speed medium. Only a few of the rays that bound the outside surface of the tube are shown for clarity. The vectors representing the surface areas, S_1 and S_2 are along the particular ray path through the center of the tube and at ranges R_1 and R_2.

the solid angle $\Delta\Omega$ that they enclose is constant, and the perpendicular areas of the ends of the ray tube are

$$S_1 = (\Delta\Omega)R_1^2 \tag{2.23}$$

and

$$S_2 = (\Delta\Omega)R_2^2 \tag{2.24}$$

Since ideal rays are normal to wavefronts, no energy passes through the sides of the ray tube. Thus, if there is no sound absorption or scattering, the energy passing through the area S_2 equals the energy which passed through S_1.

Now consider the inhomogeneous, lossless (non-absorbing, non-scattering) medium, in which c and ρ_A depend on position. Put the energy (2.12) in terms of the local pressures p_1 and p_2, and replace ΔS by S_1 or S_2. The equality of the energy entering and leaving the section of the ray tube gives

$$\frac{S_1}{\rho_{A1}c_1} \int_{\frac{R_1}{c_1}}^{\frac{R_1}{c_1}+t_g} [p_1(t - R_1/c_1)]^2 \, dt$$
$$= \frac{S_2}{\rho_{A2}c_2} \int_{\frac{R_2}{c_2}}^{\frac{R_2}{c_2}+t_g} [p_2(t - R_2/c_2)]^2 \, dt \tag{2.25}$$

where the subscripts on $\rho_{A1}, c_1, \rho_{A2}$, and c_2 indicate the local ambient densities and sound speeds and where the acoustic pressures, p_1 and p_2 are explicitly shown as functions of time and position.

2.2.4 Sound pressures in ray tubes

The sound pressures at the surfaces S_1 and S_2 follow the $1/R$, spherical divergence, rule of (2.19) in a homogeneous, non-scattering, non-absorbing medium. In such *homogeneous* media, $\rho_{A1}c_1 = \rho_{A2}c_2$, the integral (2.25) is an accurate approximation of the energy propagation, and the wavefronts are nearly spherical. These expressions are the starting place for calculations of the sound pressure in a ray tube. The changes of variable

$$\tau_1 = t - R_1/c \quad \text{and} \quad \tau_2 = t - R_2/c \qquad (2.26)$$

in the integral (2.25) gives

$$S_1 \int_0^{t_g} [p_1(\tau_1)]^2 \, d\tau = S_2 \int_0^{t_g} [p_2(\tau_2)]^2 \, d\tau \qquad (2.27)$$

The limits of integration t_g can be chosen so that $p_1(\tau_1)$ and $p_2(\tau_2)$ are essentially zero for $\tau > t_g$. Rearrangement of (2.27) gives

$$\int_0^{t_g} \{S_1[p_1(\tau_1)]^2 - S_2[p_2(\tau_2)]^2\} d\tau = 0 \qquad (2.28)$$

When $p_1(\tau_1)$ and $p_2(\tau_2)$ are proportional, one can set the contents of the bracket $\{\ldots\}$ to zero and write

$$p_2(\tau_2) = \pm p_1(\tau_1) \sqrt{\frac{S_1}{S_2}} \qquad (2.29)$$

The $+$ sign is chosen for the tube sketched in Fig. 2.4 because the areas are simple projections on an expanding wavefront.

2.3 Ray paths in a refracting medium*

Profiles of the sound speed versus depth in oceans, lakes, and rivers are often complicated. However, there are some simple approximations. Since water tends to stratify with the density, which increases with depth, the regions of constant salinity and constant temperature and the corresponding sound speed profiles are nearly horizontal in local regions. Currents disrupt horizontal stratifications and cause boundaries, or fronts, between different types of water. Oceanic fronts, such as the boundaries of the Gulf Stream, and many coastal regions that are traversed by interior wave packets called "solitons" have very complicated and time-sensitive structures.

2.3.1 Sound speed in the ocean

The sound speed in water does not depend on the direction of the ray. Therefore, the names "sound speed" and "sound velocity" etc. often can

---

* This section contains some advanced analytical material.

be used interchangeably. *In-situ* instruments called *sound velocimeters* operate by timing a megahertz pulse in a small "sing-around" water circuit in the instrument. Speed accuracies of 0.1 m/s are claimed. However, bubbles can cause deviations of several meters per second.

When the sound speed is to be determined from temperature, salinity, and depth, the best values are obtained by using the empirical formulation of Del Grosso (1974). But that equation has 19 terms, 18 of which have coefficients with 12 significant figures each. The following much simplified formula (Medwin, 1975) has less than 0.2 m/s error compared with Del Grosso for $0 < T\,^\circ\text{C} < 32$ and $22 < \text{Salinity ppt} < 35$ for depths under 1000 m:

$$c = 1449.2 + 4.6T - 0.055T^2 + 0.00029T^3$$
$$+ (1.34 - 0.01T)(S - 35) + 0.016z \tag{2.30}$$

where c = sound speed (m/s); T = temperature ($^\circ$C); S = salinity (‰; i.e., parts per thousand); and z = depth (m).

Mackenzie (1981) gives a longer formula, which claims a standard error of 0.07 m/s and is not restricted to depths less than 1000 m or a narrow range of salinities:

$$c = 1448.96 + 4.591T - 5.304 \times 10^{-2}T^2 + 2.374 \times 10^{-4}T^3$$
$$+ 1.340(S - 35) + 1.630 \times 10^{-2}z + 1.675 \times 10^{-7}z^2 \tag{2.31}$$
$$- 1.025 \times 10^{-2}T(S - 35) - 7.139 \times 10^{-13}Tz^3$$

2.3.2 Refraction at an interface

Ray paths obey Snell's Law as the waves propagate in water layers of different speeds (Fig. 2.5):

$$\frac{\sin\theta_0}{c_0} = \frac{\sin\theta_1}{c_1} = \frac{\sin\theta_z}{c_z} = a = \text{constant} \tag{2.32}$$

where c_0 is the initial sound speed, θ_0 is the initial incident angle with the vertical, and θ_1 is the refracted angle. The constant a is known as the *ray parameter*. Suppose we start a ray at a depth z_0 and sound speed c_0. By Snell's Law, the incident angle is θ_z at a depth z, where the sound speed is c_z.

2.3.3 Rays through constant sound speed layers

Now we follow a transmitted ray as it progresses through a continuous medium that is, *for convenience of calculation*, split into several constant speed layers. Since the sound speed in each layer is constant, the ray path within each layer is a straight line. Ray trace calculations use Snell's Law at the interfaces between layers to follow rays through a medium. The

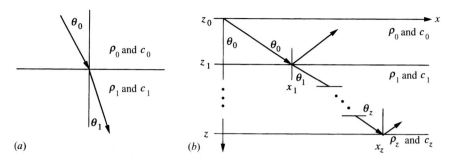

Fig. 2.5. Snell's Law and ray traces. (*a*) Refraction at an interface. (*b*) Refraction in a multilayered medium. The incident, reflected, and transmitted rays are shown.

calculations are usually done on a computer, because the same algorithm is applied repeatedly.

Let the ray leave the source at an angle θ_0 with the normal. Then the "ray parameter" is

$$a = \sin \theta_0 / c_0 \tag{2.33}$$

At the $(n + 1)$th interface, $\sin \theta_n$ is,

$$\sin \theta_n = ac_n \tag{2.34}$$

and the other trigonometric functions are

$$\cos \theta_n = \sqrt{1 - (ac_n)^2} \quad \text{for} \quad ac_n < 1 \tag{2.35}$$

and

$$\tan \theta_n = \frac{ac_n}{\sqrt{1 - (ac_n)^2}} \tag{2.36}$$

The ray parameter a is convenient because $\sin \theta_0 / c_0$ is computed once, and the rest of the functions are computed using a in (2.34) through (2.36). Refering to Fig. 2.5, we write an expression for computing the horizontal range of the ray trace to its intersection at the Nth interface

$$x_N = \sum_{n=0}^{N-1} (z_{n+1} - z_n) \tan \theta_n \tag{2.37}$$

The travel time for the ray to travel to the Nth interface is

$$t_N = \sum_{n=0}^{N-1} \frac{(z_{n+1} - z_n)}{c_n \cos \theta_n} \tag{2.38}$$

For ranges less than a few hundred meters, and near vertical angles of incidence, the ray paths are nearly straight lines. We can use the constant sound-speed approximation in most sonar applications where the ranges are less than a few hundred meters. *Note*: Many people who do ray tracing prefer to use these equations in terms of the complementary grazing angle ϕ, in which case all $\sin\theta$ are written $\cos\phi$, and $\cos\theta$ are written $\sin\phi$.

2.3.4 Rays through slowly changing sound-speed layers

Assume the sound speed profile, $c(z)$, is a continuous function of depth. Along a ray, the differential distance ds and time dt are

$$ds = dz/\cos\theta \quad \text{and} \quad dt = ds/c(z) = dz/[c(z)\cos\theta] \qquad (2.39)$$

Let the ray be in the x–z plane and measure the horizontal range as distance r. The differential displacement dr is

$$dr = dz\tan\theta \qquad (2.40)$$

where, using the ray parameter a in Snell's Law, the trigonometric functions are given by (2.33) through (2.36) and where c_n becomes $c(z)$. The travel duration and path length are

$$t_f - t_i = \int_{z_i}^{z_f} dt = \int_{z_i}^{z_f} dz \frac{1}{c(z)\sqrt{1 - a^2 c^2(z)}} \qquad (2.41)$$

and

$$r_f - r_i = \int_{z_i}^{z_f} dr = \int_{z_i}^{z_f} dz \frac{ac(z)}{\sqrt{1 - a^2 c^2(z)}} \qquad (2.42)$$

The sound speed in the nth layer is given by

$$c(z) = c(z_n) + b_n(z - z_n) \quad \text{for} \quad z_n \le z \le z_{n+1} \qquad (2.43)$$

where the sound speed gradient, with units $[(m/s)/m] = [s^{-1}]$, is

$$b_n = \frac{c(z_{n+1}) - c(z_n)}{z_{n+1} - z_n} \qquad (2.44)$$

The interfaces are at z_n and z_{n+1}. A change of variables from z to w facilitates the integration (see Fig 2.6)

$$w_n = z - z_n + \frac{c(z_n)}{b_n} \qquad (2.45)$$

where

$$dw = dz \quad \text{and} \quad c(w) = b_n w \qquad (2.46)$$

For downgoing rays (the positive z and w direction), the integrals (2.41) and (2.42) are

$$t_{n+1} - t_n = \int_{w_n}^{w_{n+1}} \frac{dw}{b_n w(1 - a^2 b_n^2 w^2)^{1/2}} \qquad (2.47)$$

$$r_{n+1} - r_n = \int_{w_n}^{w_{n+1}} \frac{ab_n w\, dw}{(1 - a^2 b_n^2 w^2)^{1/2}} \qquad (2.48)$$

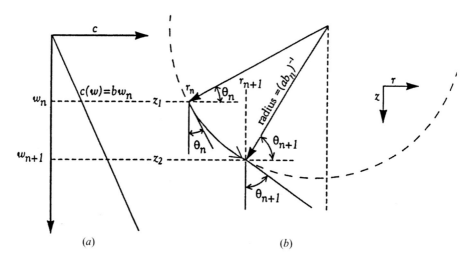

Fig. 2.6. (a) Sound-speed profile in w coordinates. (b) Circular ray path for linear dependence of speed on depth between z_n and z_{n+1}. Radius is $(ab_n)^{-1} = w_n/(\sin\theta_n)$, where b_n is the speed in the nth layer.

Integral tables give

$$t_{n+1} - t_n = \left|\frac{1}{b_n}\right| \ln \frac{w_{n+1}[1 + (1 - a^2 b_n^2 w_n^2)^{1/2}]}{w_n[1 + (1 - a^2 b_n^2 w_{n+1}^2)^{1/2}]} \tag{2.49}$$

and

$$r_{n+1} - r_n = \left|\frac{1}{ab_n}\right| \left[\left(1 - a^2 b_n^2 w_n^2\right)^{1/2} - \left(1 - a^2 b_n^2 w_{n+1}^2\right)^{1/2} \right] \tag{2.50}$$

The travel time and range through a set of layers are

$$t = \sum_{n=0}^{N-1} t_{n+1} - t_n \quad \text{and} \quad r = \sum_{n=0}^{N-1} r_{n+1} - r_n \tag{2.51}$$

If the sound speed increases such that the ray becomes horizontal, and then turns upward, the turning depth w_t is given by

$$1 - a^2 b_n^2 w_t^2 = 0 \tag{2.52}$$

If $c(z)$ is not a function of range, the upward trace is the reversal of the downward ray trace.

2.3.5 Examples of ray paths

The path of a ray depends on the sound speed structure of the ocean and the initial conditions at the source. In a "horizontally stratified ocean," the sound speed $c(z)$ is the same everywhere at depth z: this

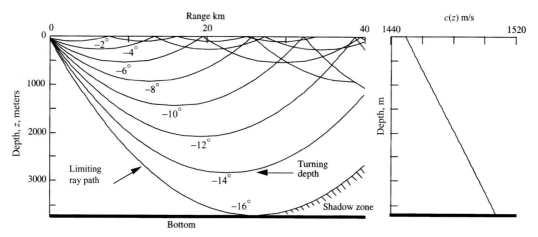

Fig. 2.7. Refracted and surface-reflected ray paths in isothermal Arctic water. Sound-speed profile is at right. The sound speed increases linearly with depth. Following a common convention for nearly horizontal ray paths, the rays are identified by the initial grazing angle (degrees) with the horizontal, where *grazing angle* $= \theta - 90°$. Each of the rays has a turning depth where the tangent to the ray path is horizontal. Steeper ray paths (omitted here, for simplicity) will reflect from the bottom as shown in Fig. 2.8.

assumption is used in all of our examples. Actual sound speed profiles are time- and space-dependent. Typical examples of ocean stratifications illustrate the range of possibilities as follows.

Arctic ocean
The temperature of the water under the sea ice is nearly constant. Although below 200 m to 4000 m the temperature ranges from $+0.5$ to $-1.0\,°C$ and the salinity is about 35‰, for a simple example, we choose $T = 0\,°C$ and salinity $= 35‰$ (parts per thousand). Then, from (2.31), the sound speed is approximately

$$c(z) \approx 1449 + 1.630 \times 10^{-2}z + 1.675 \times 10^{-7}z^2 (\text{m/s}) \qquad (2.53)$$

Since the gradient of the sound speed is constant in our example, one can use the integrated (2.49) and (2.50) to compute travel times and range for a ray having the incident starting angle θ_0. A set of ray paths for this profile is shown in Fig. 2.7.

Each ray becomes horizontal and turns at its *turning depth* w_t, given by (2.52). The ray that just grazes the bottom is a limiting ray path. The geometrical *shadow zone* is just beyond this ray path, and no direct path can get there in the ray description. In reality sound waves penetrate into the shadow region but are attenuated exponentially. The steeper paths

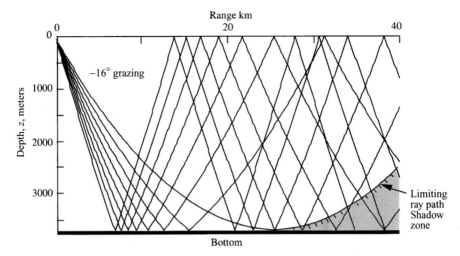

Fig. 2.8. Reflected ray paths in an isothermal Arctic water sound-speed profile shown in Fig. 2.7. The limiting ray path has an angle of incidence of 74° (−16° grazing angle). Angle increments are 2°. These steeper angles are reflected at the bottom and at the surface. The rays at nearer grazing are shown in Fig. 2.7.

are reflected at the bottom as shown in Fig. 2.8. The reflection paths that can get into the shadow region have at least two bottom reflections. As angles of incidence tend to normal, ray paths tend to straight lines.

North Atlantic ocean

A "typical" North Atlantic Profile is from the seminal research of Ewing and Worzel (1948). The sound-speed profile has a minimum sound speed, which creates a *sound channel axis* at 1300 m depth. As shown in Fig. 2.9, the ray paths pass through the axis of the sound channel (dashed line). If the ray paths do not reflect from the surface or bottom, and are not blocked by features on the ocean floor, sound channel transmissions are very efficient for long-range (thousands of kilometers) communications in the ocean. It is likely that some whales use these paths for communications.

A sound source near the surface excites a different phenomenon. Figure 2.10 shows a few rays from a shallow source. The ray paths go deep, turn, and return to the surface. At about 90 km range, the ray paths appear to converge in a region named the *convergence zone*. A magnification of the convergence zone is shown in the inset. Because many ray paths converge, the acoustic pressure is large in a convergence zone and smaller elsewhere. The region where the rays overlap is called a *caustic*.

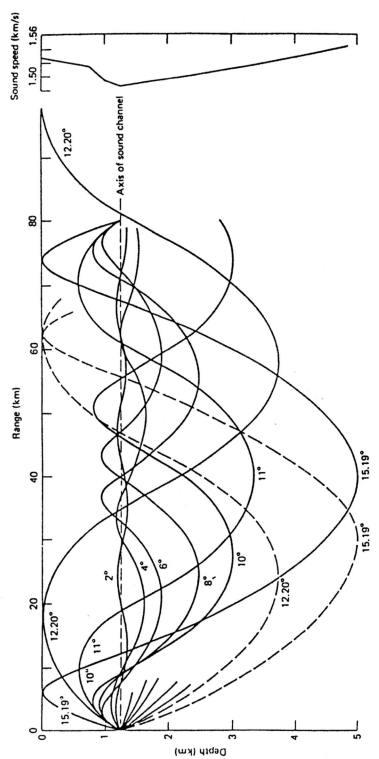

Fig. 2.9. Ray diagram for an Atlantic Ocean sound channel. Source is on the channel axis at 1.3 km depth. Sound speed profile is at the right. The grazing angles at the axis of the channel are given in degrees. Note that rays at angle $\geq$ 12.20° will be scattered at the surface. (Ewing and Worzel, 1948.)

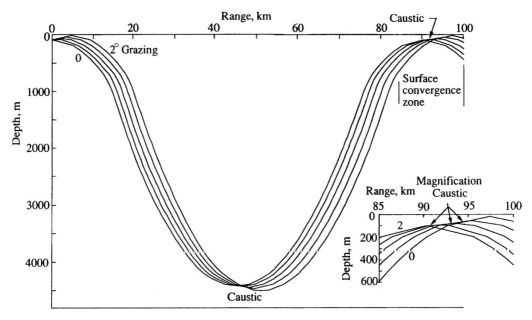

Fig. 2.10. Convergence-zone ray paths for a source at 50 m depth.

2.3.6 General comments on ray methods

Classical ray-path theory simply uses Snell's Law to determine the ray paths; the phases of the signals are not considered. However, when there is a longer duration signal, the correct wave description shows that the pressures of two or more of the simultaneous, crossing paths can "interfere" constructively or destructively, depending on their frequencies and phases at the positions of overlap. In such cases the ray treatment must be supplemented, or replaced, by a wave treatment in order to determine the correct acoustic pressures at the places and times of overlap.

2.4 Attenuation

2.4.1 Attenuation of plane waves

In general, waves attenuate (become weaker as they propagate) because of divergence, energy scatter (by inhomogeneities), and energy absorption (conversion to heat). In this section we consider the change of acoustic pressure of plane waves as they propagate in a homogeneous medium. In such a region a plane wave would experience a reduction of acoustic pressure, dp, proportional to the original acoustic pressure, p, and proportional to the distance traveled, dx. We call the proportionality constant α_e and have the equation

$$dp = -\alpha_e p \, dx \qquad (2.54)$$

When (2.54) is integrated, one obtains the natural logarithm to base $e = 2.71828$ – that is,

$$\ln (p/p_0) = \alpha_e x \qquad (2.55)$$

or

$$\alpha_e = \left(\frac{1}{x}\right) \ln \left(\frac{p_0}{p}\right) \qquad (2.56)$$

The familiar decay law follows from (2.55) – that is,

$$p = p_0 e^{-\alpha_e x} \qquad (2.57)$$

where we evaluated the constant of integration by letting $p = p_0$ at $x = 0$.

The spatial rate of amplitude decay, the amplitude decay coefficient α_e, has SI units of nepers/unit distance. (The unit *neper*, abbreviated Np, was named after John Napier (1550–1617), Scottish inventor of logarithms, which were in everyday use before the advent of digital computers.) The *distance* is usually given in meters or kilometers.

The plane-wave exponential attenuation rate, α_e, which depends on the medium, is a function of frequency. Commonly the attenuation per wavelength is very small, which permits attenuation to be included simply as a factor not only for plane waves but also in the pressure spreading from a point source. For example, including plane-wave attenuation in a lossy medium, the acoustic pressure (2.4) for a spherically diverging wave could be written as a function of $(t - R/c)$

$$p(R, t) = p_0(t - R/c)\frac{R_0}{R}e^{-\alpha_e R} \qquad (2.58)$$

where $t \geq R/c$.

In the 1920s, a new decay coefficient, advocated by the Bell Telephone system and based on $\log_{10}$ of the *relative power* in lossy communication circuits, came into common usage. The "bel" was named to honor Alexander Graham Bell (1847–1922), and one-tenth bel, the *decibel* (abbreviated dB), became much more popular than the neper. To distinguish the attenuation coefficient in decibels per unit distance, we call it simply α, without a subscript. The plane-wave attenuation coefficient is defined in terms of the *relative sound intensity*, i_1/i_2, at two points of a plane wave,

$$\alpha = (1/x)[10 \log_{10}(i_1/i_2)] \text{ dB/distance} \qquad (2.59)$$

or, since the intensity is proportional to the square of the pressure in a plane wave (or at long range in a spherical wave), one writes

$$\alpha = (1/x)[20 \log_{10}(p_1/p_2)] \text{ dB/distance} \qquad (2.60)$$

Comparing with (2.56), we find

$$1 \text{ neper} = 8.68 \text{ dB} \tag{2.61}$$

or

$$8.68\alpha_e = \alpha \tag{2.62}$$

2.4.2 Absorption losses in sea water: viscosities and molecular relaxation

Coefficients of viscosity

We now consider the effects of the *dynamic (or absolute) coefficient of shear viscosity, μ,* and the *dynamic bulk viscosity, μ_b.* The shear viscosity, which is generally called simply "the viscosity," is defined as the ratio of the shearing stress to the rate of strain. Each component of the stress is due to a shearing force F parallel to an area A, caused by a gradient of velocity at the surface. For example, for velocity components u, v, and w in the x, y, and z direction, the shearing stress in the x direction, F_x, is proportional to the shearing rate of strain, which is caused by a velocity gradient in the y direction:

$$\frac{F_x}{A} = \mu\left(\frac{\partial u}{\partial y}\right) \tag{2.63}$$

The lesser known, but important, bulk viscosity appears only when there is a compressible medium (e.g., an acoustic wave in water). It is not found in traditional, incompressible hydrodynamics but does appear in the Navier–Stokes viscosity equations as a term which is proportional to the rate of change of density.

Molecular relaxation

Consider the physical origins of bulk viscosity. It takes a finite time for a real fluid to respond to a pressure change, or to relax back to its former state after the pressure has returned to normal. The process is called *relaxation.* Chemical relaxation, which occurs in sea water, involves ionic dissociation that is alternately activated and deactivated by sound condensations and rarefactions. Surprisingly, magnesium sulfate and boric acid are the two predominant contributors to sound absorption in sea water, even though their contribution to water salinity is very much less than that of common salt (sodium chloride).

Although molecular relaxation in sea water affects the speed of propagation only very slightly, it is the source of a most important factor in the attenuation of sound.

The attenuation at sea due to absorption by relaxation processes is sensitive to the temperature and ambient pressure (depth) of the water.

It turns out that the dispersive speed, c_r, differs from c by only about 1 percent in sea water, so we henceforth assume for simplicity

$$c \cong c_r \tag{2.64}$$

We now define the molecular relaxation frequency in terms of a relaxation time for a change of molecular configuration

$$f_r = \frac{1}{2\pi \tau_r} \tag{2.65}$$

and obtain the characteristic form for the attenuation rate owing to molecular relaxation in a fluid – that is,

$$\alpha_e = \frac{(\pi f_r/c)f^2}{f_r^2 + f^2} \text{ nepers/distance} \tag{2.66}$$

For attenuation rate α in decibels/distance, use the conversion one neper $= 20 \log_{10} e = 8.68$ dB, where e is the base of the natural logarithm. The plane-wave attenuation rate for a relaxation process is therefore of the form

$$\alpha = \frac{Af_r f^2}{f_r^2 + f^2} \text{ dB/distance} \tag{2.67}$$

where $A = 8.68\pi/c$.

It is important to notice that the attenuation rate in a relaxation process reduces to

$$\alpha = Af_2 = \text{ constant} \qquad \text{for} \qquad f \gg f_r$$
$$\alpha = \frac{A}{f_r} f^2 \sim f^2 \qquad \text{for} \qquad f \ll f_r \tag{2.68}$$

These are the behaviors that dominate the attenuation in sea water, as seen in the next section.

A few more words about the physics of the relaxation process may be helpful. When the effect is expressed in terms of attenuation per wavelength, $\alpha\lambda$, one finds that

$$\alpha\lambda = (Acf_r)\frac{f}{f^2 + f_r^2} \tag{2.69}$$

which describes an effect that approaches zero for very high or very low frequencies. This is because, at very high frequencies, the relaxing molecules cannot respond fast enough to be effectively activated. At the other extreme, when the frequency is very low, the molecular relaxation follows in step with the sound wave, and there is no evidence that relaxation is taking place. However, when the frequency is approximately equal to the relaxation frequency, the activated molecules will dump energy from a condensation into a rarefaction, and some of the ordered energy of the sound wave is thereby transformed into random, thermal motion of the medium.

In *M&C* it is shown that the macroscopic descriptors of the medium, μ and μ_b, and the molecular relaxation description are related by,

$$\tau_r = \frac{(4/3)\mu + \mu_b}{\rho_A c^2} \tag{2.70}$$

Consider fresh water at 14 °C. Designating fresh water by the subscript F, the conventional constants are: $\mu_F = 1.17 \times 10^{-3}$ N-s/m²; $\rho_F = 1000$ kg/m³; and $c_F = 1480$ m/s. From the experiments in acoustic streaming (Chapter 4), it is known that the bulk viscosity for water is 2.8 times the shear viscosity – that is, $\mu_b = 2.8\,\mu_F$. Therefore, (2.70) yields

$$\tau_r = 2.1 \times 10^{-12} (s) \tag{2.71}$$

Recalling (2.68), this very small relaxation time means that the attenuation in fresh water is proportional to frequency squared for an enormous range of frequencies propagating in fresh water – that is, for

$$f \ll 1/(2\pi \times 2.1 \times 10^{-12} \text{ seconds}) \qquad f \ll 10^{11} \text{ Hz} \tag{2.72}$$

This frequency squared component of the total absorption will be seen in the straight lines that have slope 2 :1, in the log–log graph plotted in Fig. 2.11.

Molecular relaxation in sea water

In the 1950s, the mysteriously large attenuation of sound in sea water, observed at frequencies around 20 kHz during World War II, was finally explained by a series of careful laboratory experiments by Leonard *et al.* (1949) and his students Wilson (1954) and Bies (1955). The technique used was to drive a water-filled sphere, of diameter approximately 30 cm, into its modes of oscillation, and to determine the damping constants caused by the various sea salts added in proper amounts to the water. The startling realization that the relatively small amounts of magnesium sulfate salts were causing a molecular relaxation phenomenon changed, forever, the naïve thought that only shear viscosity was important. Note the region from 30 kHz to 300 kHz in Fig. 2.11.

In the 1970s, similar laboratory experiments by Fisher and Simmons (1977) and experiments by Mellen and Browning (1977) showed that there was another relaxation phenomenon, due to boric acid in sea water, moderated by the pH of the sea water. This causes a strong effect at frequencies from 500 Hz to 5 kHz (see Fig. 2.11).

The dependence of relaxation effects on temperature, shear viscosity, and bulk viscosity is discussed in somewhat more detail in Section A3.2 of Clay and Medwin (1977). The result of much research in the laboratory and at sea have been summarized by François and Garrison (1982a, b).

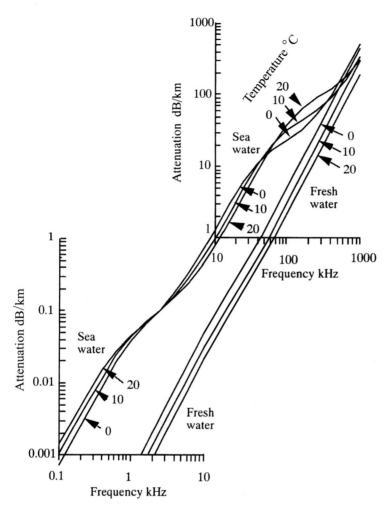

Fig. 2.11. Sound pressure attenuation rate in dB/km in fresh and sea water at temperatures 0°, 10°, and 20 °C. Calculated from François and Garrison (1982a, b) with parameters pH = 8; S = 35 ppt; depth, z = 0 m. See (2.73) and (2.78).

It is their empirical formula that will be used in this text. They give the attenuation in sea water as the sum of the two relaxation terms and the viscosity component:

$$\alpha = \frac{A_1\, P_1\, f_1\, f^2}{f^2 + f_1^2} + \frac{A_2\, P_2\, f_2\, f^2}{f^2 + f_2^2} + A_3 P_3 f^2 \quad \text{dB/km} \qquad (2.73)$$

where α is the total absorption coefficient in dB/km. The coefficients are expressed in terms of z = depth, (m); T = temperature, (°C); S = salinity (parts/1000); and the relaxation frequencies f_1 for boric acid and f_2 for magnesium sulfate. The constants are:

Boric acid component in sea water

$$A_1 = \frac{8.68}{c} 10^{(0.78\,pH-5)} \text{ dB km}^{-1} \text{ kHz}^{-1}$$
$$P_1 = 1 \tag{2.74}$$
$$f_1 = 2.8 \left(\frac{S}{35}\right)^{0.5} 10^{[4-1245/(273+T)]} \text{ kHz}$$

Magnesium sulfate component in sea water

$$A_2 = 21.44 \frac{S}{c}(1+0.025T) \text{ dB km}^{-1} \text{ kHz}^{-1}$$
$$P_2 = 1 - 1.37 \times 10^{-4}z + 6.2 \times 10^{-9}z^2 \tag{2.75}$$
$$f_2 = \frac{8.17 \times 10^{[8-1990/(273+T)]}}{1+0.0018(S-35)} \text{ kHz}$$

Pure water (Shear viscosity) component for $T \le 20\,^\circ C$

$$A_3 = 4.937 \times 10^{-4} - 2.59 \times 10^{-5}T + 9.11 \times 10^{-7}T^2$$
$$- 1.50 \times 10^{-8}T^3 \text{ (dB km}^{-1} \text{ kHz}^{-2}) \tag{2.76}$$

For $T > 20\,^\circ C$

$$A_3 = 3.964 \times 10^{-4} - 1.146 \times 10^{-5}T + 1.45 \times 10^{-7}T^2$$
$$- 6.5 \times 10^{-10}T^3 \quad \text{(dB km}^{-1} \text{ kHz}^{-2}) \tag{2.77}$$
$$P_3 = 1 - 3.83 \times 10^{-5}z + 4.9 \times 10^{-10}z^2 \tag{2.78}$$

2.4.3 Scattering losses

In carrying an acoustical message through a ray tube, any energy that is scattered out of the tube causes attenuation, just as effectively as absorption and divergence of sound. Later we consider the scattering of energy by bodies and bubbles, which is the third source of propagation loss at sea.

2.5 "The SONAR Equation": SL, SPL, and TL

We now assume that our signal is a sinusoid, so that we can include absorption simply as a frequency-independent factor in the sound transmission equations. From (2.58) the rms pressure is first written to include spherical divergence and attenuation,

$$P = P_0(R_0/R)10^{-aR/20} \tag{2.79}$$

"The SONAR Equation," which became famous before the era of computers made calculations simple, restates the pressure in equations such as (2.79) by relating it to a reference pressure. Then all terms are re-expressed in logarithmic form so that the decibel notation can be used (see below). The proclaimed virtue of the decibel notation is that additions and subtractions of quantities that are given in decibels can replace multiplications and divisions of the physically measurable quantities. Traditionally, both experimental and theoretical studies have then referred the amplitudes of the sound pressures to a "reference pressure," and have ignored the phases in the logarithmic operation.

Unfortunately, in performing the decibel operation, the physical units of the quantities disappear so that one tends to forget the physical character of the measurement. Also, unfortunately, there are several different "common" reference pressures, and authors occasionally neglect to tell the reader which reference is being used. Furthermore, the important subtleties of interferences of superimposed sound wave fields in the real world are ignored in the popular SONAR equations. Indeed, one might question the continuing use of logarithmic quantities such as the decibel which were invented to satisfy a need in the eighteenth century but which are rapidly becoming obsolete due to the common availability of digital calculators in the twenty-first century!

Consider the transmission loss in the case of CW sinusoidal sound spherically spreading from a point source in a homogeneous, attenuating medium, as expressed in decibel notation.

By dividing both sides by the rms reference pressure P_{ref} and taking the logarithm of both sides, we obtain the deceptively simple "SONAR Equation" for the sound pressure level in decibel notation:

$$\text{SPL (dB)} = \text{SL (dB)} - \text{TL (dB)} \qquad (2.80)$$

The interpretation of the symbols is

$$\text{SPL (dB)} = 20 \log_{10}(P/P_{ref}) \qquad (2.81)$$

where P is the rms sound pressure at the field position, range, R, and P_{ref} is the reference pressure, usually one newton/m^2. Also, the "source level" is defined as

$$\text{SL (dB)} = 20 \log_{10}(P_0/P_{ref}) \qquad (2.82)$$

where P_0 is the "source pressure," i.e., the sound pressure at the source reference distance of 1 meter. Finally,

$$\text{TL (dB)} = 20 \log_{10}(R/R_0) + a(R - R_0) \qquad (2.83)$$

is the transmission loss between the source and the field positions. Notice the well-known (to acoustical engineers!) transmission loss of 6 dB per double distance, for $(R/R_0 = 2)$, if one assumes that the rate of attenuation due to absorption, a, is negligible.

Above, we used the symbol, P_{ref}, for the reference pressure. Unfortunately, there are several different sound pressure values that are commonly used as reference pressures in acoustics. The generally accepted reference in underwater acoustics is either the "pascal," 1 pascal $=$ 1 newton/m², or the micropascal, 1 µPa $= 10^{-6}$ pascals, which is used throughout this text. Readers should be aware that, in the older publications, underwater acousticians used the reference 1 microbar $= 0.1$ Pa $= 10^5$ micropascal (µPa). Note also that acousticians who work in air generally use 20 µPa as their reference, because 20 µPa is the nominal threshold of human hearing at 1000 Hz.

2.6 Doppler frequency shift

2.6.1 Doppler theory

Motions of a source, receiver, or scattering object change the frequency of received signals. We assume here that the water is still, and that the propagating signal moves with sound speed c, regardless of the velocities of the source or receiver or scattering objects. A brief review of the Doppler effect follows.

Initially, we consider *source and receiver motions* along the $+x$ axis. The source moving with velocity v_s transmits either a continuous wave or a very long source ping of frequency f_s. When the source advances on its own waves, the waves are shortened, and the apparent wavelength along x is

$$\lambda_a = (c - v_s)/f_s \qquad (2.84)$$

Also, assume that the receiver has velocity v_r away from the source. In a unit time it detects f_r crests, where

$$f_r = (c - v_r)/\lambda_a \qquad (2.85)$$

Solve for f_r and obtain

$$f_r = \frac{f_s(c - v_r)}{(c - v_s)} \qquad (2.86)$$

For both actions, the frequency at the receiver tends to decrease because of the receding receiver (the numerator) and to increase because of the approaching source (denominator). Switching the sense of the movements will change the signs in (2.86), of course.

Now consider a *fixed transducer* radiating frequency f_s. A scattering object moving away with velocity v_0 receives a lesser number of crests per second. Therefore, the frequency of sound observed by the receding object is

$$f_0 = f_s(c - v_0)/c \qquad (2.87)$$

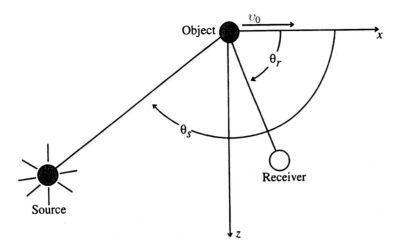

From the point of view of a fixed receiver at the source location, the scattering object becomes a receding source, and the frequency at the transducer is

$$f_r = \frac{f_s(c - v_0)}{(c + v_0)} \tag{2.88}$$

In other directions, the velocities in (2.88) become the *components* of v_0 along the directions to the source and receiver. For the case in Fig. 2.12,

$$f_r = \frac{f_s(c + v_0 \cos \theta_s)}{(c - v_0 \cos \theta_r)} \tag{2.89}$$

The Doppler shift may be positive or negative, depending on the geometry and the direction of motion.

2.6.2 Doppler measurements of particle motion

Doppler velocimeters have been used successfully in studies of mixed-layer and upper-ocean dynamics (Pinkel and Smith, 1987; Vagle and Farmer, 1992), internal waves (Pinkel, Plueddemann, and Williams, 1987), and tidal motion (Lhermitte, 1983). The platform for the device must either be fixed (e.g., the Floating Instrument Platform, FLIP, de-signed by F. Fisher and operated by the Scripps Institution of Oceanog-raphy) or have its motion referenced to an inertial frame.

The scatterers that produce the Doppler shift are plankton and de-tritus, which are found everywhere in the sea. Bubbles are additional Doppler scatterers near the sea surface. These bodies and bubbles, when they are entrained, act as tracers of the moving medium.

Consider the pulse-to-pulse coherent sonar. Two sinusoidal signals are generated at the same frequency but out of phase by 90° (the sine and cos signals). Either one of these is then gated and transmitted. The backscattered Doppler-shifted signal is multiplied by each of these reference signals, and they are then low-pass-filtered. The resulting pair of signals represent the real and imaginary components, which determine the angle of the Doppler phase shift. In the pulse-to-pulse coherent technique, the change of phase from one pulse to the next, at fixed range, is used to calculate the component of the scatterer motion parallel to the beam direction. The great virtue of this technique is that, by pinging millisecond duration pulses many times per second, drift velocities as small as 1 cm/s can be readily measured.

One flaw in the coherent sonar technique is that, if the scatterer were to move by an integral number of wavelengths, the equipment would show no apparent phase shift, and therefore would imply no motion. Also troubling is the possibility that an echo may have come from a previous pulse, scattered from a larger, more distant body. Signal coding can eliminate that problem.

It is possible to quantify the ambiguity of a pulse-to-pulse *coherent Doppler system*. If the time between pulses is t, the maximum unambiguous range is

$$R_m = c(t/2) \tag{2.90}$$

Also, to avoid aliasing, the maximum round-trip distance must change by at most $\lambda/2$ (where λ is the wavelength) between pings. This represents a maximum drift of the medium of $D_m = \lambda/4$. The maximum unambiguous drift velocity for the pulse-to-pulse coherent sonar is then

$$V_m = D_m/t = \lambda/(4t) \tag{2.91}$$

The conditions for R_m and V_m are an example of the contradictory uncertainty in their measurements. To decrease the uncertainty in drift velocity, we could increase t, but this would *increase* the uncertainty in position. In fact, the two conditions can be summarized in the resolution product,

$$V_m R_m = c\lambda/8 \tag{2.92}$$

Frequencies of around 100 kHz to 1 MHz are commonly used. The higher frequencies provide smaller values of the resolution product $V_m R_m$. But this improved resolution is achieved at the cost of reduced range, because of the greater attenuation at higher frequencies.

The pulse-to-pulse *incoherent Doppler system* follows the backscatter of a single pulse as its range increases. Doppler shift is estimated from the rate of change of phase with time. The spectrum of the echo is slightly shifted in frequency compared with the outgoing pulse, and the spectrum is rather broad, so that the peak is unclear. Although longer transmitted pulses can be used to improve the narrowness of the spectrum, by the same action the range resolution is deteriorated. In practice, the incoherent Doppler finds its greatest use because of the large ranges possible (kilometers), although the spatial resolution may be as large as many meters (Lhermitte, 1983).

Doppler frequency shifts from the ocean surface

Signals scattered at the moving sea surface are shifted in frequency. For a fixed transducer, the amount of the frequency shift is given by (2.89). In the backscatter direction, for $\theta_s = +\theta_r = \theta$ (see Fig. 2.12),

$$f_r = \frac{f_s(c + v \cos \theta)}{(c - v \cos \theta)} \tag{2.93}$$

Nominally, the Doppler shift from a "smooth" horizontal surface is zero in the specular (mirror) direction. However, when a beam insonifies a real ocean surface, in addition to specular reflection from horizontal facets, some of the scattered components reach the receiver from other parts of the ocean wave system, and these show a Doppler shift. Roderick and Cron (1970) give comparisons of theoretical and experimental Doppler shifts from ocean wave surfaces.

The Doppler shift is a measure of the component of wave velocity along the axis of the sonar system. Usually there are bubbles, zooplankton, and so forth in the water, and these objects scatter sound back to a Doppler sonar. The orbital motions of the water carry the objects, which have the velocities given by previous equations. Clearly, the results of a Doppler measurement are strongly controlled by the location and size of the region being measured.

2.6.3 Doppler navigation

Doppler navigation systems use the frequency shift of backscattered sound signals to measure the velocity of the ship relative to the bottom, or stationary objects within the water. The algebra is the same as for (2.89), and, since $v \ll c$, the frequency shift of the backscattered signal is approximately

$$\Delta f \approx \frac{2v f_s \cos \theta}{c} \tag{2.94}$$

Illustrative problems

Illustrative problem 2.1

The water temperature in the Arctic is nearly isothermal. The sound speed increases with depth because of the pressure effect. Assume a surface temperature of $0\,°C$ and salinity of 35 ppt, determine the initial angle, range, and travel time for the ray that starts at the surface, turns at 2 km depth, and returns to the surface.

Solution. Use

$$c(z) \simeq 1449 + 0.016z \text{ m/s}$$

$$c(0) = 1449 \text{ m/s}$$

$$c(2000 \text{ m}) = 1481 \text{ m/s}$$

Applying Snell's Law

$$\sin\theta_1 = \frac{1449}{1481} = 0.9784$$

$$\cos\theta_1 = 0.2068$$

$$a = \frac{\sin\theta_1}{c(z_1)} = 6.752 \times 10^{-4} \text{ s/m}$$

At the turning depth, 2000 m

$$\sin\theta_2 = 1 \quad \cos\theta_2 = 0$$

Calculate $w = z + \frac{c(0)}{b}$ where $b = 0.016 \text{ s}^{-1}$ and $w = 9.056 \times 10^4 + z$ (m).
The ray is traced from

$$w_1 = 9.056 \times 10^4 \text{ m}$$

to

$$w_2 = 9.256 \times 10^4 \text{ m}$$

The surface-to-surface travel time is

$$t = 26.22 \text{ s}$$

and surface-to-surface distance is

$$r = 38.28 \times 10^3 \text{ m}$$

The path is the arc of a circle.

Illustrative problem 2.2

To compare the pressure levels due to an omnidirectional source and a directional source in the ocean (see next chapter), we will need to first compute the rms pressure for an omnidirectional sound source operating in sea water. The values are source power $= 1 \text{ kW}$, frequency $= 100 \text{ kHz}$, and distance $R = 1 \text{ km}$. The sound speed is assumed to be constant, $c = 1473 \text{ m/s}$.

Solution. We use the Sonar Equation. The signal level is

$$\text{SPL} = \text{SL} - \text{TL}$$

(a) The source level for the omnidirectional source is conveniently written by letting $DI_t = 0$.

$$SL = 10\log_{10} \Pi + 50.8$$
$$SL = 10\log_{10}(1000) + 50.8$$
$$SL = 80.8 \text{ dB re 1 Pa} = 200.8 \text{ dB re 1 } \mu\text{Pa (at 1 m)}$$

(b) The transmission loss for an isotropic medium is given by $TL \simeq 20\log_{10}(R/R_0) + \alpha(R - R_0)$. From Fig. 2.11 the approximate absorption for sea water is found to be

$$\alpha = 2 \times 10^{-2} \text{ dB/m}$$
$$\text{For } R = 10^3 \text{ m}$$
$$TL = 60 + 20 = 80 \text{ dB}$$

(c) On combining (a) and (b), we obtain the SPL at the field position

$$SPL \simeq 80.8 - 80 \text{ dB}$$
$$\simeq 0.8 \text{ dB re 1 Pa} = 120.8 \text{ dB re 1 } \mu\text{Pa}$$

or

$$P \simeq 1.1 \text{ Pa} \quad \text{at} \quad R = 1 \text{ km}$$

Problems and some answers

2.1 A point source radiates 1000 W of acoustic power. Plot a graph of the rms pressure as a function of range from 1 to 10^4 m. A log–log graph is suggested.

2.2 Ray trace calculation. The speed–depth profile for a point in the South Pacific has been approximated by three linear segments. The parameters of the four points defining the segments are as follows:

	Speed (m/s)	Depth (m)
z	1495	0
z_1	1495	500
z_2	1485	1000
z_3	1520	4000

The problem is to compute the ray paths and travel times of rays incident at $z = 0$ at angles with the normal of $10°$, $30°$, $60°$, $70°$, $80°$, and $90°$. The travel paths are from z to z_1, z_1 to z_2, and z_2 to upper and lower turning depths.

2.3 The source is above an interface in a medium having sound speed c and the receiver is below the interface where the sound speed is c_2. The source and receiver are displaced a horizontal distance x

from each other. Use calculus to show that the minimum time for a ray path to go from the source to a point on the interface then to the receiver yields the Snell's Law path.

2.4 Determine the attenuation constants (dB/m) for 0 °C sea water at the following frequencies: $f = 1, 12.5, 25$, and 50 kHz.

2.5 Determine the attenuation constants (dB/m) for 18 °C fresh water at the following frequencies: $f = 1, 12.5, 25, 50$, and 100 kHz.

2.6 You have equipment that can transmit omnidirectionally at the power of 1000 W. Assuming a received signal of 1 Pa, over approximately what distance would you transmit in sea water for the frequencies of Problem 2.4?

2.7 The same equipment as in Problem 2.6 is used in fresh water. For a 1 Pa signal, what are the ranges for the frequencies of Problem 2.5?

2.8 Using the attenuation constants found in Problem 2.4, calculate and tabulate for comparison the loss due to spherical divergence from 1 to 10, 100, 1000, and 10 000 m; then compare with the loss due to absorption to the same distances for the frequencies 1, 12.5, 25, and 50 kHz.

2.9 Calculate the decibel correction required to convert a signal level from dB re 1 µbar to dB re 1 Pa.

2.10 Prove that the loss per wavelength, $\alpha\lambda$, is a peak when $f = f_{rm}$ in the $MgSO_4$ relaxation process. Is the situation different for boric acid?

2.11 Calculate and graph the variation of α_s with temperature for the range 0 to 30 °C at frequency 10 kHz.

Some answers to problems
2.2 For $\theta = 10°$ ray

Layer	Δx, m	Δt, s
1	88.16	0.340
2	87.86	0.343
3	531.72	2.028

2.4 $7.22 \times 10^{-5}, 1.98 \times 10^{-3}, 6.78 \times 10^{-3}, 1.88 \times 10^{-2}$ dB/m

2.6 10 km, 4.1 km, 2.1 km, 1.1 km

2.8 At 50 kHz, absorption losses are 0.18, 1.86, 18.8, 188 dB; divergence loss is greater except at 10^4 m.

2.9 dB re 1 µbar = dB re 1 Pa + 20 dB

More problems

2.12 Calculate the time and relative pressure of the received signal as in Fig. 2.3 for a point source radiating a single cycle of a sinusoid of frequency 200 Hz. Assume source depth 5 m, receiver depth 15 m, water depth 20 m, and horizontal range 60 m.

2.13 Assume that the source amplitude in the previous problem was 10^4 pascals. Calculate the duration of the interfering signal at the receiver.

2.14 The sound speed profile for a region in the deep sea is approximated by the three linear segments defined by

depth (m)	speed (m/s)
0	1495
500	1495
1000	1485
4000	1520

Compute the paths and travel times of rays leaving the surface at grazing angles 80°, 60°, 30°, 20°, and 10°.

2.15 A point source radiates 100 watts of CW acoustic power in homogeneous water. Plot a graph of the rms pressure from 1 m to 1000 m, taking into account attenuation due to spherical divergence and absorption. Assume that the frequency is (*a*) 1000 Hz, (*b*) 10 kHz, (*c*) 100 kHz. Identify the ranges where divergence is the major source of attenuation for each frequency and the regions where energy absorption is the major attenuation process.

2.16 Near estuaries the salinity varies from zero, in the stream, to close to 35 ppt in the ocean away from the outlet. Plot a curve of the variation of absorption rate with salinity for frequencies 200 Hz, 10 kHz, and 100 kHz.

2.17 A 100 kHz sonar is directed horizontally to insonify a fish. Transmitter and receiver are side by side. The fish is swimming toward the sonar at 12 cm/s. Compute the Doppler shift.

2.18 Calculate the frequency shift of a 100 kHz signal aimed at a horizontally swimming fish at 12 cm/s as viewed from the sonar at 10 m below the fish. Assume that the fish is tracked as it approaches, passes overhead, and then receeds from the transducer. Plot a graph of Doppler shift as a function of angle with the vertical.

2.19 When using a 150 kHz upward-looking sonar at a grazing angle of 30°, a signal with a frequency shift of 1370 Hz is backscattered from a corrugated surface wave. What is the velocity of the surface wave along the direction of the sonar?

Further reading

Tolstoy, I. and Clay, C. S. (1987). *Ocean Acoustics: Theory and Experiment in Underwater Sound.*

An Acoust. Soc. Am. paperback reprint of the original McGraw Hill, New York (1966) publication, supplemented and updated by six new appendices. This is a broad, authoritative book sometimes used as a graduate level text. Available from Acoustical Society of America, 2 Huntington Quadrangle, Melville, NY, 11747–4502.

Fisher, F. H. and Simmons, V. P. (1977). Sound absorption in sea water. *J. Acoust. Soc. Am.* **62**, 558–64.

The extension of the data on attenuation in sea water down to frequencies of a few kilohertz and below, where a boron component of molecular absorption becomes dominant. The results of this laboratory research supplemented the magnesium-sulfate relaxation process in sea water, which is dominant at frequencies of tens of kilohertz and above, discovered in the 1940s by R. W. Leonard and his students.

Pinkel, R. and Smith, J. A. (1987). Open ocean surface wave measurements using Doppler sonar. *J. Geophys. Res.* **92**, 12967–73.

One of several early publications by Pinkel and colleagues who have led the way in the use of underwater sound measurements to determine ocean particle velocities and ocean surface wave velocities.

Chapter 3
Sound sources and receivers

Summary

All experimental acoustical research in the ocean starts with the selection of a sound source if the study is to require active probing, and an appropriate sound receiver (hydrophone), whether the task is active or passive probing. Both types of source/receiver are called "transducers" because they lead across from the electrical world to the world of mechanical oscillations or vice versa. Transducers can be selected to be optimum as receivers, or optimum as senders, or they may be used in both modes. They can be designed to be effective for narrow or broad frequency bands, and for rapid or slow temporal response, and to produce narrow or wide beam patterns. Most transducers can be purchased as "off-the-shelf" equipment, or designed by the transducer manufacturer for a specific usage, or constructed inexpensively by the researcher (often, a Ph.D. candidate!) who knows enough about materials and designs to "do his own thing." In this chapter we discuss the fundamentals so that the student will know the options for purchase or construction, and calibration of underwater sound transducers.

For theoretical developments, it is often assumed that the source is a pulsating sphere. Typically, it is also assumed that the sound receiver is a device sensitive to acoustic pressure. In fact, sources and receivers, "transducers," have been designed with a wide range of physical, geometrical, acoustical, and electrical characteristics. Proper selection of a transducer element, or an array of elements, can provide increased sensitivity to certain frequencies or to specific directions of propagation. Furthermore, the effectiveness of either a hydrophone or sound source is a function of its physical mounting and the electrical circuit to which it is connected.

Contents

3.1 Transducer elements

3.1.1 The pulsating sphere: the acoustic monopole

At this point we make the "connection" between a very simple source and the sound field that it radiates. Consider the pulsating sphere. The instantaneous radial velocity u_r at the sphere surface, $R = a$, is described by

$$u_r]_{R=a} = U_a\, e^{i\omega t} \tag{3.1}$$

where U_a is the amplitude of the sphere radial velocity.

From (1.121), the isotropic radiated pressure at range R is

$$p = \frac{P_0 R_0 \exp\left[i(\omega t - kR)\right]}{R} \tag{3.2}$$

We also derived, the radial particle velocity of the acoustic field. At $R = a$, the particle velocity is

$$u_r]_{R=a} = \frac{P_0 R_0 \exp\left[i(\omega t - ka)\right]}{a\rho_A c}\left(1 - \frac{i}{ka}\right) \tag{3.3}$$

Equate the two expressions for u_r at $R = a$ and thereby obtain an expression for $P_0 R_0$. Insert that value into (3.2) to give the acoustic pressure in the "monopole" field in terms of the source parameters

$$p = \frac{ik(\rho_A c)\dot{V}\, \exp\left[i(\omega t - kR + ka)\right]}{4\pi R(1 + ika)} \tag{3.4}$$

where $\dot{V} = 4\pi a^2 U_a$.

---

* This section contains some advanced analytical material.

One concludes that

(1) for a point source with constant radial velocity amplitude, the CW radiated pressure is proportional to the sound frequency of the CW source;
(2) for a given CW frequency, the radiated acoustic pressure is proportional to the amplitude of the source rate of volume flow $4\pi a^2 U_a$ (m^3/s);
(3) for the same volume flow and sound frequency, the radiated sound pressure is proportional to the impedance $\rho_A c$ of the medium. Consequently, since the $\rho_A c$ of water is about 5000 times the $\rho_A c$ of air, an underwater sound source is an ineffective source of sound in air.

3.1.2 Generalized sources of sound

The pulsating sphere is a prototype of some simple underwater sound sources called monopoles. In fact Sims (1960) describes a low frequency, high-power, gas-filled bag that is a realization of the pulsating sphere described in the previous section. In addition, as will be seen in Chapter 6, when bubbles are formed at sea, they act as natural pulsating spheres.

The mechanical sources for acoustic waves are monopoles, dipoles, and quadrupoles; specifically they consist of:

(1) the acceleration of mass per unit volume (a monopole such as (3.4));
(2) the spatial rate of change (divergence) of force per unit volume exerted on the medium (a dipole as described in Section 3.1.3);
(3) a third (quadrupole) term: which is the double divergence of what hydrodynamicists designate as the "Reynold's stress tensor." This term is acoustically important only when the source of noise is due to turbulence. It is particularly important in air, where it explains the noise caused by the turbulence of a jet aircraft exhaust.

3.1.3 The acoustic dipole

A dipole is a combination of two equal-amplitude, out-of-phase monopoles with a small separation (re λ) between them. The dipole occurs, for example, in the case of an unbaffled oscillating membrane which simultaneously creates a condensation at one face and a rarefaction at the other. When the membrane oscillation reverses, the condensation is replaced by a rarefaction, and the rarefaction by a condensation. When the membrane is in a rather small baffle, $kl \ll 1$, where l is the effective separation between the two sides, a dipole exists. Physically, it can be seen that the two out-of-phase radiations will completely cancel each

Fig. 3.1. The idealized dipole source. Two point sources of equal strength but opposite phase are separated by a distance l, much less than a wavelength. The far-field directionality is a three-dimensional figure-eight pattern.

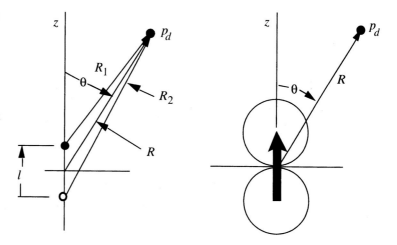

other on a plane perpendicular to the line joining the two poles; the radiation from the two poles will partially cancel each other everywhere else. The idealized dipole is shown in Fig. 3.1.

Call the summation, the dipole pressure $p_d = p_+ + p_-$, and obtain it by using two out-of-phase components of the pressure of a spherical wave from a point source. The addition is

$$p_d = P_0 R_0 \left\{ \frac{\exp[i(\omega t - kR_1)]}{R_1} - \frac{\exp[i(\omega t - kR_2)]}{R_2} \right\} \qquad (3.5)$$

For ranges large compared with the separation l, use the Fraunhofer approximation (1.30) to write

$$R_1 \simeq R\left(1 - \frac{l}{2R}\cos(\theta)\right) \quad \text{and} \quad R_2 \simeq R\left(1 + \frac{l}{2R}\cos(\theta)\right) \qquad (3.6)$$

As expected for an interference effect, the small differences between R_1 and R_2 in the denominators of (3.5) are not important, and each may be approximated by R. But the terms kR_1 and kR_2 in the exponentials, which determine the phases and the interference, are crucial. After factoring out the common terms, $1/R$ and $\exp[i(\omega t - kR)]$, and expanding the remaining imaginary exponentials for the dipole condition, $kl \ll 1$, we obtain

$$p_d = \frac{P_0 R_0}{R} e^{i(\omega t - kR)} \{ikl \cos\theta\} \qquad (3.7)$$

which is recognized as the monopole pressure multiplied by $\{ikl \cos\theta\}$.

When attenuation is included, this becomes

$$p_d = \frac{P_0 R_0}{R} e^{i(\omega t - kR)} [ikl \cos\theta] 10^{-\alpha R/20} \qquad (3.8)$$

The dipole is characterized by two significant effects compared with the individual monopole radiations. First, the radiated pressure is reduced because of the factor kl. Second, the radiation pattern is no longer isotropic; it now has a directionality given by $\cos\theta$, where θ is the angle with the dipole axis. That is, there is a maximum (but very much reduced) acoustic pressure along the line of the dipole, and there is zero sound pressure in the central plane perpendicular to the dipole line. The directionality is sometimes called a figure-eight pattern because of the way the cosine looks when it is plotted in polar coordinates (Fig. 3.1).

3.1.4   Materials and mechanisms

Now a few words about the electro-acoustical materials that are being used for modern transducers. Experimentalists who have specific needs in selecting or designing their transducers should consult a transducer manufacturer or a book on the subject.

The most common material used for underwater transducers is the polycrystalline, "piezoelectric" material barium titanate ($BaTiO_3$), which was discovered in the 1940s. A piezoelectric material shows a voltage across electrodes when subjected to pressure (sound receiver), and changes in dimension when a voltage is applied (sound source).

In the manufacture, the granular $BaTiO_3$ material is fused into a ceramic-like block (which looks like the ceramic of a coffee mug). Electrodes are applied to selected surfaces of blocks of any desired shape. The blocks are heated to a temperature (Curie Point, approximately 120 °C) at which the minute crystals become cubic. A DC polarizing voltage is applied, and, as the element is cooled, the block becomes an assemblage of tetrahedron crystals with a preferred axis. Depending on the electrode orientation, this causes the material to be piezoelectric for shear or compression. The addition of other chemicals to the barium titanate (e.g., lead zirconate) can lead to specific design advantages as a source or receiver.

Because the electrical polarization properties of the material follow a hysteresis loop (resembling a ferromagnetic material such as iron) when an electric field is applied and reversed, those materials are called ferroelectrics. A major advantage of the ferroelectric ceramic is that it can be formed into a large variety of sizes and shapes to fit the particular array design. Transducer manufacturers (e.g., Channel Industries, Santa Barbara, California) have developed elements in various sizes and in the forms of cylinders, rods, tubes, disks, plates, hemispheres, and so forth.

Magnetostrictive transducers depend on the physical phenomenon of certain ferromagnetic materials such as nickel, which expand or contract when a magnetic field is applied. Since the change of dimension (contraction or expansion) is independent of the direction of the current that is creating the magnetic field, there is a frequency doubling when an AC signal is applied. (Note that a noisy 60-cycle power transformer radiates a hum of frequency 120 Hz.) This may be avoided by superimposing the AC signal over a larger DC polarizing field. Magnetostrictive devices have been particularly effective as low-frequency sound sources.

A large variety of piezoelectric plastics (e.g., polyvinylidene fluoride, PVDF) and new electrostrictive, ferroelectric, and magnetostrictive materials and composites are now being used as transducer elements. There are unusual configurations such as benders and segmented cylinders, as well as mushroom-shaped *Tonpilz* (the German word translates to "sound mushroom") mechanical oscillators for low frequencies. A promising fiber-optic flextensional hydrophone has been patented (Brown, 1994). The latest copies of the technical literature must be researched for these rapidly changing developments.

3.2 Line arrays of discrete sources

The sound pressure produced by an array of discrete sources such as in a multi-element sound source is readily adapted to computer summations. For example, assume there are N sources evenly spaced over a distance W in a straight line along the y-axis (Fig. 3.2). The separation between adjacent elements is

$$b = \frac{W}{N-1} \tag{3.9}$$

The pressure p_n of the nth source at distance R and angle θ is

$$p_n = \frac{a_n P_0 R_0}{R_n} \exp\left[i(\omega t - kR + \frac{nkW\sin\theta}{N-1})\right]10^{-\alpha R/20} \tag{3.10}$$

where a_n is a dimensionless amplitude factor of the nth source; θ is the angle with the z axis, which is perpendicular to the line array (Fig. 3.2); and α is the rate of attenuation in the medium.

In the spirit of the Fraunhofer far-field approximation, let R_n in the denominator become simply R. Also at long range, since $\omega t - kR$ is

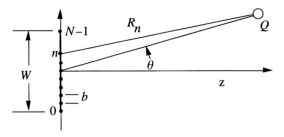

Fig. 3.2. Geometry for the directivity of a straight line of discrete sources with separations b, and array opening W.

common to all signals, we factor it out and obtain the pressure in terms of the transducer directional pressure response D_t:

$$p = D_t \frac{P_0 R_0}{R} \exp\left[i(\omega t - kR)\right]10^{-\alpha R/20} \quad R \gg W \qquad (3.11)$$

where D_t represents the "directional response," the "directivity function," or, simply, the "directivity" of a transducer or an array of transducers:

$$D_t \equiv \sum_{n=0}^{N-1} P_n \exp\left(i \frac{nkW \sin \theta}{N-1}\right) \qquad (3.12)$$

The complex quantity can be written in the form $D_t = A + iB$, where

$$A = \sum_{n=0}^{N-1} a_n \cos\left(\frac{nkW \sin \theta}{N-1}\right) \qquad (3.13)$$

and

$$B = \sum_{n=0}^{N-1} a_n \sin\left(\frac{nkW \sin \theta}{N-1}\right) \qquad (3.14)$$

Then the magnitude of the array directivity is found from

$$D_t = A + iB \quad \text{and} \quad |D_t| = (A^2 + B^2)^{1/2} \qquad (3.15)$$

The a_n are relative amplitudes of the transducer elements. It is often convenient to normalize the a_n by setting

$$\sum_{n=0}^{N-1} a_n = 1 \qquad (3.16)$$

Proper choices of the contributions by elements of the array whose amplitudes are P_n can improve array performance. The simplest, equally driven, arrays have

$$a_n = 1/N \qquad (3.17)$$

However, other choices of the contributions by elements of the array can improve array performance. Generally, the improvements desired are either to reduce the side lobes or to narrow the central lobe.

When the functional dependence on θ is included, the directivity of a symmetrical array with an odd number of elements is

$$D_t(\theta) = a_0 + 2\sum_{n=1}^{N_h} a_n \cos\left(\frac{nkW \sin \theta}{N-1}\right) \qquad (3.18)$$

where

$$N_h = \frac{N-1}{2} \qquad (3.19)$$

Two common weightings of elements (see Fig. 3.3) are

$$\text{triangular}: a(n) = a(-n) = (N_h - n)/N_h \qquad (3.20)$$

$$\text{cosine}: a(n) = a(-n) = 1 + \cos(\pi n/N_h) \qquad (3.21)$$

where (3.21) and (3.22) can be normalized by using (3.16).

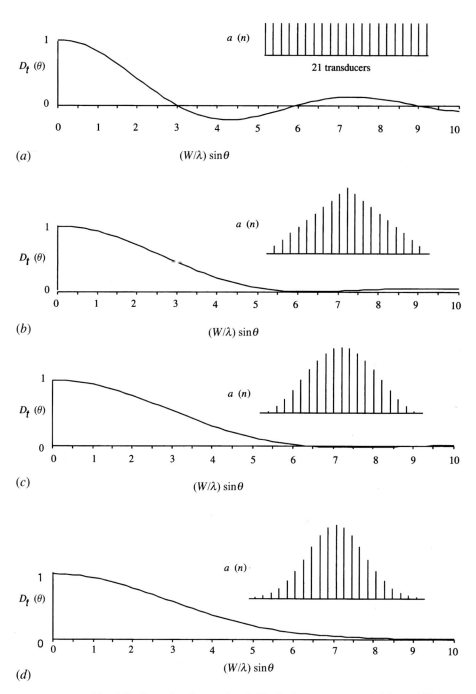

Fig. 3.3. Examples of array directivities for four common weightings of 21 equivalent spaced transducers: (*a*) uniform, (*b*) triangular, (*c*) cosine, and (*d*) Gaussian. Angle θ is measured with respect to the normal to the line of elements, the transducer axis *z*. The array aperture width is *W*.

Another weighting choice that is very popular because it often makes theoretical computations more tractable and convergence more rapid, and because it has no side lobes, is the so-called Gaussian directivity:

$$D_{tG}(\theta) = \exp\left[-(kW_G)^2(\sin^2\theta)/4\right] \qquad (3.22)$$

where W_G is the width parameter. One determines the $\theta_{1/2}$ for an array directivity $D(\theta)$ at its "half-power" response, $(1/2)^{1/2} = 0.707$, and solves for W_G in the expression

$$\exp\left[-(kW_G)^2\sin^2(\theta_{1/2})/4\right] = D(\theta_{1/2}) = 0.707 \qquad (3.23)$$

These types of "weighting" can also be applied to rectangular, circular, and other arrays.

The examples of linear, triangular, cosine, and Gaussian array directivities shown in Fig. 3.3 demonstrate the rule that, for the same number of elements, weightings that decrease the side lobes, unfortunately, also widen the main beam.

3.3 Directivity of continuous line sources

The integration of a continuous distribution of point sources along a straight line yields a line source. As shown in Figs. 3.2 and 3.3(*a*), let the array be many identical discrete *sourcelets* at very close spacing over a line of length W. But now replace the summation by an integration (Fig. 3.4) over a continuous line source. The strength of a differential sourcelet is proportional to dy/W. Let p be the resulting sound pressure in the local field region.

In the Fraunhofer plane wave approximation, the path difference from the sourcelet at origin 0 and that at y is $y\sin\theta$ (see Fig. 3.4). The differential pressure dp due to a sourcelet at y is

$$dp = (dy/W)P\,\exp\left[i(\omega t - kR + ky\sin\theta)\right] \qquad (3.24)$$

The pressure is calculated by the integration from $-W/2$ to $W/2$ of

$$p = PD_t\,\exp\left[i(\omega t - kR)\right] \qquad (3.25)$$

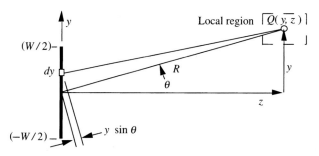

Fig. 3.4. Construction for a continuously distributed source at large range.

where

$$D_t = \frac{1}{W} \int_{-W/2}^{W/2} \exp(iky \sin\theta)\, dy \qquad (3.26)$$

Integration gives

$$D_t = \frac{1}{W} \frac{\exp(iky \sin\theta)}{ik \sin\theta} \Bigg|_{-W/2}^{W/2} \qquad (3.27)$$

Evaluation for the limits of integration gives

$$D_t = \frac{1}{W} \frac{\exp[(ikW \sin\theta)/2] - \exp[(-ikW \sin\theta)/2]}{ik \sin\theta} \qquad (3.28)$$

The expression for D_t reduces to

$$D_t = \frac{\sin[(kW \sin\theta)/2]}{(kW \sin\theta)/2} \qquad (3.29)$$

The preceding expression has the form $(\sin F)/F$, which is sometimes called the *sinc* function. Since the *sinc* is indeterminate as F tends to zero, the directivity D_t at zero is evaluated by forming the derivatives

$$\lim_{F \to 0} \frac{(d/dF)\sin F}{(d/dF)F} = \lim_{F \to 0} \frac{\cos F}{1} = 1 \qquad (3.30)$$

The *sinc* function is shown in Fig. 3.5(*a*), in a linear graph, and in Fig. 3.5(*b*) in polar coordinates for the case $F = (kW \sin\theta)/2$ when the source extent $W = 4\lambda$. In the polar pattern, the length of the radius vector is proportional to the directional response at that angle and, although the

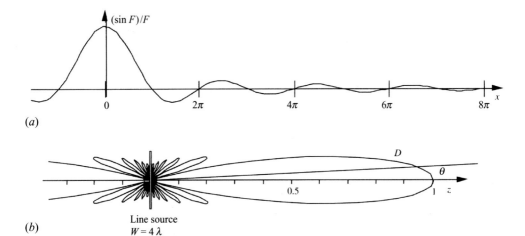

(*a*)

(*b*)

Fig. 3.5. Far-field radiation from a uniform line source (*a*) the $(\sin F)/F$ function, (*b*) polar radiation pattern of the directional response, $|D|$ versus angle, where $F = (kW \sin\theta/2)$ with $W = 4\lambda$. The angle θ is measured with respect to the normal to the line source.

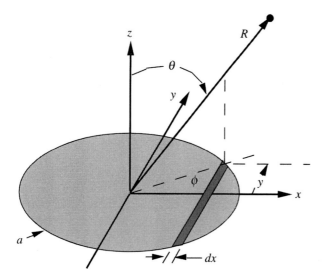

Fig. 3.6. Geometry for calculation of radiation from a circular piston transducer.

source is sketched in, the directivity is applicable only at long ranges from the source. The important sinc function specifies the directional response not only of a line source but of a line receiver as well.

We have given this analysis for point sources aligned in the y direction. We can do the same for sources in the x direction. If each of the dy sources on the y axis were a line of sources along the x axis, a rectangular source would be formed. The resulting directional response would be the product of the directional responses for the x and y directions.

Such a source is called a rectangular "piston source" when the individual elements within the source plane are very close together and have the same amplitude and phase. The function is the same for a receiver. The sinc function comes up again in the scattering of sound from a rectangular plane segment.

3.4 Circular piston source*

3.4.1 Far-field directivity

The geometry for a circular piston transducer is given in Fig. 3.6. We are interested in the directivity so we suppress the absolute pressure term and the propagation term $\exp\left[i(\omega t - kR)\right]$ and consider only the dependence on the "latitude" θ. It is assumed that the piston is in an infinite baffle (or enclosed), so that there is no back radiation, which would produce a dipole term.

---

* This section contains some advanced analytical material.

The field pressure will depend on an areal integration of the elements $2y\,dx$ in the plane of the source (rather than along the length for the line source). In this Fraunhofer far-field approximation, the range to any elements is approximately R. When spherical divergence and medium attenuation are included,

$$p = \frac{P_0 R_0}{R} 10^{-\alpha R/20} \frac{2}{\pi a^2} \int_{-a}^{a} y \cos(kyx \ \sin\theta)\, dx \qquad (3.31)$$

Define the transducer directional response D_t as

$$D_t \equiv \frac{2}{\pi a^2} \int_{-a}^{a} y \cos(kx \ \sin\theta)\, dx \qquad (3.32)$$

Change (3.31) to polar coordinates $x = a\cos\phi$ and $y = a\sin\phi$ and obtain

$$D_t = \frac{2}{\pi} \int_0^\pi \cos(\zeta\phi)\, \sin^2\phi\, d\phi \qquad (3.33)$$

where we use $\zeta = ka\sin\theta$ to agree with the form of the cylindrical Bessel Function in Abramowitz and Stegun (1964), which is written

$$J_1(\zeta) = \frac{\zeta}{\pi} \int_0^\pi \cos(\zeta \ \cos\phi)\, \sin^2\phi\, d\phi \qquad (3.34)$$

Using our value of ζ, (3.33) and (3.34) give the directional response for a circular piston source:

$$D_t = \frac{2J_1(ka \ \sin\theta)}{ka \ \sin\theta} \qquad (3.35)$$

The complete expression for the radiated pressure for a circular piston source has the same form as for a rectangular piston source, but with the different directivity factor as given by (3.35). Table 3.1 presents numerical values of the relative pressures and intensities of the circular piston directivity factor.

A linear graph of (3.35) is shown in Fig. 3.7. The central lobe of the radiation is down to fractional pressure $2^{-1/2} = 0.707$ (i.e., half-intensity) at $ka \ \sin\theta \cong 1.6$. The central lobe reaches zero at $ka \ \sin\theta = 3.83$. The first side lobe is $180°$ out of phase with the central lobe and shows a peak of sound pressure and intensity at $ka \ \sin\theta \cong 5.0$.

Examples of the directivity functions of a circular piston are shown in polar coordinates in Fig. 3.8 and in linear and polar graphs in Fig. 3.9. The pattern begins to resemble a "searchlight beam" only when ka is quite large.

3.4.2 Near-field

The circular piston–radiated pressure field at great distance ($kR \gg 1$), which was derived in the previous section, has been known for a long

Table 3.1 *Values of the circular piston pressure directivity function D_t, and intensity directivity D_t^2 (in terms of $z = ka \, \sin \theta$)*

z	$\frac{2J_1(z)}{z}$	$\left[\frac{2J_1(z)}{z}\right]^2$	z	$\frac{2J_1(z)}{z}$	$\left[\frac{2J_1(z)}{z}\right]^2$
0.0	1.0000	1.0000	7.0	−0.0013	0.00000
0.2	0.9950	0.9900	7.016	0	0
0.4	0.9802	0.9608	7.5	+0.0361	0.0013
0.6	0.9557	0.9134	8.0	0.0587	0.0034
0.8	0.9221	0.8503	8.5	0.0643	0.0041
			9.0	0.0545	0.0030
1.0	0.8801	0.7746	9.5	0.0339	0.0011
1.2	0.8305	0.6897			
1.4	0.7743	0.5995	10.0	+0.0087	0.00008
1.6	0.7124	0.5075	10.173	0	0
1.8	0.6461	0.4174	10.5	−0.0150	0.0002
			11.0	−0.0321	0.0010
2.0	0.5767	0.3326	11.5	−0.0397	0.0016
2.2	0.5054	0.2554	12.0	−0.0372	0.0014
2.4	0.4335	0.1879	12.5	−0.0265	0.0007
2.6	0.3622	0.1326			
2.8	0.2927	0.0857	13.0	−0.0108	0.0001
			13.324	0	0
3.0	0.2260	0.0511	13.5	+0.0056	0.00003
3.2	0.1633	0.0267	14.0	−0.0191	0.0004
3.4	0.1054	0.0111	14.5	−0.0267	0.0007
3.6	0.0530	0.0028	15.0	−0.0273	0.0007
3.8	+0.0068	0.00005	15.5	0.0216	0.0005
3.832	0	0			
			16.0	0.0113	0.0001
4.0	−0.0330	0.0011	16.471	0	0
4.5	−0.1027	0.0104	16.5	−0.0007	0.00000
5.0	−0.1310	0.0172			
5.5	−0.1242	0.0154	17.0	−0.0115	0.00013
6.0	−0.0922	0.0085	17.5	−0.01868	0.00035
6.5	−0.0473	0.0022			

time. But it was the advent of high-speed computers that permitted the calculation of the very complicated acoustic field closer to the source. Figure 3.10 is an early three-dimensional plot of the pressure field near a circular piston source of size $a/\lambda = 2.5$. There is an inner circle of very high pressure at the face of the transducer and two somewhat

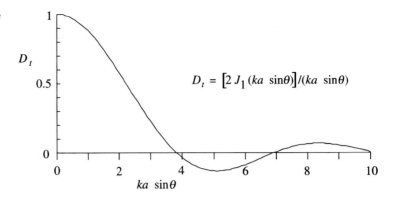

Fig. 3.7. Directional pressure response of a circular piston transducer.

$$D_t = [2 J_1 (ka\ \sin\theta)]/(ka\ \sin\theta)$$

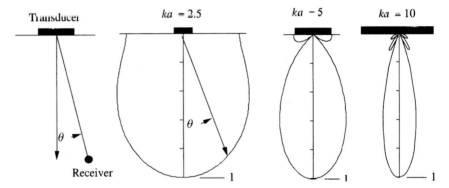

Fig. 3.8. Polar coordinate graphs of directivity patterns of circular piston transducers for values of *ka*.

lesser-pressure amplitude rings out toward the periphery of the radiating disk. Along the axis of the source the pressure amplitude is an oscillating function of range until it reaches a maximum, which turns out to be at $R \cong a^2/\lambda$. Beyond that range the pressure decreases into the far field. An increase in the ratio a/λ increases the number of pressure maxima and minima over the face of the source and along the axis. The maxima of these pressures on the face of the source can be very much higher than the average, so that destructive cavitation (see Section 4.2) may take place at local "hot spots," although the average pressure implies that there is no problem.

The complex "near field" is caused by constructive and destructive interferences of the radiation from different sub-areas of the transducer face. Near the source, these radiating wavelets may differ in travel path by $\lambda/2$ and cause almost complete cancellation of the sound pressure. Along the axis of the transducer, a critical range R_c exists where it is

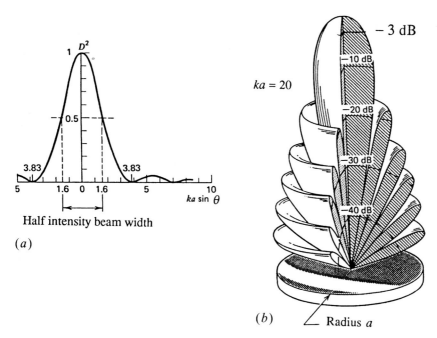

(a)

(b)   Radius a

Fig. 3.9. Directional response of a circular piston transducer of radius a in linear graph as a function of $ka \sin \theta$ (*left*) and in logarithmic plot for $ka = 20$ (*right*). The half-intensity beam-width angle where $D^2 = 0.5$, or -3 dB with respect to the axial value is indicated. (Eckart, 1968.)

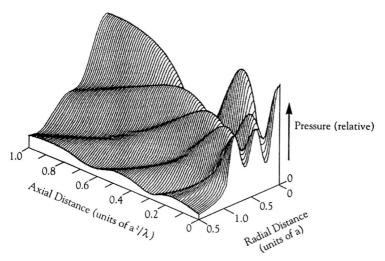

Fig. 3.10. Three-dimensional plot of relative pressure amplitude in the near field of a circular piston of size $a/\lambda = 2.5$. (Lockwood and Willette, 1973.)

no longer possible for wavelets traveling the longest path (from the rim of the piston) to interfere destructively with those traveling the shortest path (from the center of the piston). It is left as a problem for the reader to show that, for a circular piston source, this path difference, beyond which there can be no near-field axial minimum, is

$$R_{\text{rim}} - R_{\text{axis}} \cong a^2/(2R_c) < \lambda/2 \tag{3.36}$$

from which a critical distance is

$$R_c \cong a^2/\lambda \tag{3.37}$$

The range at which the experiment is "safely" in the far field, so that the pressure varies "essentially" as R^{-1}, is somewhat arbitrary. Often the far field for the circular piston is defined as beginning at the greater critical range, $\pi a^2/\lambda$.

3.5 Transducer descriptors

3.5.1 Total power radiated

The total acoustic power output by a CW point source radiating into 4π radians of solid angle is given in Section 2.1.4. Computation of the total radiated power when the pressure amplitude directional response is $D_t(\theta, \phi)$ is a simple adaptation of that calculation.

The "message" power passing through an increment of surface, ΔS in terms of the pressures $P(R)$ or $P_{ax}(R)$ at the field range, R, is

$$\Delta \Pi_M = \frac{P^2(R)}{2\rho_A c} \Delta S(R, \theta, \phi) \tag{3.38}$$

where

$$P(R) = P_{ax}(R)D_t(\theta, \phi) \tag{3.39}$$

or, in terms of the reference pressure P_0 at axial reference range R_0,

$$\Delta \Pi_M = \frac{P_0^2 R_0^2}{2\rho_A c R^2} D_t^2(\theta, \phi)10^{-\alpha R/10} \Delta S(R, \theta, \phi) \tag{3.40}$$

where the medium attenuation is now included.

Using the coordinates sketched in Fig. 3.6, the increment of area ΔS can be written as

$$\Delta S = R^2 \sin\theta \, d\theta \, d\phi \tag{3.41}$$

An integration over all angles gives the total radiated power Π_M. Typically the transducer is enclosed so that there is no back radiation. Then the θ integration limits are 0 to $\pi/2$,

$$\Pi_M = \frac{P_0^2 R_0^2}{2\rho_A c} 10^{-\alpha R/10} \int_0^{\pi/2} \sin\theta d\theta \int_0^{2\pi} D_t^2(\theta, \phi) \, d\phi \tag{3.42}$$

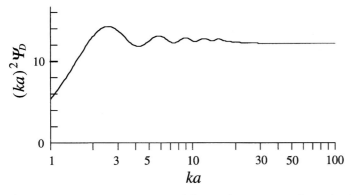

Fig. 3.11. Integrated beam pattern of the circular piston transducer of radius a as a function of ka. It is customary to plot $(ka)^2 \Psi_D$ as the ordinate. The $ka > 10$, Ψ_D is approximately $4\pi/(ka)^2$.

The double integral over the angles appears often. It is called the *integrated beam pattern* ψ_D. For an enclosed transducer ψ_D is defined as

$$\psi_D \equiv \int_0^{\pi/2} \sin\theta\, d\theta \int_0^{2\pi} D_t^2(\theta, \phi)\, d\phi \tag{3.43}$$

In general, lacking an analytical expression, the integrated beam pattern is evaluated numerically. The example of the theoretical circular piston transducer is shown in Fig. 3.11.

In terms of the integrated beam pattern Ψ_D, using SI units, the total radiated power over all angles is

$$\Pi_M = \left[\frac{P_0^2 R_0^2}{2\rho_A c} 10^{-\alpha R/10}\right]\psi_D = \left[\frac{P_{ax}^2 R^2}{2\rho_A c}\right]\psi_D \quad \text{watts} \tag{3.44}$$

For comparison to omnidirectional radiation, recall the expression for total power (2.16). In that case, the point source, $D_t = 1$ and $\Psi_D = 4\pi$. It is strongly recommended that the output of transducers be reported in total watts of radiated power. It is easy, using (3.44) to go from peak pressure on the axis of the transducer to total radiated power. However, notice that, because of near-field interferences, P_0 should not be *measured* at one meter from the source. It should be obtained by extrapolation back from a measurement at a far-field range.

The total radiated power is proportional to the integrated beam pattern. In order to keep the same sound intensity (proportional to p^2) along the axis of a transducer when there is an increase of beam width (e.g., from $ka = 10$ to $ka = 5$ in Fig. 3.8) the total radiated power must be increased.

If one knows the total radiated power at range R and the integrated beam pattern, one can solve (3.44) to obtain the reference peak pressure one meter from the transducer on the radiation axis:

$$P_0^2 = \frac{2\Pi_M \rho_A c}{\psi_D R_0^2} 10^{\alpha R/10} \tag{3.45}$$

3.5.2 Descriptors of radiation pattern

The need to comprehend the complicated radiation patterns (and receiving patterns) and to compare transducers in some simple way, has led to several single-number descriptors.

Beam directivity factor

The directivity factor of a transmitter, Q_t, is the ratio of the transducer sound intensity in the axial beam direction ($\theta = 0$) to the intensity, at that same far-field range, that would be caused by a point source radiating the same total power omnidirectionally (over 4π steradians).

$$Q_t = \frac{\text{Beam axial intensity at } R}{\text{Point source intensity at } R} = \frac{P_{ax}^2/(2\rho_A c)}{\Pi_M/(4\pi R_{ax}^2)} \tag{3.46}$$

Substitute Π_M from (3.44) and obtain the transmitter directivity factor

$$Q_t = 4\pi/\psi_D \tag{3.47}$$

Beam directivity index

The directivity index is obtained by taking the logarithm of Q_t

$$DI_t = 10 \log_{10} Q_t \quad \text{dB} \tag{3.48}$$

Special cases of the directivity index for piston sources are:
a rectangle of dimensions, $L \cong W$:

$$DI_t = 10 \log_{10} \frac{k^2 L W}{\pi} \quad kL \gg 1 \tag{3.49}$$

and a circle of radius a

$$DI_t = 10 \log_{10} (k^2 a^2) \quad ka \gg 1 \tag{3.50}$$

Beam width

The *half-intensity beam width* is the angle $\Theta_{\text{beam}} = 2\theta$ measured from $-\theta$ where the acoustic intensity is half the axial value, past the axis, to the angle $+\theta$ where the acoustic intensity is again half the axial value (see Fig. 3.9). It is also called the "half-power beam width," an unfortunate misnomer which comes from electric circuit theory.

For example, from Table 3.1, the half-intensity value of the circular piston beam is at $ka \sin\theta = 1.6$. Therefore, for $ka = 20$, the half-intensity angle is $\theta = \sin^{-1} 0.08 = 4.6°$ and the half-intensity beam width is $\Theta_{\text{beam}} = 2\theta = 9.2°$.

3.5.3 Equivalence of source/receiver directivity

The same analytic expressions that were derived for the pressure field due to an array of sources can be used to describe the directional response of an array of pressure-sensitive receivers.

In many applications, the same array is used to transmit sound pressures and then switched to receive sound pressures. These arrays of transducers are reciprocal devices. The directional response of the array as a transmitter is the same as its directional response as a receiver.

3.6 Free-field calibration of transducers

A relatively simple, absolute method for the calibration of transducers is the "free-field" reciprocity technique, which is performed in a region free of scatterers. Needed are two small, "identical," "point" hydrophones, A and B (one of which can be used as a source as well), and a third transducer, C, that acts solely as a source (Rudnick and Stein, 1948).

Transducers A and B should be "identical" in shape and size to minimize differential sensitivities due solely to different diffraction fields around the transducers. Transducers A and B should be "point" transducers (size $\gg \lambda$) so that pressure is inversely proportional to the separation range R between source and receiver.

The free-field condition can be satisfied by using a ping or an impulse short enough to complete the measurement before extraneous reflections are received at the hydrophone. When an impulse is used, the Fourier Transform of the pulse yields a broad spectrum that allows many frequencies to be calibrated simultaneously.

Transducer C is a source of far-field pressures (inversely proportional to the separation) used to determine the relative sensitivities of A and B as hydrophones. The procedure is simple (see Fig. 3.12). Transducer A is placed at a position of the sound field produced by C; its open-circuit voltage, V_{AC}, is measured. Then B is placed at the same position and its voltage, V_{BC}, is obtained. This yields the ratio of sensitives, K_{VPB}/K_{VPA}.

The final two measurements are the current I_B, drawn by B when it is used as a source, and the open-circuit voltage V_{AB} across A at the same time. When these voltages, current, and separation are known, the sensitivity of hydrophone A is given by

$$K_{VPA} = \left(\frac{2R V_{AC} V_{AB}}{\rho_A c V_{BC} I_B} \right)^{1/2} \text{volts/pascal} \qquad (3.51)$$

where ρ_A = water density and c = speed of sound in water.

The reciprocity technique thereby provides an absolute calibration of the sensitivity of hydrophone A by measurements of electrical quantities and distance alone. Furthermore, the sensitivity of hydrophone A is

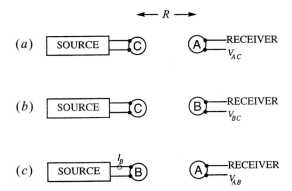

Fig. 3.12. Scheme for obtaining free-field reciprocity calibrations. In (a) C is the source, and the voltage at transducer A is measured to yield V_{AC}. In (b) C is the source, and the voltage at B is V_{BC}. In (c) B is the source, and the current to B is I_B when the voltage across A is V_{AB}.

the key to unlock the sensitivity of B because we know the relative sensitivities, K_{VPB}/K_{VPA}. Finally, by using either hydrophone, we have measured the output of source C or any other transducer that we can specify in terms of pascals produced at the measurement range/source voltage input. A precaution: we have assumed that the transducers are not driven beyond their linear range and that they do not produce non-linear distortion.

3.7 Self-reciprocity calibration of transducers

A transducer can be calibrated by pointing it toward a large reflecting surface (Carstensen, 1947), such as the smooth water–air surface above the transducers in a water tank. Assume that the separation from the mirror is distance d, and the frequency of the short duration sinusoidal signal is f.

Define $K_{VP} = V_m/p_m$, the voltage generated by pressure p_m at the transducer, where V_m is the open-circuit voltage after mirror reflection of the signal, and p_m is the pressure after mirror reflection.

Also define $S = p_1/I_s$, the source response, pressure per ampere input, where p_1 is the output pressure at range 1 meter when the source draws a current, I_s amperes.

The ratio of the transducer sensitivity (as a hydrophone) to its response (as a source) is called the reciprocity parameter, J:

$$J = K_{VP}/S = (V_m/p_m)/(p_1/I_s) \qquad (3.52)$$

The calibration is performed in a spherically divergent field, so that the pressures are inversely proportional to the distances:

$$p_m/p_1 = R_0/(2d) \qquad (3.53a)$$

Combining 3.52 and 3.53 and setting $R_0 = 1$ m, we find the transducer sensitivity when used as a hydrophone,

$$K_{VP} = V_m \left(\frac{2d}{p_m p_1}\right)^{1/2} \qquad (3.53b)$$

and the same transducer output pressure at 1 meter, when used as a source,

$$S = \frac{1}{I_s}(2d p_m p_1)^{1/2} \qquad (3.53c)$$

This simple, nearly magical, calibration of the transducer as a source or receiver is predicated on several conditions: the field is spherically divergent so that the source is effectively a point at the range of the calibration; the reflecting surface is large and smooth (compared with the wavelength); the transducer axial propagation is perpendicular to the reflector; and the source "ping" is long enough to identify the frequency of the CW but short enough to avoid scatter from nearby objects.

3.8 Transducer dependence on mounting

A warning: a transducer's sensitivity and radiation pattern are functions of the geometry and the material of the mounting and its surroundings. A sound source or receiver that is imbedded in a real marine environment such as in a whale may have a spectral sensitivity and polar output quite different from those in an infinite rigid baffle.

Transducer calibrations should be performed *in situ* or in a setting that duplicates its intended physical environment at sea.

3.9 Other types of sonars

Comparisons of mapping and object location operations that use radar in air or sonar in water demonstrate the large differences between the use of electromagnetic waves in air and sound waves in water. For radar the electromagnetic wave velocity is 3×10^8 m/s. The pulse travel time for a range of 30 km is 2×10^{-4} s, and a simple radar systems can send, receive, and display in a very short time. The time required to make a $360°$ image at $1°$ increments can be less than 0.1 s. Thus radar systems can use a single rotating dish to give good images. Consider an airborne radar. In 0.1 s, an aircraft moving at a little less than the speed of sound in air (about 600 miles/h or 1000 km/h) moves only about 30 m. However,

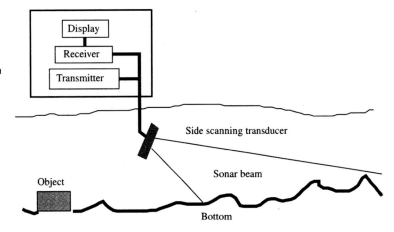

Fig. 3.13. Side-scanning sonar. The sonar looks to the side of the ship and makes an echo sounding record as the ship moves. The time of return of a pulse is interpreted as the range to the bottom feature that caused the scatter. Display software converts the "raw image" to a map of features on the bottom.

the attenuation of electromagnetic waves in sea water is very large, and radar does not have a useful working range in the ocean.

For sonar (sound speed = 1500 m/s), the time required to range to 30 km is 40 s. In a sequential data acquisition system that takes one echo measurement at a time, several hours at one location would be needed to make one 360° image. A ship moving at 9 km/h (2.5 m/s) moves 100 m during the time for a single echo ranging measurement. One technological solution has been to acquire sonar data in parallel by transmitting and receiving in many directions at the same time.

In this section we describe some of the ingenious technological solutions to this problem.

3.9.1 Side-scan sonars

The side-scan sonar is an echo sounder that is pointed sideways (Fig. 3.13). However, although the design concepts are the same as for the simple vertical piston echo sounder, the sending transducer produces a fan-shaped beam, and the receiver has a time-variable gain to compensate for range. Side-scanning sonars are used to give images of rough features on the sea floor. The instruments are also used to locate and identify objects such as sunken ships (see figure in Prologue).

3.9.2 Multibeam sonars

Figure 3.14 shows an example of a multibeam sonar for sea-floor mapping. This system is intended to map a swath of depths along the ship track. Since these systems are usually mounted on the hull of the ship, the receiving array points in different directions as the ship pitches and

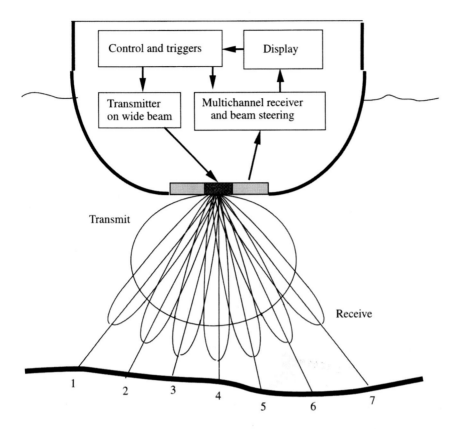

Fig. 3.14. Multichannel sonar system using preformed beams. A cross-section of the ship is shown. The transmission is a broad beam. By adjusting time delays of the receiving elements, the multi-element receiving array is preformed to a set of narrow beams that look from port to starboard and measure the depths to various positions such as 1 to 7. As the ship moves, the computer makes a contour plot of the depths. The use of color coding yields a highly revealing picture.

rolls. The data-reduction system must compensate for the ship motions and the direction in which the receiving array is pointing when the echoes arrive.

3.9.3 Doppler sonars

Doppler sonars are used to determine the velocities of ships relative to the water or the sea floor. They may also be used to measure the motion of the ocean surface or swimming objects, or internal waves, within the volume. The theory of operation is given in Section 2.6.

Another application of the Doppler phenomenon is the oceangoing "portable" Doppler velocimeters, as seen in Figs. 3.15 and 3.16.

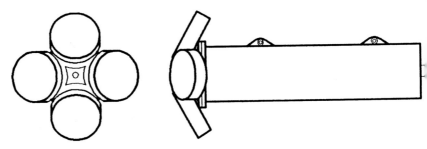

Fig. 3.15. Physical configuration of an acoustic Doppler current velocimeter. The four transducers here are oriented 30° off the cylinder axis with 90° azimuthal spacing. The instrument body is a cylinder of length 81 cm and diameter 18 cm. When used with 75 kHz transducers, the four transducers span 76 cm measured perpendicular to the instrument axis. At 1200 kHz, with smaller transducers, the span is 22 cm. (Courtesy RD Instruments, San Diego, California.)

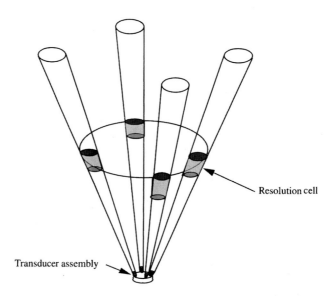

Fig. 3.16. Beam pattern of the vertically oriented Doppler velocimeter. The velocity measurement region is the space bounded by the four beams. A short duration pulse ("ping") is used. The resolution cell is defined by the transducer beam pattern and the ping duration. Up to 128 cells are available in the instrument shown.

3.9.4 Passive listening systems

Passive acoustical systems may range in complexity from a single hydrophone to an elaborate, steered array of hydrophones.

Buckingham *et al.* (1992) coined the phrase "acoustic daylight" to describe a very different application of passive acoustics. The basic idea is sketched in Fig. 3.17. Physical, biological, and man-made sounds at sea can be used as an acoustic analog of daylight – that is, they illuminate

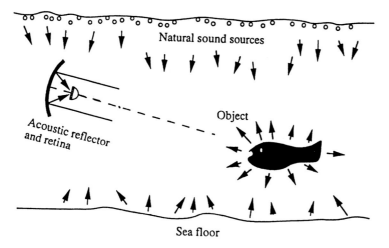

Fig. 3.17. Object insonified by acoustic daylight. The sketch shows natural sources from near the sea surface and scattered sound from the sea floor, illuminating an object that scatters sound toward a receiver. The directional receiver is sketched as an acoustic reflector that is focused on an acoustic retina.

(more correctly, insonify) unknown objects in the sea. Since the ocean water between the surface and the bottom is acoustically transparent, the situation is the acoustic analog of an optically transparent atmosphere. The natural sounds of the sea insonify objects in the water, and the objects scatter these waves. Therefore, if one were to scan with the proper acoustic equipment, one should be able to identify the waves scattered by objects in the ocean, and thereby identify the objects themselves (Buckingham and Potter, 1994).

Using acoustic daylight to sense objects at sea is much more difficult than using our eyes to identify objects on a sunny day. Since the natural sounds are extremely variable in frequency spectrum, amplitude, and phase, some kind of multibeam, acoustic lens-retina or focusing reflector-retina system is needed to compare the relative sounds from many directions at the same time, and to identify the scattering body. One acoustical advantage may be that identifiable sounds in the sea cover a vast frequency range from fractions of a cycle to megahertz. This is to be compared with the optical frequencies to which the eye is sensitive – frequencies that cover a range (violet to red) of only 2 to 1.

3.9.5 Steered arrays

Transmitting or receiving arrays of transducers are steered by adding the signals from each transducer after proper time delays. The same analysis applies to send or to receive; we give the analysis for a receiving array.

Consider the array of transducers in a line perpendicular to the direction $\phi = 0°$ (Fig. 3.18). To electronically/digitally steer the array, we insert appropriate time delays in each y_n-channel.

Let the signal at the 0th hydrophone be $p(t)$ and the channel amplification factor be a_0. From the geometry in Fig. 3.18 the plane wavefront

Fig. 3.18. Electronically/ digitally steered array for a plane wave entering the line of transducer array elements at angle ϕ. To steer the array, the elements at positions y_0, y_1, and so forth are given time delays T_0, T_1, and so forth that depend on the angle. The combination of the transducer sensitivity, A/D conversion, and amplifiers for the individual elements have the values a_n. The concept works for sources as well as receivers.

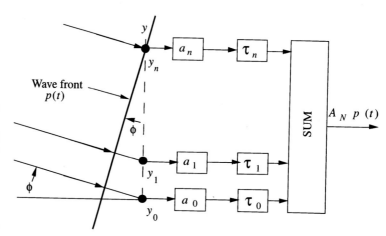

arrives at the nth hydrophone at advance Δt_n before reaching the 0th hydrophone, where

$$\Delta t_n = (y_n \sin\phi)/c \tag{3.54}$$

The signal at hydrophone n is $p(t - \Delta t_n)$. The A/D conversion and amplification are in the a_n. The time delay τ_n is inserted to produce the signal $p(t - \Delta t_n + \tau_n)$.

The sum signal for N channels is

$$A_N p_N(t) = \sum_{n=0}^{N-1} a_n\, p(t - \Delta t_n + \tau_n) \tag{3.55}$$

where A_N is an amplitude factor.

Now, if τ_n is chosen to equal Δt_n, the signals add in phase for that direction ϕ, and we have

$$A_N p_N(t) = p(t) \sum_{n=0}^{N-1} a_n \quad \text{for} \quad \Delta t_n = \tau_n \tag{3.56}$$

This method of array steering is called "delay and sum." The only assumption is that the signals in each channel are the same except for their time delays. Delay and sum processing works for any $p(t)$. The directional response of a steered array in other directions can be computed by choosing an incoming angle ϕ' and letting

$$\tau_n = (y_n \sin\phi')/c \tag{3.57}$$

Then

$$\Delta t_n - \tau_n = y_n(\sin\phi - \sin\phi')/c \tag{3.58}$$

The directional response of the array as a function of ϕ depends on the value of ϕ'.

We have assumed that the incident sounds are plane waves. This is equivalent to assuming that the curvature of the wavefront is small over the dimensions of the array (e.g., less than $\lambda/8$). The plane-wave assumption is effective for small arrays or distant sources.

Arrays are built in many configurations: cylinders, spheres, and so on. The multibeam sonar described in Fig. 3.14 is one example. By using time delays, almost any shape can be steered to receive signals of any curvature from any direction. However, when the arrays are built around a structure, diffraction effects can modify and may deteriorate the performance.

3.9.6 Dual-beam sonars

The amplitude of an echo from a body depends on its location in the sending and receiving sonar beams. To measure the apparent scattering length or the backscattering cross-section, we must know the object's position in the sonar beams. If the object is uncooperative, like a freely swimming fish, some ingenuity is needed.

In 1974, Ehrenberg suggested the potential effectiveness of the dual-beam sonar system as a means to locate a scatterer. Traynor and Ehrenberg (1990) showed that, in addition to being simple, the method is practical. As shown in Fig. 3.19, the system has a wide-beam source transducer and a narrow-beam transducer. The transducers are circular, and their beam patterns are symmetrical about the axis.

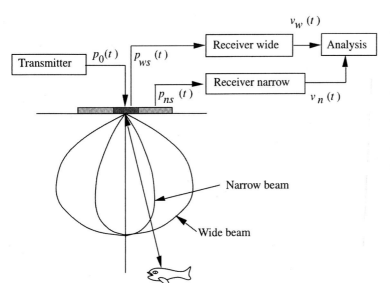

Fig. 3.19. Dual-beam transducer system. The response of the wide beam is approximately the reciprocal of the response of the narrow beam. The ping is transmitted on the wide beam, and the echo is received on both transducers. The echo amplitudes are compared in the analyzer.

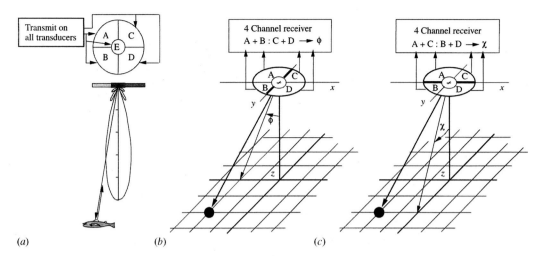

(a)   (b)   (c)

Fig. 3.20. Split-beam sonar and dual-beam sonar. The sonars transmit a narrow beam by using transducers $(A + B + C + D + E)$ together. For dual-beam echo processing, the output of transducer E gives the wide beam. For split-beam echo processing, the transducers A, B, C, and D are the receivers. (a) A fish is in the sonar beam at the range R and angles ϕ and χ. (b) The comparison of the phases of $(A + B)$ and $(C + D)$ gives the angle ϕ. (c) The comparison of the phases of $(A + C)$ and $(B + D)$ gives the angle χ. (Based on Traynor and Ehrenberg, 1990.)

Generally one assumes spherical divergence from the source and spherical divergence from the scatterer, that there are no refraction effects, and that there is no incidental scattering between the source and the scattering object. Transmission can be from either the wide- or narrow-beam transducer.

3.9.7 Split-beam sonars

The receiving transducer has four sectors and four receiving channels (Fig. 3.20). The first step in the signal processing is to identify an echo in all four receiver channels. The accepted echo is processed to determine its direction. The conversion of measured phase shifts to implied directions requires acoustical calibrations of the system. Commonly, split-beam systems transmit a narrow beam by using all transducers on transmission. Details of the echo processing are in the software.

Illustrative problem

Assume that the directional transducer has dimensions 0.12×0.06 m. For the same parameters as in illustrative problem 2.2 of Chapter 2, how much would the SPL be increased along the axis of the transducers?

$$k = 2\pi f/c$$
$$= 2\pi (10^5/1473)\ \text{m}^{-1} = 427\ \text{m}^{-1}$$
$$H = 0.06\ \text{m}, \quad W = 0.12\ \text{m}$$
$$DI_t = 26.2\ \text{dB}$$
$$SL = SL(\text{omni}) + DI_t$$
$$\simeq 80.8 + 26.2 = 107.0\ \text{dB re 1 Pa} = 227.0\ \text{dB re 1 μPa (at 1 m)}$$
$$SPL = 107.0 - 80 = 27\ \text{dB re 1 Pa} = 147\text{dB re 1 μPa at 1000 m}$$
$$= 20\log(P/P_r)$$
$$P = 22\ \text{Pa at } R = 1\ \text{km}.$$

The effect of the directionality of the transducer is to increase the axial signal by 26.2 dB, that is an acoustic pressure about 20 times greater than for the omnidirectional source.

Problems and some answers

3.1 Compare graphically the radiation patterns of a circular piston of diameter $2a$ and a rectangular piston of width $W = 2a$.

3.2 Plot the polar radiation pattern of a piston transducer of radius 10 cm when radiating frequency 15 kHz in the ocean.

3.3 Plot the polar radiation pattern of a square transducer of side W, at the "corner" angle $\phi = \chi$.

3.4 Calculate the directional response of a 26 cm, square piston transducer in the plane $\chi = 0$. The radiated frequency is 30.0 kHz. Plot in decibels, using polar coordinates.

3.5 Assuming that the circular transducer of Fig. 3.9 is radiating sound of frequency 20 kHz, what is the piston radius?

3.6 The approximation is sometimes made that the directional response of a square piston of width W, in a plane parallel to one edge, is about the same as for a circular piston of diameter $2a = W$. Calculate the directional responses of pistons of $k(2a) = kW = 33$.

3.7 A calibration is needed for the output of a circular piston source of radius 20 cm over the frequency range 5 to 50 kHz. Where should the hydrophone be placed to ensure that it is safely out of the near field.

3.8 Verify the directivity patterns for a 21-element line source (as shown in Fig. 3.3) for those four cases: (a) uniform (piston) point source weighting; (b) triangular weighting; (c) cosine weighting; (d) Gaussian weighting.

3.9 Calculate the central lobe beam widths for the cases in Fig. 3.3.

3.10 Calculate the relative pressure of the first side lobe peak, compared with the axial value, for the four beam patterns in Problem 3.8.

3.11 State the far-field directivity function for a rectangular piston source of length L in the x direction and width W in the y direction, propagating in the z direction.

3.12 Determine the far-field directivity in the xz plane and yz plane of a rectangular piston source of dimensions L and W. Plot the directivity as a function of $0 < \phi < 2\pi$ for the cases in which (a) $L = W$; (b) $L = 2W$; (c) $L = 10W$.

3.13 Use a "reasonable" criterion to derive the critical range for the far-field directivity of a rectangular piston source of dimensions L and W, propagating in the xz plane, or yz plane.

3.14 Calculate the half-intensity beam widths in the xz and yz planes for a rectangular piston source of $L = 2W$ and $L = 10W$.

3.15 Calculate the half-intensity beam widths of the triangular, cosine, and Gaussian line sources in Fig. 3.3.

Answers to some problems

3.2 Graph: At $5°$, -0.3 dB; at $10°$, -1.3 dB; at $15°$, -3.0 dB; at $20°$, -5.6 dB

3.4 Graph: At $5°$, -3.2 dB; at $10°$, -19.5 dB; at $15°$, -13.6 dB

3.5 $a = 24$ cm

Further reading

Lighthill, J. (1978, 1979). *Waves in Fluids*. Cambridge, UK: Cambridge University Press.
A *tour-de-force* presentation of all forms of waves including an elegant exposition of the sources of sound, consisting of monopoles, dipoles, quadrupoles. Lighthill was the first to quantitatively attribute the acoustic power and directionality of the noise of jet aircraft engines to quadrupole sources generated at the boundary between the high speed exhaust and the quiescent air adjacent to the exhaust. The concept is applicable also in water, of course.

Wilson, O. B. (1985). *An Introduction to the Theory and Design of Sonar Transducers*. Peninsula Publishing, Box 867, Los Altos, CA. 94023
A good survey of ocean transducer theory and design as taught at the US Naval Postgraduate School.

Chapter 4
Intense sounds: non-linear phenomena

Summary

A naïve ocean scientist would expect that the acoustic pressure radiated from an electromechanical underwater sound source will be proportional to the applied voltage, and that the wave form of the sound pressure will be a replica of the wave form of the applied voltage. Not so. Even if the transducer and its electrical circuitry are designed to meet these objectives, the reality of the ocean medium prevents this idealization from happening. Consequences of intense sound fields in the ocean include: generation of harmonics which may cause sinusoidal waves to become repeated shock waves; creation of cavities of water vapor at positions of intense sounds; generation of sum and difference sound frequencies when there are two, coexistent, intense, simple sinusoids; acoustic streaming (jetting and water circulation) near intense sound sources; sound radiation pressure which creates forces on nearby objects. Some of these effects prevent the successful use of sound. Others give ocean scientists and engineers opportunities to create unique applications of sounds in the sea.

Contents

4.1 Harmonic distortion and shock waves

The approximations that lead to infinitesimal amplitude acoustics, which
have been assumed up till this point, obscure the extraordinary behaviors
that occur when sound pressures are large. It is important to understand
these effects whenever high-intensity sound sources are used underwater.
The first correction that we consider occurs because the sound speed is
a function of ambient pressure.

Figure 4.1 sketches the difference between the "equations of state"
of air and water.

The physical relation between incremental changes of pressure and
density, the slope in the Fig. 4.1, determines the speed of sound in the
medium at that ambient pressure as first described in (1.55):

$$c^2 = \frac{\Delta p}{\Delta \rho} = \frac{E}{\rho_A} \tag{4.1}$$

To calculate the speed, we need an analytic expression for the equation
of state of the fluid, $p_A = p_A(\rho_A)$. An appropriate form is the adiabatic
relation

$$\left(p_A \rho_A^{-\Gamma} \right)_s = K \tag{4.2}$$

where K and Γ are empirical constants. The subscript s identifies
the condition of zero heat exchange in a reversible process (constant

Fig. 4.1. Schematic sketch of
equations of state for air and
water, $p_A = p_A(\rho_A)$. The
curves are not to scale. The
steep slope for water shows its
relative incompressibility. The
values p_{AO} and ρ_{AO} are the
standard conditions (one
atmosphere). The correct
slopes at standard conditions
for water are about 23 times
greater than they are for air.

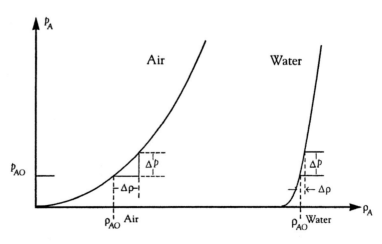

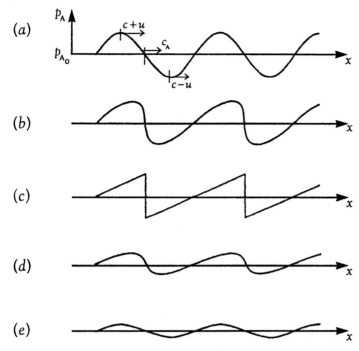

Fig. 4.2. Stages of a finite-amplitude sinusoidal sound wave. (*a*) Close to a source with large sinusoidal pressure swings. The local speeds are indicated. (*b*) Non-linear distortion after propagation away from source. (*c*) Fully developed repeated shock wave (sawtooth wave) away from source. (*d*) Aging, repeated shock wave after greater loss of higher-frequency components. (*e*) Infinitesimal nearly sinusoidal amplitude of a former shock wave.

entropy). This is generally a good assumption for sound propagation at sea, except in bubbly water.

To calculate the signal speed $u + c$ at a point where the excess density is Δ_ρ, use (1.59):

$$u = \left(\frac{\Delta \rho}{\rho_A} \right) c_A \qquad (4.3)$$

Then the rate of advance of a point on a wave depends on the sum

$$u + c = c_A \left(1 + \beta \frac{\Delta \rho}{\rho_A} \right) \qquad (4.4)$$

where $\beta = 1 + B/(2A)$. The ratio B/A is called the "parameter of non-linearity." For water B/A $\simeq$ 5.

Consider the *finite* sinusoidal wave sketched in Fig. 4.2(*a*). The wave's crest is at higher pressure than the undisturbed medium; therefore the speed of sound at the crest is $c > c_A$. The particle velocity at the crest is u. The crest advances at speed $c + u$. In a rarefaction, u is negative,

$c < c_A$, and the wave trough advances at the lesser speed $c - u$. The net result is that the crests advance relative to the axial positions, and the troughs lag behind the axial positions. As the effect continues, the wave distorts to a form resembling Fig. 4.2(b). Finally, if the pressure swing is large enough for a sufficient number of wavelengths, the finite amplitude wave becomes a "sawtooth" wave or "repeated shock wave," (Fig. 4.2(c)). The progressive growth construction cannot continue beyond this point; to do so would be to imply double values of pressure at various parts of the wave.

After the shocks are formed, the phenomenon of harmonic distortion proceeds into "old age" as the higher frequencies, which define the corners of the shock-wave fronts, dissipate too rapidly to be compensated by harmonic growth. In that condition the corners of the shock wave erode and become rounded. The shock wave thereby returns toward a sine wave form, but with greatly reduced amplitude (Fig. 4.2(d) and (e)).

Fourier analysis of the various forms of the sawtooth wave reveals that energy has been taken from the fundamental of frequency f (the original sinusoidal frequency) and redistributed into the second, third, and higher harmonics as the wave distorts. In the ultimate repeated shock, the pressure amplitude of the second harmonic (frequency $2f$) is half that of the fundamental, the third harmonic ($3f$) has one-third the amplitude of the fundamental, and so on. The cascading redistribution of energy from the fundamental to the upper harmonics is accompanied by an increased real loss of acoustic energy because the newly generated higher frequencies dissipate into thermal energy at a much faster rate than the fundamental (Chapter 2).

It is useful to compare non-linear propagation in air with that in water. For the same pressure amplitude of the fundamental, the second harmonic will grow to the same magnitude in a distance only $1/5000$ as far in air as in water. Even though the sound is diverging from the source, the sensitive listener often hears the popping, crackling sound of mini-shock waves from nearby jet aircraft during takeoff.

The effect that we have described limits the energy that can be put into a sound beam by a source. As the input power increases, more and more power goes into the higher harmonics which, with their larger attenuation rate, act to dissipate the new power. Consequently an intensity is reached where additional input of energy at the source does *not* produce increased sound pressures in the field. The limit is called "saturation."

The saturation effect for a spherically divergent beam in water is illustrated in Fig. 4.3. Where the response is linear, the 45° line shows that field pressure is proportional to source pressure. The ultimate saturation effect is increasingly evident at the larger ranges. For the four greatest

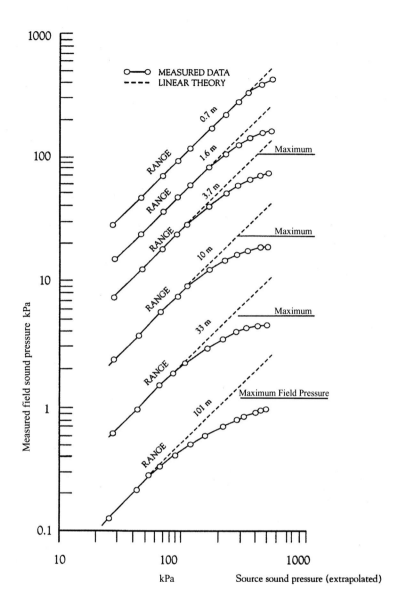

Fig. 4.3. Amplitude response curves in kilopascals showing the reduced field pressures due to extra losses caused by non-linear propagation at increased source pressures for six ranges in water. Piston source diameter = 7.6 cm, $f = 454$ kHz. The horizontal lines at the right are the asymptotic saturation pressures. (Data from Shooter, *et al.*, 1974.)

ranges, the saturation pressure is indicated by the horizontal lines to the right.

A simultaneous effect of non-linear harmonic distortion is the degradation of the beam radiation pattern. Since the axial part of the beam has the greatest intensity, it suffers more than the off-axis field. As a result, the lesser attenuation at greater angles off-axis causes a broadening of the beam. This also causes the side lobe intensities to be closer to the intensity of the strong central part of the beam.

4.2 Cavitation

Near sea level, minute bubbles of micron or submicron size are always present in the ocean. When the rarefaction (tension) phase of an acoustic wave is great enough, the medium ruptures or "cavitates." For sound sources near the sea surface, the ever-present cavitation nuclei permit rupture to occur at pressure swings of the order of 1 atm (0.1 MPa), depending on the frequency, duration, and repetition rate of the sound pulse. Cavitation bubbles may also be produced by Bernoulli pressure drops associated with the tips of high-speed underwater propellers. Natural cavitation is created by photosynthesis and the life processes of some marine animals.

Several extraordinary physical phenomena are associated with acoustic cavitation. Chemical reactions can be initiated or increased in activity; living cells and macromolecules can be ruptured; violently oscillating bubbles close to a solid surface can erode the toughest of metals or plastics; light may be produced by cavitation (sonoluminescence). The high pressures and high temperatures (calculated to be 30 000 Kelvin) at the interior during the collapsing phase of cavitating single bubbles can cause emission of a reproducible pulse of light of duration less than 50 picoseconds and have been proposed as a vehicle for controlled nuclear fusion.

Of direct importance to the use of sound sources at sea is the fact that, as the sound pressure amplitude increases, ambient bubbles begin to oscillate non-linearly, and harmonics are generated. At sea level, the amplitude of the second harmonic is less than 1 percent of the fundamental as long as the pressure amplitude of the fundamental of a CW wave is less than about 0.01 atm rms (1 kPa) (Rusby, 1970). This increases to about 5 percent harmonic distortion when the signal is about 10 kPa.

When the peak pressure amplitude is somewhat greater than 1 atm, the absolute pressure for a sound source at sea level will be theoretically less than zero during the rarefaction part of the cycle. In using CW below 10 kHz, this negative pressure, or tension, is the trigger for a sharply increased level of harmonic distortion and the issuance of broadband noise. Any attempt to increase the sound pressure amplitude appreciably beyond the ambient pressure will cause not only total distortion but also the generation of a large cloud of bubbles which will actually *decrease* the far-field acoustic pressure.

The detailed bubble activities during cavitation have been studied in several laboratories. Acousticians have identified gaseous cavitation resulting in streamers of hissing bubbles that jet away from regions of high acoustic pressure swings, and vaporous cavitation, which radiates shock waves of broadband noise.

The nuclei for cavitation often are bubbles caught in crevices of solid particles. If the acoustic pressure swing is great enough, the bubbles grow by a process called "rectified diffusion." During a cycle of this action, more gas diffuses inward from the liquid to the bubble during the expansion part of the cycle when the bubble is larger, than moves outward during the contraction part of the cycle, when the bubble surface is smaller. After growth to a critical radius, the bubble will expand explosively.

The acoustic pressure threshold for cavitation depends on the criterion selected to define its existence. In the laboratory it is possible to use visual observation. More commonly, the broadband noise of the cavitating bubbles is evidence of the onset. Additionally, the non-linear generation of subharmonics of half-frequency is one criterion that defines the onset of cavitation.

The sound field of an oceangoing transducer consists of regions close to the source and on the face of the source where the near field sound pressure is either greater than, or less than, the calculated pressure at the one meter reference range. Furthermore, both acoustic radiation pressure and streaming (see Section 4.6) move cavitating bubbles to new positions where they experience different growth rates. Therefore there are "hot spots" where cavitation activity is significantly greater than the prediction from the average reference pressure. The sound intensities that can be effectively radiated deep below the surface are higher than those that can be radiated at the surface, because there are fewer cavitation nuclei, the radii of these nucleii are smaller, and the ambient pressure is greater.

The dependence of cavitation onset and harmonic distortion on pulse duration is shown in Fig. 4.4. For pulse durations less than 100 ms, the

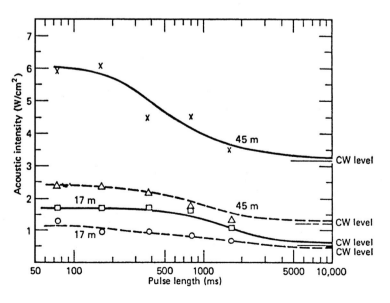

Fig. 4.4. Pulse-length dependence of the onset of cavitation (solid lines) and of 10 percent harmonic distortion (dashed lines) in sea water for 7 kHz sound at depths of 17 m and 45 m. (Rusby, 1970.)

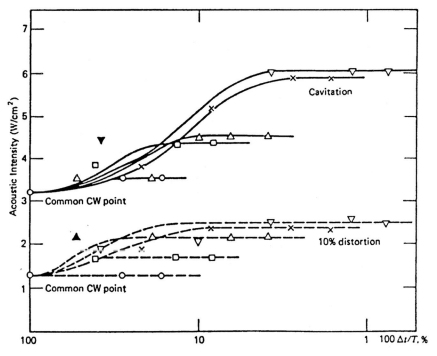

Fig. 4.5. Duty-cycle dependence of onset of cavitation and 10 percent harmonic distortion at 45 m depth in sea water for different pulse lengths of 7 kHz sound. Pulse durations are: 74 ms, ∇; 165 ms, x; 380 ms, Δ; 800 ms, $\square$; 1650 ms, $\bigcirc$. (Rusby, 1970.)

average acoustic intensity required for 10 percent distortion is much greater than for CW.

When the sound is alternately turned on and off, the allowable acoustic intensity, before distortion or cavitation, increases with decreasing percentage of on-time, $\Delta t / T$, from CW (100 percent) to lower values as shown in Fig. 4.5. An asymptotic value is reached that is higher with shorter pulse duration.

4.3 Parametric, difference frequency, sources

Section 4.1 described the harmonic distortion that characterizes a single, intense sound beam. In 1963, Westervelt pointed out that if two intense sound beams are coaxial, the non-linearity of the medium also creates entirely new propagating frequencies; these are sum and difference frequencies of the original "primary" sound frequencies. For example, when two primary beams of frequencies 500 kHz and 600 kHz are superimposed, a secondary beam of 100 kHz and another of

1100 kHz will be generated. So-called "parametric" or virtual sources will be distributed all along the intense part of the interacting primary beams.

The difference frequency is particularly attractive for technical applications because it has an extremely narrow beam at relatively low frequency. In the example given, the difference frequency of 100 kHz will have approximately the narrow beam width of a 550 kHz source radiating from the same transducer. The volume-distributed parametric source acts as if it were a highly directional end-fire array, so it is sometimes called a "virtual-end fire array." Furthermore, the bandwidth of the difference frequency of the parametric source is very much greater than it is for a linear transducer.

For a simple description of the non-linear action, consider how sum and difference frequencies can be generated at a point traversed by CW sounds of two different frequencies. Assume plane waves carrying the primary frequencies ω_1 and ω_2 described by

$$p_1(t) = P_1 \cos(\omega_1 t) \tag{4.5}$$

and

$$p_2(t) = P_2 \cos(\omega_2 t) \tag{4.6}$$

Assume that these two primary waves are traveling in the same direction and that they are both very intense. Each beam will be modulated by the other. For example, the amplitude of the p_1 wave will be modulated by the presence of p_2 and will become $P_1(1 + m \cos(\omega_2 t))$, where $m = P_2/P_1$ is the modulation amplitude. When the modulated p_1 is added to (4.6), we obtain the sum

$$p(t) = P_1 \cos(\omega_1 t) + P_2 \cos(\omega_2 t) + P_1 m \cos(\omega_1 t) \cos(\omega_2 t) \tag{4.7}$$

The third term is the non-linear interaction and $P_1 m$ is a measure of its strength. To recognize that sum and difference frequencies have been generated, rearrange the product of the cosines by using the trigonometric relation

$$2 \cos x \cos y = \cos(x + y) + \cos(x - y) \tag{4.8}$$

and p becomes

$$p(t) = P_1 \cos(\omega_1 t) + P_2 \cos(\omega_2 t) + \frac{P_1 m}{2}[\cos(\omega_\Sigma t) + \cos(\omega_\Delta t)] \tag{4.9}$$

The third term is a function of the sum and difference frequencies;

$$\begin{aligned} \omega_\Sigma &= \omega_1 + \omega_2 \\ \omega_\Delta &= \omega_1 - \omega_2 \end{aligned} \tag{4.10}$$

Fig. 4.6. Geometry for the generation of a difference frequency from two intense coaxial high-frequency beams. The new frequency arises from a continuous distribution of virtual sources throughout the beam. It is observed at Q.

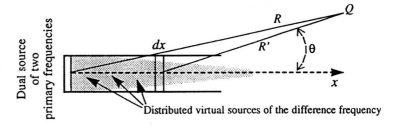

Sum and difference frequencies such as these would be generated at all points of intense interaction along the beams and would constitute a volume distribution of "virtual" secondary sources. For a sound beam they become equivalent to a properly phased array of simple real sources in an end-fire configuration (Fig. 4.6).

The difference frequency can be exploited. To show this, we use a Huygens type of calculation for a distribution of sources to determine the directional dependence of the difference tone component, ω_Δ. Refer to Fig. 4.6.

Assume that the cross-sectional diameter of the volume is small compared with the wave length of the difference tone. The element of volume is $S_0\, dx$, where S_0 is the cross-sectional area of the high-frequency primary beams. The secondary wavelets diverge spherically with the relatively low sound-pressure attenuation rate of α_Δ (nepers/m). The sound pressure at Q is proportional to the integral over all sources,

The development in Chapter 5 of $M \& C$ shows that the

$$\text{beam width} = (2\,\theta_\Delta) \cong 4(\langle\alpha\rangle/k_\Delta)^{1/2} \tag{4.11}$$

Not only is the parametric difference tone beam width much narrower than would be expected from a piston of the size used or the primaries, but analysis predicts and experiments verify that there will be virtually no side lobes in the radiation pattern of the difference frequency. In addition, one notes that, under the assumptions of the development, the beam width of the difference tone does not depend on the size of the primary sources. A further advantage of the parametric source is that its beam width is relatively insensitive to the value of the difference frequency.

Figure 4.7 shows the beam patterns of two primary waves – one at 418 kHz and the other at 482 kHz, and the difference tone produced at 64 kHz. The difference tone has a central beam width like those of the high-frequency primaries, and its side lobes appear to be down by about 40 dB compared with the axial value.

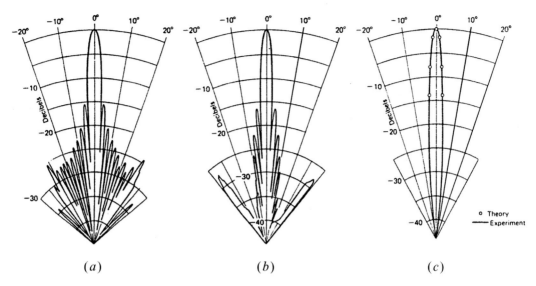

Fig. 4.7. Directivity patterns of a parametric source. (*a*) Primary 418 kHz. (*b*) Primary 482 kHz. (*c*) Beam pattern of difference frequency, 64 kHz. The solid lines are experimental; values obtained by a few theoretical calculations are shown by open circles. (From Muir, 1974.)

Another advantage of the parametric array is the increase in bandwidth. The bandwidth of a primary source is normally some percentage of the central frequency, say ±5 percent. When the primaries are operated within this percentage bandwidth, the difference frequency will range over this same absolute bandwidth, centered on its lower frequency. For example, if the primaries are 500 ± 5 percent (i.e., 475 to 525 kHz) and 600 ± 5 percent (i.e., 570 to 630 kHz), the difference frequency will have the relatively wide bandwidth of 100 ± 25 kHz.

Unfortunately the great virtues of the parametric sources are bought at the cost of very low power efficiency which is generally a fraction of one percent. There are three apparent ways to increase the efficiency: increase the difference frequency, increase the primary power, decrease the primary beam width. Only increased primary power offers direct promise for increased efficiency without sacrificing the advantage of the parametric source. However, that avenue is restricted by saturation effects and beam broadening, as well as cavitation at high intensities.

When gas bubbles are present in an intense sound beam, they may be driven into large-amplitude, non-linear oscillations at or near their resonance frequencies. This effect can substantially increase the power of the difference frequency, but with some loss in the radiation directionality.

Even though the efficiencies are very low, parametric sources are in everyday use. They are one solution to the need for transducers which have a narrow beam width and a uniform response over a wide range of low frequencies. Because their low frequency permits them to penetrate several layers of sediment, they are particularly desirable as seismic transducers for profiling lower layers of the ocean bottom.

4.4 Explosives as sound sources

4.4.1 The shock wave

An underwater chemical explosion (e.g., TNT, Pentolite, Tetryl) is an effective source of sound for experiments at sea. The energy available from the explosion is approximately 4.4×10^3 joules/g (TNT). The explosion starts with a very rapid chemical reaction which creates product gases at temperatures of the order $3000\,°C$ and pressures of about 50 000 atm.

There are then two sources of sound in the water surrounding a deep-water explosion: the shock wave, which carries about half of the energy of the explosion and which initially propagates spherically at speeds greater than the conventional 1500 m/s; and the huge oscillating gas bubble that follows.

First consider the shock wave. An approximate description is that the shock shows an instantaneous rise in pressure to a maximum value P_m. The pressure then decays exponentially (Fig. 4.8). Studies during and since World War II have shown that both the peak pressure and the time constant of the decay can be scaled according to a universal parameter, $(w^{1/3}/R)$, where w is the weight of the explosive and R is the range to the measurement position. See Cole, *Underwater Explosions* (1948), for a discussion of theory and early experiments. Rogers (1977) has used weak shock theory to derive accurate expressions for the peak pressure and decay time constant as a function of the charge-range parameter.

The most common explosives in current experiments at sea are 0.82 kg SUS (Signal, Underwater Sound) charges that are composed principally of TNT. A great deal of effort has been expended to determine the source strength and decay time constant of SUS charge. The best empirical equations appear to be those of Chapman (1985), who conducted extensive trials at sea and verified a relation originally by Slifko (1967) for the peak shock-wave pressure of the 0.82 kg SUS charge,

$$P_m = 50.94(w^{1/3}/R)^{1.13} \text{ MPa} \qquad (4.12)$$

where P_m is given in megapascals when the charge weight w is in kilograms and the propagation range R is in meters. The range dependence is $R^{-1.13}$ rather than R^{-1}, because of excess attenuation at the shock front.

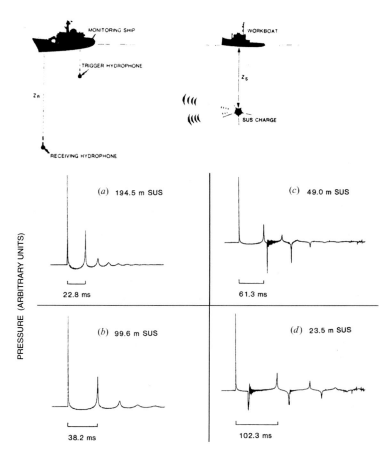

Fig. 4.8. Sound pressures from an explosive at various depths and ranges at sea. The geometry is above. The first sound impulse is the shock wave (note the steep rise in pressure, and exponential decay), which is observed at all depths. At 194.5 m depth, the shock is followed by several oscillations of the bubble pulse. The pressure graph is similar at 99.6 m depth, but there is a greater time between the shock and the first bubble pulse, owing to the depth dependence of the bubble period. At explosive depths 49.0 m and 23.5 m, one can identify the shock wave and the several bubble pulses and their phase-reversed reflections from the ocean surface. (From Chapman, 1985.)

The early exponential decay immediately after shock passage is given by

$$p = p_m \exp\left(-t/\tau_s\right) \tag{4.13}$$

where τ_s is in seconds.

Chapman's redetermined expression for the decay time constant for the shock from a 0.82 kg SUS charge is

$$\tau_s = 8.12 \times 10^{-5} w^{1/3} (w^{1/3}/R)^{-0.14} \text{ seconds} \tag{4.14}$$

4.4.2 Shock front propagation; the Rankine–Hugoniot equations

The propagation equations that were derived for infinitesimal (acoustic) waves in Chapter 1 are not valid for shock waves. However, the appropriate equations of conservation of mass, momentum, and energy at a shock front, the "Rankine–Hugoniot equations," are easily derived (Fig. 4.9).

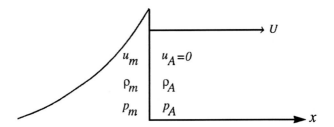

Fig. 4.9. Conditions at a plane shock front that moves with velocity, U. In front of the shock, the particle velocity is zero, pressure is p_A, and density is ρ_A. At the shock front the maximum pressure, density, and particle velocity are p_m, ρ_m and u_m, respectively.

Conservation of mass

Let an observer move with the front. The mass per unit area per unit time entering from the right is $M = \rho_A U$. The mass per unit area per unit time leaving the front must be the same, $M = \rho_m(U - u_m)$. Equating, we get the Rankine–Hugoniot equations for conservation of mass;

$$M = \rho_A U = \rho_m(U - u_m)$$

$$u_m = \left(\frac{\rho_m - \rho_A}{\rho_m}\right) U \tag{4.15}$$

Conservation of momentum

For conservation of momentum, notice that the net momentum per unit area per unit time delivered to the entering fluid is $\rho_A U^2 - \rho_m(U - u_m)^2$. The net pressure acting toward the right is $p_m - p_A$. Equating the two gives the conservation of momentum equations

$$p_m - p_A = \rho_A U^2 - \rho_m(U - u_m)^2 \tag{4.16}$$

simplify, using (4.15),

$$p_m - p_A = \rho_A u_m U \tag{4.17}$$

from which we obtain the shock front propagation speed, U:

$$U = \sqrt{\frac{(p_m - p_A)\rho_m}{(\rho_m - \rho_A)\rho_A}} \tag{4.18}$$

Since water is only slightly compressible, $\rho_A/\rho_m \cong 1$. Therefore, the shock speed U depends on the *average* slope $(p_m - p_A)/(\rho_m - \rho_A)$ in the p, ρ graph, e.g., Fig. 4.1. On the other hand, the speed of sound at the peak, c_m, depends on the *local incremental* slope, $(\Delta p/\Delta\rho)_{p_m\rho_m}$, which is clearly larger than the average slope. Therefore $c_m > U > c_A$.

The speed of the shock front can be significantly greater than the conventional speed of sound in sea water.

4.4.3 The gas globe

The huge gas globe which encloses the products of the chemical components of the explosion, as well as vaporized water, contains about half the
total energy of the explosion. Sound is radiated during acceleration of
the bubble radius. After the initial acceleration, the great bubble deceler-
ates, expands past the point where the internal bubble pressure equals the
ambient pressure, slows, and reaches a maximum radius. At maximum
radius, the internal gas pressure is less than the ambient pressure, and
the bubble contracts. The oscillating gas bubble is responsible for the
bubble pulses, following the shock wave in Fig. 4.8.

The period of the gas bubble oscillation is a function of the energy
available to it after the shock has been formed, and the ambient density
and pressure of the water that surround it. To calculate the approximate
period, assume that the bubble is spherical and determine the partition
of non-shock energy, Y.

At the maximum radius, a_m, the kinetic energy is zero, and the
internal energy is much less than the potential energy. Therefore we can
assume that the non-shock energy, which contains about half of the total
explosion yield, is all potential energy:

$$Y = (4/3)\pi a_m^3 p_A \tag{4.19}$$

or

$$a_m = \sqrt[3]{\frac{3Y}{4\pi p_A}} \tag{4.20}$$

To calculate the approximate period of the motion, assume that the
bubble oscillates spherically as an ideal bubble (see Section 6.5.2) at its
depth, where the ambient conditions are p_A, ρ_A. The period of oscillation
of a spherical bubble is given by

$$T = 2\pi a_m \sqrt{\frac{\rho_A}{3Y p_A}} \tag{4.21}$$

Substitute a_m from (4.19) to obtain the approximate period,

$$T = KY^{1/3}\rho_A^{1/2} p_A^{-5/6} \tag{4.22}$$

where dimensionless $K \cong 2$ depends on our assumptions. The units for
Y are joules.

This equation is important for our purposes because it specifies the
approximate dependence of bubble period on depth (recall Fig. 4.8)
and explosive yield. The reality is that, because of the large differential
pressure from its top to bottom, large explosive bubbles are not spherical
but are flattened or dimpled at the bottom. Furthermore, the gas globe
often splits into two because of that dimpling. Also, the non-harmonic

pulsation does not generate a sinusoid, and the effective period changes as the bubble rises due to buoyancy. Nevertheless, (4.22) is a useful initial predictor.

4.4.4 Interaction with the ocean surface

When the explosive is close to the ocean surface, phase-reversed signals (caused by reflection from the water–air interface) will be received after the direct received signals. Figure 4.8(c), with the source at depth 49.0 m, shows the reflected shock wave just after the first bubble pulse.

Note the "noise" that follows the reflected shock in Fig. 4.8. When the positive shock pressure is reflected, it is phase-shifted, and it becomes a tension. This negative pressure can have a greater absolute value than the ambient pressure. As a result, the water cavitates (Section 4.2). The newly created microbubbles of many sizes near the surface will reradiate damped oscillations at many frequencies. This will give the appearance of random noise radiated from the surface. The reflected shock loses significant energy as it ploughs through the water and causes cavitation. Note, however, that the reflected first *bubble pulse* in Fig. 4.8(c) does not appear to have a large enough tension to give evidence of cavitation. The signal shown for explosive depth 23.5 m should now be easy for the reader to explain.

In experiments at sea, it is often the case that the cacophony of multiple sources of sound that comes from an underwater explosive is not desirable. To avoid the bubble pulses and its sea surface reflections, experimenters can set off the explosive close enough to the surface to cause the bubble gases to vent before an oscillation can take place. The isolated shock wave is then easier to use.

4.5 Acoustic radiation pressure

A sound beam transports acoustic energy and acoustic momentum. The average acoustic momentum carried through unit area in unit time causes an *acoustic radiation pressure*. The *Langevin* radiation pressure P_{RL} is the time average of the product of the momentum per unit volume, $\rho_A u$, by the particle velocity, u:

$$P_{RL} = \frac{\text{momentum transfer}}{\text{area} \quad \text{time}} = \langle \rho_A u^2 \rangle \qquad (4.23)$$

The Langevin radiation pressure is the difference between the pressure in a beam perpendicular to a wall and the static pressure in the fluid behind the wall. For additional explanations, see, for example Beyer (1975).

The quantity in (4.23) is also the average acoustic energy per unit volume, $\langle \varepsilon \rangle$, at the point. It is equal to the average acoustic intensity divided by the speed of sound,

$$P_{RL} = \langle \rho_A u^2 \rangle = \langle \varepsilon \rangle = \frac{\langle I \rangle}{c_A} \qquad (4.24)$$

Therefore, if one determines the Langevin radiation pressure, the energy density is obtained, and the intensity of the beam is readily calculated. The technique is frequently used in ultrasonic measurements where the insertion of a hydrophone would disturb the acoustic field that is being measured. It can also be used in calibration of a high-frequency underwater source. To use the technique, a diaphragm is inserted perpendicular to the beam, and the radiation force is measured.

Strictly speaking, radiation pressure is not solely a high-intensity effect; it exists for infinitesimal sound amplitudes, as well. However, because the radiation pressure depends on the *square* of the acoustic pressure it is very much easier to measure at high intensities than at low intensities.

4.6 Acoustic streaming

In addition to harmonic distortion, intense sound waves cause a unidirectional flow called "streaming." Acoustic streaming is characterized by an outward jetting of the medium in front of any transducer that propagates high-intensity sounds. In the early days the phenomenon was called the "quartz wind" because quartz crystals were used as underwater sound sources. In air, the effect is experienced most easily by turning up the gain of an audio system and placing one's hand in front of the loudspeaker to feel the breeze.

In the case of a piston transducer propagating a sound beam in the open sea, the acoustic streams move *away* from the source on the sound axis. The axial stream velocity increases somewhat beyond the source where the higher attenuation of the newly generated frequencies (due to harmonic distortion) increases the momentum transfer from the sound wave to the bulk medium (see, for example, Starritt *et al.*, 1989). At much greater ranges, the stream velocity decreases due to the weaker acoustic field and fluid mixing. Nevertheless, the streams jet out for distances of the order of meters and gradually circulate back to the side of the transducer. The acoustic streams start to flow in tenths of seconds.

The many different acoustic streaming patterns that occur in intense sound fields depend on the geometry of the region and the form of the acoustic field. The basic reason for the flow can be easily understood

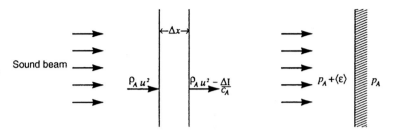

Fig. 4.10. Acoustic radiation
pressure and acoustic
streaming in a beam. *Left,*
sound attenuation through
the path Δx causes a decrease
of acoustic momentum along
the path $\Delta I / c_A$; this is taken
up by a streaming momentum
of the bulk medium. *Right,*
Langevin radiation pressure of
a sound beam measured at a
wall is the average energy
density in the beam.

as momentum transfer from the attenuating sound wave to the bulk medium. Following an argument of Fox and Herzfeld (1950), consider a section of a plane wave beam (Fig. 4.10) carrying sound in the $+x$ direction.

For an attenuating medium such as water, the spatial change of intensity is given by

$$\Delta \langle I \rangle = -2\alpha_e \langle I \rangle \Delta x \qquad (4.25)$$

where α_e is the rate of attenuation of acoustic pressure, nepers per unit distance (Section 2.4).

The spatial decrease of energy density, ε, due to the absorption of energy in distance Δx is

$$\Delta \langle \varepsilon \rangle = \frac{\Delta \langle I \rangle}{c_A} = \frac{-2\alpha_e \langle I \rangle \Delta x}{c_A} \qquad (4.26)$$

The spatial rate of decrease of acoustic momentum per unit area per unit time is compensated by the change of static pressure:

$$\Delta p_A = \frac{2\alpha_e \langle I \rangle \Delta x}{c_A} \qquad (4.27)$$

If that pressure is unsupported, it will cause the fluid to drift. Otherwise stated, the loss of momentum from the acoustic beam is taken up by a gain of momentum of the fluid mass; thus conservation of momentum is fulfilled.

A more general study of the streaming phenomenon (Medwin and Rudnick, 1953) shows that the streaming *vorticity* (*vorticity* $= \nabla \times u_2$ where u_2 is the streaming velocity) is generated at all points where there is sound absorption. If the acoustic field is completely away from boundaries, the volume absorption rate, α_e, determines the stream vorticity. If the sound field is adjacent to a boundary, the absorption in the *acoustic boundary layer* is the source that drives streams in the medium. The acoustic boundary layer is the viscous region within which a grazing plane-wave particle velocity decreases exponentially from the plane-wave value to zero at a rigid wall. It was originally calculated by Lord

Rayleigh (1877); see also Pierce (1981). The thickness to e^{-1} is given as

$$l = \left(\frac{2\mu}{\omega\rho_A} \right)^{1/2} \tag{4.28}$$

The detailed flow patterns in specific situations depend on the local absorption of energy, the geometry of the region in which flow is driven by these distributed acoustical sources, and the sound intensity. For example, very strong eddies form around a bubble attached to a wall – say, at a sound transducer (Elder, 1959). Potentially destructive streaming velocities thousands of times greater than the acoustic particle velocity have been observed.

The first specific example of streaming in free space was solved by Eckart (1948), who considered a non-divergent beam of radius a within a tube of larger radius.

A streaming experiment was then used by Liebermann (1949) to calculate the sea water bulk viscosity, μ_b a quantity whose numerical value was unknown at the time (see Section 2.4.2). By this calculation he showed that the great difference between the theoretical and experimental values of the attenuation of sound in sea water that existed prior to the 1950s was due to the lack of knowledge of the macroscopic quantity, bulk viscosity (or the microscopic descriptions, the molecular relaxation, or the structural relaxation).

The streaming pattern is sensitive to the form of the primary acoustic field. When the correct divergent beam pattern is used and the physical constants of the medium are known, predictions of the streaming velocity can be within 2 percent of the measured values (Medwin, 1954). The streaming technique is a practical means to determine the bulk viscosity of fluids from simple acoustical and drift measurements. An excellent early summary of acoustic streaming was written by Nyborg (1965). Continuing research on the subject can be found in the triennial *Proceedings of the International Symposium on Non-linear Acoustics*.

Illustrative problem

(a) Given a difference tone generation due to primary frequencies 418 kHz and $f_2 = 482$ kHz, calculate the 3 dB beam width.

(b) Calculate the size of piston that would have been required to produce the same half power beam width.

(c) Calculate the radius of the transducers used. (Assume piston radiation for the primaries.)

(d) Calculate the reduction in source radius when using the end-fire array for the same beam width.

(a) Write

$$\text{avg} f = \frac{418 + 482}{2} = 450 \text{ kHz}$$

At 450 kHz, the attenuation is 0.15 dB/m or 1.7×10^{-2}Np/m.

$$f_d = 482 - 418 = 64 \text{ kHz}$$

$$k_d = \frac{\omega d}{c} \simeq 268 \text{ m}^{-1}$$

$$2\phi_d \simeq 4 \left(\frac{\alpha_e}{k_d} \right)^{1/2}$$

$$\simeq 0.032 \text{ rad} \quad \text{or} \quad 1.8°$$

(b) For a circular piston the -3 dB angle is given by

$$k_d a \sin \phi_d = 1.6$$

Therefore we require

$$a = 1.6 \, (k_d \sin \phi_d)^{-1}$$

$$a = 37 \text{ cm}$$

(c) The -3 dB angle in Fig. 4.7 is approximately $2°$. The -3 dB angle for a circular piston is at $ka \sin \phi = 1.6$. Therefore for $f = 450$ kHz,

$$a(\text{primary}) = \frac{1.6}{(18.8)(0.035)} = 2.6 \text{ cm}$$

(d) The radius has been reduced to $2.6/37 = 7\%$

Problems

4.1 Calculate the second harmonic generated in an intense unattenuating sinusoidal plane wave as shown in Fig. 4.2 by following a scheme first used by Black (1940). Show that in time Δt the advance of a crest is $\Delta x = \beta(\Delta p/\rho_A)c_A \Delta t$. Then assume that the distorted wave is simply the sum of a fundamental plus second harmonic, $p = P_1 \cos kx + P_2 \sin 2kx$, and show that the zero-slope peak $(dP/dx) = 0$ results in $P_2 \cong P_1 k \Delta x/2 = \beta(\Delta p/\rho_A)kx P_1/2$.

4.2 Use Fig. 4.3 to determine the rate of attenuation of the 454 kHz non-linear wave propagating to range 101 m if the source pressure is 100 kPa, 200 kPa, and 300 kPa. Compare with the attenuation rate α in Chapter 2.

4.3 Based on Fig. 4.5 and assuming that the density of cavitation nuclei remains constant with depth, plot a graph of onset of cavitation dependence on duty cycle for pulse duration 74 ms at transducer depth 5 m.

4.4 At zero axial distance Fig. 3.10 shows how the near-field pressure varies across the face of a piston transducer. Replot those pressures on a two-dimensional graph and compare the hot-spot pressures where cavitation will most likely occur with the average acoustic pressure over the face of the transducer.

4.5 Plot a graph of peak explosive shock pressure versus range for a deep explosion from a 0.82 kg SUS charge.

4.6 Sketch the shock pressure p versus t for the first 200 ms after the detonation of a 0.82 kg SUS charge at depths 10 m, 100 m, and 500 m.

Further reading

Westervelt, P. J. (1963). Parametric acoustic array. *J. Acoust. Soc. Am.*, **35**, 535–7.
This landmark paper desribes how the non-linear effects of intense sound waves produced by two frequencies could be used to produce a narrow beam source at the difference frequency of the two components. The implications have been far-reaching in transducer design.

Cole, R. H. (1948). *Underwater Explosions*. Princeton, NJ: Princeton University Press.
An orderly and detailed presentation of the results of WWII research by many researchers on the effects of underwater explosions. Unfortunately, the book is marred by many (easily recognized and easily corrected) errors.

Chapter 5
Interpreting ocean sounds

Summary

Signals are the messages that we want to receive. Noises are the sounds that we don't want to receive. We all know the saying "Beauty (or ugliness) is in the eye of the beholder." The comparable acoustical maxim is "Signals (or noise) are in the ear of the listener."

The ocean is filled with a great variety of natural "noises" and "signals" of physical and biological origin which range in frequency from less than 1 hertz to many megahertz. In this chapter we consider signal processing techniques which allow us to isolate, characterize, identify and interpret the sources of this cacophony. The descriptions of these miscellaneous ocean sounds are in terms of the duration or intermittency of the spectral and temporal components, and their acoustical powers and geometrical radiation patterns.

Passive acoustical measurements reveal specific information about natural sound sources. These observations permit one to transmute ocean "noises" to "signals" that provide unique knowledge about the sounds emitted by the great whales, fishes, and lesser marine life for species communication and foraging as well as the physical processes of rainfall, surface waves, internal waves, ocean turbulence, and seismic activity.

For example, the passive analysis of sounds heard at a remote ocean buoy easily distinguishes between wind "noise" and precipitation "noise." It is the basis for mathematical inversions, which yield detailed, quantitative meteorological interpretation of *in situ* wind speed and rainfall. When there is no precipitation, the sound at a buoy hydrophone is a quantitative measure of the local wind speed and gustiness. During precipitation, the sound reveals the local raindrop sizes and numbers per second, per unit surface area. These valuable details go far beyond the primitive information gathered by traditional rain gages used by weather observers. Deductions of meteorological information from underwater acoustical measurements are much more accurate and locally specific than those from satellite radar.

There are two types of ocean sound that concern us: signals and noises. The types of signals include impulses and CW tones of short or long duration and constant or varying frequencies; they also include complicated coded messages and random sequences.

There are many examples: the sonar operator searching for the "signal" of a submarine will call the sounds of whales and dolphins "noise." Needless to say, the marine mammal seeking to communicate or locate food would characterize man-made sounds as "noise." Some sounds that were called "noise" for many years are now recognized as containing information that qualifies them as "signals"; for example, the sound of rainfall at sea is now used to measure the size and number of raindrops per square meter per second.

Motion of the medium relative to the transducer, flow noise, electrical circuit noises, and the 60 or 50 Hz electrical interferences from power lines are generally regarded as noise by everybody.

Traditionally, underwater ambient noise has been specified in terms of the sound measured at a convenient hydrophone, some distance from the sources. The origins of the sound are often a mystery. We initiate a different approach. In Section 5.5 we will survey the acoustic power, source pressure, directionality, and intermittency of physical, biological, and man-made ocean sounds *at their source*. When this information is known, our knowledge of propagation in the ocean allows us to calculate the ambient sound at any location.

Contents

---

[*] This section contains some advanced analytical material.

5.1 Sampling rules

Practically all underwater sound signals and noise are recorded digitally, and the results of analysis are displayed on computers. The acoustician uses signal acquisition and digitizing equipment, signal processing algorithms, and graphic display software to make this happen.

The noisy signal that enters a pressure-sensitive hydrophone becomes an analog signal. Hydrophone signals are sampled to convert them from analog to digital format in order to enter a digital computer. We must sample signals properly or we get garbage. Sampling rules apply to both temporal and spatial sampling of the oceanic environment. When the rules are obeyed, the original continuous signal can be recovered from the sampled signal with the aid of an interpolation procedure. If the rules are not obeyed, and sampling is too sparse, the original signal cannot be recovered.

The Nyquist sampling rules are

(1) *Space Domain Rule. The spatial sampling interval must be less than half of the shortest wavelength of the spatial variation.*
(2) *Time Domain Rule. The time interval between samples must be less than half of the shortest period in the signal.* Otherwise stated, *the sampling frequency must be greater than twice the highest frequency component in the signal.*

5.1.1 Spatial sampling

In traditional marine biology, samples are taken by towing a net through the water. A net is lowered to the depth, opened, and towed for a specified distance. The following example of spatial sampling uses a two-dimensional map (or depth profile) of a marine biological survey in which the results of typically sparse *net* sampling and detailed *acoustical* sampling are compared. The details of the data acquisition are not important here.

Figure 5.1 shows a simulation of sampling of the density of organisms or "particles" in the ocean, using nets. In this simulation, the length of the tow is 1.5 km. Simulated sampling stations are occupied at approximately 5 km spacings, and samples are simulated at 50 m depth intervals. The inadequately sampled data do not show much structure or pattern.

For comparison with Fig. 5.1, the almost continuous acoustical sampling of the biomass is shown in Fig. 5.2. This figure shows strong features including regions with many scatterers, indicating large biomass (black spots), and regions with few or no scatterers (white).

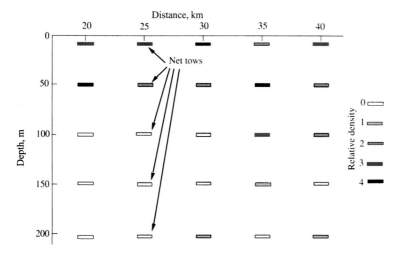

Fig. 5.1. Simulation of net sampling of marine organisms. The net is assumed to be towed at a set of stations at different ranges and depths. Each net tow is over a 1.5 km distance. The relative densities of organisms captured in the simulated net tows are indicated by shaded rectangles using the relative density code at the right, 0, 1, 2, 3, and 4 (Clay, 1998).

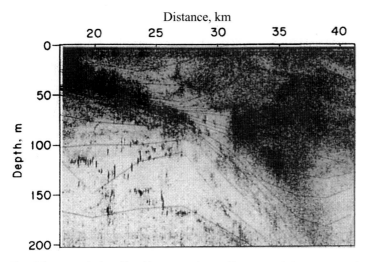

Fig. 5.2. Acoustical profile of biomass. The profile was made by continuously recording the backscatter output of a downward-looking sonar as the ship traveled. The simulation of net sampling shown in Fig. 5.1 was made from this profile. (From Nero *et al.*, 1990.) *Note*: The details of the profile depend on the frequency of the sonar.

5.1.2 Temporal sampling

The electrical signal $x(t)$ is to be sampled by an analog-to-digital converter to create a sequence of numbers (Fig. 5.3(a)). The clock gives

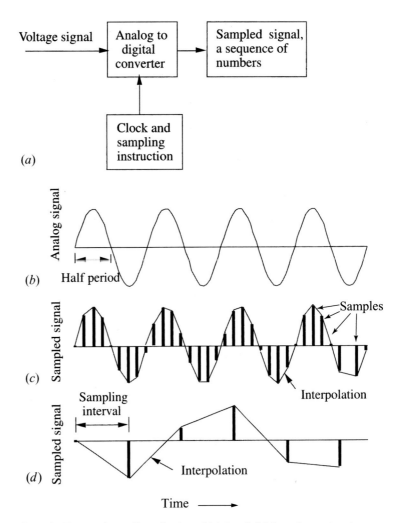

Fig. 5.3. Temporal sampling of a sinusoidal signal. (*a*) Sampling system to change an analog voltage into a sequence of numbers. (*b*) Input analog voltage. (*c*) The result of sampling four times during each half-period. The vertical lines represent the magnitudes of the sampled voltages. The straight lines between the ends of the vertical lines are interpolations. (*d*) The distorted result of sampling at times greater than the half-period. Compare the interpolated waveforms of (*c*) and (*d*) to the analog signal of (*b*) (Clay, 1998).

the sampling instruction. The sample is the instantaneous value of the signal voltage at the clock time (Fig. 5.3(*b*)). No information is recorded between samples where straight lines are drawn.

Examples of data taken at two different sampling intervals are shown. In Fig. 5.3(*c*) there are four samples in a half-period. In Fig. 5.3(*d*) the sampling interval is larger than the half-period. Reconstruction of the inadequately sampled signal, Fig. 5.3(d), does not resemble the original signal, whereas Fig. 5.3(c) does.

A common rule of thumb is to sample at intervals less than the period/3 for approximate reconstruction of the original signal. Communication textbooks give more elaborate methods to get excellent reconstructions based on the Nyquist theorem.

5.2 Frequency filtering

Electrical filters were originally introduced into electronic systems by radio and telephone engineers to separate the signals they wanted from those that they didn't. We use the filters that are built into our radio or television set when we select a channel that tunes in our desired station's carrier frequency and rejects others. Audio amplifiers also have frequency filters (equalizers, bass, and treble controls) to modify the amplitudes of the input frequencies and to enhance the quality of the sound coming from the speakers.

Digital communication engineers have developed the digital equivalent of the analog filter. The incoming analog electrical signal is digitized by an analog-to-digital converter, and the filter operations are done by a computer. In many systems, the filtered digital sequence of numbers is converted back to an analog signal for listening and display.

5.2.1 Frequency filtered response

The frequency response of a filter is the ratio of the output to input voltages for a (long-duration) sinusoidal input signal (the oscillator). Two measurements are sketched in Fig. 5.4. To record signals digitally, we need a low-pass filter to prepare the signal for the digitizing operation. The filter shown in Fig. 5.4(b) is a "low pass," "anti-aliasing filter." It is adjusted to pass frequencies that are less than half the sampling frequency of the analog-to-digital converter and to reject higher frequencies. It thereby can prevent higher-frequency components from appearing as *alias* signals. The action of a bandpass filter is sketched in Fig. 5.4(c). It is used to pass a signal within the designed frequency range and to reject unwanted signals and noise.

5.2.2 Time domain view of frequency filtering

Figure 5.5(a) shows a short-duration signal, $x_1(t)$, which then passes through an appropriate bandpass filter to give the output $y_1(t)$ in Fig. 5.5(b). The high- and low-pass settings on the filter were chosen to pass the signal with an acceptable amount of distortion of waveform.

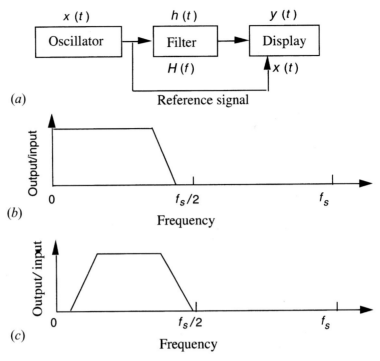

Fig. 5.4. Frequency filters and their responses. (*a*) Block diagram for a typical filter response measurement. (*b*) Response of an anti-aliasing, low-pass filter that is used ahead of analog-to-digital conversions at sampling frequency f_s. (*c*) Response of a bandpass filter (Clay, 1998).

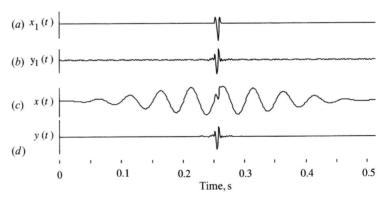

Fig. 5.5. Filter operation shown in the time domain. (*a*) Signal input is a 150 Hz ping having a duration of 0.01 s. (*b*) Signal out of a 50–150 Hz bandpass filter. (*c*) Input 150 Hz ping and a 20 Hz whale song. (*d*) Filtered signal output using the 50–150 Hz bandpass filter (Clay, 1998).

A longer-duration, low-frequency whale song $x_2(t)$, is emitted during this same time so that the sum of the two signals at the input (Fig. 5.5(c)) is

$$x(t) = x_1(t) + x_2(t) \tag{5.1}$$

The output of the bandpass filter $y(t)$ is shown in Fig. 5.5(d). The bandpass filter effectively removes the interfering whale song (a "noise" in this case) and reveals the short-duration 150 Hz ping.

5.2.3 Frequency domain analysis of finite duration signals

Most spectrum analyzers combine digital sampling and computers to do the spectral analysis. The digital spectrum analyzers often can have the equivalent of more than 1000 very narrow bandpass filters. Examples of spectral analysis are shown in Fig. 5.6(b) and (c). The spectrum of the 150 Hz ping is in Fig. 5.6(b) and the spectrum of the short-duration ping and the longer-duration whale song are presented in Fig. 5.6(c).

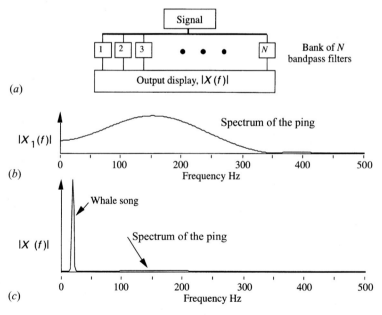

Fig. 5.6. Frequency domain analysis of the signals of Fig. 5.5 using a digital spectral analysis. The bandwidths of the equivalent bandpass filters are 2 Hz. (a) Block diagram of an analog spectrum analyzer that uses a bank of bandpass filters. (b) The digitally calculated spectrum of the 150 Hz ping in Fig. 5.5. (a) The digitally calculated spectrum of the ping and whale song in Fig. 5.5(c). The spectral amplitude factor of the ping is 1/40 that of the longer duration, narrow frequency band, whale song (Clay, 1998).

5.3 Time-gated pings

The spectrum of a signal depends on its time-domain waveform. Consider some pings and their spectra. These comparisons display the relation of periodicity and duration in the time domain, to the peak frequency and bandwidth of the frequency spectrum.

For these examples, the ping has a slow turn-on and turn-off. The signal $x(t)$ is

$$x(t) = 0.5[1 - \cos(2\pi t/t_p)] \sin(2\pi f_c t) \quad \text{for} \quad 0 < t < t_p \qquad (5.2)$$

and

$$x(t) = 0 \quad \text{otherwise} \qquad (5.3)$$

where t_p is the total (non-zero) ping duration, and f_c is the (carrier) frequency. The amplitude factor in the [...] gives a spectrum with very small side lobes. This signal is similar to the sound-pressure radiated by many sonar transducers and some marine animals. The envelope of the sine wave is tapered from zero to a maximum and then back to zero.

5.3.1 Dependence of spectrum on ping frequency

Figure 5.7 shows the dependence of the frequency spectra (b), on the carrier frequencies of three pings of different frequencies and equal durations in (a). The pings were chosen to be long so that the widths of spectral peaks are narrow. The 50 Hz signal has a spectral peak at approximately 50 Hz. The other signals have their spectral peaks at 100 and 150 Hz.

5.3.2 Dependence of spectrum on ping duration

Figure 5.8 illustrates the dependence of the widths of the spectral peaks on the durations of the pings. The effective durations of the signals t_d are a little less than the t_p in (5.2) because the turn-ons and turn-offs are very gradual. The same frequency, 100 Hz, was used for all examples. To define the bandwidth Δf, we use the half-power width, given by the two frequencies where the amplitude is 0.707 of the peak amplitude. The spectra shown in Fig. 5.8(b) are the moduli, the absolute amplitudes. The widths of the spectra decrease as the signal duration increases, and are approximately the reciprocals of the durations of the signals. The

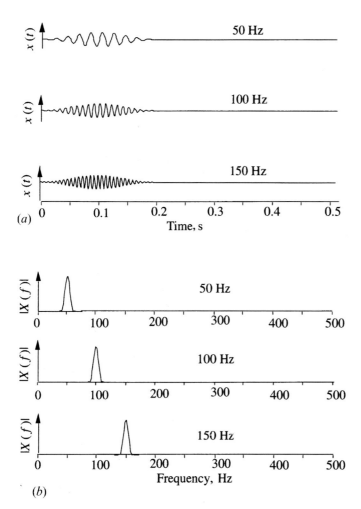

Fig. 5.7. Signals of same duration but different frequencies. (*a*) Time domain presentation of pings with carrier frequencies of 50, 100, and 150 Hz. (*b*) Spectral amplitudes of these pings. The spectra were computed using digital algorithms (see Chapter 6, *M&C*).

comparisons are

t_p, s	t_d, s	$1/t_d$, Hz	Δf, Hz	
0.01	0.0085	118.0	120	
0.02	0.017	58.8	60	(5.4)
0.04	0.034	29.4	30	
0.10	0.080	12.5	13	

These comparisons follow a rule of thumb:

$$(\Delta f)(t_d) \geq 1 \tag{5.5}$$

We use the $\geq$ sign for $(\Delta f)(t_d)$ because many signals have durations

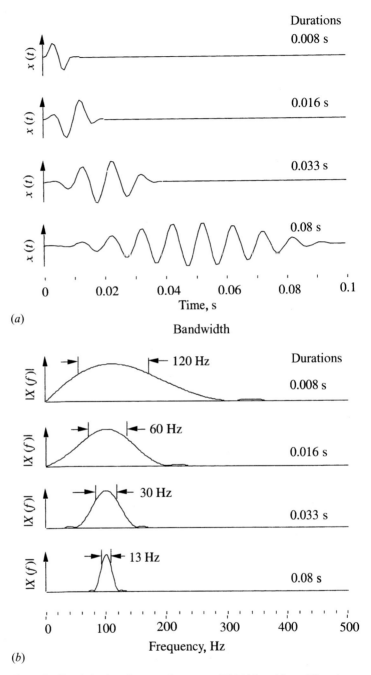

Fig. 5.8. Signals having the same frequency (100 Hz) and four different durations. (*a*) Signals in the time domain. (*b*) Spectral amplitudes in the frequency domain. The bandwidths, bracketed by arrows, were measured at the half-power points (i.e., at 0.707 of the peak amplitudes) (see Chapter 6, *M&C*).

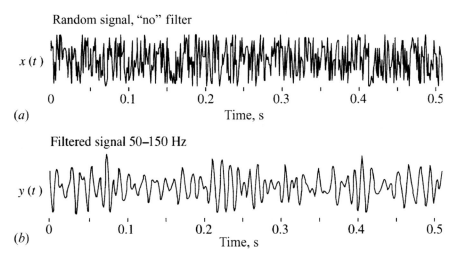

$x(t)$

Random signal, "no" filter

0   0.1   0.2   0.3   0.4   0.5

(a)   Time, s

Filtered signal 50–150 Hz

$y(t)$

0   0.1   0.2   0.3   0.4   0.5

(b)   Time, s

Fig. 5.9. (*a*) Random signal created by using a random number generator. The sampling interval was 1 ms. (*b*) Result of bandpass filtering the signal in (*a*) through a 50–150 Hz bandpass filter.

greater than $1/\Delta f$. The time t_d gives the *minimum* duration of a signal for a sonar system to have a bandwidth Δf.

5.4 Power spectra of noise*

Sound pressures that have random characteristics are often called noise, whether they are acoustical, or electrical that are cleverly created as "pseudo noise," or are the result of random and uncontrolled processes in the ocean.

5.4.1 Signals with random characteristics: spectral density

In their simplest form, signals that have random characteristics are the result of some process that is not predictable. In honest games, the toss of a coin and the roll of a die give sequences of random events. Natural processes such as earthquakes are generators of random signals. For simulations and laboratory tests, computers and function generators are used to make sequences or sets of random numbers. Many of the algorithms generate sequences that repeat, and these algorithms are known as *pseudorandom number generators*. Programming languages usually include a function call such as rnd( ) in the library of functions (ee Fig. 5.9).

Simulation of a random signal starts with a random function generator which gives a sequence of random numbers: $x(0)$, $x(1)$, and so on.

---

* This section contains some advanced analytical material.

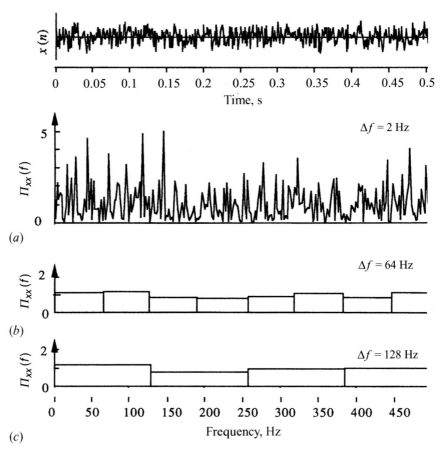

Fig. 5.10. Smoothing of power spectra by filtering. The top trace is a random signal $x(n)$ of $x(t)$. Filter bandwidths are (a) $\Delta f = 2$ Hz, (b) $\Delta f = 64$ Hz, and (c) $\Delta f = 128$ Hz.

Figure 5.9 shows a sequence where the numbers have been connected by interpolation lines. The result of bandpass filtering the input random signal gives a new random signal, e.g., the output $y(t)$ in Fig. 5.9(b).

The operations of bandpass filtering, squaring the signal, and summing or integrating the squared signal are indicated in Fig. 5.10. The effective number of independent trials is

$$N_{it} = t_d \, \Delta f_m \tag{5.6}$$

where t_d is the duration of the signal, and Δf_m is the filter bandwidth.

Assume the filtered signal is $y_m(n)$, where the subscript indicates the filtering by the mth filter. The signal is squared, summed, and averaged over t_d to give the power. Since the mean square output (e.g., volt^2) is proportional to the filter bandwidth and the duration of the signal, it is customary to define the "power" spectral density,

$$\Pi_{xx}(f_m) = \frac{1}{N\Delta f_m} \sum_{n=0}^{N-1} [y_m(n)]^2 \qquad (5.7)$$

where N is the number of samples. The first step in deriving an equivalent integral expression, for continuous functions of time, uses the multiplication and division by t_0:

$$\Pi_{xx}(f_m) = \frac{1}{N t_0 \Delta f_m} \sum_{n=0}^{N-1} [y_m(n)]^2 t_0 \qquad (5.8)$$

Let $N t_0$ become t_d, the duration of the signal, and t_0 become dt. The summation becomes the integral:

$$\Pi_{xx}(f_m) = \frac{1}{t_d \Delta f_m} \int_0^{t_d} [y_m(n)]^2 dt \qquad (5.9)$$

If $x(n)$ has the units of volts, the so-called "power" spectral density has units of (volts)^2/Hz. True power spectral density would require division by a load resistance in an electrical circuit to give watts/Hz.

Since a hydrophone output in volts is proportional to the acoustic pressure, when $x(n)$ has the units of pascals (Pa), the spectral density has units of (Pa)^2/Hz. The true *intensity spectral density* requires division by $\rho_A c$, to give (Pa^2/$\rho_A c$)/Hz = (watts/m^2)/Hz.

Acoustic spectra are often reported in dB relative to one $(\mu Pa)^2$/Hz, so that the Intensity Spectrum Level

$$\text{ISL} = 10 \log_{10}\{[\Pi_{xx}(f_m)]/[(\mu Pa)^2/Hz]\} \qquad (5.10)$$

The spectrum levels depend on the reference sound pressure, which is sometimes unclear. It is very much better to use SI units such as $(Pa)^2$/Hz or (watts/m^2)/Hz, rather than quantities without a reference.

5.4.2 Spectral smoothing

Consider the following example of spectral analysis. A random signal is constructed of 512 magnitudes at separation $t_0 = 0.001$ s and duration 0.512 s. Figure 5.10 shows the results of processing the signal by the equivalents of very-narrow, wide, and very-wide bandpass filters.

The output of the narrow 2 Hz filter, Fig. 5.10(a), is extremely rough. Figure 5.10(b) shows the result of using a wider filter, $\Delta f = 64$ Hz. Here the number of independent samples is 32. The spectrum is much smoother and has less detail. An increase of the filter width to $\Delta f = 128$ Hz is shown in Fig. 5.10(c). Another random signal would have a different spectrum. *These examples show the basic trade-off between*

resolution and reduction of roughness or variance of the estimate of the
spectral density of noises.

5.4.3 Traditional measures of sound spectra

The measurement of underwater sounds has inherited the instrumentation and the vocabulary that were developed for measurements of sounds
heard by humans in air. The principal areas of interest to humans have
been acoustic pressure threshold for hearing; acoustic threshold for damage to hearing; threshold for speech communication in the presence of
noise.

The vast amount of data required to evaluate human responses, and
then to communicate the recommendations to laymen, forced psycho-
acousticians and noise control engineers to adopt simple instrumentation
and a simple vocabulary that would provide simple numbers for complex
problems. Originally this was appropriate to the analog instrumentation.
Unfortunately, even digital measurements are now reported according to
former constraints. For example, the "octave band," which is named for
the eight notes of musical notation that corresponds to the 2:1 ratio of
the top of the frequency band to the bottom, remains common in noise-
control work. For finer analysis, one-third octave band instruments are
used; they have an upper- to lower-band frequency ratio of $2^{1/3}$, so that
three bands span one octave.

The use in the ocean of instruments and references that were designed
for air has caused great confusion. The air reference for acoustic pressure
level in dB was logically set at the threshold of hearing (approximately
20 μPa at 1000 Hz) for the average adult human. This is certainly *not*
appropriate for underwater measurements, where the chosen reference
is 1 μPa or 1 Pa.

Furthermore, plane-wave intensity (of CW) is calculated from (1.65),
where intensity $= P_{rms}^2/(\rho_A c)$ (where P_{rms}^2 is the mean squared pressure;
$\rho_A =$ water density; and $c =$ speed of sound in water). Therefore, the
dB reference for sound intensity in water is clearly different from that in
air because the specific acoustic impedance $\rho_A c$ is about 420 kg/m²s for
air compared with 1.5×10^6 for water. This ratio corresponds to about
36 dB, if one insists on using the decibel as a reference.

The potential for confusion in describing the effects of sounds on
marine animals is aggravated when physical scientists use the decibel notation in talking to biological scientists. Confusion will be minimized if psychoacoustical characteristics of marine mammals – such as
thresholds of pain, hearing, communication perception, and so forth – are
described by the use of SI units, i.e., pascals (acoustic pressure at a

receiver), watts/m^2 (acoustic intensity for CW at a receiver), and joules/m^2 (impulse energy/area at a receiver). Likewise, only SI units should be used for sources – that is watts (power output of a continuous source) and joules (energy output of a transient impulse source). The directivity of the source should always be part of its specification. All of these quantities are functions of sound frequency and may be expressed as *spectral* densities (i.e., per 1 Hz frequency band).

5.5 Sounds at sea

Ambient sound in the ocean is caused by a large number of physical and biological elements that are indigenous to the sea, as well as by powerful man-made sources. Each contributes to the local sound pressure by an amount that depends on the source characteristics and the attenuation between the source and receiver. Since low-frequency sounds have smaller attenuations, even distant sources of low-frequency sound are important at any measurement position. The sound often reaches the receiver by paths through the bottom, or scatter from the coastal shore, as well as by multipaths through the water. Here, where possible, we give an outline of source characteristics such as the acoustic power radiated, the source pressure, directivity pattern, source spectrum, intermittency, and location.

Much of this essential information about natural sound sources at sea is simply not known at this time, and marine scientists have accepted (by default) any hydrophone measurements they can get. Unfortunately, much published data on sounds at sea do not reveal the conditions of the experiment that affect the quoted numbers – for instance, specifics such as hydrophone height above the ocean floor or nearby topography at the hydrophone position, nearby scatterers, depth of water, distance from coastal reflectors, source and hydrophone distances from the ocean surface, and roughness of the ocean surface. Sometimes experimenters have buried these details into a single transmission number which they then use to "calibrate" the site of the experiment.

When the source is known, algorithms such as the ray-path procedures (Chapter 2) or waveguide propagation (Chapter 8) can be used to predict the sound at any location.

During World War II, and for many years thereafter, the US Defense Department sponsored a huge number of measurements of sound levels in the sea. (Many of these reports have been declassified and may be purchased from the US Department of Commerce, National Technical Information Service, Springfield, VA 22151.) However, most experiments were, for expedience, limited to the frequency range of military interest of the moment, and unfortunately they are reported in systems of units that are no longer in use.

Historically, several rms reference pressures P_{ref} have been used to calculate SPL $= 20 \log(P/P_{ref})$ dB – for example, P_{ref} may be the rms values 0.0002 dyne/cm^2, 1 dyne/cm^2, 1 mbar, 1 Pa, 1 μPa. In order to present a simple and consistent system of units, we convert original values in dB to the SI units of pascals, watts, and joules. However, to facilitate reference to the literature, we present both the original values in dB and the converted values in SI units. Converted numbers are rounded to one or two significant figures as seems appropriate.

More recently, the civilian community has recognized that fisheries research vessels must be acoustically designed to minimize the effect on marine life. Furthermore, the International Council for the Exploration of the Sea has given recommendations for the proper acoustical parameters and units that should be measured during such research (Mitson, 1995).

5.5.1 Natural physical sounds

Figure 5.11 is an updated traditional presentation of spectra of non-biological sounds sensed by hydrophones at sea. The reader should be aware that these curves are simple, approximate representations that may be incorrect by large factors (many decibels) at specific times or places. The traditional curves have been supplemented by recent measurements of the sound of rain for different wind and cloud conditions. The left ordinate in Fig. 5.11 is sometimes incorrectly called *power spectral density* (similarly for the *level* in dB on the right). In fact, the left ordinate is *proportional* to the intensity spectral density [(watts/m^2)/Hz], which can be obtained by dividing Pa^2/Hz by $\rho_A c \cong 1.5 \times 10^6$ (kg/m^{-3})(m/s), as described previously.

In general, ocean sound at frequencies from 0.1 Hz to 5 Hz is attributed to the non-linear interactions of surface waves, "wave–wave interactions," and microseisms, and is dominated by sound propagating through the bottom layers (Kibblewhite and Wu, 1991). It is a function of depth and the sea surface spectrum of heights.

From 5 to 20 Hz there is again a strong correlation of noise with wind speed (Nichols, 1987). The *natural* sound-producing mechanisms that have been proposed at frequencies between 20 and 500 Hz are wave-turbulence interactions and oscillating bubble clouds (Cary and Bradley, 1985; Prosperetti, 1988). However, particularly near shipping lanes, the noise in the 10–150 Hz band is due largely to the machinery of distant ships (Ross, 1987).

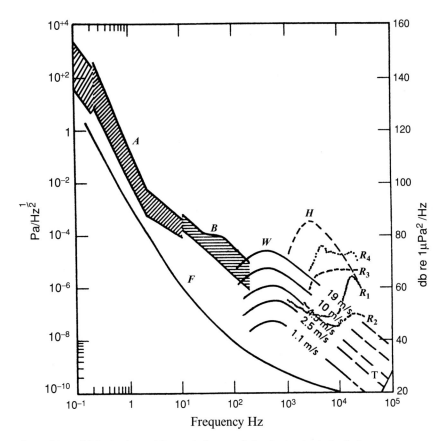

Fig. 5.11. Published values of "intensity" spectral density levels of physical sounds measured at various locations by many researchers. The heavy solid line at the bottom is the empirical minimum at sea. A, "seismic noise" due to earthquakes and wind; B, ship noise; F, thermal noise; H, hail; W, sea surface sound at five different wind speeds; R_1, drizzle ($\approx$1 mm/h) with 0.6 m/s wind over lake; R_2, drizzle with 2.6 m/s wind over lake; R_3, heavy rain (15 mm/h) at sea; R_4, very heavy rain (100 mm/h) at sea; T, thermal noise.

Knudsen Sea State Sound

Ocean sound in the band 500–20 000 Hz has been called "wind noise," "sea state noise," or "Knudsen noise" because, during World War II, Vern O. Knudsen discovered that it correlated very well with wind speed (Knudsen *et al.*, 1948).

The depth dependence due to the attenuation of sea surface sound by near-surface bubble layers and bubble plumes cannot be ignored. The sound from radiating surface bubbles is modified by inactive older bubbles below the surface (Chapter 6) and by absorption attenuation (Chapter 2). For these reasons, when measured several meters below the surface, the traditional straight lines of the Knudsen spectra in Fig. 5.11

actually droop (at about $f > 5$ kHz) to lower levels. Nevertheless, the sound due to breakers has been shown to be sufficiently well correlated with winds at sea to permit estimates of wind speed from ambient sound measurements on the sea floor. The distribution of these overhead breakers can be determined from the *variation* of the underwater ambient sound (Farmer and Vagle, 1989). (See Chapter 10.)

Summary of Knudsen Sea State Sound

In conclusion, Knudsen noise spans the frequency range from 500 Hz, probably to 50 kHz. Microsource description: dipoles near ocean surface each of individual peak source pressures 0.1 to 1.0 Pa. These microsources are transient, damped oscillations of microbubbles radiated immediately after their creation by breakers (Medwin and Beaky, 1989; Updegraff and Anderson, 1991). Local breaking depends on surface chemistry, sea swell, and local winds. The sound level increases with increasing wind speed partly because there are more waves breaking simultaneously at higher wind speeds than at lower wind speeds. Ignoring absorption and local variations, the average spectral intensity is independent of depth. There is potential for determining underwater sound from satellite photos of surface foam coverage (Monahan and O'Muircheartaigh, 1986).

Rainfall sound

The intensity spectral density during rainfall at sea (Fig. 5.11) is an example of "noise" that has yielded information about the source. For light rainfall, "drizzle" (≈ 1 mm/h), at very low wind speeds over "calm" seas, this most prominent peak (curve R_1) is due to transient, exponentially decaying oscillations of bubbles created by normal incident raindrops 0.8 mm < diameter < 1.1 mm. At higher wind speeds the peak spectrum level lowers, broadens, and shifts to higher frequencies (curve R_2) because these small raindrops enter at *oblique* incidence (Medwin, Kurgan, and Nystuen, 1990).

During heavy rainfall (>7.6 mm/h), there are also raindrops of diameter up to several millimeters falling at terminal speeds up to 10 m/s. The larger drop kinetic energy creates larger bubbles which radiate strongly for greater durations at the lower frequencies from 2 to 10 kHz. Each drop diameter has a distinctive energy spectrum. The larger drops produce a broader spectrum (curves R_3 and R_4) than those shown for light rain (<2.5 mm/h) in Fig. 5.11, partly because the larger drops generate splash hydrosols that create bubbles when they fall back into the sea. The type of rainfall (from stratus or cumulus clouds) can be readily deduced from this shape difference of the underwater sound spectrum.

(For details, see Chapter 11.) In summary, radiated power and source level depend on number and distribution of raindrop sizes and local angle of drop incidence (see Fig. 5.11). The significant frequency range is 1 to 25 kHz. Individual sources are dipoles perpendicular to surface. The "energy" spectral density per raindrop, up to 10^{-7} Pa^{2}s/Hz. Rainfall sound is composed of a broadband impulse source and (usually) much more energetic damped bubble oscillations. Knowledge of the rainfall drop size distribution and the ocean surface slope permits calculation of the underwater sound spectrum which, except for sound absorption, is essentially independent of depth. By matrix inversion, measurement of the underwater sound spectrum permits calculation of the number of (drops/m^2)/s within raindrop diameter bands (e.g., 100 μm bands), which are summed to give the total rainfall rate.

Free-drifting ocean buoys reporting signals of rainfall sound via satellite can be inverted to yield rainfall description in remote ocean regions (Nystuen, 1998). (See Chapter 11.)

5.5.2 Natural biological sounds

Figure 5.11 omits biological sounds, not because they are unimportant but because of their great diversity and complexity, and because so little is known about them. Probably the first identification of the noise of marine animals was by Knudsen *et al.* (1948), who observed that "colonies of certain species of snapping shrimp close their pincers with a loud audible click." In some coastal regions and in the frequency range of 200 to 20 000 Hz, snapping shrimp "intensity" spectral densities range from 1 to 100 Pa^2/Hz (the corresponding spectrum levels range from 60 to 80 dB re 1 μPa^2/Hz).

A remarkable sound radiation specialization is shown by the "Plain-fin midshipman" fish, *Porichthys notatus*. The 15 cm long Type I male can actively resonate its swimbladder to produce a loud, mating-attractive 100 Hz sound for many minutes duration, or transiently "grunt" 50 ms duration impulses at a repetition rate of two per second, for threat display (Brantley and Bass, 1994). A brain "motor volley" of that same frequency drives the surrounding bladder muscles, which act as a pacemaker to maintain the radiation (Bass and Baker, 1990).

An extensive review of how animals use sound, and their anatomy for sound production, is in the symposium volume of Busnel and Fish, *Animal Sonar Systems* (1980), and in *The Sonar of Dolphins* (Au, 1993). Sound recordings have been made of many species of odontocete (toothed) animals, including dolphins and porpoises, as well as baleen (non-toothed) whales. The vast diversity of mammalian sounds has been described as "clicks," "whistles," "screams," "barks," "moans," "rum-bles," "chirps," "growls," "ratchets," "shrieks," "raucous screeches," "horn blasts," "grunts," and "songs," depending on the poet, naturalist, or scientist, as well as the animal itself. Some more recent research has been very specific in terms of the sound's source level, directivity pattern, temporal character, frequency spectrum, and intermittency. A dolphin's train of ultrasonic pulses is shown in Fig. 5.12. (Also see Chapters 12–17.)

Our purpose in this section is simply to give a "flavor" of the radiation from marine mammals. We use the term "apparent" source power when it is calculated by extrapolation from far-field sound pressure measurements. Similarly, peak sound pressure measurements at the hydrophone are extrapolated back to $R_0 = 1$ m to give the peak "apparent" source pressure P_0. More recent research uses recorders strapped to, or embedded in, the animal.

The apparent source power in watts (5.11) is calculated from the acoustic radiation (3.44), after taking account of the integated beam pattern Ψ_D (3.43) (which depends on the far-field directivity) and the sound transmission loss between the source and the receiver,

$$\Pi_M = \frac{(P_0 R_0)^2}{2\rho_A c}\Psi_D \qquad (5.11)$$

The frequency range of sounds produced by baleen whales can be as low as tens of hertz for "moans," to kilohertz for "songs," to tens of kilohertz for "clicks" (see Cummings and Holliday, 1987). The radiation has been extrapolated from hydrophone back to the animal to imply sound source pressure amplitudes ranging from 3 Pa to 3000 Pa referred to 1 m (130 to 190 dB re 1 μPa). We use these sound pressures and (5.11) to estimate the acoustic radiation power. Assuming that the whale radiates omnidirectionally and that $\Psi_D = 4\pi$, the acoustic radiations of whales range from 4×10^{-5} to 40 watts. The combination of low frequencies and high source powers implies that these animals can communicate for large ranges across the oceans.

Obviously, different species of dolphins emit different types of sounds. Some dolphin transmissions are frequency-modulated continuous tonals (whistles), from 5 to 30 kHz, which last for several seconds. These appear to be for communication. Others, at frequencies near 100 kHz, appear to be for identification of objects. Beam widths (3 dB down) of different species of dolphins have been measured or estimated from 6° to 16°. The monograph by Au (1993) gives results of extensive research on the *Tursiops truncatus* sonar. Au (1993, Sect. 6.5.1) states that the radiation is equivalent to that of a circular transducer of radius $a = 5.7$ cm. Note that Au's figures show the (+) peak to (−)

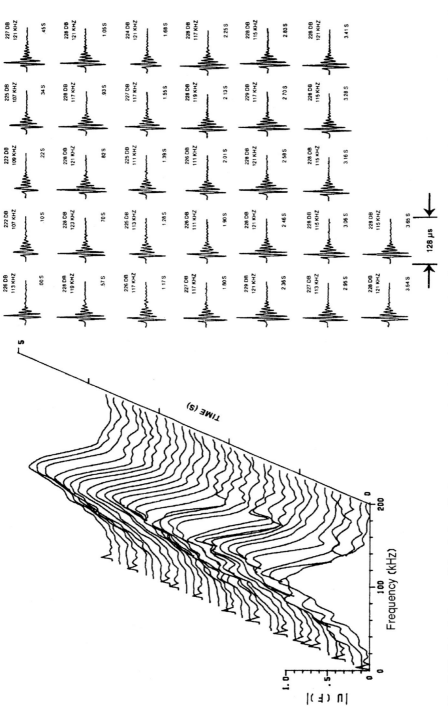

Fig. 5.12. A typical sonar click train for a *Tursiops truncatus* performing a detection task in the open sea near Hawaii. The spectra of the sequence of clicks are at the left. The ordinates are relative spectral amplitudes. The sound pressure $p(t)$ of individual signal clicks is at the right. The central frequency and source level are above each click, and its clock time is below. The (−) peak to (+) peak (at maximum) pressures are indicated in dB re 1 µPa at 1 m. To get the vertical coordinate scale of the signals, these should be converted to sound pressures in Pa. The average time between clicks is about 0.120 ms. (From Au, 1993, Fig. 5.2.)

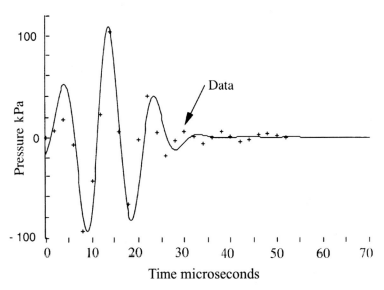

Fig. 5.13. Dolphin click and an approximate analytical function. The dolphin is a *Tursiops truncatus.* The data points (+) were read from Au (1993, Fig. 5.3), who aligned and averaged (stacked) the 32 clicks shown in Fig. 5.12. The smooth curve is where $P_0 = 110$ kPa: $f_0 = 100$ kHz, $T_0 = 13$ ms, $\Delta T = 33$ ms, and $\phi = -\pi/4$. (From *M&C*, Chapter 6.)

peak sound-pressure level in dB re 1 µPa; we convert to kilopascals in Fig. 5.13.

The maximum of the average click, Fig. 5.13, has "apparent" peak-to-peak sound pressures of about 200 kPa referred to 1 m (peak-to-peak SPL = 226 dB re 1 µPa). Such a very high-intensity sound would be expected to cause cavitation if it occured continuously at sea level. But the signal is a damped sinusoid of duration only 50 µs, which may not cause cavitation (see Section 4.2). An approximate analytical function for the clicks is

$$p(t) = P_0 \sin (2\pi f_0 t + \phi) \exp \left[-\pi^2 (t - \tau_0)^2 / \Delta \tau^2\right] \qquad (5.12)$$

Clay's comparison of (5.12) and the data is shown in Fig. 5.13.

To compute the acoustic energy radiated by a dolphin, recall (2.12). The message energy is

$$E_m = \frac{\Psi_D R_0^2}{\rho_A c} \int_{R/c}^{R/c+t_g} dt \left[p_0 \left(t - \frac{R}{c} \right)^2 \right] \qquad (5.13)$$

where the 4π is replaced by the integrated beam pattern, Ψ_D, defined in (3.43), and t_g is the duration of the click.

For $a = 5.7$ cm and $f_0 = 100$ kHz, ka is 11. The integrated beam pattern, $\Psi_D \approx 12/(ka)^2 = 0.09$. A numerical evaluation of (5.13), has been made by Clay using the points shown in Fig. 5.13. One click has the apparent message energy

$$E_m \approx 2.6 \times 10^{-4} \text{ joules} \qquad (5.14)$$

The character of repeated "clicks" can be seen in Fig. 5.12. The waveforms of the messages are not quite the same.

The dolphin *Tursiops truncatus* appears to alter its sound output according to its environment. The peak sound pressures were about 30 to 300 times higher when measured in the open waters than in tanks, and the output frequencies were near 115 kHz in open waters compared with nearly 50 kHz in tank experiments.

5.5.3 Ship noise

Ship noise generally overwhelms biological sounds over a very large area surrounding the ship, e.g., see Chapter 16.

The principal sources of ship noise are (*a*) radiation of engine noise from the ship hull; (*b*) "blade passage" tones from the propeller; and (*c*) cavitation from propeller blades. The engine noise is common to all rotating machinery. When the engine is not isolated from the ship, its vibration drives the hull; thereupon, sound is radiated according to the modes of vibration of the hull panels. The blade passage tones are caused by the rotating dipole created by the pressure in front of the propeller being positive, while the pressure behind is negative (Gray and Greeley, 1980). The frequency of these tones is the product of the number of blades times the shaft rotation frequency. Hydrodynamic cavitation is the creation of bubbles by the lowered pressures along the back edge, particularly near the tip, of the rotating blades. It is strongly dependent on the propeller tip speed. (See also Section 4.2 for acoustic cavitation).

The complexity of the multiple sources of ship noise is rendered even more complex by the fact that these noise sources are close to the ocean surface. For a smooth surface, the radiation is a function of the ratio of source depth/sound wavelength, which may cause the lower frequencies to radiate as dipoles, while the higher-frequency sounds act as monopoles (see Sections 3.1.2 and 3.1.3, which describe multipoles). Finally, when the rms height of the ocean surface is large compared with the sound wavelength, the sound at the surface is scattered rather than reflected, and the dipoles become relatively strong monopoles.

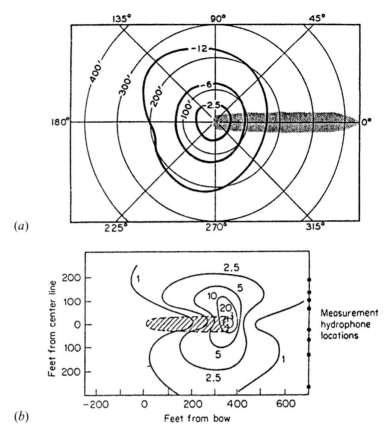

Fig. 5.14. (*a*) Average noise directivity in 200–400 Hz band for 15 World War II freighters. Contours are sound pressure levels in dB re 1 dyne/cm². From Eckart (1946, 1968). (*b*) High-frequency (2500–5000 Hz) ship noise "butterfly" pattern. (From Urick, 1983, Fig. 10.9.) The original caption is "Equal pressure contours on the bottom in 40 ft of water of a freighter at a speed of 8 knots. Contour values are pressures, in dynes per square centimeter in a 1-Hz band, at a point on the bottom, measured in the octave band 2500 to 5000 Hz." To translate the US Navy units of 1940s to SI units, use 1 knot = 1.15 miles/h = 1.85 km/h; 1 ft = 0.3095 m; 1 dyne/cm² = 10⁵ µPa = 0.1 Pa; intensity spectral density = (watts/m²)/Hz = (Pa²/Hz)/(ρ$_A$c.) The ship outlines are shaded.

Radiation at low frequencies of 200 to 400 Hz (Fig. 5.14(*a*)) looks somewhat omnidirectional. But rms pressure contours for the frequency band 2500 to 5000 Hz are closer to dipole rather than monopole (Fig. 5.14(*b*)). They were drawn from the values measured by eight sea floor hydrophones as a freighter steamed by. The sources of ship noise are spread along the ship. However, the radiated energy is generally greatest abeam and appears to come principally from the propeller. The

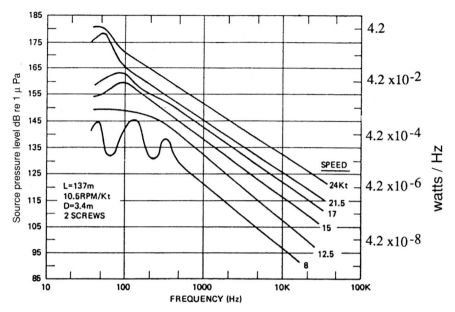

Fig. 5.15. Radiated spectra of noise maxima referred to 1 m from the British cruiser *Cardiff*, considered as a point source, as measured during World War II. The two ship propellors shafts of diameter 3.4 m rotated at 10.5 rpm/knot. The column on the right is the apparent spectral power, watts/Hz. (Adapted from Ross, 1987.)

propeller noise power radiated astern is reduced by being absorbed and scattered by the bubbly wake. It is reduced forward by diffraction by the ship body. The ocean surface reflection of the dominant dipole source results in radiation not unlike that of convected, distributed quadrupole sources in a jet engine exhaust. The net effect is that the sea-floor measurement of ship radiation is typically a "butterfly" pattern as shown in Fig. 5.14. This figure, here reprinted from Urick (1983), originally came from *U.S. Naval Ordnance Laboratory Report* 7333. (See also Chapter 16.)

The "butterfly" pattern, which is presented as spectral pressures (1 Hz band) in Fig. 5.14(*b*), was calculated from the measured squared pressure in the octave band, 2500 to 5000 Hz. This reduction to spectral values is sometimes done by assuming equal intensity in each 1 Hz band and simply dividing by the bandwidth, 2500 Hz. Possibly the stated values were calculated by recognizing that there is an f^{-2} dependence of radiated intensity in the typical ship noise spectrum (see Fig. 5.15).

The spectral pressure levels and spectral power of a WWII cruiser are shown as a function of ship speed in Fig. 5.15.

Further reading

See Chapters 11, 15, and 17, for current observations and interpretations of rain noise, marine mammal and fish sounds.

Chapter 6

Sound radiated or scattered by prototype marine bodies and bubbles

Summary

This chapter provides the basic concepts and analytical tools needed for active and passive acoustical studies of underwater animals, plankton, and other bodies and bubbles in the sea. It is clear that acoustical techniques are essential because optical visibility is so very limited and net sampling, or observations from remote underwater vehicles, or fixed observatories in the vast ocean volumes, are so sparse. As a result, man's search for identifiable life and obstacles at sea is largely dependent on the wise use of sound as an active or passive probe. Generally, a pressure sensitive hydrophone measures the sound pressure radiated or scattered by the objects.

First, we study sound generated or backscattered from single spherical bodies and free bubbles. These extreme models are the elementary sources and scatterers. Information about assemblages of real bodies or bubbles in the ocean can then be deduced from the frequency-dependent echoes, excess attenuation, volume reverberation, and the effect on sound speed dispersion.

The study of sound scattering by marine bodies has its roots in the classical physics research of Rayleigh, Helmholtz, Kirchhoff, and Born. However, ocean bodies do not have simple geometries that can be solved analytically; therefore, often one uses numerical techniques based on those classical formulations to describe the interaction of sound with objects in the sea.

To expand on this chapter, readers should see Chapters 10 through 24 for descriptions of recent and current acoustical measurements of marine bodies and multiple bubbles. Those chapters include on-going studies of droplets and bubbles produced by physical processes as well as detailed passive and active acoustical evaluations of biological presence and activity in shallow water and in the deep sea. Current, specific references are given in later chapters.

In an underwater acoustics experiment, generally a pressure-sensitive hydrophone measures the scattered sound pressure. The scattered sound pressures are the "observables." They can be compared with the results of theoretical and numerical calculations to yield a characterization of the scattering body.

We begin with brief descriptions of scattering measurements, which define the meanings of the scattering parameters. These same parameters are used in theoretical calculations.

Contents

6.1 Scattering of plane and spherical waves

An object is *effectively* insonified by plane waves when its dimensions are less than the diameter of the first Fresnel zone (Fig. 1.18). Within the first Fresnel zone, an actual spherical wavefront can be approximated as *being a plane wave.*

Oscilloscope traces of incident and scattered sound pressures at range R are sketched in Fig. 6.1. The incident sound pressure is assumed to be a ping having a high carrier frequency f and the duration $t_p \gg f^{-1}$.

---

* This section contains some advanced analytical material.

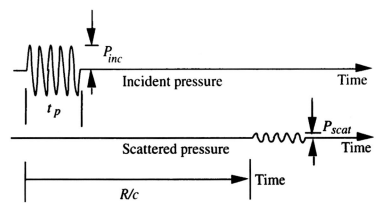

Fig. 6.1. Sketches of incident pressure and scattered sound pressure at range R. The amplitudes of the envelopes of the incident and scattered sound pressures are P_{inc} and P_{scat}. The travel time for the sound to scatter to the pressure-sensing hydrophone is R/c.

For very simple and approximate measurements, we use the peak pressures of the pings to measure the incident and scattered sound pressures. The pressure of the incident sound pressure at the object is $p_{inc}(t)$. It has the peak value P_{inc}. The time dependence is

$$p_{inc}(t) = P_{inc}e^{i2\pi ft} \quad \text{for} \quad 0 \le t \le t_p \quad \text{and} \quad p_{inc}(t) = 0 \quad \text{otherwise} \quad (6.1)$$

where t_p is the ping duration.

The sign of the exponential was chosen to match that used in most scattering theories.

Including attenuation, the scattered sound pressure is the time-dependent

$$p_{scat}(t) = P_{scat} \exp[i2\pi f(t - R/c)]10^{-\alpha R/20} \quad (6.2)$$

where t is the ping duration for $R/c \le t \le R/c + t_p$.

The theoretical solution shows that there is a shadow due to the destructive interference of the diffracted sound and the incident sound, that arrive behind the object (Fig. 6.2) at essentially the same time, with differently phased amplitudes. When the incident wave is a pulse of sufficiently short duration, it is experimentally possible to isolate and measure the scattered sound in that region.

6.1.1 Scattering length: the acoustical "size" of a scatterer

At a large distance from the object, the amplitudes of the scattered waves decrease as $1/R$ by spherical divergence and $10^{-\alpha R/20}$ by energy absorption (and scattering) in the medium. We assume that the backscattered

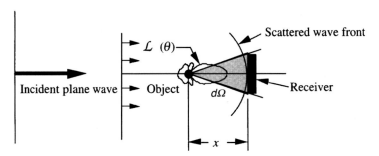

Fig. 6.2. Measurement of sound scatter from a sphere. In the forward direction, a close receiver senses the pressure waves over the subtended solid angle, $d\Omega$, where there are interferences. The sketched scatter directivity $\mathcal{L}(\theta)$ is for a rigid sphere having $ka = 5$.

energy is separated from the incident. Using the peak amplitudes of the incident and scattered sound pressures of a ping, we define a *complex acoustical scattering length*, $\mathcal{L}(\theta, \phi, f)$, which is the apparent "size" as determined by the experiment. At this time, we use the absolute value $|\ldots|$ and ignore the phase

$$P_{scat} = P_{inc} |\mathcal{L}(\theta, \phi, f)| (10^{-\alpha R/20})/R \tag{6.3}$$

or

$$|\mathcal{L}(\theta, \phi, f)| = RP_{scat}[10^{\alpha R/20}]/P_{inc} \tag{6.4}$$

Although the acoustical scattering length has the dimensions of length, it is different from any actual dimension of the body.

The scattering expression (6.4) is for a "point" receiver. Actually a receiving transducer integrates the sound pressures over its area ΔS. The dimensions of the transducer, which covers angles $\Delta\theta$ and $\Delta\phi$, limit the angular resolution of measurements of the scattering length (see Fig. 6.2).

6.1.2 Differential scattering cross-section

The concept of the differential scattering cross-section comes from quantum physics, where it describes the scattering of particles at a given angle. In acoustical measurements, the differential acoustical scattering cross-section at angle θ, ϕ is the absolute square of the acoustical scattering length

$$\Delta\sigma_S(\theta, \phi, f) = |\mathcal{L}(\theta, \phi, f)|^2 = \frac{P_{scat}^2(\theta, \phi, f)R^2}{P_{inc}^2} 10^{\alpha R/10} \tag{6.5}$$

where ϕ is the angle out of the plane.

The SI dimensions of $\Delta\sigma_s$ are m². The functional dependence in (6.5) explicitly indicates that, in general, both $\Delta\sigma_s(\theta, \phi, f)$ and $P_{scat}^2(\theta, \phi, f)$ depend on the geometry of the measurement and the carrier frequency of the ping. The measurement is called *bistatic* scatter when the source and receiver are at different positions. When they are at the same position, it is called backscatter, or sometimes *monostatic* scatter.

Specialization to the backscattering direction,

$$\theta = 0 \quad \text{and} \quad \phi = 0 \tag{6.6}$$

gives the differential backscattering cross-section,

$$\Delta\sigma_S(0, 0, f) = \Delta\sigma_{bs}(f) = |\mathcal{L}(0, 0, f)|^2 \tag{6.7}$$

This very important function is sometimes called the *backscattering cross-section*, $\sigma_{bs}(f)$ (Clay and Medwin, 1977). The Δ notation is used here to emphasize the differential character and to parallel the general case (6.5),

$$\sigma_{bs}(f) \equiv \Delta\sigma_{bs}(f) \tag{6.8}$$

The effective acoustic backscattering length is defined as

$$\mathcal{L}_{bs}(f) = \mathcal{L}(0, 0, f) \tag{6.9}$$

(Note: For typographical simplicity, we often omit functional dependencies when they are obvious.)

6.1.3 Total cross-sections for scattering, absorption and extinction

The theoretical total scattering cross-section is the integral of $\Delta\sigma_s(\theta, \phi, f)$, over the entire 4π of solid angle

$$\sigma_s(f) \equiv \int_0^{2\pi} d\phi \int_0^{\pi} \Delta\sigma_s(\theta, \phi, f) \sin\theta \, d\theta \tag{6.10}$$

The total scattering cross-section can also be defined by

$$\sigma_s = \Pi_{scat}/I_{inc} \tag{6.11}$$

where Π_{scat} is the total power scattered by the body, and I_{inc} is the incident intensity. The two definitions are equivalent, as can be easily proved.

The total power lost from the incident wave due to absorption by the object, Π_{abs}, determines the absorption cross-section

$$\sigma_a = \Pi_{abs}/I_{inc} \tag{6.12}$$

The total power removed from the incident beam, $\Pi_{scat} + \Pi_{abs}$, is used to define the *extinction cross-section*,

$$\sigma_e = (\Pi_{scat} + \Pi_{abs})/I_{inc} \qquad (6.13)$$

The sum of the total scattering cross-section and the total absorption cross-section is the total extinction cross-section,

$$\sigma_e = \sigma_s + \sigma_a \qquad (6.14)$$

If there is no body absorption, the total scattering and extinction cross-sections are equal,

$$\sigma_e = \sigma_s \quad \text{(no absorption)} \qquad (6.15)$$

Experimentally, it is difficult to isolate the integral scattered sound power in regions where there are interferences with the incident sound wave. As an alternative, the differential extinction cross-section could be measured (Fig. 6.2).

The receiver, which is at distance x behind the scattering object, senses the sum of the scattered and the incident wave pressures. This raw measurement depends on the value of x and $\Delta\sigma_s(\theta, \phi, f)$ within the solid angle $d\Omega$. Appreciable curvature of the wavefront at the receiver can complicate the measurement if the dimensions of the receiver are many wavelengths in extent.

Wave scattering depends on the object's size, call it a, relative to the acoustic wavelength. As we will see later in this chapter, for $a/\lambda \ll 1$, a simple assumption of spherical wave divergence scatter is often a useful approximation. At $a/\lambda \gg 1$, ray approximations give good results. At intermediate values, $0.5 \ll a/\lambda \ll 20$, solutions require detailed evaluations based on wave theory.

Some authors compute a *total acoustic scattering cross-section*, σ_s, by multiplying the differential backscattering cross-section, $\Delta\sigma_{bs}$ by 4π. This procedure is correct only in special cases where the scattering *is known to be isotropic (omnidirectional)* – for example, scattering from a spherical gas bubble, near resonance, is known to be isotropic.

6.1.4 Target strength of a scatterer

The target strength is a logarithmic measure (to the base 10) of the differential cross-section. It depends on the geometry of the measurement,

$$TS(\theta, \phi, f) = 10 \, \log[\Delta\sigma_s(\theta, \phi, f)/(1 \, \text{m}^2)] \quad \text{(dB)} \qquad (6.16)$$

where the reference area is $1 \, \text{m}^2$.

The most common usage is for backscatter. Then

$$TS(f) = 10 \log[\Delta\sigma_{bs}(f)/(1 \, \text{m}^2)] \quad \text{(dB)} \qquad (6.17)$$

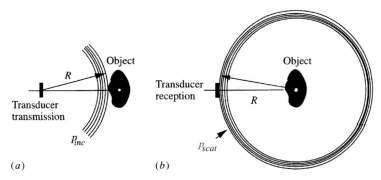

Fig. 6.3. Sound scattered by a small object. (*a*) Outgoing ping from a transducer. (*b*) Ping scattered back to the transducer. It is assumed that the sound spreads spherically from the transducer and also spreads spherically from the object, and that the Kirchhoff approximation holds.

In terms of the general statement of the scattering size, the target strength referred to 1 m is

$$TS(\theta, \phi, f) = 20 \log[|\mathcal{L}(\theta, \phi, f)/1\,\mathrm{m}] \quad (\mathrm{dB}) \qquad (6.18)$$

or, more particularly, in terms of the backscattering size,

$$TS(f) = 20 \log[|\mathcal{L}_{bs}(f)|/1\,\mathrm{m}] \quad (\mathrm{dB}) \qquad (6.19)$$

The measured cross-sections or scattering lengths are simpler to use and closer to the physical concept than their logarithmic functions, target strength. For most bodies, the backscattering length, backscattering cross-section, and target strength are also functions of the angles of entry of the sound at the scatterer.

6.1.5 Backscatter measurements

A typical backscattered sound measurement is sketched in Fig. 6.3. Using Fig. 6.3, the incident sound pressure at the object is

$$P_{inc}\, e^{i2\pi ft} = \frac{P_0 e^{i[2\pi f(t-R/c)]}\, R_0}{R} 10^{-\alpha R/20} \qquad (6.20)$$

where the reference sound pressure for the source P_0 is at the reference distance R_0, usually 1 m.

The sound scattered back to the source–receiver, is

$$P_{scat}(f) = P_{scat}\, e^{i2\pi ft} = \frac{P_0 e^{i[2\pi f(t-2R/c)]}\, R_0 \mathcal{L}_{bs}(f)}{R^2} 10^{-2\alpha R/20} \qquad (6.21)$$

The travel time to go from the source to the scattering object and back to the receiver is $2R/c$.

The Kirchhoff approximation assumes that the pressure reflection coefficient, $\mathcal{R}$, and pressure transmission coefficient, $\mathcal{T}$, that would be *derived for reflection and transmission of an infinite plane wave at an infinite plane interface can be used at every point of a rough surface interface.*

The Kirchhoff approximation is often called the *geometrical optics approximation* because the ray description is assumed to represent the reflected and transmitted waves at the point where the ray strikes the plane interface.

6.1.6 Scattering of a spherical wave at a plane facet

The ratio of the reflected pressure from a finite square $(w_x = w_y)$ facet to that from an infinite plane $|P(w_x, w_y)|/|P(\infty, \infty)|$ is shown in Fig. 6.4. The Fraunhofer (incident plane-wave) region is approximately $w_x/(\lambda R)^{1/2} < 0.5$ where R is the distance from source to reflector. In this region the relative backscattered pressure is proportional to the area of the facet.

What is called "the Fresnel region" is characterized by amplitude oscillations above and below unity at approximately $w_x/(\lambda R)^{1/2} > 0.5$.

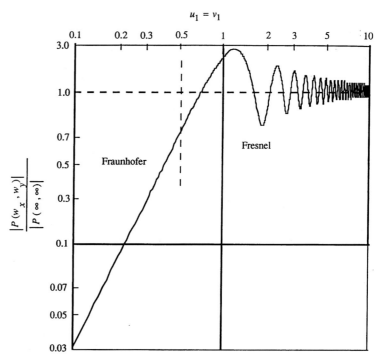

Fig. 6.4. Relative pressure of a sinusoidal wave backscattered from a square facet $(w_x = w_y)$ compared to an infinite plane. The sound waves are vertically incident. The parameters are $u_1 = w_x/(\lambda R)^{1/2}$ and $v_1 = w_y/(\lambda R)^{1/2}$ where R is the range. (From Clay, Chu, and Li, 1993.)

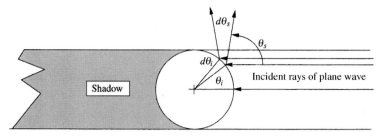

Fig. 6.5. Ray acoustics for geometrical scatter from a fixed, rigid sphere at high frequencies, $ka \gg 1$.

The ratio tends to the constant pressure "image" region of unity at large $w_x/(\lambda R)^{1/2}$; there the finite facet is large enough to give the same reflected pressure as an infinite plane. Figure 6.4 is solely due to the interference effect when a spherical wave is incident on a square plane facet; edge diffraction has been ignored. These theoretical results have been verified by a set of laboratory experiments (Clay, Chu, and Li, 1993).

6.2 Scattering from a sphere*

The marine acoustician is interested in spheres because scattering from these simple models has been well studied, and the results of this research are applicable to many forms of marine life. Furthermore, an acoustically small and compact *non-spherical* body, whose dimensions are much less than those of the sound wavelength, scatters in about the same way as a sphere of the same volume and same average physical characteristics.

Scattering from a sphere is an ideal vehicle to demonstrate the contributions by the three components of scatter from a fluid body: reflection, diffraction, and transmission. In Section 6.2.1 we use the rigid sphere to provide one reference point, the reflection component, which dominates in "geometrical scatter" ($ka \gg 1$). In Section 6.2.2 we introduce "Rayleigh scatter" ($ka \ll 1$), in which the weaker wave diffraction around the body dominates. In Rayleigh scatter the ratio of the backscatter cross-section to the geometrical cross-section is proportional to the very small quantity $(ka)^4$.

6.2.1 Geometrical scatter from an acoustically large, rigid sphere

The behavior when $ka \gg 1$ is described as "geometrical," Kirchhoff, or "specular" (mirrorlike) scatter. In the Kirchhoff approximation, a plane wave reflects from a small area as if the local, curved surface is a plane. The situation can be understood by using rays as illustrated in Fig. 6.5.

---

* This section contains some analytical material.

In the ray description, scatter consists of a spray of reflected rays that follow the Kirchhoff reflection approximation. That is, each reflected ray obeys the simple law of reflection, in which the ray is reflected with its angle of reflection equal to its angle of incidence. It is as if the reflection point is a plane that is tangent to the sphere at that point. Diffraction effects, mainly from the edge of the shadow and behind the sphere, are ignored in this approximation.

We now calculate the scattering from a fixed, rigid, perfectly reflecting sphere at very high frequencies, $ka \gg 1$. The incident sound is a plane wave of intensity I_{inc}. There is no energy absorption in the medium. No energy penetrates into the sphere.

First, calculate the incoming power at angle θ_i, for ring increments $d\theta_i$, on the sphere surface (Fig. 6.5).

The surface area increment is

$$dS_i = 2\pi(a \sin \theta_i)a \, d\theta_i \tag{6.22}$$

The component perpendicular to the surface is

$$dS_\perp = dS_i \cos \theta_i \tag{6.23}$$

The input power in the ring is

$$d\Pi_{inc} = I_{inc} \, dS_\perp = I_{inc}(2\pi a^2 \sin \theta_i \cos \theta_i \, d\theta_i) \tag{6.24}$$

Therefore

$$d\Pi_{inc} = I_{inc}\pi a^2 \sin(2\theta_i) \, d\theta_i \tag{6.25}$$

Next, calculate the scattered power. The rays that are within angular increment $d\theta_i$ at angle θ_i are scattered within increment $d\theta_s = 2d\theta_i$ at angle $\theta_s = 2\theta_i$. The increment of *geometrically scattered* power, measured at range R, is

$$d\Pi_{gs} = I_{gs}2\pi(R \sin \theta_s)R \, d\theta_s \tag{6.26}$$

or,

$$d\Pi_{gs} = I_{gs}2\pi R^2(\sin 2\theta_i)(2d\theta_i). \tag{6.27}$$

Assume that there is no loss of power, $d\Pi_{inc} = d\Pi_{gs}$. Therefore

$$I_{gs} = I_{inc}\frac{a^2}{4R^2} \quad \text{or} \quad P_{gs} = P_{inc}\frac{a}{2R} \tag{6.28}$$

or, in terms of the scattering length,

$$P_{gs} = P_{inc}\frac{|\mathcal{L}_{gs}|}{R} \quad \text{where} \quad |\mathcal{L}_{gs}| = a/2. \tag{6.29}$$

Also,

$$|\mathcal{L}_{gs}|/(\pi a^2)^{1/2} = 1/(2\pi^{1/2}) = 0.28 \tag{6.30}$$

In this geometrical scatter approximation, the derivation shows that scattered power is not a function of θ_i. All differential geometrical cross-sections, including the backscattering cross-section, are equal. A high-frequency acoustical wave gets a "wall-eyed" view of the sphere. From (6.5), assuming no attenuation in the medium, the differential geometrical cross-section is

$$\Delta\sigma_{gs} = |\mathcal{L}_{gs}|^2 = (a^2/4) \quad \text{for} \quad ka \gg 1 \tag{6.31}$$

The total geometrical scattering cross-section is

$$\sigma_{gs} = 4\pi\,\Delta\sigma_{gs} = \pi a^2 = A \quad \text{for} \quad ka \gg 1 \tag{6.32}$$

Note that, at very high sound frequencies, $ka \gg 1$, σ_{gs} is equal to the actual cross-sectional area, A.

The fact that $\mathcal{L}_{gs}$ and σ_{gs} are approximately independent of frequency for $ka > 10$ has important practical applications. This independence has been exploited by acousticians for many decades in laboratory calibrations of sonars, because the high-frequency backscattered sound pressure is almost a delayed replica of the transmitted signal or message.

The ray treatment is deceptively simple. In fact, it is an incomplete description of the problem because it ignores the complicated wave interferences between direct and scattered sound in the shadow region.

6.2.2 Rayleigh scatter from an acoustically small sphere ($ka \ll 1$)

When the sound wavelength is very much greater than the sphere radius, the scatter is due solely to diffraction. There are two simple conditions that then cause scatter. (1) If the sphere bulk elasticity $E_1 (= compressibility^{-1})$ is less than that of the water, E_0, the incident condensations and rarefactions compress and expand the body, and a spherical wave is reradiated. This monopole reradiation occurs with opposite phase when $E_1 > E_0$. (2) If the sphere density, ρ_1, is much greater than that of the medium, ρ_0, the body's inertia will cause it to lag behind as the plane wave swishes back and forth. The motion is equivalent to the water being at rest and the body being in oscillation. This action generates dipole reradiation (Section 3.1.3). When $\rho_1 < \rho_0$, the effect is the same but the phase is reversed. In general, when $\rho_1 \neq \rho_0$, the scattered pressure is proportional to $\cos\theta$, where θ is the angle between the scattered direction and the incident direction.

The simplest sphere model is a small, fixed, incompressible sphere that has no waves in its interior. There will be monopole scatter because the body is incompressible. There will also be dipole scatter because it is fixed. To illustrate this, we use a simple derivation that shows

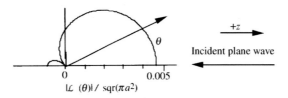

Incident plane wave

$|\mathcal{L}\ (\theta)|\,/\ \text{sqr}(\pi a^2)$

Fig. 6.6. Polar scattering pattern for a fixed, rigid sphere, at $ka = 0.1$, i.e., circumference is one-tenth of the wavelength. The reference length is $(\pi a^2)^{1/2}$.

the essential physical concepts. The same result can be derived formally (*M&C*) by using the first two terms, which are the largest terms, of the mode solution for $ka \ll 1$.

An incident plane wave moves from right to left in the $-z$ direction, Fig. 6.6. The fixed, rigid sphere scatters sound to the receiver at range R and angle θ. The sphere is so small relative to λ that, at any instant, its entire surface is exposed to the same incident acoustic pressure at that time. The sound pressure on the sphere $Pe^{i\omega t}$ is the sum of the incident plane wave pressure of amplitude P_{inc} and the scattered pressure wave of amplitude P_{scat}. Dropping the common time dependence

$$P = P_{inc}e^{ikR\cos\theta} + P_{scat} \tag{6.33}$$

where

$$z = R\cos\theta \tag{6.34}$$

The boundary condition on the surface of a rigid sphere is that the normal components of displacement and particle velocity are zero at all times. The normal (radial) component of particle velocity is related to the pressure by

$$u_R = -\frac{1}{ik\rho_A c}\left(\frac{\partial P}{\partial R}\right) \tag{6.35}$$

We use (6.33) to find the velocity relation at $R = a$, where $u_R = 0$,

$$\frac{\partial P}{\partial R}\bigg]_{R=a} = ikP_{inc}\cos\theta\, e^{ika\cos\theta} + \frac{\partial P_{scat}}{\partial R} = 0 \tag{6.36}$$

The gradient of the scattered component is

$$\frac{\partial P_{scat}}{\partial R}\bigg]_{R=a} = -ikP_{inc}\cos\theta\, e^{ika\cos\theta} \tag{6.37}$$

Therefore, from (6.35), the radial component of the scattered particle velocity at $R = a$ is

$$u_{scat}] = \frac{P_{inc}}{\rho_A c}\cos\theta\, e^{ika\cos\theta} \quad \text{at} \quad R = a \tag{6.38}$$

Expansion of the exponential for small ka gives

$$u_{scat} = \frac{P_{inc}}{\rho_A c}(\cos\theta + ika \cos^2\theta) \quad \text{at} \quad R = a \qquad (6.39)$$

Equation (6.39) gives the scattered radial particle velocity that is necessary to make the total fluid particle velocity equal to zero on the sphere. We will show that this scattered particle velocity is equivalent to radiation from a source composed of a monopole and a dipole.

Monopole component

To calculate the equivalent monopole rate of volume flow, $\dot{V}_m$, (m^3/s), we integrate u_{scat} over the surface of the sphere,

$$\dot{V}_m = \int_A u_{scat}\, dA = 2\pi a^2 \int_\pi^0 u_{scat} \sin\theta d\theta \qquad (6.40)$$

The integral over the first term of (6.39) is zero. The integral over the second term gives the monopole flow through the surface that surrounds the sphere,

$$\dot{V}_m = \frac{ika}{\rho_A c}\frac{4\pi a^2}{3} P_{inc} \qquad (6.41)$$

From (3.4), the monopole sound pressure at range R is given in terms of its volume flow,

$$P_m = -\frac{ik\rho_A c \dot{V}_m}{4\pi R(1 + ika)}e^{-ik(R-a)} \qquad (6.42)$$

Using (6.41), at large range $kR \gg 1 \gg ka$, the pressure due to the monopole is

$$P_m = \frac{(ka)^2 a}{3}\frac{P_{inc}}{R}e^{-ikR} \qquad (6.43)$$

Dipole component

To calculate the acoustic dipole, notice that the first term in (6.39) is equivalent to alternating flow in the z direction (because of the time factor that we have suppressed). This dipole flow is obtained by integrating over the sphere:

$$\dot{V}_d = \frac{2\pi P_{inc} a^2}{\rho_A c}\int_{\pi/2}^0 \cos\theta \sin\theta d\theta = \frac{\pi P_{inc} a^2}{\rho_A c} \qquad (6.44)$$

Using (6.44) in (3.4), the long-range dipole solution (3.7) has the dipole axis, $l = 2a$, so that the dipole pressure is

$$P_d = (ka)^2 \frac{a}{2}\cos\theta \frac{P_{inc}e^{-ikR}}{R} \qquad (6.45)$$

Scattered pressure

The scattered pressure at large R is the sum of the monopole and dipole components of pressure,

$$P_{scat} = P_m + P_d = \left[\frac{2(ka)^2}{3} \left(1 + \frac{3}{2} \cos \theta \right) \frac{a}{2} \right] \frac{P_{inc} e^{-ikR}}{R} \qquad (6.46)$$

A low-frequency (small-body) polar radiation pattern was shown in Fig. 6.6.

The scattering length and differential scattering cross-section for the small, fixed, rigid sphere are calculated from $(P_{scat})^2/(P_{inc})^2$, as in (6.5), and are extrapolated back to $R = 1$ m,

$$|\mathcal{L}| = a \frac{(ka)^2}{3} \left(1 + \frac{3}{2} \cos \theta \right) \qquad (6.47)$$

and

$$\Delta\sigma_s(f, \theta) = |\mathcal{L}|^2 = \frac{(ka)^4}{9} \left(1 + \frac{3}{2} \cos \theta \right)^2 a^2 \qquad (6.48)$$

Although the analytical derivation above was for $ka \ll 1$, (6.47) and (6.48) are useful for ka as large as 0.5 (see Fig. 6.7). Backscattering is obtained by setting $\theta = 0°$. The results are

$$|\mathcal{L}_{bs}| = a \frac{5(ka)^2}{6} \qquad (6.49)$$

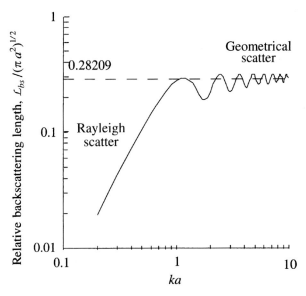

Fig. 6.7. Relative backscattering lengths for a rigid sphere. The Rayleigh scattering region where relative backscattering lengths are proportional to $(ka)^2$ starts at approximately $ka < 0.5$. The dashed line is the high-frequency asymptote, $1/(4\pi)^{1/2}$.

and

$$\sigma_{bs} = a^2 \frac{25(ka)^4}{36} \qquad (ka \ll 1) \tag{6.50}$$

The relative scattering cross-section, obtained by division by πa^2, is proportional to $(ka)^4$. This dependence characterizes what is widely known in optics and acoustics as "Rayleigh scatter." It is Rayleigh scatter that accounts for the blue of the sky. In optics, it is the intensities that are directly measured. In acoustics, pressures are directly measured.

The acoustical scattering cross-section in Rayleigh scatter is very much smaller than the geometrical cross-section because the sound waves bend around, and are hardly affected by, acoustically small non-resonant bodies. Ocean waves diffract around small rocks, and very low frequency (long-wavelength) swell waves diffract around small islands, in similar ways.

When the small sphere is an elastic fluid, the Rayleigh scatter region remains. But then the scattering depends also on the relative elasticity and the relative density, compared with the water medium surrounding it. This was first demonstrated by Lord Rayleigh (1896, Sec. 335), who found that for a fluid, elastic sphere,

$$\Delta\sigma_S(f, \theta) = (ka)^4 \left[\frac{e - 1}{3e} + \frac{g - 1}{2g + 1} \cos\theta \right]^2 a^2 \quad \text{for} \quad ka \ll 1 \tag{6.51}$$

where

$k = 2\pi/\lambda$ = wave number in the surrounding medium
$g = \rho_1/\rho_0$ = ratio of density of sphere to that of the medium
$h = c_1/c_0$ = ratio of sound speed in sphere to that in the medium
$e = E_1/E_0$ = ratio of elasticity of sphere to that of the medium
$c^2 = E/\rho$
$e = gh^2$
θ = angle between incident and scatter directions

The backscattering cross-section for a small, fluid sphere follows by setting $\theta = 0°$ in (6.51).

The relative magnitude of the monopole component of the polar scattering pattern depends on the first term in the square brackets of (6.51); the magnitude of the dipole component depends on the second term. Most bodies in the sea have values of e and g which are close to unity. On the other hand, for a gas bubble, $e \ll 1$ and $g \ll 1$. In that case, the scatter is omnidirectional because the elasticity term dominates.

Highly compressible bodies such as bubbles are capable of resonating in the $ka \ll 1$ region. Resonant bubbles produce scattering cross-sections several orders of magnitude greater than for a rigid sphere of the same size.

The total scattering cross-section for the small fluid sphere is obtained by integrating (6.51) over all angles,

$$\sigma_s(f) = \pi a^2 \int_0^{4\pi} \Delta\sigma_s d\Omega$$

$$= 4\pi a^2 (ka)^4 \left[\left(\frac{e-1}{3e} \right)^2 + \frac{1}{3} \left(\frac{g-1}{2g+1} \right)^2 \right] \qquad ka \ll 1$$

(6.52)

Light follows essentially the same backscattering laws as sound. One big difference is that the wavelength of visible light is of the order 5×10^{-5} cm. Therefore, almost all scattering bodies in the sea, even the everpresent ocean "snow," have optical cross-sections equal to their geometrical cross-sections. However, the same particles are very much smaller than the wavelength of *sounds* in the sea (almost always 1 cm or more). These bodies are therefore in the Rayleigh region, where they scatter sound very weakly. This is the major reason why the sea is turbid for light but transparent to sound.

6.2.3 Scatter from a rigid sphere

The theoretical solution for a rigid sphere is frequently used to display many of the characteristics of scattering phenomena. One reason is that it gives a fair approximation to a very dense sphere in water, which is often used for calibration in experimental work.

The backscattering length is sometimes expressed relative to $a/2$ (Anderson, 1950). Some researchers also use the name "form function" for that ratio. The form function for spheres is

$$F_\infty(ka) = |\mathcal{L}_{bs}|/(a/2) \qquad (6.53)$$

The backscattering length relative to $(\pi a^2)^{\frac{1}{2}}$ for a rigid sphere is presented in logarithmic scale in Fig. 6.7. The figure demonstrates the $(ka)^2$ dependence (slope 2:1) in the Rayleigh scattering region $(ka \ll 1)$, the asymptotic constant value of geometrical scatter $(ka > 1)$, and the oscillations at $ka \gtrsim 1$ due to diffracted waves. The relative backscatter length is shown on a linear scale in Fig. 6.8. The peaks and troughs of backscatter at $ka > 1$, which are caused by interference between the waves that diffract around the periphery and the wave reflected at the front surface of the sphere, occur with a periodicity due to their path difference.

The directional scattering at various values of ka is shown in Fig. 6.9. As ka increases, the sound tends toward isotropic, particularly for $\theta < \pi/2$. The number of diffraction side lobes increases as ka increases, and at large ka they oscillate around the ultimate isotropic solution for this component. The increasingly large forward scattered lobe is out of phase

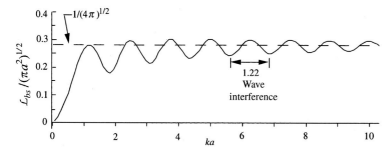

Fig. 6.8. Linear presentation of the relative backscattering length as a function of ka for a rigid sphere. The periodicity of the peaks and troughs, due to interference between the diffracted wave and the front-face reflection, occurs at regular intervals of $ka = 1.22$.

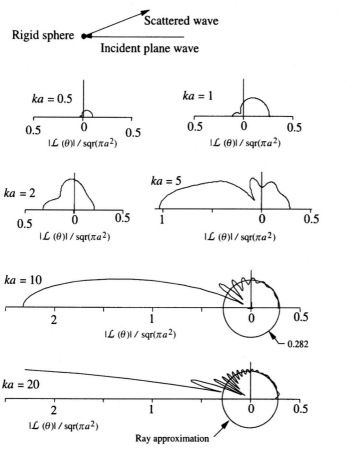

Fig. 6.9. Polar directivity of relative acoustic scattering length for a rigid, fixed sphere at a large range. The sound is incident from the right. The circles on cases $ka = 10$ and $ka = 20$ are the ray approximation described in Section 6.2. In the large ka cases, the strong forward-scattered lobe is out of phase with, and is largely cancelled by, the incident wave.

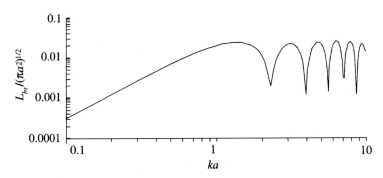

Fig. 6.10. Relative backscattering length for a fluid-filled sphere, with $g = 1.06$ and $h = 1.02$. These values of g and h are typical of marine animals.

with the coexistent incident sound so that, at high frequencies, the sum gives the shadow behind the obstacle.

6.2.4 Scatter from a fluid sphere

Fluid spheres are often used to represent small marine animals such as zooplankton, and sometimes even larger animals when the sound frequency is low ($ka \ll 1$). The sound speeds and densities of sea life are nearly those of water. For example, Clay has calculated the backscattering magnitude and directivity patterns for a fluid sphere with characteristics similar to fish flesh, $g = 1.06$ and $h = 1.02$, where

$$g \equiv \frac{\rho_1}{\rho_0}, \quad h \equiv \frac{c_1}{c_0}, \tag{6.54}$$

Figure 6.10 shows the relative sound backscattering length for a fluid-filled sphere as a function of ka. Compare with Fig. 6.7 for a rigid sphere. These curves, following Anderson's equations, do not include viscous and thermal losses. The incident plane wave interferes with the scattered wave as described in the previous section. For details, see Neubauer, *Acoustic Reflection for Surfaces and Shapes* Naval Research Lab, 1986.

6.3 A brief history of ocean bubble measurements

In the late 1950s, several studies introduced oceanographers and sonar users to the great importance of bubbles at sea. Urick and Hoover (1956) were concerned about the military implications of sound scatter from the rough sea surface; they discovered that much of the scatter came from *below* the surface, presumably from bubbles created by breaking waves. About the same time, Blanchard and Woodcock (1957) were interested in the airborne salt nuclei, generated from breaking waves, that affect thunderstorm activity at sea and climate throughout the world; they waded out from shore, scooped up a jar of bubbly water in the surf, and obtained the first measurements of bubbles caused by breaking waves. In the same year, LaFond and Dill (1957) wrote an internal US Navy Laboratory memorandum with the provocative title "Do Invisible Bubbles Exist in the Sea?". Starting from the evidence that sea slicks were formed over turbid surface water convergence zones, or were correlated with internal waves that had brought minute particulate matter to the surface, they concluded that bubbles may indeed be present in some parts of the ocean.

All of this research activity ran counter to early laboratory experiments, and intuition. The belief at the time was that any bubbles that may be created would soon disappear either by buoyant action, which would bring them popping to the surface, or by gas diffusion forced by bubble surface tension, which would squeeze the gas out of any small bubble. The flaws in this reasoning are now recognized: (1) ocean bubbles are not clean but may have solid or dissolved material on the surface which would inhibit gas diffusion, or they may exist in crevices; (2) ocean currents will create a friction drag that can overcome buoyant forces; (3) bubble populations at sea are continually replenished by the myriad bubble source mechanisms. Most of the bubbles found near the surface of the open sea appear to be continually generated by spilling or plunging breakers, or during rainfall. Particularly in coastal regions, the sources of ocean bubbles also include those entrained by continental aerosols that drop into the sea, generated by photosynthesis of marine plants or life processes of marine animals, or the decomposition of organic material, or released from gas hydrates on or below the ocean floor.

The immense number of microbubbles per unit volume that have been identified in the ocean are a major factor in near-surface sound propagation. They have also proved to be a unique tool in the study of near-surface ocean characteristics.

The statistical size distribution of bubbles caused by breaking waves was first measured in a laboratory flume by Glotov *et al.* (1962). The first photographic evidence of bubbles in a quiescent sea was that of Barnhouse *et al.* (1964), and the first measurements in the coastal ocean were by Buxcey *et al.* (1965), who used acoustical techniques. Research from the latter two Master's theses was reported by Medwin (1970). The dependence of ocean bubble densities on depth, season of the year, time of day/night, wind speed, and presence of sea slicks was revealed by using various acoustical techniques (Medwin, 1977) and extensive photography (Johnson and Cook, 1979). See *The Acoustic Bubble* by T. G. Leighton, Academic Press, 1994 and Chapter 21.

The omnipresence of optically unidentified particles, sometimes called "*detritus*" or "snow," makes it difficult for simple photography to positively identify bubbles of radius less than about 40 microns. It required laser holography (O'Hern *et al.*, 1988) to prove, unequivocally, that there can be as many as 10^5 to 5×10^6 bubbles per cubic meter e.g., at radii between 15 and 16 microns near the ocean surface even during calm seas, and that the inverse acoustical determinations of the preceding 20 years had been essentially correct. It is generally assumed that the peak density is somewhere around radius 10 to 15 μm, depending on ocean chemistry, and that the number density decreases for larger bubbles.

The principal practical techniques for bubble identification and counting at sea have proved to be inversions of linear acoustical measurements. Determinations of bubble densities have been based on acoustic backscatter, excess acoustic attenuation, and differential sound speed, as well as on non-linear behavior and acoustic doppler shift. The acoustical techniques have also included passive listening at sea to the sound under breaking waves, or during rainfall, to determine the number of newly created bubbles. (See Chapters 10 and 11.)

The direct consequence of bubbles at sea has been demonstrated to result in near-surface excess attenuations as great as 60 dB/m and speeds of sound that are tens of meters per second less than the approximately 1500 m/s that would be measured by a sound velocimeter, or assumed from a calculation based on temperature, salinity, and depth alone.

6.4 Scattering from a spherical gas bubble

6.4.1 Scattering directivity

The rigid, impenetrable sphere is one extreme limit of the physical properties of a simple body. The other is the gas bubble.

The realistic gas-filled sphere has an important low-frequency, omnidirectional breathing-mode resonance and higher-order, directional resonances. The relative directional scattering lengths of a gas-filled sphere insonified by a plane wave are shown in Fig. 6.11. These curves were calculated from Anderson's (1950) equations; they do not include viscous and thermal losses. They show that for small ka, the scatter is essentially omnidirectional.

6.4.2 Backscattering length and backscattering cross-section

The backscatter from a sea-level gas bubble, is shown in Fig. 6.12, where it is compared with the backscattering length of the rigid sphere. The sea-level bubble has a very large response to a CW plane wave – that is the major resonance at $ka = 0.0136$. There are also very much

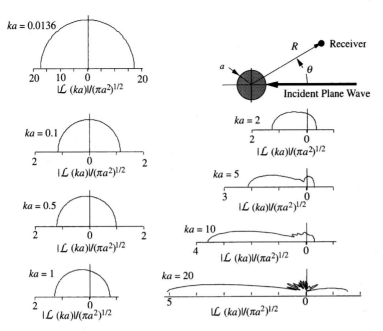

Fig. 6.11. Directionality of the scattering from an air-filled bubble as a function of *ka*. Plane wave incidence. Note particularly the case of virtually omnidirectional scatter at $ka = 0.0136$ for a sea-level air bubble at resonance.

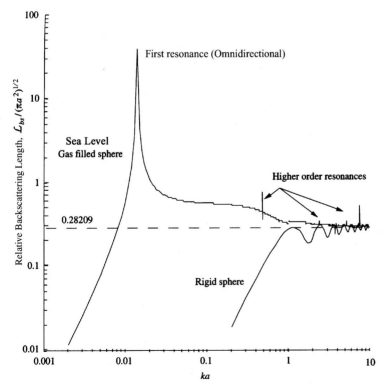

Fig. 6.12. Relative backscattering length of a spherical air bubble at sea level compared with that of a rigid sphere. Inclusion of viscous and thermal damping would lower the bubble peak and broaden the resonance.

weaker resonances at higher frequencies. The calculation does not include the significant bubble damping effects due to viscous and thermal losses. When these are included, they decrease the height and broaden the resonance curve of backscatter.

6.5 Single pulsating bubbles*

6.5.1 Stiffness and equivalent mass of a pulsating bubble

The sea-level resonance peak at $ka = 0.0136$ (Fig. 6.12) has been known since the original research on the musical sounds of running water (Minnaert, 1933). There is a relatively simple way to derive this "breathing frequency" at resonance.

In the $ka \ll 1$ condition, the bubble is very effectively driven by the virtually uniform acoustic pressure over its surface. Furthermore, a pulsating (breathing) bubble radiates more effectively than in any other mode (Section 3.1.1).

When $ka \ll 1$, the acoustical parameters of an oscillating bubble can be "lumped" into an equivalent mass, stiffness, and mechanical resistance, so that the acoustical system resembles the mechanical system of a mass on a spring. This type of analog is very commonly used by engineering acousticians; see, for instance, Olson (1947). Lighthill (1978) calls these "acoustically compact" regions. The equivalent mass is due to the inertia of the adjacent layer of water that envelops the bubble and that has essentially the same radial displacement as the bubble surface. The spring stiffness is determined by the compressibility of the bubble volume, plus the surface tension effect for very small bubbles. The equilibrium bubble radius, a, takes the place of the reference position of the spring at rest, and the radial displacement, $da = \xi$, corresponds to the linear displacement of the spring. The acoustical damping due to reradiation, viscosity, and thermal conductivity will later be described, in the lumped constant approach, as a mechanical resistance. For the moment, we assume that there is no damping.

The mechanical equation of motion of a mass–stiffness system is

$$m\frac{\partial^2 \xi}{\partial t^2} + s\xi = 0 \qquad (6.55)$$

The solution is of the form

$$\xi = \xi_0 e^{i\omega_b t} \qquad (6.56)$$

Substitution into (6.55) leads to the system natural frequency,

$$\omega_b = 2\pi f_b = \sqrt{\frac{s}{m}} \qquad (6.57)$$

We now determine the equivalent s and m for an oscillating bubble.

---

* This section contains some advanced analytical material.

Bubble stiffness

Consider a spherical gas bubble of volume $V = (4/3)\pi a^3$, surface area $S = 4\pi a^2$, and interior gas pressure $p_{int} \cong p_A$, where p_A is the static ambient pressure at that point in the ocean. Assume the bubble experiences an incremental interior pressure change, $dp_{int} \ll p_A$, and that the gas follows the adiabatic relation, $p_{int} V^\gamma = \text{constant}$, where γ is the ratio of specific heats of the bubble gas.

Differentiating yields

$$\frac{dp_{int}}{dV} = \frac{-\gamma p_A}{V} \qquad (6.58)$$

Use $dV = 4\pi a^2 \xi$, where ξ is the small radial surface displacement. For the present, assume that there is no surface tension force, so that the restoring force is

$$4\pi a^2 dp_{int} = -(12\pi \gamma p_A a)\xi \qquad (6.59)$$

This is a form of Hooke's Law, with stress proportional to strain. The proportionality constant is the stiffness of the bubble,

$$s = 12\pi \gamma p_A a \qquad (6.60)$$

Equivalent bubble mass

The inertial force experienced by the radiating bubble is calculated to determine the equivalent mass. It is not the mass of the bubble gas but rather the entrained water next to the bubble that comprises the mass of the pulsating system.

The pressure radiated omnidirectionally by a pulsating bubble is

$$p = \frac{P_a a}{R} \exp[i(\omega_b t - kR)] \qquad (6.61)$$

where P_a is the pressure amplitude at $R = a$.

For radial motion, the acoustic force is

$$\rho_A \frac{\partial^2 \xi}{\partial t^2} = -\frac{\partial p}{\partial R} \qquad (6.62)$$

where ρ_A is the density of the water.

Use (6.61) to obtain

$$\rho_A \frac{\partial^2 \xi}{\partial t^2}\Bigg]_{R=a} = \frac{P_a a}{R^2}(1 + ikR)\exp[i(\omega_b t - kR)]_{R=a} \qquad (6.63)$$

Since $ka \ll 1$, (6.63) simplifies to

$$\rho_A \frac{\partial^2 \xi}{\partial t^2}\Bigg]_{R=a} = \frac{p}{a}\Bigg]_{R=a} \qquad (6.64)$$

The inertial force at the surface is a form of Newton's Second Law,

$$F_m]_{R=a} = -4\pi a^2 p]_{R=a} = -4\pi a^3 \rho_A \frac{\partial^2 \xi}{\partial t^2}\Bigg]_{R=a} \qquad (6.65)$$

so we identify the effective mass as

$$m = 4\pi a^3 \rho_A \tag{6.66}$$

It is interesting to observe that, in this low-frequency approximation, the effective mass of the pulsating bubble is equivalent to that of an oscillating shell of water three times the volume of the bubble itself.

6.5.2 Simple pulsation; breathing frequency

Inserting the mass (6.66) and the stiffness (6.60) into (6.57) gives the simple harmonic breathing frequency of a small bubble ($ka \ll 1$) under the assumption of this derivation (no surface tension, adiabatic gas oscillations, no energy absorption):

$$f_b = \frac{1}{2\pi}\sqrt{\frac{s}{m}} = \frac{1}{2\pi a}\sqrt{\frac{3\gamma p_A}{\rho_A}} \tag{6.67}$$

The ambient pressure can be written in terms of the depth z by expressing

$$p_A = p_{A0} + \rho_A g z \cong 10^5(1.01 + 0.1z) \quad \text{pascals} \tag{6.68}$$

where p_{A0} = sea level atmospheric pressure = 1.01×10^5 pascals; $g = 9.8$ m/s^2; $\rho_A \cong 1030$ kg/m^3; $\gamma = 1.4$ for an air bubble; z = depth (meters).

For a spherical air bubble in water we get the simplified expression for the breathing frequency:

$$f_b = \frac{3.25\sqrt{(1+0.1z)}}{a(\text{meters})} = \frac{3.25 \times 10^6}{a(\text{microns})}\sqrt{(1+0.1z)} \tag{6.69}$$

At sea level, (6.67) may be written $ka = 0.0136$. Some people use the simple mnemonic that a bubble of radius 60 μm resonates at a frequency of about 60 kHz near the sea surface.

It turns out that the simple (6.67) is an acceptable predictor for the resonance frequency of most of the bubbles in the sea. Exceptions are: (a) bubbles of radii less than 5 μm (important in cavitation); (b) dirty bubbles (with a skin of debris); (c) bubbles in crevices; (d) bubbles near surfaces; and (e) non-spherical bubbles.

6.5.3 Damping constants

The inclusion of thermal conductivity and shear viscosity, as described in (M&C) causes very little change in the resonance frequency but it produces significant changes in the bubble damping at resonance. It therefore causes major corrections to resonant bubble scattering. The result is

$$\frac{P_{scat}}{P_{inc}} = \frac{-a/R_0}{[(f_R/f)^2 - 1] + i[ka + (d/b)(f_R/f)^2 + 4\mu/(\rho_A \omega a^2)]} \quad (6.70)$$

where d, b are thermal constants.

Even without these microscopic details it is useful to look at the "big picture."

The imaginary term in the square brackets is called the *total damping constant*,

$$\delta = \delta_r + \delta_t + \delta_v \quad (6.71)$$

which is the sum of the reradiation (scattering) term $\delta_r = ka$; the thermal damping term δ_t, which is a function of d/b; and the viscous damping term δ_v, which is a function of the kinematic coefficient of shear viscosity of the water, μ/ρ_A.

At resonance $f_R = f$, the scattered wave amplitude P_{scat} is limited only by the square-bracket imaginary term in (6.71). Now define the *damping constants at resonance* within that term,

$$\delta_R = \delta_{Rr} + \delta_{Rt} + \delta_{Rv} \quad (6.72a)$$

where $\delta_{Rr} = k_R a$ = resonance damping constant due to reradiation (scattering); $\delta_{Rt} = (d/b)$ = resonance damping constant due to thermal conductivity; $\delta_{Rv} = 4\mu(\rho_A/\omega_R a^2)$ = resonance damping constant due to shear viscosity.

Figure 6.13 shows the damping constants *at resonance* for a small, clean air bubble in fresh water at sea level. One notes that the resonance thermal damping δ_{Rt}, is significant over a very wide range of frequencies > 1 kHz; the resonance viscous damping, δ_{Rv}, is important above 100 kHz, and the resonance radiation damping, δ_{Rr} is important below 1 kHz.

A simple approximation to the resonance damping constant curve from 0.1 kHz to 100 kHz is

$$\delta_R \cong 0.0025 f^{1/3} \quad (6.72b)$$

with f in Hz.

6.5.4 Acoustical cross-sections

When the small bubble is insonified by a plane wave, its total *acoustical scattering cross-section*, σ_s, is defined as the total scattered power divided by the incident plane-wave intensity:

$$\sigma_s(f) = \Pi_s/I_{inc} = \frac{4\pi R_0^2 (P_{scat}^2/\rho_A c)}{(P_{inc}^2/\rho_A c)} = \frac{4\pi a^2}{[(f_R/f)^2 - 1]^2 + \delta^2(f)} \quad (ka < 1)$$

$$(6.73)$$

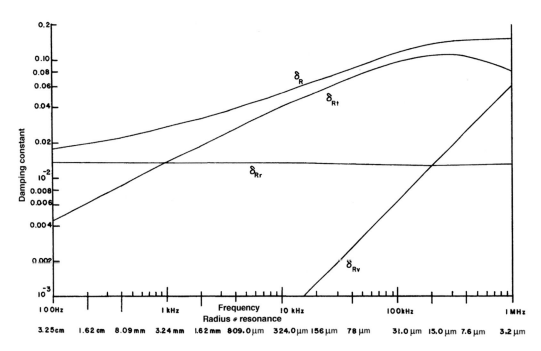

Fig. 6.13. The damping constants *at resonance* for a small, clean, air bubble in fresh water at sea level.

At resonance, the total scattering cross-section is

$$\sigma_s(f_R) = \frac{4\pi a^2}{+\delta_R^2} \quad \text{when} \quad f = f_R \tag{6.74}$$

Notice that the total scattering cross-section becomes $4/\delta_R^2$ greater than the geometrical cross-section. For example, from Fig. 6.13 we see that a 30 micron radius bubble with resonance frequency of about 100 kHz will have a damping constant of about 0.1, and therefore a total scattering cross-section 400 times larger than its geometrical cross-section! Because the scattering is omnidirectional when $ka \ll 0.1$, at resonance the differential scattering cross-section in any direction is simply

$$\Delta\sigma_S = \sigma_S/(4\pi) = a^2/\delta_R^2 \tag{6.75}$$

The backscattering length $|\mathcal{L}_{bs}|$ is the square root of $\Delta\sigma_s$, and the relative backscattering length is

$$|\mathcal{L}_{bs}|/(\pi a^2)^{1/2} = (\sigma_s)^{1/2}/(2\pi a) \quad ka < 1 \tag{6.76}$$

The comparison of the resonant bubble at sea level, where $ka = 0.0136$, with the rigid sphere of the same size reveals the immense selectivity of bubble resonance. This was shown in Fig. 6.12. The rigid sphere has a relative scattering length that decreases as $(ka)^2$ for frequencies below $ka = 1$. Therefore, at $ka = 0.0136$, the resonating bubble has a relative scattering length of about 10^5 greater than for a rigid sphere of the same size. For this reason, an acoustical scattering experimenter has a very easy task distinguishing a rigid sphere from a resonating bubble of the same size.

The attenuation of a plane wave depends on both the bubble scattering cross-section, σ_s, which removes sound from the beam, and the absorption cross-section, σ_a, which converts sound energy to heat. The sum of the two is the extinction cross-section, σ_e,

$$\sigma_e = \sigma_a + \sigma_s \tag{6.77}$$

The extinction cross-section is calculated directly from the rate at which the incident pressure does work on the bubble, divided by the incident plane-wave intensity,

$$\sigma_e = \frac{\frac{4\pi a^2}{T} \int_T \left[p_{inc} \frac{\partial \xi}{\partial t} \right]_{R=a} dt}{I_{inc}} \tag{6.78}$$

The integration yields

$$\sigma_e = \frac{4\pi a^2 (\delta/\delta_r)}{[(f_R/f)^2 - 1]^2 + \delta^2} \tag{6.79}$$

Using (6.77), we find the absorption cross-section by subtraction of (6.73) from (6.79). The result is

$$\sigma_a = \frac{4\pi a^2 \left(\frac{\delta_t + \delta_v}{\delta_r} \right)}{[(f_R/f)^2 - 1]^2 + \delta^2} \tag{6.80}$$

Comparing these equations, we obtain relations between the cross-sections in terms of the damping constants,

$$\frac{\sigma_a}{\sigma_s} = \frac{\delta_t + \delta_v}{\delta_r} \tag{6.81}$$

and

$$\frac{\sigma_e}{\sigma_s} = \frac{\delta}{\delta_r} \tag{6.82}$$

Again we see that, at resonance, the acoustical cross-sections are very much larger than the geometrical cross-section. This fact, combined with the simple, direct connection between resonance frequency and bubble radius, is the key to the advantages of acoustical measurements in determining bubble populations at sea. By comparison, when light scatters from a bubble or particle, the optical cross-section is at most equal to the geometrical cross-section.

The physical reason for this magnified reaction to a sonic wave is easy to explain. Using the concept of specific acoustic impedance (the ratio of the acoustic pressure to the acoustic particle velocity), the bubble at and near resonance is effectively a "hole" of very low impedance compared with the water. This hole distorts the incoming acoustic field over a large volume surrounding the bubble. The distorted field causes a power flow toward the bubble center from a section of the incoming plane wave that is far beyond the bubble cross-section. When this is "viewed" acoustically from a great distance, the absorption, scatter, and extinction give the impression of a body that is far larger than its true size.

The left side of Fig. 6.14 shows the scattering and extinction cross-sections and the right side the absorption cross-sections as functions of bubble radius for three frequencies, where $ka < 1$. The very prominent peaks at resonance are clearly seen. For bubbles of radius smaller than the resonance size, the cross-sections are approximately proportional to a^6. However, notice that larger bubbles have large extinction and scattering cross-sections. Therefore, if there is a mixture of bubble radii, those of radius much greater than the resonance radius may make a significant contribution to the measurement of total scatter or extinction if there are enough of them. An absorption measurement is not contaminated by larger bubbles. See Fig. 6.15 and Section 6.6.2 for an estimate of this effect.

6.5.5 Pulsations in a sound field

We now rewrite the equation of the lumped mechanical motion of a bubble, (6.55), to include damping when the bubble is insonified. In addition to the inertial force and the stiffness restoring force, there is a damping force, $R_M(\partial \xi / \partial t)$, which is assumed to be proportional to the radial velocity. The mechanical resistance, R_M, represents energy losses caused by sound reradiation, shear viscosity, and thermal conductivity. The equation describing the radial displacement, ξ, of the insonified bubble is

$$m\frac{\partial^2 \xi}{\partial t^2} + R_M\frac{\partial \xi}{\partial t} + s\xi = 4\pi a^2 P_{inc}e^{i\omega t} \tag{6.83}$$

The solution of this equation, which may be demonstrated by substitution, is

$$\xi = \frac{-i4\pi a^2(P_{inc}/\omega)e^{i\omega t}}{R_M + i(\omega m - s/\omega)} \tag{6.84}$$

It can be shown (M&C) that the lumped mechanical resistance, R_M, is simply related to the damping constant, δ, as follows.

$$R_M + i(\omega m - s/\omega) = i4\pi a^3 \rho_A \omega \left\{ \left[\left(\frac{f_R}{f}\right)^2 - 1 \right] + i\delta \right\} \tag{6.85}$$

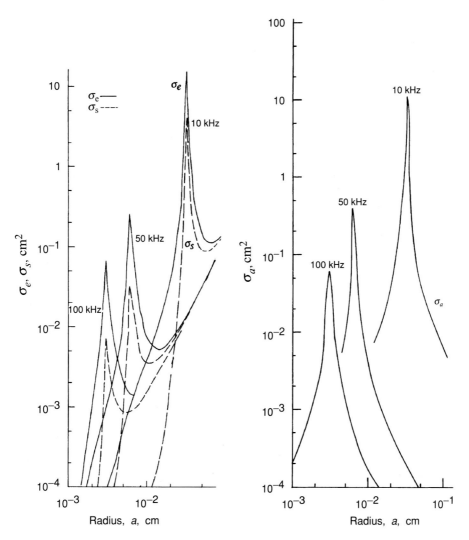

Fig. 6.14. *Left,* extinction (solid line) and scattering (dashed line) cross-sections of small sea-level air bubbles in fresh water, insonified by 10, 50, or 100 kHz plane waves. *Right,* absorption cross-sections for the same parameters. (From Medwin, 1977.)

Equating the reals gives the connection between R_M and δ,

$$R_M = 4\pi a^3 \rho_A \omega \delta = \omega m \delta \qquad (6.86)$$

where we have used (6.66) for the mass m.

6.5.6 Other effects: proximity of ocean surface

The bubbles created by a breaking wave or by a raindrop are close to the ocean surface – call it depth, z. When a pulsating bubble is close

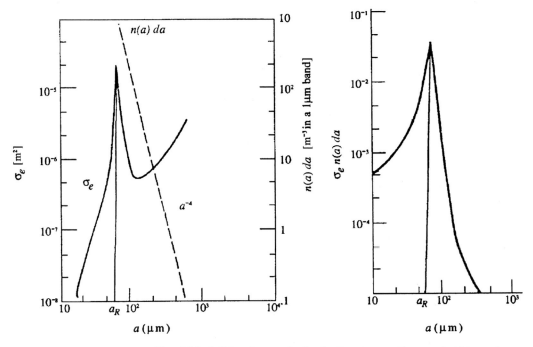

Fig. 6.15. *Left,* log–log graph of extinction cross-sections per bubble, σ_e, for 50 kHz insonification (ordinate at left side) and $n(a)da$ bubble density proportional to a^{-4} (ordinate at right side). *Right,* the product of σ_e and $n(a)da$ is calculated from the graph on the left. The area under the curve in a linear–linear graph would be S_e (Medwin, 1977).

to the "smooth" ocean surface, radiation from the bubble will reflect from the surface as a spherically diverging wave. For a "smooth" ocean surface, the reflected wave will have a virtual source strength that is equal to that of the real pulsating bubble and a phase that is shifted by 180°. The combination of the source and its out-of-phase reflection at a separation $l = 2z$ constitutes a dipole as described in Section 3.1.3. (See Strasberg, 1956.)

Non-sphericity
Strasberg (1956) also showed that an oscillating, ellipsoidal bubble of the same volume as a spherical bubble will have weaker radiation and a higher fundamental resonance frequency.

The change in volume resonance, when a bubble is non-spheroidal, has been studied also by Weston (1967) and Feuillade and Werby (1994). For the same volume of gas, as the aspect ratio of the prolate ellipsoid, or cylindrical volume with endcaps, increases from that of a spherical bubble (unity aspect ratio), the resonance frequency increases, the magnitude of the scattered energy decreases, and the breadth of the resonance curve increases. Both of these cases are possible models of swimbladders of fish.

Non-linearity

Sometimes the pulsations of the bubble are so large that shape oscillations take place, in addition to the radial pulsation. Longuet-Higgins (1992) has proved that under this condition the second harmonic of a shape oscillation can be resonant to the pulsation frequency, and an excess dissipation results. This continues until the oscillation amplitude is lowered to a level where only the omnidirectional pulsation remains, with its predictable theoretical damping.

The complex non-linear oscillations of bubbles have also been studied optically, most recently by Stroud and Marston (1994), Asaki and Marston (1995), and Feuillade and Werby (1994). Because of the difficulty of observing volume fluctuations optically, the work is usually conducted with larger bubbles. It has been found that clean water bubbles of radius greater than 800 microns show damping two to four times larger than expected in linear oscillations.

Salinity

Different damping constants in the real ocean would cause different expressions for the scattering, absorption, and extinction cross-sections. These would, in turn, affect the bubble densities obtained by the inversion of those equations.

There is evidence that the bubble damping constant in salt water *is* different from that in fresh water. Scofield (1992) found that the damping constants of rainfall-produced bubbles of radius from 140 to 300 microns ($23 > f_R > 11$ kHz) were about 10 percent greater in 35 ppm salt water than in tap water. Kolaini (personal communication, 1997) has found similar differences for bubbles generated in salt water, compared with fresh water.

Remote sensing

Since most bubbles are close to the ocean surface, a factor that affects attempts to obtain bubble densities by remote backscatter, is the proximity of the bubble to the sea surface and the roughness of the surface. The total picture of backscatter includes consideration of four paths: (1) direct backscatter from the bubble region; (2) bubble scatter that is then backscattered from the sea surface to the source; (3) insonification of the bubble by sound scattered from the sea surface; (4) the path from surface scatter to bubble, back to surface, and finally back to sound source. The answer, which depends on the assumption of the surface roughness effect, has been considered by Clay and Medwin (1964), McDaniel (1987), and Sarkar and Prosperetti (1994).

A further important factor is that the insonified volume must not be large if the backscatter is to be inverted to yield bubble density with good spatial resolution. Sonars that are far from the scattering region can provide only gross average values of the bubble densities.

6.6 Multi-bubble effects: scatter and absorption, attenuation and dispersion*

6.6.1 Backscatter from an ensemble of bubbles, volume reverberation

When widely spaced bubbles are insonified, the acoustical cross-sections of the individuals simply add. By comparison with attenuation and dispersion experiments in which the air-to-water volume fraction, U, approached 10^{-2}, Feuillade (1995) has proved that, for realistic ocean distributions of bubble sizes and random spacings, the classical multiple scattering development for attenuation or dispersion of sound in a random-sized bubbly mixture, as used in this chapter, is appropriate.

Clouds of bubbles are produced naturally when large numbers of individual bubbles from breakers are entrained by the turbulence and Langmuir circulations under a breaking wave (Thorpe, 1982). These bubble clouds have been offered as an explanation of the significant backscatter of low-frequency sound ($f < 2$kHz) from the sea surface (Prosperetti et al., 1993) as well as the ambient low-frequency noise ($f < 500$ Hz) at sea. In the latter case it is postulated that the volume of the bubble cloud has a density and stiffness that allows it to oscillate as a huge "pseudo-bubble" with low-frequency modes.

When bubbles of random sizes are in a separated, close-packed, random, three-dimensional configuration as in the wake of a ship, there will be significant reflection at the pressure-release face of the bubble cluster. Consider the general situation:

In the ocean there is a continuous distribution of bubble sizes, so

$$S_{bs} = \left[\sum_N \Delta\sigma_n\right] = \int_a [n(a)\,da][\Delta\sigma_s(a, f)] \qquad (6.87)$$

where S_{bs} is the backscattering cross-section per unit volume for all bubbles contained in the sampled volume and $n(a)\,da =$ the number of bubbles of radius between a and $a + da$, per unit volume. Insert the differential cross-section for small ($ka < 1$) bubbles, $\Delta\sigma_s(a) = \sigma_s(a)/4\pi$ from (6.73), to get

$$S_{bs} = \left[\sum_N \Delta\sigma_n\right] = \int_a [n(a)\,da]\left[\frac{a^2}{[(f_R/f)^2 - 1]^2 + \delta^2(f)}\right] \quad (ka < 1)$$

$$(6.88)$$

---

* This section contains some advanced analytical material.

When $n(a)\,da$ is known, one can numerically integrate (6.88) for a given frequency, f.

A simple analytical integration of this type of equation was first provided by Wildt (1946), who reported on World War II studies of sound scattering and attenuation in ship wakes. In his solution it is assumed that the scatter is dominated by the resonant bubbles, and that neither δ nor $n(a)\,da$ change appreciably over the important narrow band of radii around the resonance radius. Then $n(a)$ may be factored out of the integration. Also, the substitution is made

$$q = (f_R/f) - 1 = (a/a_R) - 1 \qquad (6.89)$$

where q is a small number, and $dq = da/a_R$.

For the convenience of using a known definite integral, the integration over q is extended from $-\infty$ to $+\infty$, and we obtain the backscattering cross-section per unit volume, due to bubbles,

$$S_{bs} = (a_R)^2 n(a) \int_{-\infty}^{+\infty} \frac{a_R\,dq}{(2q)^2 + \delta_R^2} = \frac{\pi a_R^3 n(a)}{2\delta_R} \quad (ka < 1) \qquad (6.90)$$

A similar integration is used for the *extinction* cross-section. Numerical integrations have been compared with the approximation (6.90) by Medwin (1977), Commander and McDonald (1991), and Sarkar and Prosperetti (1994). The accuracy of the approximate integration depends on the assumed ocean variation of bubble density with radius, and the frequency of interest. In the cases studied, the errors are a few percent. Proper calculation of the backscatter for bubbles larger than $ka = 0.1$ would require integration over the directional radiation patterns as well.

6.6.2 Attenuation due to bubbles

The presence of upper ocean bubbles, each with absorption and scattering cross-sections, causes extra attenuation beyond that described in Section 2.4.

Bubbles of one size
Assume that we have water containing N bubbles of radius a per unit volume. Assume, also, that the bubbles are separated enough so that there are no interaction effects. Effectively, this is true when the separation is greater than the square root of σ_e. (Sometimes a stronger criterion is used – separation greater than the wavelength.)

If the incident plane-wave intensity is I_{inc}, the power absorbed and scattered out of the beam by each bubble is $I_{inc}\sigma_e$, where σ_e is calculated from (6.79). The spatial rate of change of intensity is

$$\frac{dI}{dx} = -I_{inc}\sigma_e N \qquad (6.91)$$

Integrating,

$$I(x) = I_{inc}\exp(-\sigma_e Nx) \tag{6.92}$$

After traveling a distance x, the change in intensity level will be

$$\Delta IL(\text{dB}) = 10\log_{10}\left(\frac{I(x)}{I_{inc}}\right) = -10\sigma_e Nx \log_{10}e \tag{6.93}$$

The spatial attenuation rate due to bubbles is

$$\alpha_b(\text{dB/distance}) = \frac{\Delta IL}{x} = 4.34\sigma_e N \tag{6.94}$$

where the units are generally α_b in dB/m, σ_e in m^2, and N in m^{-3}.

Bubbles of many sizes

When there are bubbles of several sizes, the number per unit volume can be defined in terms of the number in a radius increment (typically, one micron) as

$$n(a)\,da = \frac{\text{number of bubbles of radius between } a \text{ and } a + da}{\text{volume}} \tag{6.95}$$

The extinction cross-section per unit volume, S_e, for sound traversing a random mixture of non-interacting bubbles in the range $ka < 1$ is calculated by using (6.79) in the integration

$$S_e - \int_0^\infty \sigma_e n(a)\,da - \int_0^\infty \frac{4\pi a^2(\delta/\delta_r)n(a)\,da}{[(f_R/f)^2 - 1]^2 + \delta^2} \tag{6.96}$$

This S_e replaces $\sigma_e N$ in (6.94) to give the expression for the attenuation rate due to a mixture of bubbles of $ka < 1$,

$$\alpha_b = 4.34 S_e \text{ dB/m} \tag{6.97}$$

In some bubble determinations at sea, one seeks the bubble density $n(a)\,da$ from the results of an attenuation measurement by inverting (6.97) and (6.96). The simplest approximation is to assume that the major contribution to the attenuation is near the resonance frequency. As shown in the derivation of (6.90), Wildt (1946, p. 470) assumes that only bubbles close to resonance contribute to S_e, and that the bubble density and the damping constant are constant over that interval. In our notation, the approximate value is

$$S_e = \frac{2\pi^2 a_R^3 n(a_R)}{\delta_R} \tag{6.98}$$

Therefore, from (6.97) and using (6.98), in an extinction experiment an approximate value of the bubble density at the resonance frequency $n(a_R)$ is obtained from

$$\alpha_b = \frac{85.7 a_R^3 n(a_R)}{\delta_R} \tag{6.99}$$

Numerical integration will give a better value than this approximate theory. See Sarkar and Prosperetti (1994) for some possibilities.

A graphical solution can also be used to find S_e when one knows the dependence of bubble density on the bubble radius. At sea, it turns out that dependences from a^{-2} to a^{-4} are common.

A graphical calculation was shown in Fig. 6.15. Assume that the insonification is by a plane wave of frequency 50 kHz. In the graph at the left there is a plot of the single bubble extinction cross-section, σ_e, as a function of bubble radius (the left ordinate). In this same graph, there is a plot of bubble density $n(a)\,da$ that is assumed to vary as a^{-4} (the right ordinate). In the graph on the right the product of the curves, $\sigma_e n(a)\,da$ is drawn. The total values S_e will be given by the areas under the curves (when replotted on a linear graph, of course). The strong dominating contribution of $\sigma_e n(a)\,da$ near the resonance radius is quite clear. The dependence on $n(a)\,da$ has been discussed by Commander and McDonald (1991).

6.6.3 Sound speed variation with frequency and void fractions

Single radius bubbles

From (1.55), the speed of longitudinal waves in a fluid medium may be written as

$$c_A^2 = \frac{E_A}{\rho_A} = \frac{1}{\rho_A K_A} \tag{6.100}$$

where $K_A = 1/E_A$ is the compressibility of the ambient medium (fractional change of volume per unit applied pressure, pascals^{-1}) and ρ_A is its density. The presence of a small void-fraction of bubbles (i.e. fractional volume of gas in bubble form to volume of water) $< 10^{-5}$ has a negligible effect on the density, but it has a very significant effect on the compressibility.

In bubbly water the compressibility is made up of a part due to the bubble-free water, K_0, and a part due to the compressibility of the bubbles themselves, K_b:

$$K_A = K_0 + K_b \tag{6.101}$$

The compressibility for the bubble-free water is a function of the speed in bubble-free water,

$$K_0 = \frac{1}{\rho_0 c_0^2} \tag{6.102}$$

For now, assume all bubbles have the same radius a, and that $ka < 1$. The compressibility of the bubbles is found in terms of the displacement when insonified by an incident plane wave,

$$K_b = \frac{(\Delta v/v)}{\Delta P} = \frac{N\Delta v}{\Delta P} = \frac{NS\xi}{P_{inc}e^{i\omega t}} = \frac{NS^2}{m\omega^2[(-1 + \omega_R^2/\omega^2) + i R_M/(\omega m)]} \tag{6.103}$$

where N is the number of bubbles per unit volume, Δv is the change of volume, $S = 4\pi a^2$ is the surface area of each bubble, and ξ is the radial displacement of the bubble surface.

To simplify, we use $\delta = R_M/(\omega m)$ and also define the frequency ratio,

$$Y = \omega_R/\omega \tag{6.104}$$

After multiplying numerator and denominator by the complex conjugate, we obtain

$$K_b = \frac{N4\pi a(Y^2 - 1 - i\delta)}{\rho_A\omega^2[(Y^2 - 1)^2 + \delta^2]} = K_0(A - iB) \tag{6.105}$$

where

$$A = \frac{(Y^2 - 1)}{(Y^2 - 1)^2 + \delta^2} \frac{4\pi a N c_0^2}{\omega^2} \quad \text{and} \quad B = \frac{\delta}{(Y^2 - 1)^2 + \delta^2} \frac{4\pi a N c_0^2}{\omega^2} \tag{6.106}$$

From (6.100) and (6.105), the speed in bubbly water is

$$c_A = \left(\frac{1}{\rho_A K_A}\right)^{1/2} = \left(\frac{1}{\rho_A(K_0 + K_b)}\right)^{1/2} = \frac{c_0}{(1 + A - iB)^{1/2}} \tag{6.107}$$

where we assumed that $\rho_A \cong \rho_0$.

The propagation constant is

$$k_A = \frac{\omega}{c_A} = \frac{\omega(1 + A - iB)^{1/2}}{c_0} \tag{6.108}$$

Assume that the correction is small compared with unity, so that the first terms of the Taylor expansion may be used to replace the square root,

$$k_A = k_0\left(1 + \frac{A}{2} - i\frac{B}{2}\right) \tag{6.109}$$

The plane wave propagating through the bubbly medium is

$$p_{inc} = P_{inc}e^{ik_Ax} = P_{inc}\exp(-k_{Im}x)\exp[i(\omega t - k_{Re}x)] \tag{6.110}$$

where

$$\begin{aligned} k_{Im} &= k_0(B/2) \\ k_{Re} &= k_0(1 + A/2) \end{aligned} \tag{6.111}$$

The imaginary part of the complex propagation constant represents the attenuation of the wave: this was called α_b earlier. The real part is the wave number for the propagation of constant phase surfaces at the dispersive speed ω/k_{re}.

Calculations of α_b and Re$\{c\}$ are given by Farmer *et al.* (1998) for an assumed bubble density.

The speed in bubbly water is a function of frequency:

$$\text{Re}\{c\} = \frac{\omega}{k_{Re}} = c_0 \left\{ 1 - \left[\frac{(Y^2 - 1)}{(Y^2 - 1)^2 + \delta^2} \right] \left[\frac{2\pi a N c_0^2}{\omega^2} \right] \right\} \qquad (6.112)$$

It is useful to write the speed in fractional terms, volume of gas in bubble form divided by volume of water. This is called the void fraction, U:

$$U = N \left(\frac{4}{3} \pi a^3 \right) \qquad (6.113)$$

$$\text{Re}\{c\} = c_0 \left\{ 1 - \left[\frac{(Y^2 - 1)}{(Y^2 - 1)^2 + \delta^2} \right] \left[\frac{3 U Y^2}{2 a^2 k_R^2} \right] \right\} \quad (ka < 1) \qquad (6.114)$$

where $k_R = \omega_R / c_0$ is the value of k_0 at resonance.

Graphs of the relative change in sound speed predicted by (6.114) are plotted in Fig. 6.16 for two different densities of bubbles of the same radius. The dispersion curves show that at incident frequencies below

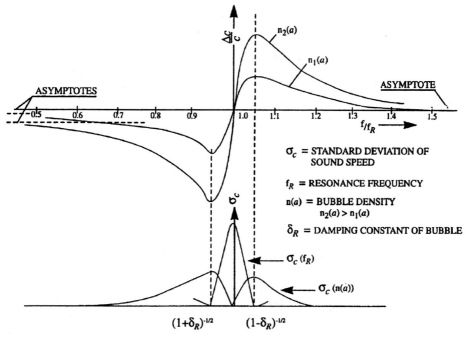

Fig. 6.16. *Above,* sound speed dependence on frequency for two different bubble densities in a liquid containing only one bubble size. *Below,* standard deviation of the sound speed dispersion as a function of change in bubble density, $\sigma_c[n(a)]$, and as a function of resonance frequency change due to variation of ambient pressure above the bubble population, $\sigma_c(f_R)$. (From Medwin *et al.*, 1975. See also Wang and Medwin (1975) for the theory.)

the bubble resonance frequency, the relative differential sound speeds, $\Delta c/c_0$, are below the value for bubble-free water, and approach a negative asymptote. Incident sounds of frequencies above the bubble resonance frequency reach a peak sound speed and then decrease to approach the speed in bubble-free water as the frequency becomes very large.

The limit of the dispersion curves (Fig. 6.16) at high frequencies is

$$c_{hf} = c_0 \left[1 + \frac{3UY^2}{2a^2k_R^2(1+\delta^2)} \right] \to c_0 \quad f \gg f_R \qquad (6.115)$$

Therefore, bubbles do not affect the sound phase speed *if the frequency is high enough*. This explains why commercial *sound velocimeters*, which use megahertz frequency pulses to measure the travel time over a fixed short path, do not recognize the dispersive effect of bubbles. Sound velocimeters provide $c_A = c_0$ even in bubbly water because at 1 MHz, $f \gg f_R$ for all significant bubble fractions in the sea. However, the instruments show "drop-outs" in bubbly water.

The limit at low frequencies is

$$c_{lf} = c_0 \left(1 - \frac{3U}{2a^2k_R^2} \right) \quad f \ll f_R \qquad (6.116)$$

Since ak_R is a constant for a given gas at a given depth (for an air bubble at sea level $ak_R = 0.0136$ from Section 6.4), the low-frequency asymptotic speed depends only on the void fraction, U. The simple approximations above are acceptable for $U \lesssim 10^{-5}$ and $ka \lesssim 1$.

Multiple radii bubbles
Using the same approximations, the generalization to the multiradii bubbles of the ocean medium is accomplished by replacing N by $n(a)\,da$ and U by $u(a)\,da$ in the previous development. Because all contributions to the compressibility are very small quantities, they add linearly, and the speed of sound in the bubbly region can be written in terms of the integral over all radii:

$$\mathrm{Re}\{c\} = c_0 \left\{ 1 - \int_a \frac{(Y^2-1)\dfrac{3Y^2u(a)da}{2a^2k_R^2}}{(Y^2-1)^2 + \delta^2} \right\} \qquad (6.117)$$

For values of U greater than 10^{-4}, the approximation of (6.109) is inadequate, and consequent predictions of (6.114) are incorrect. Hall (1989, Appendix 2) and R. Goodman (personal communication, 1997) have proposed alternative expressions that work for $U > 10^{-4}$ provided that $n(a)\,da$ is known over the entire range of the integration. The low-frequency asymptotic value can be obtained by use of Wood's Equation (see below), which is valid for *all* void fractions and does not require a knowledge of $n(a)\,da$.

Void fractions in bubbly water: Wood's Equation and void fractions
When $n(a)\,da$ is not known, there is a simple technique to find U for any void fraction because the low-frequency asymptotic value of the speed in Fig. 6.16 is dependent only on the fluid "zero frequency" compressibility due to the presence of bubbles.

Following Wood (1955), let ρ_b and ρ_w represent the densities of air in bubbles and in water, respectively, and let E_b and E_w be the bulk moduli of elasticity of air and water. Since U is the fraction of air by volume, $(1-U)$ is the fraction of water by volume. For bubbly water, we first express the average density, ρ_A, and the average elasticity, E_A, which are needed to determine the speed at low frequencies, as follows:

$$\rho_A = U\rho_b + (1-U)\rho_w \tag{6.118}$$

$$\frac{1}{E_A} = \frac{U}{E_b} + \frac{1-U}{E_w} \tag{6.119}$$

$$c_{lf} = \sqrt{\frac{E_A}{\rho_A}} = \sqrt{\frac{E_b E_w}{[UE_w + (1-U)E_b][U\rho_b + (1-U)\rho_w]}} \tag{6.120}$$

For bubble-free water, E_w is calculated from ρ_w and c_w given in (2.30). For the bubble gas, one uses $c^2 = \Delta p/\Delta \rho$ with $p\rho^{-\gamma} =$ constant for adiabatic propagation in a gas to obtain $E_b = \gamma P_A$ where γ is the "ratio of specific heats."

When the low-frequency asymptote of the sound speed is experimentally determined in any mixture of bubbles, U can be calculated from (6.120), which is "Wood's Equation."

The approximate (6.116) for non-interacting bubbles agrees with Wood's Equation when $U \leq 10^{-5}$. Feuillade (1996) has shown that, although interactions are important if the bubble sizes are identical, there is no significant interaction effect for a mixture of bubble radii, as exists in the sea.

6.7 Active measurements of bubbly water

6.7.1 Techniques for linear bubble counting

Remote sensing by backscatter
Because bubble density is sensitive to ocean parameters, its measurement can reveal a wealth of information about physical, chemical, and biological processes in the sea and on the sea floor.

Determinations of bubble density by backscatter sometimes use an upward-looking source and receiver on the sea bottom and follow an analytical development such as in Section 6.6.1. Sometimes the transducers are on a ship, underwater vehicle, or buoy (e.g., Løvik, 1980; Vagle and

Farmer, 1992). Segments of the backscattered sound are received by opening and closing an electronic gate in the receiving system. The advantage of remote sensing is that the bubbly medium is undisturbed by the intrusion of equipment. A major disadvantage is that the insonified volume is generally quite large and poorly defined. The axial extent is limited by the windowing procedure; the cross-sectional extent depends on the divergence of the sound beam and the range. Therefore the backscatter comes from a spherical cap which covers more than one depth. Since bubble populations are sensitive functions of depth, the divergence of the beam produces an undesirable averaging over a range of depths. Another disadvantage is that it is necessary to correct for the specific attenuation of the signal before and after it acts in the scattering region.

In situ *sensing*

Several techniques, originally used in physical acoustics laboratories to determine the ultrasonic characteristics of liquids or solids, have been brought to sea to measure the acoustical characteristics of bubbly water. "Bubble counting" depends on the distinctive dispersive sound speed or attenuation or backscatter in the bubbly parts of the ocean.

The pulse–echo technique

Several cycles of a high-frequency sound are emitted from a piston transducer and returned by reflection from a nearby plane-rigid plate (Fig. 6.17, left) (e.g., Barnhouse *et al.*, 1964; Buxcey *et al.*, 1965; and Medwin, 1970.). After sending, the source transducer is switched automatically to receive, and the decreasing amplitude of the series of reflections of the ping is measured (Fig. 6.18). The attenuation due to normal absorption and diffraction of the ping is first calibrated in a tank

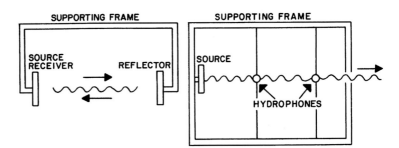

Fig. 6.17. Experimental techniques for measuring bubble densities and void fractions at sea. The set-up at the left is drawn for a pulse–echo system. When the source and receiver are at separation less than the plate diameters, the device may also be used as a standing wave resonator system. The scheme at the right is for measuring the amplitude and phase changes for propagation past two or more hydrophones. The device is oriented vertically or horizontally.

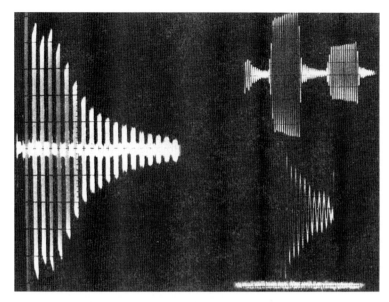

Fig. 6.18. Oscillograms of pulse–echo signals. The multiple echoes at the left are for sound frequency 200 kHz. Evidence of the near field is seen in the first three or four echoes. The right top shows two echoes of a 30 kHz signal, with the transient backscatter between them. Right center is the amplified backscatter between echoes. Right bottom is the background noise. (From Medwin, 1970.)

of non-bubbly water. The increased attenuation at sea is attributed to bubble absorption and scatter of sound out of the system (Section 6.6.2), and the average bubble density is calculated for the region traversed by the pulse. The number per unit volume of many different sizes can be "counted" by using several frequencies.

An added feature of the pulse–echo technique is that backscatter can be detected separately, and measured, between the times of the echoes (Fig. 6.18). The backscatter occurs while the ping is traveling away from the transducer. It has the appearance of a tapered signal because the solid angle viewed by the receiver becomes smaller as the ping moves toward the reflector.

A third feature of the pulse-echo system is that the time between pulses can be used to determine the speed of the pulse, provided the phase shifts at the transducer and reflector have been calibrated.

In the pulse–echo system it is necessary to have clean pulses that are replicas of the applied voltage, without the transient growth and decay characteristic of most transducers. The desirable low Q source used in the Buxcey *et al.* (1965) experiments was a "home-made" Mylar electrostatic transducer. This has an output acoustic pressure proportional to the applied voltage, and a receiving voltage sensitivity that is inversely proportional to the acoustic pressure at the face over a wide range of frequencies. When used in the send–receive mode, its electro-acoustical transducer performance was therefore independent of frequency up to about 200 kHz in this version.

The resonator technique

If the source–receiver plate and reflector (Fig. 6.17) are close compared with their diameters, and the sound is CW, the source and reflector act as a leaky one-dimensional multifrequency resonator. When a "white noise" CW sound that covers a broad range of frequencies is radiated from the source, a series of standing waves (Section 1.5.5) is formed in the resonator. The frequencies of the standing waves are a nearly harmonic series. The amplitudes and breadths of the standing wave responses at these frequencies, measured at the plate pressure antinode, depend on the system attenuation and the bubbly medium absorption between the plates. The resonance Q at sea is compared with the laboratory calibration to obtain the bubble density within the resonator at that frequency (see Fig. 6.19). Breitz and Medwin (1989) describe the equipment and the calculation. Medwin and Breitz (1989) give experimental results and comparisons with other measurements at sea. In addition to the bubble density determination, it is also possible to calculate the dispersion of the sound speed from the absolute frequency of the harmonics and the corrected separation of the plates.

Cartmill and Su (1993) have used the resonator technique in a very large wave-making facility. They show that the number density of bubbles from 35 to 1200 μm radius produced by breaking waves in salt water is an order of magnitude greater than in fresh water for the same wave system. Farmer et al. (1998) have mounted five resonators (depths 0.7, 1.3, 1.9, 3.3, and 5.5 m) on a vertical frame to obtain simultaneous sound speed dispersion and void fractions for radii 20 to 500 μm during 12 m/s winds at sea.

Fig. 6.20 is an example of the range of bubble densities that have been measured under breaking waves at sea. Even in quiescent coastal waters, densities as great as 10^3 to 10^5 bubbles within a 1 micron radius (say, between 30 and 31 microns) per cubic meter are not uncommon near the sea surface and in coastal waters.

The direct transmission technique

The simplest of all techniques uses a source to radiate sound to two or more nearby hydrophones (Fig. 6.17, right). The signal received at each hydrophone is analyzed by Fast Fourier Transform to yield the amplitude and phase. The ratio of the amplitudes gives the attenuation; the difference of the phases, corrected for the whole number of periods,

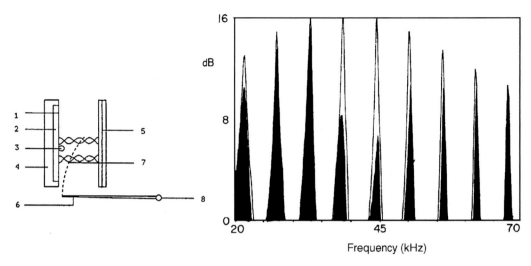

Fig. 6.19. *Left*, schematic of the standing wave during calibration of the resonator at the fourth "harmonic" as used by Breitz and Medwin (1989). The source is a solid dielectric flat plane capacitor, 1, backed by an aluminum plate, 2, in a plexiglas base, 4. The receiver is a probe hydrophone, 3. The reflector consists of a constrained damping layer between two aluminum plates, 5. A counted array of bubbles, 7, from a capillary, 6, is swept through the standing waves of the resonator by oscillating the capillary tube. *Right*, nine resonance peaks from 20 to 70 kHz during calibration with bubbles (blackened) and without bubbles (outlined). The changes of heights and breadths of the peaks are used to determine the sound absorption at these several frequencies; for the calibration, the number of bubbles is calculated and compared with the visually counted bubbles.

provides the speed of propagation. When the propagation experiment uses a multi-frequency signal such as a sawtooth wave, the bubble densities and dispersion can be evaluated simultaneously for several bubble radii. See Medwin (1977).

Comments on in situ *bubble-counting techniques*

The direct transmission technique involves measurements of both extinction (scatter and absorption) and dispersion. The pulse–echo technique has a lesser dependence on scatter than direct transmission because some of the scattered energy is retained between the plates. The nearly closed resonator leaks only a small fraction of the scattered energy and is thereby almost a pure bubble absorption device. The resonator largely avoids the "noise" due to scattering from bubbles that are larger than the resonating bubbles (see Fig. 6.14, left).

All bubble-counting techniques must be calibrated by use of a known bubble stream, which may be obtained by electrolysis or from a pipette

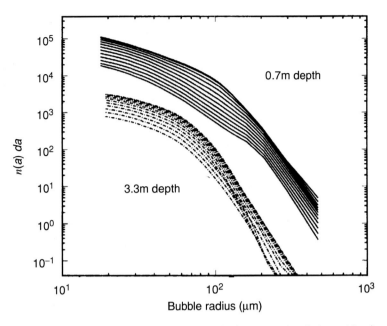

Fig. 6.20. Resonator determinations of bubble densities at sea during a 12 m/s wind in the Gulf of Mexico. The ordinate is number of bubbles/m^3 in a 1 micron radius increment. Simultaneous measurements were made at 36 frequencies to yield 36 bubble densities for radii from 15 μm to 450 μm every 2 seconds during a 2-minute interval at each of five depths (not all results are shown). Data at the two depths show increasing air fractions due to breaking waves as time progressed. The curves are smoothed between the data points. (Farmer and Vagle, 1997.)

fed with pressurized gas. Results of field measurements are more trust-worthy when two of the bubble effects are measured and compared at the same time – for example, scatter and extinction, absorption and dispersion, extinction and dispersion.

6.7.2 Dependence of bubble densities on physical, biological, and aerosol sources

Bubbles are generated by many physical, biological, and chemical actions. A very large range of bubble densities have been measured near the ocean surface under different conditions. Measurements, to date, have demonstrated the dependence of bubble densities on water depth, bubble depth, time of day or night, wind speed, rainfall, cloud cover, season of year, presence of sea slicks. Far less is known about bubbles with biological origins.

Results obtained in coastal waters are: (*a*) increased bubbles caused by increased breaking waves at higher wind speeds; (*b*) seasonal dependence of increased biological activity in coastal waters; (*c*) more plentiful smaller bubbles in daylight hours due to photosynthesis; (*d*) more common larger bubbles at night, possibly caused by offshore winds ("sea breeze") dropping continental aerosols, which trap bubbles when they fall into the sea, or biological activity on the sea floor.

There are several acoustic consequences of bubbles at sea. Bubbles cause frequency-dependent ray refraction near the sea surface. Because the bubble density generally increases as rays approach the surface, low frequencies are refracted and surface-reflect at more nearly normal incidence. There is a lesser effect in the opposite direction and for higher frequencies. Waveguide modes (Chapter 8) are distorted by the varying dispersion near the ocean surface.

The temporal variation of bubble densities and bubble radii near the ocean surface causes fluctuations in sound speed. An example is in Fig. 6.21. When the number of bubbles of a given radius changes, the dispersion curve for that bubble radius moves up and down. Additionally, when the pressure over a bubble changes (e.g., during passage of a wave crest or trough), the bubble resonance frequency changes, and the speed may move from the positive to the negative region of differential speed. The frequency spectrum of the sound fluctuations mimics the frequency spectrum of the ocean surface wave displacements. That is, the spectral slope of the sound fluctuation is F^{-5} where F is the ocean wave frequency in consonance with the Pierson–Moscovitz ocean wave height spectrum (Medwin, 1974).

The temporal fluctuation of near-surface sound speed has been measured for several frequencies as shown in Fig. 6.21. At frequencies over 25 kHz, the speed is already close to the high-frequency asymptotic value. At lower frequencies, the changing speeds, which are substantially below the bubble-free speed, are evidence of temporal changes in the void fraction.

6.7.3 Collective bubble oscillations and low-frequency noise

When waves break, particularly in the case of plunging breakers, the cloud of bubbles that is produced is a region of much reduced compressibility and slightly reduced density. The bubble cloud is capable of resonating with frequencies that depend on the shape, dimensions, average compressibility, and average density, just as single bubbles can. The theory that ambient noise in the region $f < 500$ Hz (Section 5.5.1) may be partly due to collective oscillations of bubble patches (e.g., Prosperetti,

Fig. 6.21. *Above*, temporal variation of sound speed ΔC and void fraction U, obtained by direct measurement at seven frequencies at depth 50 $\pm$ 10 cm during 8 m/s wind. (From Lamarre and Melville, 1994.) For the lower frequencies, 6, 10, 15, 20, and 25 kHz, the asymptotic low speeds were as much as 120 m/s below the value for bubble-free water. A record low sound speed of 700 m/s and void fraction of 1.6 $\times 10^{-4}$ was observed at 5 kHz at another time. *Below*, seven rapidly changing dispersion curves obtained by resonator technique during 30 seconds at depth 0.7 m with a 12 m/s wind, measured by Farmer *et al.* (1998).

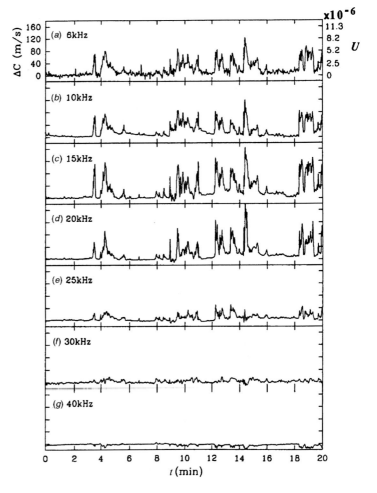

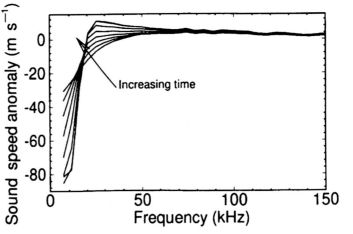

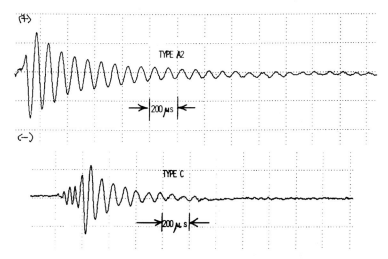

Fig. 6.22. Pressure radiated by damped microbubble oscillations observed in laboratory study of breaking waves. In Type A2 (*above*), the pulsation frequency was 10.4 kHz (radius 312 μm) and there were two decay rates. The fact that the first four cycles decayed at a higher than normal rate was explained by Longuet-Higgins (1992) as due to shape oscillations. Initial amplitude was 0.36 Pa, on axis at 1 m. There are two proposed explanations for the early part of Type C (*below*). Possibly it was a bubble that was rapidly moving away from the surface; this would initially make its amplitude increase owing to the dipole axis becoming larger, and then cause the frequency to decrease due to its increasing distance from the surface (see Section 6.5). Another possibility is that the bubble was initially non-spherical and that, as it became spherical, its radiation became more efficient and its frequency became lower. Initially frequency was 25.6 kHz; final frequency, 11.5 kHz. Peak pressure on axis at 1 m was 0.33 Pa. (From Medwin and Beaky, 1989.)

1988) has been supported by several experiments in the laboratory and at sea (e.g., Loewen and Melville, 1991).

6.8 Sea surface microbubble production

6.8.1 Bubbles from breaking waves

The ambient underwater sound that has been known since World War II as "Knudsen sea noise" and which, for 40 years, had been attributed to miscellaneous turbulent actions caused by wind at the sea surface is now known to be comprised of the damped radiations from newly formed microbubbles created by spilling breakers (Medwin and Beaky, 1989; Loewen and Melville, 1991). Figure 6.22 shows two examples of damped acoustic pressures due to what have been poetically called "screaming infant microbubbles" created by a laboratory, fresh water-spilling breaker. The bubble radius is determined by reading the zero crossings on an oscilloscope trace. The observed damping rate, D_b, is generally within 10 percent of the theoretical damping rate for fresh water. It is calculated from the δ_R in Fig. 6.13.

At the moment of their creation, the infant microbubbles are shock-excited by the sudden radial inflow of water and the simultaneous application of surface tension. They then show damped oscillations. Sometimes there are two damping rates for a bubble; a higher rate of damping due to shape oscillations is followed by the theoretically predicted lower rate. Several other bubble types have been identified, as well as larger bubbles spawning smaller bubbles and bubbles that split into nearly equal parts whose radiation interferes to cause amplitude pulsations called "beats."

By placing a hydrophone close to the breaking wave, an individual oscillation can be isolated, and the bubble radius can be determined from the frequency and the damping constant. This was done during intermittent spilling breakers of a wind-driven laboratory fresh water surface by Medwin and Beaky (1989) and at sea by Updegraff and Anderson (1991). By using two calibrated hydrophones, the differential time of arrival can be employed to deduce the position and orientation of the bubbles, so that their axial source pressures can be calculated (Daniel, 1989).

One can duplicate the 5 dB/octave slope of what is called the Knudsen "wind" noise by analyzing sound from spilling breakers generated by a *plunger* with no wind at all; a rather conclusive proof that Knudsen noise is not directly attributable to wind, although it is fairly well correlated to wind speeds! (See Fig. 6.23.)

Because it is easy in the laboratory to identify the individual damped oscillations, one can actually count the bubbles within a 1 micron radius increment, produced per square meter of water surface (Medwin and Daniel, 1990). (See Fig. 6.24.) Supported by similar work at sea, it is now clear that the Knudsen sea-noise spectrum from 500 Hz to probably 50 000 Hz is due to the cumulative sound of damped individual bubble oscillations of a large range of sizes, intermittently created by breaking waves.

In the fresh water study, breaker bubbles of radii from 0.050 mm to 7.4 mm were observed with resonance frequencies 65 kHz to 440 Hz, respectively (Fig. 6.24). Since it is very close to the surface, an oscillating bubble combines with the phase-inverted reflection from the pressure-release surface to radiate as a dipole.

The peak production was found to be about 6 bubbles per m^2 surface area in a 1 micron radius increment at radius 150 mm for these small "spillers." The total gas encapsulated by the laboratory breaking wave was 23 cm^3/m^2, calculated simply from the number of bubbles and their volumes. This number is of interest to those concerned with the important question of gas exchange at the water surface.

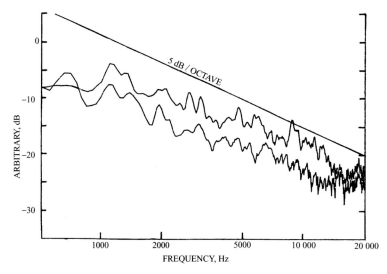

Fig. 6.23. Laboratory breaking wave spectrum (jagged lines), compared with Knudsen sea-noise spectral slope 5 dB/octave. The lower graph was obtained during continuous recording, the upper graph during intermittent recording when breakers were seen over the hydrophones. Breakers were driven by a plunger source with the hydrophone 24 cm below the surface. The *height* of the comparison straight line Knudsen spectrum is arbitrary. (From Medwin and Beaky, 1989.)

Loewen and Melville (1991) were able to duplicate the surface sound spectrum of Fig. 6.23 by combining the bubble production data of Fig. 6.24, with the assumption that the relative bubble pulsation amplitudes (ξ/a) were 0.015 and the depth of the bubbles was 1.0 cm. This is presented in Fig. 6.25. Furthermore, they were able to show that, with certain simplifying assumptions, the inverse problem could be solved to yield the number of bubbles in a given radius increment calculated from the spectral sound of the surface.

6.8.2 Bubbles from rainfall

Precipitation is an additional source of bubbles near the sea surface. The impact of rain or hail will first create an impulse of radiation (Fig. 6.26). More important, this is often followed, after a few milliseconds, by the birth of one or more bubbles, which then radiate sound during their damped oscillations. The newly created microbubble is typically the source of *dipole* radiation because its wavelength is generally large compared with its proximity to the ocean surface (Section 3.1.3).

A dependence on salinity, temperature, and water surface slope has been demonstrated by Jacobus (1991) and Miller (1992).

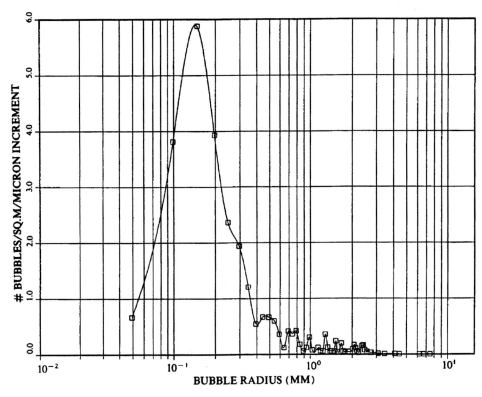

Fig. 6.24. Bubbles produced by a spilling breaker in fresh water. About 500 bubbles were counted from 10 breakers covering surface areas of an average of 320 cm² for this graph. (From Medwin and Daniel, 1990.)

Fig. 6.25. Comparison of a predicted sound spectrum (thick line), calculated from measurements of bubble production, with the measured laboratory sound spectrum (dashed line) for breaking waves. The model assumed that the fractional radial displacement to radius of the bubbles, ξ / a, was 0.015 and the depth of the bubbles was 1 cm. (From Loewen and Melville, 1991.)

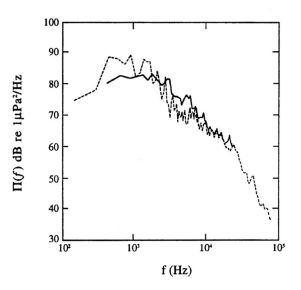

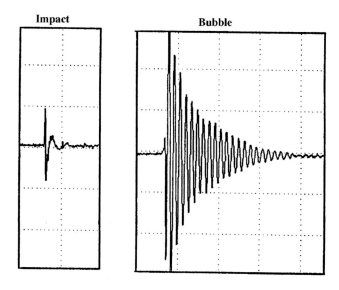

Fig. 6.26. The pressure signal for the impact sound (*left*) and bubble sound (*right*) caused by a small water drop of diameter 0.83 mm entering the water at normal incidence and terminal velocity. The time between impact peak and bubble peak was 17.7 ms. The spacing of the time grid is 400 μs. (From Medwin *et al.*, 1992.)

The shape of the underwater sound spectrum provides a characterization of the type of cloud, stratiform or cumuliform, from which the rain came. The meteorological characterization is not really mysterious: stratus are low clouds that produce predominantly small drops of rain, i.e., "drizzle." These cause small oscillating bubbles and a narrow spectrum with a predominant peak around 15 kHz. Cumulonimbus clouds have strong vertical air currents that allow raindrops to grow to large diameters before they fall into the sea, where they create larger bubbles and radiate a broad sound spectrum.

One can obtain the drop size distribution (and total rainfall rate) by inversion from the spectrum of the underwater sound (Nystuen, 1998). See Chapter 11 by Nystuen for details of more recent research.

6.9 Bubbles in sediments

By the 1950s, seismic exploration geophysicists had extended their activities to near offshore and lakes. In some areas, instead of getting the usual reflection records that showed subsurface structure, the records "rang like a bell" (Werth, Liu, and Trorey, 1959; Levin, 1962). Levin's extensive measurements in Lake Maracaibo, Venezuela, showed that there was near-perfect reflection from a pressure-release bubbly layer at the ocean floor. Independently, signal theorists in competing laboratories described the ringing as due to the upward traveling reflection signals being multiply reflected in the water layer. Clay (1991) gives an elementary treatment of optimum methods used to sort out the signal. Sonar detection of bubbles rising through the water column (McCartney and Bary, 1965) awakened geophysicists and acousticians to the fact that there are significant inclusions of bubbles in ocean and lake sediments. Several thousand gassy areas have now been identified in the sediments of the Gulf of Mexico, Southern California coastal regions, and the Bering Sea (Anderson and Bryant, 1989); undoubtedly, many others will be found as geophysical prospecting continues in the oceans of the world.

The acoustical consequences of bubbly sediments are substantial: the sound velocity can be significantly lower than for the water medium, and even more divergent from the expected sediment velocity; the attenuation, principally due to scattering through the sediment, can be great enough to prevent sound penetration in geophysical oil exploration. It is speculated that gas hydrates elected from the sea floor are a significant source of bubbles and thereby a marine source of atmospheric gases.

See the definitive paper by Lyons *et al.* (1996) which lays out the reason for bubble formation and the effects on sound propagation.

Illustrative problem

A body, presumably a big fish, is detected at a range of 1 km by backscatter of 20 kHz sound that has an $\text{SPL}_{bs} = +80$ dB re 1 μPa. Assume isothermal water and that the sonar source level is 220 dB re 1 μPa. What is the target strength of the fish?

Solution. To calculate the target strength we need the total attenuation loss. From Fig. 2.11 we get $\alpha \simeq 3 \times 10^{-3}$ dB/m. Therefore the transmission loss is

$$
\begin{aligned}
\text{TL} &= 20 \log \frac{R}{R_0} + \alpha R \\
&= 20 \log \frac{10^3}{1} + (3 \times 10^{-3})(10^3) \\
&= 63 \text{ dB}
\end{aligned}
$$

The target strength is

$$
\begin{aligned}
TS &= \text{SPL}_{bs} - SL + 2\text{TL} \\
&= +80 - 220 + 2(63) \\
&= -14 \text{ dB re 1 m}^2
\end{aligned}
$$

Problems

6.1 Show that the peak backscattering cross-section at bubble resonance is greater than the value at high frequencies ($f \gg f_R$) by
the factor $(1 + \delta_{RF}^2)/\delta_{RF}^2$.

6.2 A broadband sound signal is backscattered principally at 5 kHz
from 100 m depth. Assuming that the principal scatterer is an air-
filled, spherical swimbladder of a fish, and that the fish is neutrally
buoyant:

    (a) What is the equivalent radius of the bladder?

    (b) What is the estimated weight of the fish?

6.3 Calculate the backscattering cross-section of Sergestidae at 30 kHz,
assuming a density ratio $g = 1.03$ and an effective radius $a' = $
6 mm.

6.4 A near-surface experiment at 300 kHz yields $S_v = -95$ dB. Simultaneous sampling reveals 10^7 diatoms per cubic meter and 1
copepod per cubic meter. Assuming that the diatoms have a radius
of 10 μm, the copepods $a' = 1$ mm, and $e \simeq 1.25$, $g = 1.02$ for
both plankton, decide which scatterer is the probable cause of the
S_v and how many are actually present in a cubic meter.

6.5 Assume that surface tension is a significant restoring force, and
recalculate the stiffness and the resonance frequency. Plot the effect
as a function of bubble radius.

6.6 How will a bubble skin of detritus affect the resonance frequency?

6.7 Plot the dependence of bubble breathing frequency on depth. Calculate the correction needed for resonance radius, when a constant
frequency is used at various depths.

6.8 Compare the peak backscattering length at resonance, given in
Fig. 6.12, with the correct value obtained by considering viscous
and thermal damping. Make the comparison for the three sample
frequencies 1 kHz, 10 kHz, and 100 kHz.

6.9 Fig. 6.15 shows a graphical calculation of the effect of an a^{-4}
bubble distribution. Complete the calculation by determining S_e
for the example given. Warning: Use a linear graph.

6.10 Redraw Fig. 6.15 for the case of 100 kHz insonification. Comment.

Answers to some problems

6.1 At resonance $\sigma_s \text{ (res)} = a^2/\delta_{RF}^2$

       At $f \gg f_R$ $\sigma_s \, f_R = a^2/(1 + \delta_{RF}^2)$

       ratio $= (1 + \delta_{RF}^2)/\delta_{RF}^2$

6.2 (a) $a = 3.25(1 + 0.1z)^{1/2}/f_R'$, $z = 100$ m

       $a = 2.2 \times 10^{-3}$m, 0.22 cm

(*b*) Assume fish $\sim$ 10% bone and 90% flesh.
Using bone and flesh density, $\rho_{\tilde{f}} \tilde{=} 1.1\rho_0$.

6.3 $\sigma_{bs} \simeq 8.5 \times 10^{-9}$ m^2

6.4 Copepods. Approximately 0.05 per m^3.

Further reading

See Chapters 10, 16, and 17 for descriptions of recent and current acoustical measurements of marine bodies and bubbles. The chapters include on-going studies of droplets and bubbles produced by physical processes as well as detailed passive and active acoustical evaluations of biological presence and activity in shallow water and in the near-surface deep sea.

Blanchard, D. C. and Woodcock, A. H. (1957). Bubble formation and modification in the sea and its meteorological significance. *Tellus* **9**, 145–58.
A landmark paper describing the first near-shore measurements of bubbles produced by breaking waves, with discussion of the significance of bubbles, and the droplets that they produce, in the "seeding" of clouds.

Leighton, T. G. (1994). *The Acoustic Bubble*, 613 pp., Academic Press.
A "*Tour de force*" monograph on the significance of bubbles in its very many aspects including hydrodynamic and acoustic cavitation, biological effects of ultrasound, ocean "populations," sonochemistry, and industrial applications.

Chapter 7
Ocean bioacoustics

Summary

It has been many decades since innovative, commercial fishermen learned to use sound in their active sonar hunt for schools of fish. More recently, curious "ocean tourists" have listened, excitedly but without understanding, to the "moans" and "clicks" of the great whales. Nevertheless, at the beginning of the twenty-first century, the vast biological world beneath the sea surface is largely unknown. Each year brings revelations of the lives and activities of plankton, fishes, and sea mammals. The scientific use of sound frequencies from a few hertz to several megahertz, at ranges from centimeters to thousands of kilometers, is just beginning to enhance the marine biologist's understanding of life in the sea. A diverse catalog of sonars is now available for these studies.

Following this introductory chapter, readers specifically interested in the uses of sound in marine biology may wish to skip to Chapters 12–17, written by some of the world's leading ocean scientists, where we present a sampling of the remarkable progress currently being made by actively probing with, and passively listening to, sounds in the seas.

Contents

* This section contains some advanced analytical material.

7.1 Criteria for marine bioacoustical detection

When physicists and engineers study the interaction of sound with marine life, they create simple acoustical models to "explain" or "understand" or "interpret" the results of measurements. Their plankton, fish, and sea mammals are constructed of geometrical elements such as spherical bodies, or straight or curved cylindrical bodies, which may or may not enclose spherical or cylindrical swimbladders. In this way, a minimum number of elements are used to match a set of experimental data and extrapolate to unmeasured realms. (The Austrian physicist–philosopher Ernst Mach, whose name is enshrined in descriptions of the speed of supersonic vehicles, called this method "economy of thought.") Abiding by this philosophy, the following simple presentation (Table 7.1) shows the effective and optimum marine bioacoustical detection frequencies at **sea level**. It is based on the actual length, L, of the mammals and the fishes, and the equivalent radius of the plankton (column two). Those bodies that do not contain resonating bubbles are shown in column three and those that do, in column 4.

When the fish or plankton does not contain a bubble, or the cavity is lipid-filled, the minimum frequency for significant backscatter of sound from the marine body (third column) depends on the equivalent spherical radius of the body. In that case, the Rayleigh Criterion (Section 6.2.2) for significant wave scatter, $ka \geq 1$, (where a is the effective radius and k is the wave number $= 2\pi f/c$,) will determine whether there can be easily detectable backscatter of the frequency $f = kc/(2\pi)$. For $ka \geq 1$ the total acoustical scattering cross-section, σ_S, is approximately equal to the geometrical cross-section, πa^2, of a spherical body. The parameters of Table 7.1 have been calculated in terms of the actual animal length, L; equivalent spherical radius, $a_{es} = 0.1L$; swimbladder equivalent cylindrical radius, $a_{ec} = 0.04L$; swimbladder length $= 0.3L$; assuming swimbladder volume $= 5\%$ of fish body volume.

Detection frequencies for those bodies that contain bubbles which may resonate to the incoming sound are in the fourth column. Recall Section 6.2. Note that when the body contains a swimbladder the possibility and frequency of bubble resonance depends on whether the animal cavity is gas-filled, and the depth of the fish. For example, Pacific hake, $L = 40$ cm contains a swimbladder which would be resonant at 200 Hz

Table 7.1 *Marine bioacoustical detection criteria*

Sound frequencies for active detection of marine animals and plankton

Marine animals and plankton	Length, L, or equivalent radius, a	Effective detection frequencies for non-resonant bodies	Optimum detection frequencies for bodies with resonant bubbles
Largest mammals, whales	$L \geq 2$ meters	$f \geq 120$ Hz	$f \geq 1$ Hz
Larger fishes: cods, tuna, grenadiers, hake, salmon	2 meters $\geq L \geq 20$ cm	1.2 kHz $\leq f \leq 12$ kHz	15 Hz $\leq f \leq 600$ Hz
Smaller fishes: anchovy, shrimp	20 cm $\geq L \geq 2$ cm	12 kHz $\leq f \leq 120$ kHz	150 Hz $\leq f \leq 6000$ Hz
Zooplankton: euphausiids, siphonophores	$L < 2$ cm; $a \leq 1$ cm	120 kHz $\leq f \leq 1200$ kHz	1.5 kHz $\leq f \leq 60$ kHz

near the sea surface and 900 Hz at 200 m depth (R.W. Aleso, personal communication).

Mammals, fish, and plankton that carry gas bubbles for flotation are particularly easy to detect if the search frequency is close to the bubble resonance frequency. The vast number of bubble-carrying species of marine animals was summarized many years ago by Marshall (1970) when he pointed out that a gas-filled swimbladder is present in the adult stage of species comprising about a third of the mesopelagic fauna (*ca.* 150 to 1000 m depth), notably in myctophids (*ca.* 180 to 200 species), stomiatoids (*ca.* perhaps 250 species), gonostomatids (excluding cyclothone, spp.), hatchetfish, most trichiuroids (*ca.* 40 species), and some melamphaids. He also suggested that the swimbladder is either absent or regressed in the adults of all (*ca.* 150) species of the bathypelagic fishes (1000 to 4000 m depth). Marshall declared that, except for certain squaloid shars, chimaeroids, aleocephalids, and ateleopids, fishes of the benthopelagic fauna (living near the bottom) have a well-developed, gas-filled swimbladder, and that of some 750 species, the main groups are rattails (Macrouridae, *ca.* 300 species), deep-sea cods (Moridae, *ca.* 70 species), and brotulids (*ca.* 250 species).

Marshall's respected catalog has been out of date for some time and there are continuing efforts to update it.

In 2002, the "Consortium for Oceanographic Research" (CORE) began to coordinate a 10 year "Census of Marine Life" (CoML) to "promote and fund research assessing and explaining the diversity, distribution, and abundance of species throughout the world's oceans." This very large, intercontinental, interdisciplinary program is expected to use acoustics as one of its most important tools. See internet address: www.coreocean. org.

7.2 Operation of a simple sonar

The introduction of digital recording and data analysis broadened the range of usefulness of sonars so that a single instrument may be able to do several related tasks. Digital software has replaced many of the analog operations in sonar systems, and digital signal processing has improved the adaptability of a system to new tasks.

Collaborations between marine biologists and acousticians have had profound effects on the way marine biologists work and visualize their data. These collaborations have changed the bulk of sonar applications from military to civil. There are more sonars on sport and commercial fishing boats than there are on naval vessels.

Traditionally, bioacousticians make sonar transects and measure backscatter. A qualitative interpretation is that high-frequency sonar transects are sensitive to the distributions of small animals, whereas low-frequency transects show the large animals. But quantitative interpretation of sonar data requires a more detailed knowledge of the scattering of sound waves by animals. Laboratory studies of the scattering of sound by single animals are bridges that take us from mere volume scattering measurements to the goal of size and density distributions obtained by inverse methods. Acoustical oceanography brings new tools to marine biology and limnology and provides challenging problems for acousticians.

Although we use the terms "marine biology" and "oceanography," we mean to include studies in "fresh water" and "limnology." To biologists, the salt water (marine) environment supports life forms that are different from those in fresh water lakes and rivers. To acousticians, the main differences between salty and fresh water are in the physical parameters that affect backscatter: sound absorption, sound speed, and water density.

Sonar transducers may be pointed in any direction and placed at any depth. The fish or animal may have any orientation relative to the sonar beam and may be in any part of the beam. For a consistent geometry, we use vertically downward transducers in many illustrations because almost all bio-acoustic surveys use them. The data analysis is the same for transducers that point in any direction.

A use of a sonar to measure fish echoes is sketched in Fig. 7.1. The sonar transmits a ping that has the carrier frequency f_c. The fish of length L is at the range R, angle θ, and directivity D. Assume that the fish is

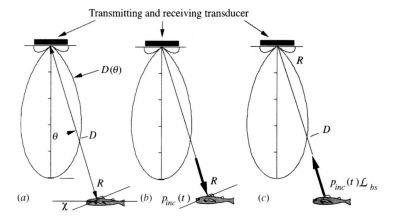

Fig. 7.1. Scattering of sound by a fish. (*a*) Fish at angle θ, directivity D of a sonar beam. (*b*) Sound pressure incident on a fish. (*c*) Fish as a source of scattered sound pressure. The sourcelet reradiates a pressure proportional to L_{bs} back to the transducer.

completely within the first Fresnel zone (Section 1.7.5):

$$L < r_1 = \sqrt{R\lambda/2} \qquad (7.1)$$

So that the local wavefront at the fish is effectively a plane wave. From Chapter 3, the incident sound pressure at time $t = R/c$ at the fish at range R is

$$p_{inc}(t) = D\frac{p_0(t - R/c)R_0}{R}10^{-\alpha R/20} \qquad (7.2)$$

The backscattered sound pressure from the fish, the "echo," is proportional to the incident sound pressure and to the *acoustical backscattering length* of the fish, $\mathcal{L}_{bs}$ (Section 6.1.1), which is a function of frequency and is assumed here, for simplicity, to be independent of the geometry. The backscattered sound radiates as if the fish is a sound source. At large ranges the scatterer reradiates spherically, and the backscattered pressure that is received from the fish is

$$p_{scat}(t) = D\frac{p_{inc}\mathcal{L}_{bs}}{R}10^{-\alpha R/20} \qquad (7.3)$$

Use (7.2) to replace the incident sound pressure and get

$$p_{scat}(t) = D^2\frac{p_0\mathcal{L}_{bs}R_0}{R^2}10^{-\alpha R/10} \qquad (7.4)$$

Scattering lengths may be measured experimentally or calculated theoretically. Scattering cross-sections are described in Section 6.1.2. We are interested in the backscattering cross-section,

$$\sigma_{bs} = |\mathcal{L}_{bs}|^2 \qquad (7.5)$$

Fig. 7.2. A night-time 70 kHz echo sound record taken in Trout Lake, Wisconsin. Some of the bottom echoes saturate the receiver electronics and cause the trace to be white. (From Rudstam, Clay, and Magnuson, 1987.)

The backscattering length and backscattering cross-section depend on the incident angle at the fish, as well as the sound frequency.

Results of an echo-sounding traverse in a lake are shown in Fig.7.2.

7.3 Sound backscattered by many bodies: volume reverberation*

To calculate the scattering by randomly spaced fish or plankton in a sonar beam one assumes that the relative positions with respect to one another and with respect to the transducers change from ping to ping. The spatial changes are assumed to be large enough to make the relative phases change by several π. Foote (1983) studied the linearity and addition theorems, under these conditions for, fisheries acoustics. The ratio (fish length)/λ ranged from 7 to 21. His experiments showed that for the frequencies used: (1) the backscattered pressures from fish add linearly; (2) backscattering acoustic cross-sections of live free-swimming fish can be determined from measurements on anesthetized samples; and (3) time integral-echo-squared processing is valid for ensembles of similar fish. The acoustically estimated and true fish densities were the same. This is important because operations on echoes from simple "objects" that will be developed here will work for the scattering from ensembles of complicated objects such as fish and zooplankton.

Based on laboratory research, and for simplicity, the following characteristics of the objects are assumed:

(1) The ratio of backscatter to incident sound is essentially the same for the fish in any part of the transducer beam.
(2) The backscatter from a fish has a broad directional pattern toward the transducer.
(3) The backscattering cross-sections of individual fish are near the mean of all backscattering cross-sections for fish in the ensemble.
(4) For the narrow frequency bandwidth of a sonar ping, the frequency dependence of the backscattering lengths is constant, and the incident ping is scattered back with nearly the same wave form.
(5) Multiple scattering and interactions between fishes are ignored.

* This section contains some advanced analytical material.

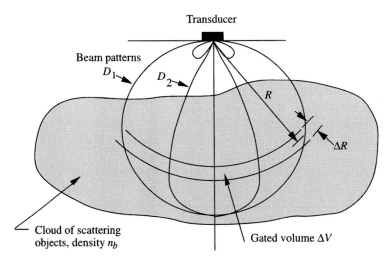

Fig. 7.3. Scatter from many objects within two beams of different directivity. A time gate in the receiver selects the echoes from the "gated volume." The time-gated volumes are the same for both the senders and receivers.

7.3.1 Randomly spaced objects in a directional beam

The instantaneous acoustic pressure of the backscattering from many objects is the algebraic sum of the pressures from the separate objects. This time-dependent sum is called "reverberation." Again, the analysis is given for co-located transmitting and receiving transducers. The geometry for a typical sonar measurement of the sound scattered from a volume containing scattering objects is shown in Fig. 7.3 for two different transducer directivity patterns. All radiation goes into the lower half-space. In our usage, the reference omnidirectional "half-space" responses are $D_1 = 1$ and $D_2 = 1$. The location of the ith object in the directional responses of the two transducer beams is indicated by D_{1i} and D_{2i}.

The calculation of the many-object backscattered energy problem requires summations of the time integral of the pressure squared. Four simplifying assumptions are used to reduce the expressions to something that can be "explained." First, within a population of scattering objects, the objects are nearly alike and can be replaced by a population of average objects. Second, the objects are uniformly and randomly distributed within layers of the selected volume. Third, the duration of the ping t_d is very small compared with the time gate that selects the shell thickness of the scattering volume, ΔR. And fourth, the attenuation of sonar energy within the ensemble can be ignored because the density of objects is small. Actually, extinction effects *are* observed in dense schools of fish (Foote, 1983).

Table 7.2 *Integrated beam pattern* Ψ_D

Source	Receiver	Ψ_D	Condition
Piston, radius a	Same	$5.78/(ka)^2$	$ka > 10$
Omnidirectional	Piston, radius a	$12/(ka)^2$	$ka > 30$
Rectangular, $L \times W$	Same	$17.4/(k^2 LW)$	$kL, kW \gg 1$
Omnidirectional	Rectangular	$\pi^2/(k^2 LW)$	$kL, kW \gg 1$
Omnidirectional	Same	4π	
Half-space	Same	2π	

From Clay and Medwin (1977).

The scattered pressure in (7.4) was for a single object. For many objects, we label the scattered pressure of the ith object $p_{i,scat}(t)$. The directional responses of the sonar beams are D_{1i} and D_{2i}. The sum over all objects within the gated volume is:

$$p_\Sigma(t) = \sum_{i=0}^{N-1} p_{i,scat}(t) \tag{7.6}$$

The summation over the positions of the objects counts the number of objects in the insonified shell and weights the amplitudes. The dependence on the beam patterns is gathered into an expression called the *integrated beam pattern* Ψ_D (Table 7.2) (see also Section 3.5.1 and (3.43)),

$$\psi_D \equiv \int_\phi d\phi \int_\theta D_1^2(\phi, \theta) D_2^2(\phi, \theta) \sin\theta \, d\theta \tag{7.7}$$

In terms of the integrated beam pattern, an *effective* sampled volume ΔV_e is

$$\Delta V_e = R^2 \Psi_D \Delta R \tag{7.8}$$

The density of scatterers n_b (number/volume) and the mean backscattering cross-sectional area σ_{bs} define the *volume backscattering coefficient:*

$$S_v(f) = n_b < \sigma_{bs} > \tag{7.9}$$

where both S_v and $\langle \sigma_{bs} \rangle$ depend on frequency.

If there are different animals, or different sizes of the same species within the scattering volume, multiple frequencies can be used to sort out the different components.

One can adapt this development to a distribution of different type bodies, or the same type of different sizes. Let the jth kind of object have the density n_{bj} and the squared scattering amplitude $|\mathcal{L}_{bs,i}|^2$. If the total number of objects of all types is N_b, the average volume backscattering coefficient is the weighted sum over all objects,

$$(S_v) \approx \frac{1}{N_b} \sum_i n_{fi} |\mathcal{L}_{bs,i}|^2 \qquad (7.10)$$

The SI units are n_{fi} in m^{-3} and $|\mathcal{L}_{bs,i}|^2$ in m^2.

Some workers prefer the logarithmic sonar equation. Then a reference, $S_{v,ref} = 1$ m^2/m^3, is used, and the *volume backscattering strength* is given as

$$S_v = 10 \log_{10}[(S_v)/(S_{v,ref})] \text{ dB} \qquad (7.11)$$

7.3.2 Scattering layers

Deep scattering layers (DSL)

Many zooplankton and fish respond to light. They move up at dusk and down at daybreak. They tend to concentrate at strong changes of properties in the water column, such as the base of the mixed layer and oceanic fronts. Much of the knowledge of the vertical migrations of plankton has been obtained by echo sounding. Sonars have identified deep scattering layers (DSL) throughout the world's oceans, down to depths of several hundred meters. Although the vertical migration of the DSL is a striking feature on the records, other groups of scatterers seem to remain at almost constant depth, probably because of temperature preferences (see Fig. 7.4).

Because plankton and nekton are sensitive to the presence of nutrients, food, light, and other physical–chemical characteristics of their environment, they may respond to the change of any of these. Many water properties change at the water mass boundaries, and the boundary is often marked by changes of the volume scattering strength. Marine animals may concentrate at the interface or behave as if the interface were a barrier.

Backscattering at two different frequencies depicts the migration of larger and smaller marine organisms just after sunset (Fig. 7.5).

The motion of an internal wave of periodicity two or three minutes is illustrated by volume acoustical backscatter in Fig. 7.6. In an ocean dumping ground where man-made underwater garbage is made "visible" by backscatter.

Near-surface scatterers

The near-surface region of the ocean plays host to a large number of scatterers, in addition to the transient presence of the DSLs that come up during the night. This is also the region that shows the greatest effects of surface wave action. The full-time residents of the near-surface region are biological and physical in origin.

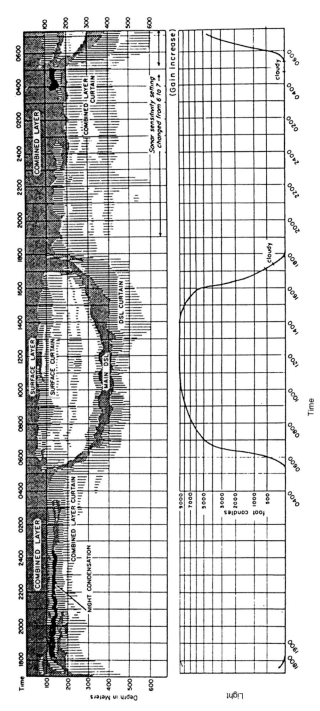

Fig. 7.4. *Above,* diagram of scattering layers prepared from a 37 hour long echogram recorded in the Bay of Bengal at 06°10′N, 93°07′E, beginning 1700 hours November 24 and ending 0715 hours November 26. A downward pointing 30 kHz sonar was used. Stippled patterns indicate heavy scattering; medium and light vertical lines indicate medium and light scattering, respectively. Lesser scattering between the "surface curtain" and main deep scattering layer (DSL), centered at 250 m during the day, corresponds in position to the intermediate layer seen as heavy scattering on numerous other echograms. *Below,* curve indicating intensities of incident light for the period the echogram was being recorded. (Bradbury *et al.* in Farquhar, 1970.)

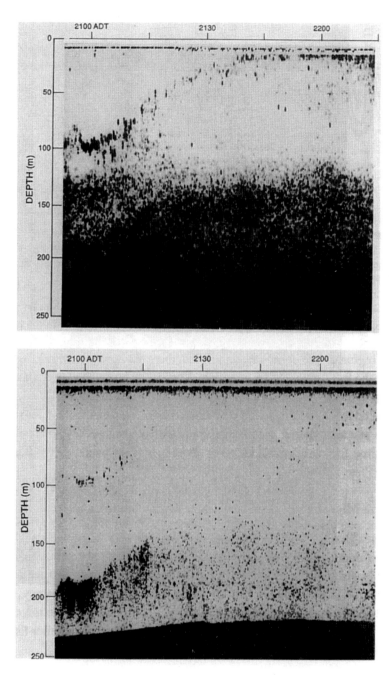

Fig. 7.5. Migrating fish populations just after sunset (2055 ADT). The use of two different frequencies, 50 kHz (*below*) and 200 kHz (*above*) permits the tracing of larger and smaller marine organisms, respectively (Cochrane *et al.*, 1991).

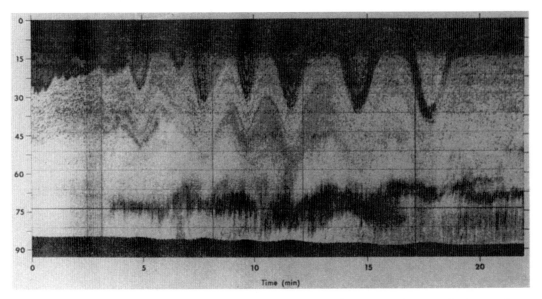

Fig. 7.6. Acoustical imaging of a non-linear internal wave packet by sound backscatter. Frequency 200 kHz. The three clearly defined layers between 15 m and 30 m depth have been attributed to temperature structure, while those below have been ascribed to neutrally buoyant marine organisms. (Orr, 1989, personal communication.)

In one experiment, the biological scatterers in the near-surface region were found to be predominantly of the phylum Arthropoda, subclass copepod. The copepods are approximately millimeters in length. To study them required the use of sound frequencies of the order of hundreds of kilohertz. For example, Barraclough *et al.* (1969) perfomed net sampling of the upper ocean from the surface to a depth of about 100 m and simultaneously measured backscatter with a 200 kHz sonar. The experiment, which was conducted over a great circle route from British Columbia, Canada to Tokyo, showed a coincidence of depth of strong echoes with depth of maximum catches that were 99 percent *Calanus cristatus,* a species of copepod. The peak depth was at about 40 m during the day. There were about 100 copepods per cubic meter, comprising a wet density of about 1.5 g/m^3.

The near-surface ocean is also a nightly home to some anchovies, which ingest a bubble to keep them floating, quietly and unobtrusively, while they rest and try to avoid the search by predators. Undoubtedly other behaviors of interest to marine biologists await the concurrent use of sonars.

As discussed in Chapter 6, the upper region of the ocean contains bubbles that are not part of any zooplankton or fish, and that cause substantial acoustical effects. These bubbles may be the vented gas of zooplankton or fish, the products of photosynthesis, or the result of break-ing waves, precipitation, cosmic rays, decaying matter, or gas hydrates associated with hydrothermal plumes at the sea floor.

7.4 Variability of fish structure and sound scatter

Acoustical models of fish are partly empirical and partly based on the anatomy of a fish. It has been known for a long time that the main contributions to sound scattering by fish come from their swimbladders (Hersey and Backus, 1962; Haslett, 1962). Early swimbladder models were simple shapes such as gas-filled spheres and prolate spheroids. Laboratory measurements of sound scattering by fish give empirical for-mulae such as those of Love (1969) and McCartney and Stubbs (1970). Many papers on biological sound scattering and research are collected in the meeting proceedings, edited by Farquhar (1970).

The swimbladder is an active organ of the fish that maintains buoy-ancy and attitude by controlling the bladder volume. For different fish species, the shapes of swimbladders are extremely variable and can be quite complicated (Whitehead and Blaxter, 1964). The swimblad-der shapes, dimensions, and tilt with respect to the axis of the animal all determine the sound backscattering of frequencies for which $ka > 1$ where a is an average radius.

Acoustical models of fish are simple representations without the fine details of the shape and anatomy of fish. The simplest models are spheres that are gas-filled to represent the swimbladders or fluid-filled to represent the average flesh of the animal. The progression of the science has been from such simple approximations to more complicated finite cylindrical or prolate spheroidal models that begin to look like the ani-mals. The introduction of multiple-frequency sonars and the acquisition of high-quality data have permitted greater sophistication of the models and correspondingly more accurate identification of unknown animals (see Chapter 13 by Horne and Jech).

Relationships between actual fish length and high-frequency target strength are important in fisheries research. (We use the terminology "high frequency" when $ka > 1$, although in many cases our developments hold for the larger region $ka > 0.2$.) The radius a is usually subscripted (e.g., a_{ec} is the equivalent cylindrical radius, and a_{es} is the equivalent spherical radius). Acousticians and biologists have used measured target strengths of single fish to construct empirical target strength–fish length formulae (McCartney and Stubbs, 1970; Love, 1971; Foote and Traynor, 1988). The earlier measurements used caged fish. More recently, *in situ* target strength measurements have been effectively compared with net captures. When local "calibrations" are used, the empirical formulae work very well.

7.5 Sound backscattered by zooplankton

Acoustical estimates of zooplankton populations can be made if the scattering lengths are known as a function of frequency and zooplankton size. In most acoustical surveys, zooplankton are too close together for echoes from individuals to be resolved. In the laboratory, the echoes from individuals are very small because the animals are small. McNaught (1968, 1969) and Greenlaw (1979) suggested using multiple-frequency sonars to separate different sizes of zooplankton populations. The methods depend strongly on the frequency dependence of scattered sound as described in Chapter 6.

At sea, there are tiny gas-bubble carrying plankton that, though only millimeters in extent, have a very large acoustical scattering effect. One example is *Nanomia bijuga,* a colonial hydrozoan jellyfish (order Siphonophora, suborder Physonectae) that has been identified as a primary cause of some sound scattering layers off San Diego, California and elsewhere. Positive identification was accomplished by simultaneous viewing from a deep-sea submersible while recording the backscattered sound.

Siphonophores consist of many specialized individuals, aligned along an axis that may be as long as 75 cm. One of the major groups of siphonophores, the Physonectae, is identifiable by bubble-carrying pneumatophores, which operate as flotation elements for the colony. This individual is approximately a prolate spheroid of dimensions about 3 mm × 1 mm and contains from 12.6 mm^3 to 0.25 mm^3 of ejectable, rechargable, carbon monoxide gas ($\gamma = 1.40$). The resonance frequencies range from about 7 to 27 kHz at 100 m depth. Some bubble-carrying aspects have been identified to depth 3000 m (S. Haddock, personal communication) (see Fig. 7.7). See Chapter 12 by Holliday and Stanton.

Problems

7.1 Calculate the target strength of a copepod of approximate equivalent radius $a' = 1$ mm when measured with sound of frequency 300 kHz. Assume that $e \simeq 1.25$ and $g = 1.02$.

7.2 (*a*) A scattering layer of volume backscattering strength $S_v = -70$ dB at 30 kHz is assumed to be due only to a high density of euphausiids of equivalent spherical radius 4.15 mm and backscattering cross-section $\sigma_{bs} = 9.26 \times 10^{-11}$ m^2. What density of euphausiids is implied by this assumption?

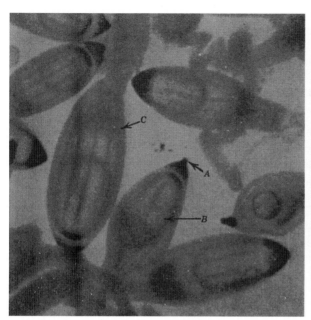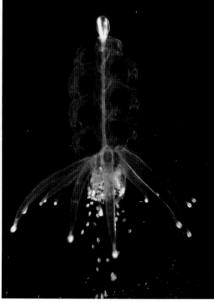

Fig. 7.7. *Left*: Laboratory photograph of a collection of pneumatophores of
Nanomia bijuga with contained gas bubbles: A, pore; B, gas gland; C,
longitudinal muscle band. (From Barham, 1963.) *Right*: A siphonophore at sea.
Note the ellipsoidal flotation gas bubble, of major axis approximately 6.5 mm.
The total length of this particular specimen was about 8 cm; it was captured at
150 m depth. Siphonophores have been seen with lengths up to 40 meters and
at depths greater than 3000 m. Siphonophore bubbles may be detected, and
non-destructively measured at a distance, by their acoustical resonance. (Photo
courtesy of Dr. S. Haddock, 2004, MBARI). (See colour plate section).

> (*b*) Assuming that the euphausiids are of 10 % greater radius, how
>   many would there be?
> (*c*) How many of 10 % lesser radius?

7.3 Perform the following calculations:

> (*a*) Given a siphonophore float of volume 4 mm^3 at sea level, cal-
>   culate f_R for the bubble.
> (*b*) Calculate the backscattering cross-section at resonance for a
>   single float.
> (*c*) If there are 10 resonant floats in a volume of 1000 m^3, calculate
>   the volume reverberation strength S_v.

7.4 A near-surface experiment at 300 kHz yields $S_v = -95$ dB. Simultaneous sampling reveals 10^7 diatoms per cubic meter and 1 copepod
per cubic meter. Assuming that the diatoms have a radius of 10μm,
the copepods $a' = 1$ mm, and $e \simeq 1.25$, $g = 1.02$ for both plankton,
decide which scatterer is the probable cause of the S_v and how many
are actually present in a cubic meter.

7.5 Integrate the directivity patterns of $ka = 2, 5, 10,$ and 20 in Fig. 6.11 to find the total scattering cross-section in each case. Compare with $ka = 1$ and $ka = 0.5$, which are closer to omnidirectional.

Further reading

See Chapters 12–17 for recent and current ocean-acoustic research on marine life, ranging from plankton through fish to mammals, conducted by several university and government laboratories active in the Atlantic and Pacific oceans. Detailed technical references are given.

Foote, K. G. and Stanton, T. K. (2000). Acoustical methods, In *ICES Zooplankton Methodology Manual*, pp. 223–58. ed. R. Haris, P. H. Wiebe, J. Lenz, H. R. Skjoldal and M. Huntley, New York: Academic Press.

Medwin, H. and Clay, C. S. (1998). *Fundamentals of Acoustical Oceanography*, New York: Academic Press.

See Sections 9.3 through 9.6 and 10.3 through 10.6. Detailed estimates of the statistical scatter from acoustical models of fish and plankton.

Chapter 8

Ocean waveguides: scattering by rough surfaces, barriers, escarpments, and seamounts

Summary

The applied scientist's approach to solving a complex problem is to employ a simple model that describes the essence of the problem. The acoustical scientist has the additional advantage of judging the importance of each spatial parameter by comparing its size to the acoustic wavelength.

For example, as far as the interaction with sound is concerned, one can use low frequencies to "reduce" a rough rigid surface to a plane rigid surface, and approximate an escarpment or seamount as a wedge, provided that the imperfecting "bumps" on the surface are small compared to a wavelength. Comparisons of ocean trials with results from physical scale models or computer models allow one to test whether the simplifications that make the problem tractable have preserved the accuracy of the predictions. By using models scaled to both the time domain and frequency domain, the characteristics of sound propagation in ocean waveguides and over escarpments and seamounts have been studied and understood at a small fraction of the cost of conducting a thorough experiment at sea.

Contents

* This section contains some advanced analytical material.

8.1 Normal modes in ocean waveguides*

People who play string instruments have a head start in understanding
the concept of "modes" in shallow water! In fact, much of the vocabulary
is the same for musicians and underwater scientists.

When a violinist draws his bow, there is a maximum motion near
the string center and almost no apparent motion at either end. The string
appears to vibrate as a half of a sinusoidal wave. A physicist would call
the musical pitch that one hears, and which corresponds to the half-wave
vibration, "the fundamental" of the complex sound. A frequency analysis
of the sound would show that there are additional components which
are whole number multiples of the fundamental frequency and whose
magnitude depends of the exact position of the bow. The physicist calls
these other components "upper harmonics." The quality of the musical
sound depends of the strengths of the several frequency components that
are generated. (The underwater acoustics analogy is the importance of
the placement of a sound source in determining the modes of propagation
in shallow water.)

If the string is now lightly touched at its midpoint, only those "modes"
of vibration that do not require motion at the midpoint will remain. The
full length of the string will appear to have been forced to vibrate into
two half wavelengths and the sound will have a pitch corresponding
to twice the frequency of the fundamental. (The physicist says we are
hearing "the second harmonic"; the musician simply says he is playing "a
harmonic" which is an octave above the fundamental.) Similarly, when
lightly touched at a one-third point, the entire vibrating string appears
to be divided into three, still-smaller, half wavelengths, and one hears
"the third harmonic." The initial distribution of energy within the modes
depends on the position where the string is bowed. When it is bowed
near the bridge, more modes are stimulated and a richer, some would
say, "more strident" sound is heard. If the string were to be bowed at its
midpoint, only modes that show motion at the midpoint would be seen
and heard, and the sound would be quite different.

---

* This section contains some advanced analytical material.

We are able to see the violin string modes because they are large transverse vibrations of a solid in air. On the other hand, the invisible sounds in the ocean have very much smaller displacements which are detected by pressure-sensitive hydrophones. Also, hydrophones measure longitudinal, not transverse, oscillations. But, with a little imagination, one can learn from the musical analogy, as follows:

A sound source in shallow water radiates energy, which underwater acousticians analyze into "normal modes of propagation" in the water waveguide. In analogy to the string, the distribution of acoustic energy within the normal modes depends on the fractional depth of the source in the ocean channel. The modes can be detected and identified by moving a hydrophone vertically through the water column even at large range, and especially when there is a continuous rather than a pulsed sound source. Laboratory scale models of the ocean have been used to verify this behavior in constant depth coastal regions as well as wedge-like, near-shore, shallow water.

We start our understanding of the "normal mode" description in a plane-bounded medium by looking at solutions of the wave equation in cylindrical coordinates. This is convenient because the structure of the shallow water medium is assumed to be only a function of depth, z, and the cylindrical coordinates are given as z, r, θ. Our solutions are for a simple harmonic source, which produces a continuous wave of angular frequency ω.

8.1.1 Wave equation in a plane waveguide

The cylindrical coordinate scalar wave equation for sound pressure is

$$\frac{\partial^2 p}{\partial r^2} + \frac{1}{r}\frac{\partial p}{\partial r} + \frac{\partial^2 p}{\partial z^2} = \frac{1}{c^2}\frac{\partial^2 p}{\partial t^2} \tag{8.1}$$

Let the pressure be expressed as proportional to the product,

$$p \sim U(r)Z(z)T(t) \tag{8.2}$$

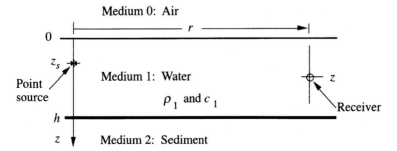

Fig. 8.1. Source and receiver in a plane ocean waveguide. For simplicity, the upper layer, medium 0, is air. The water layer is medium 1. The sediment, medium 2, may be a single medium or many layers.

Then the substitution of (8.2) into (8.1) and division by p permits the separation of the variables

$$\frac{1}{U(r)} \left(\frac{\partial^2 U(r)}{\partial r^2} + \frac{1}{r} \frac{\partial U(r)}{\partial r} \right) + \frac{1}{Z(z)} \frac{\partial^2 Z(z)}{\partial z^2} = \frac{1}{c^2 T(t)} \frac{\partial^2 T(t)}{\partial t^2} \qquad (8.3)$$

For the harmonic source, $T(t) = \exp(i\omega t)$, the right-hand side reduces to $-\omega^2/c^2$. For all values of r and z, the r-dependent term and the z-dependent term must each be equal to constants. The separation constants are designated $-\kappa^2$ for the radial term and $-\gamma^2$ for the vertical z term. The separated expressions are

$$\frac{\partial^2 U(r)}{\partial r^2} + \frac{1}{r} \frac{\partial U(r)}{\partial r} = -\kappa^2 U(r) \qquad (8.4)$$

$$\frac{\partial^2 Z(z)}{\partial z^2} = -\gamma^2 Z(z) \qquad (8.5)$$

$$k^2 = \omega^2/c^2(z) \qquad (8.6)$$

$$\kappa^2 + \gamma^2 = k^2 \qquad (8.7)$$

where c depends on z; κ is the horizontal component of the wave number; γ is the vertical component of the wave number. Since $U(r)$ is only a function of r, (8.4) implies that κ is constant for all values of z.

For the angle of incidence θ, the horizontal and vertical components of k are

$$\kappa = k \sin \theta \qquad (8.8)$$

and

$$\gamma = k \cos \theta \qquad (8.9)$$

Equations (8.6) and (8.7) show that γ depends on $c(z)$. Correspondingly, the function $Z(z)$ depends on $c(z)$ and ω.

8.1.2 Range dependence

The 0th order solution of the radial (8.4) for an outgoing wave is the Hankel function, $H_0^{(2)}(\kappa r)$ which at very large range, for an outgoing wave is simply:

$$H_0^{(2)}(\kappa r) \approx \sqrt{\frac{2}{\pi \kappa r}} e^{-i(\kappa r - \pi/4)} \quad \text{for} \quad \kappa r \gg 1 \qquad (8.10)$$

8.1.3 Depth dependence

A simplified (idealized) waveguide consists of a homogeneous fluid (water) that has pressure-release interfaces at the upper and lower boundaries. Recalling Chapter 1, reflections beyond the critical angle can be replaced by a pseudo-pressure release interface at Δz beneath the water sediment interface. Hence, a water layer over a real sediment bottom can be approximated by an idealized cylindrical waveguide. The boundary conditions require the pressure to vanish at the upper and lower interfaces. Correspondingly, the "*eigenfunction* solutions," $Z(z)$, must satisfy

$$Z(z)|_{z=0} = 0 \quad \text{and} \quad Z(z)|_{z=h} = 0 \tag{8.11}$$

Solutions of the $Z(z)$ for the boundary conditions of (8.11) are

$$Z(z) = \sin(\gamma z) = 0 \quad \text{at} \quad z = 0 \quad \text{and} \quad z = h \tag{8.12}$$

where γ is the vertical component of the wave number.

The requirements of (8.12) give the "modal equation" for the idealized waveguide:

$$\gamma_m h = m\pi \quad \text{or} \quad \gamma_m = m\pi / h \tag{8.13}$$

where m is an integer, which designates the "mode number." The γ_m are known as the *eigenvalues*. The values of κ_m are given by (8.7) and (8.13)

$$\kappa_m = (k^2 - \gamma_m^2)^{1/2} \quad \text{and} \quad k^2 \geq \gamma_m^2 \tag{8.14}$$

The requirement that κ_m be real gives the mode cut-off condition, because the γ_m increase with mode number.

In an idealized (pressure-release) waveguide, the depth-dependent *eigenfunctions*, $Z_m(z)$, are simply

$$Z_m(z) = \sin(m\pi z / h) \quad m = 1, 2, 3 \tag{8.15}$$

Recall from (8.2), the acoustic pressure is proportional to Z_m. Examples of the $Z_m(z)$ are shown in Fig. 8.2.

The top and bottom interfaces are assumed to be pressure-release boundaries. The fluid layer is assumed to be homogeneous.

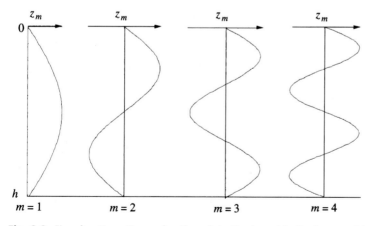

Fig. 8.2. Eigenfunctions Z_m as a function of depth z in an idealized waveguide of depth h.

8.1.4 Sound pressure as the sum of modal contributions

The sound pressure is the sum of the pressures in the modes. The sound pressure, for source and receiver in medium 1 is,

$$p\,(r, z_s, z, t) = A_s e^{i\omega t} \sum_{m=1}^{M} \frac{\rho}{v_m (2\pi \kappa_m r)^{1/2}} Z_m(z_s) Z_m(z) e^{-(i\kappa_m r - \pi/4)} \quad (8.16)$$

where ρ is the ambient density and A_s depends on source power. The summation is over the allowed modes M.

The pressure amplitude of the wave decreases as $1/\sqrt{r}$, as one would expect for a cylindrical spreading wave that is trapped in a layer. While the sound pressure has a single frequency, its dependence on range and depth is complicated because the components of sound pressures in each of the modes have different dependencies on range and depth.

In real waveguides, the absorption losses in the bottom and the water layer cause the sound pressures to decrease faster than $1/\sqrt{r}$. All of these kinds of losses are included in an empirical mode attenuation rate δ_m. Also, a mode excitation factor q_m takes account of the ability to put more energy in selected modes depending on the source position. The result is

$$p\,(r, z_s, z, t) = e^{i\omega t} \sqrt{\Pi/\Pi_0} \sum_{m=1}^{M} \frac{q_m}{\sqrt{\kappa_m r}} Z_m(z_s) Z_m(z) e^{-(i\kappa_m r - \pi/4) - \delta_m r} \quad (8.17)$$

where Π/Π_0 is the relative source power and q_m is a mode excitation factor. See Tolstoy and Clay, 1987, pp. 81–4.

8.2 Geometrical dispersion: phase velocities and group velocities*

Dispersion, in which different frequencies travel at different speeds, was described in Chapter 6 for the bubbly regions of the upper ocean. In a waveguide, in addition to the dispersion that may be caused by the physical properties of the medium, there is dispersion due to the geometry, i.e., *geometrical* dispersion. There are two different types of velocity in a waveguide – the *phase velocity* which one knows from free space propagation, and the *group velocity*.

8.2.1 Phase velocity

Let us pick a mode and phase of the sound pressure and "ride" it as it moves along the waveguide. The mth mode phase in (8.17) is given as the imaginary part of the exponential:

$$\text{Phase of mode } m = (\omega t - \kappa_m r + \pi/4) \quad (8.18)$$

---

* This section contains some advanced analytical material.

For example, a pressure maximum travels with the velocity $r/t = \omega/\kappa_m$, and this defines the phase velocity along the r coordinate. Using the subscripts r and m for the velocity along r and the mth mode, the phase velocity is

$$v_{rm} = \omega/\kappa_m \tag{8.19}$$

8.2.2 Group velocity

Group velocity is the speed of transmission of the energy in a mode. The analytical explanation is in Section 11.2.2 of (M&C). Let it suffice here to state that the condition is that the slope $(\omega t - \kappa_m r)$ be zero with respect to changes in the modal wave number κ_m. Algebraically, the condition is

$$\frac{d}{d\kappa_m}(\omega t - \kappa_m r) = 0 \tag{8.20}$$

which gives the time, t, required for frequency component ω of the signal to travel the distance r. The *group velocity*, u_{gm}, of the mth mode is therefore

$$u_{gm} = (r/t) = \frac{d\omega}{d\kappa_m} \tag{8.21}$$

Theoretical examples of the phase and group velocities for a shore in Florida are shown in Fig. 8.3. Obvious observations are (1) the highest velocities are c_2; (2) the mode cut-off (vertical segment) moves to higher frequency as the mode number increases; (3) the group velocities have minima that are much smaller than the sound speed in the water

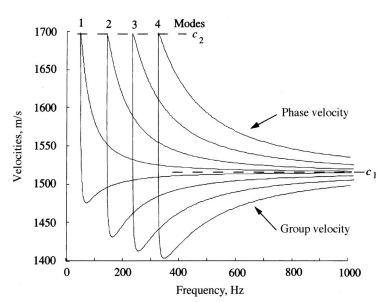

Fig. 8.3. Phase and group velocities for the Jacksonville, Florida, shoal shallow-water waveguide. The parameters are $c_1 = 1516$ m/s; $\rho_1 = 1033$ kg/m³; $c_2 = 1.12c_1$, (1698 m/s), $\rho_2 = 2\rho_1$; $h = 18$ m. The first four modes are shown. The maximum velocities are c_2, and the high-frequency limit is c_1. The values are from Ewing and Worzel (1948, Fig. 9).

layer, c_1 (these give the last arrivals in a mode); (4) the group velocities decrease as the mode number increases; and (5) at very high frequency, both the phase and group velocities tend to c_1.

8.3 Single frequency propagation in a scale model of a water-wedge waveguide

Waveguides with range-dependent depths are actually the waveguides found most often in nature. Perhaps surprisingly, normal-mode solutions which were originally calculated for plane parallel waveguides, have been successful in describing sound propagation in coastal ocean wedge waveguides, as well. The ocean shore and the continental shelf are often good examples of wedge waveguides. Solutions to this problem have been given by Buckingham (1987) as well as Biot and Tolstoy (in Tolstoy and Clay, published by the Acoustical Society of America, 1987).

Perhaps the most convincing experiment, and comparison with theory, was the laboratory model work of Tindle, Hobaek, and Muir (1987) which is described in Figs. 8.4–8.6.

The details are in the captions. The source array was constructed of line sources. The excitation was a gated 4-cycle, 80 kHz sine wave that was bandpass-filtered between 60 and 100 kHz. The excitations of the elements of the array were chosen to transmit selectively single modes.

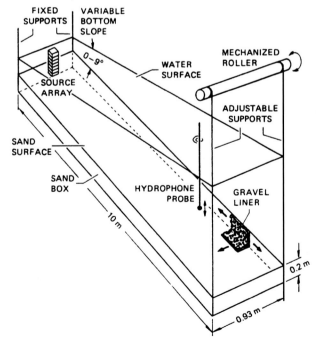

Fig. 8.4. Shallow-water model of a wedge waveguide: water over sand. Water: $c_1 = 1490$ m/s and $\rho_1 = 1000$ kg/m3. Sand: $c_2 = 1780$ m/s and $\rho_2 = 1.97\rho_1$. The water depth at the source was 120 cm. The measured attenuations in sand was not detectable. The waveguide attenuation coefficient in the water layer was $0.06f$ (m kHz). (From Tindle et al., 1987.)

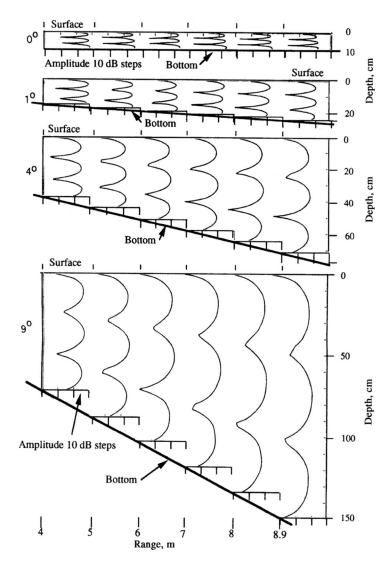

Fig. 8.5. Amplitude levels of the sound pressure in dB for the third mode. The source array was at 0 range, where the local water depth is 10 cm. The source array was adjusted to excite the third mode at 80 kHz. At each range the probe hydrophone was run from top to bottom to measure the sound pressures. Converted to dB, the amplitude levels are $20 \log[|Z_3(z, r)|]$ in dB relative to arbitrary levels. (From Tindle *et al.*, 1987.)

The receiver was a probe hydrophone. Signals from the hydrophone were recorded as a function of depth, and the receiving array processing was performed by the computer. The slope of the bottom interface was adjusted for different ocean wedge angles.

In range-dependent waveguides, it is important to use the local co-ordinates in making sound field measurements and for mode-filtering of the sound pressures. The shallow-water wedge is simple because cylindrical coordinates are its reference. The air–water interface is the upper surface of the wedge and x, the water sediment interface, is the lower plane of the wedge. In cylindrical coordinates, the intersecting planes

Fig. 8.6. Single-mode transmissions as functions of depth and slope. For all transmissions, the water depth at the source is 10 cm. The range is 8.9 m. Time is measured relative to the same time delay, after the transmission. The numbers 1, 2, and 3 are the mode numbers. (*a*) Constant depth (0° slope): the source array is vertical. The depth at the receiver is 10 cm. (*b*) Wedge 9° slope: the source array is curved to match the cylindrical coordinates. The probe receiver is moved vertically, and the waveforms arrive later as depths increase. (From Tindle *et al.*, 1987. See also Chu, 1990.)

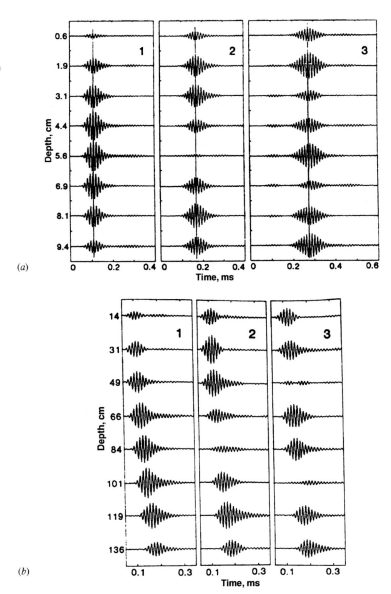

form the wedge axis; r is the radial distance from the line of the intersecting planes; and the angle θ is measured relative to the upper surface.

According to the "adiabatic approximation," the eigenfunctions $Z_m(z, r)$ should stretch as the water depth increases. Figure 8.5 shows the stretching or adjustment of the third mode to fit the bottom depth for bottom slopes of 0°, 1°, 4°, and 9°. If the amplitudes of the sound pressures were plotted instead of the levels in dB, one would see the nulls and changes of phase of $Z_m(z, r)$. The structure is the same at different ranges for the constant-depth waveguide (slope 0°). The increase of the slopes to 1°, 4°, and 9° show how the modes are stretched as they adjust to the changing depths. The wave numbers γ_m and κ_m are functions of the local depth.

Wave forms of sound transmissions for the "Pekeris" constant-depth waveguide, and for a set of wedges with 0° and 9° slopes, are shown in Fig. 8.6. The laboratory model can be scaled to real-world conditions by letting the frequency be 80 Hz, and then the depth at the source becomes 100 m and the maximum range becomes 8900 m.

The amplitudes and travel times of the wave forms depend on mode number, receiver depth, and slope. First we consider the travel times in the uniform waveguide (Fig. 8.6(a)). The delays of the arrivals depend on the group velocities of the modes. Figure 8.3 shows the group velocities for a 18 m water depth. If the laboratory waveguide depth were scaled to 18 m, the frequency would be 440 Hz. As consistent with the group velocities, the travel times increase as the mode numbers increase. The amplitude and phases of the wave forms depend on depth, as in the eigenfunctions (Fig. 8.5). Next, the laboratory world changes, and the bottom has the 9° slope. The transmitting array is curved to keep constant range from the axis of the wedge. If the adiabatic approximation works, then the wave forms should be trapped in each mode and be identifiable at the probe hydrophone depths (Fig. 8.6(b)). The arrivals maintain their identities, and the amplitudes and phases correspond to the local eigenfunctions. Wave forms at the deeper receivers have time delays relative to the receiver at 14 cm depth because the receiving hydrophone was moved vertically and the receivers were not at constant radius from the axis. Thus, the deepest receiver is 9 cm farther from the axis than the top receiver. This demonstrates the importance of measuring along the local coordinates – that is, at constant radius for a wedge waveguide. The adiabatic mode theory gives the correct arrival times and amplitude dependence on range for different bottom slopes.

8.4 Propagation in a coastal waveguide: time domain

The time domain pulse solution for the wedge problem has tremendous advantages over the normal mode wave solution. In the time domain, all of the reflection arrivals and the diffraction from the intersection of the planes are identifiable and separable. Furthermore, the use of numerical Fourier transformations can give frequency domain results for any combination of arrivals during any time interval.

The coordinates and geometry for an idealized coastal water wedge are shown in Fig. 8.7. The crest line of the intersecting planes is along the

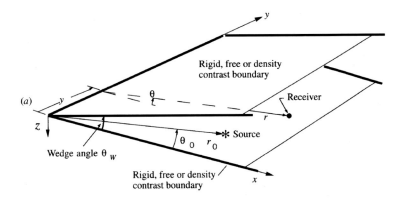

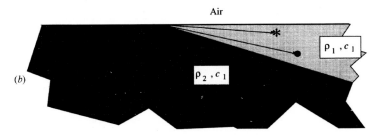

Fig. 8.7. (*a*) Coordinate system and geometries in the case of an idealized model of a beach. The line of the intersecting planes is along the y axis; all radial distances are measured normal to this line. In (*b*) the lighter shaded region represents water. The darker shaded region is the ocean bottom. The wedge is under a half-space of air.

y axis. (The line of intersection is misleadingly called the "wedge apex" in some of the literature.) The boundary planes of the wedge are normal to the x–z plane. This choice lets us use the z coordinate for depth and to have the x and y coordinates in the horizontal plane. The source is in the x–z plane at $y = 0$. The receiver may be displaced a distance y. The radial distance from the y axis is r. To simplify notation in the fluid, we use c for the sound speed in the fluid and ρ for the density in the wedge.

Consider the geometry in Fig. 8.7. There are two critical times, t_{direct} and t_0 defined by (8.22) and (8.23), which separate three distinctive domains (see Fig. 8.8).

$$t_{dir} = \frac{1}{c}[(r - r_0)^2 + y^2]^{1/2} \tag{8.22}$$

where t_{dir} is the time for a ray to travel from the source to the receiver, and

$$\tau_0 = \frac{1}{c}[(r + r_0)^2 + y^2]^{1/2} \tag{8.23}$$

where τ_0 is the time required for a ray to go from the source to the line

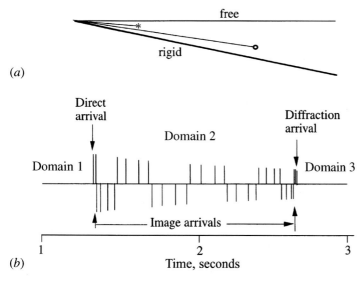

(a)

(b)

Fig. 8.8. Arrivals in a simple free/rigid, water wedge. (a) Delta function source and receiver in the same perpendicular plane of an 11° wedge. The source (asterisk) is at range 1000 m, angle 4°. The receiver (circle) is at range 3000 m, angle 8°. Sound speed is 1500 m/s. Upper boundary is free, lower boundary is rigid. (b) Pressure signals due to a pulse source. Domain 1 has no arrivals before the direct arrival. Domain 2 contains the image arrivals. Domain 3 contains the diffraction arrival. (From Chapter 11, *M&C*. See Chu, 1990.)

of the intersecting places and then to the receiver. The diffraction from the line of intersection begins after τ_0.

Domain 1 is the time before the direct arrival from source to receiver; domain 2 encompasses the reflections and the reflections of reflections etc. that occur before the final component, the diffraction arrival from the wedge crest, which defines the beginning of domain 3.

The acoustic pressures within the wedge depend on the values of t relative to t_{dir} and τ_0. Figure 8.8 was calculated by using the theory of Biot and Tolstoy (see next section), assuming that the surfaces are smooth, and the line joining the planes is straight. In the real world, the relative importance of these contributions depends on the roughness of the surfaces (compared to the sound wavelength) and the straightness (also, compared to the sound wavelength) of the line intersection where the upper plane and the bottom plane meet. Rough surface effects will be considered later.

8.5 Propagation over a rigid wedge: time domain

The unique, closed-form, time-domain description of sound propagation in a fluid-filled, wedge-like space between two rigid surfaces was given by Biot and Tolstoy in 1957. The theory was applicable not only to that geometry but also to scattering by a rigid wedge. In either case, a point source was shown to produce reflections from the two planes and diffraction from their intersection. The diffraction part of the solution was significantly different from the traditional 19th century Helmholtz–Kirchhoff ("HK") frequency-domain description of these problems as presented by Born and Wolf (1965) and Trorey (1970).

Beginning in 1978, this Biot–Tolstoy ("BT") theory was reformulated, reinterpreted, transformed to the frequency domain, and tested in well-defined laboratory model experiments conducted by students at the Ocean Acoustics Laboratory of the Naval Postgraduate School. See Bremhorst and Medwin (1978), Medwin and Novarini (1980), Jebsen and Medwin (1982), Novarini and Medwin (1985). The applicability of the theory to a great diversity of circumstances ranging from highway noise barriers to seamounts to randomly rough ocean surfaces was soon recognized. The results of these early experiments triggered research involving a wide range of problems at several laboratories e.g., Kinney *et al.* (1983), Chu (1989, 1990). Extensions and new applications of this early work are continuing to this day.

In this and later sections we look at wedges as *elements of scattering by obstacles* so that we can interpret the character of the sea floor and the sea surface when they are sensed at remote ranges.

NB Those who have studied classical optics will know that the Helmholtz–Kirchhoff (HK) technique is a popular classical implementation of Huygens' Principle for reflection and diffraction propagation of electromagnetic waves. Its virtue is that it gives one a "feel" for problems in wave propagation and its predictions can be readily verified by simple undergraduate optics laboratory experiments. On the other hand, when the HK theory is applied to the large range of frequencies common in acoustics, it has proven to be very limited, e.g., only when the incidence is nearly perpendicular to the surface and when the frequency is "high" does it yield good predictions for diffraction scatter from a plane slab such as a highway noise barrier. For the very large frequency range of acoustical wavelengths used at sea, it often turns out that the HK theory is seriously inaccurate. There have been many attempts to correct the flaws in the HK theory. These struggles became moot in the late 1970s when the BT exact solution for scattering by a wedge (or plate) was shown to agree with all critical experiments. For these reasons we will not take up more space to discuss the HK technique.

8.5.1 Implementation of the modified Biot–Tolstoy theory of scatter from wedges and plates

To provide a building block for several applications we turn to the simplified, but accurate, interpretation (Medwin, 1981) of the Biot–Tolstoy (1957) theory of scattering by a rigid, infinite wedge. Biot and Tolstoy used a "normal coordinate method" to solve the ideal, infinite-rigid, wedge problem in the time domain. The detailed derivation is in Section 11.8 of (*M&C*). The complete solution for an infinite wedge assumes radiation from a point source of volume flow $\dot{V}$,

$$p = [\rho_\mathrm{A}\dot{V}/(4\pi R)][\delta(t - R/c)] \qquad (8.24)$$

For digital implementation, one writes the impulse pressure field radiated into free space as

$$p = (P_0 R_0/R)(\Delta t)\delta_f(t - R/c) \qquad (8.25)$$

where δ_f is the source finite delta function (2.1), which is defined in terms of the impulse duration Δt.

$$\delta_f(t) = 1/\Delta t \quad \text{for} \quad 0 \le t \le \Delta t \quad \text{and} \quad \delta f(t) = 0 \quad \text{otherwise} \qquad (8.26)$$

The introduction of (8.24)–(8.26) into the BT theory (which some have called the BTM method) permitted Biot–Tolstoy's conclusion to become the following simple expression for the time-dependent diffracted pressure from a rigid wedge as a function of time, the rate of volume flow, and the source/receiver wedge geometry.

$$p(t) = \left(\frac{-\dot{V}\rho_\mathrm{A}c}{4\pi\theta_w}\right)\left(\frac{\beta_+}{rr_0\sin\eta}\right)\exp(-\pi\eta/\theta_w) \quad t \ge \tau_0 \qquad (8.27)$$

$$p(t) = 0 \quad t < \tau_0$$

Again (see Fig. 8.9) r_0 and r are perpendicular ranges from source or receiver to wedge crest, respectively; θ_0 and θ are angles from a wedge side to source or receiver, respectively; θ_w is the wedge angle measured in the fluid as shown; y is the separation of the range vectors at the wedge crest; τ_0 is the "least" (minimum) time from source to crest to receiver; ρ_A and c are the density and sound speed of the medium; and $\dot{V}$ is the source strength in m³/s.

In terms of an impulse pressure P_0 of duration Δt at source range R_0, the source strength, i.e., the rate of volume flow, is

$$\dot{V} = \frac{4\pi P_0 R_0}{\rho_\mathrm{A}}\Delta t \qquad (8.28)$$

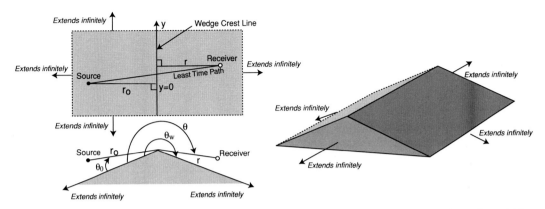

Fig. 8.9. Wedge geometry and symbols for the BT solution, shown for a rigid wedge obstacle. The unfolded geometry and the least-time path are pictured.

Also, for a rigid wedge, as given in (*M&C*) (12.2.3),

$$\beta_+ = \sum \frac{\sin\left[(\pi/\theta_w)(\pi \pm \theta \pm \theta_0)\right]}{1 - 2\exp(-\pi\eta/\theta_w)\cos\left[(\pi/\theta_w)(\pi \pm \theta \pm \theta_0)\right] + \exp(-2\pi\eta/\theta_w)}$$

(8.29)

where η is given in terms of the hyperbolic cosine,

$$\eta = \text{arc cosh}\left\{[c^2t^2 - (r^2 + r_0^2 + y^2)]/[2rr_0]\right\}$$

(8.30)

and the least time, from Fig. 8.9, is

$$\tau_0 = [(r + r_0)^2 + y^2]^{1/2}/c$$

(8.31)

For diffraction, we are interested in the time, τ, after the least time, τ_0,

$$\tau = t - \tau_0$$

(8.32)

In terms of τ and τ_0, we have

$$\eta = \text{arc cosh}\left[\frac{c^2(2\tau_0\tau + \tau^2)}{2rr_0} + 1\right]$$

(8.33)

or

$$(\sinh\eta)^{-1} = \left[\left(\frac{c^2\tau_0\tau}{rr_0} + \frac{c^2\tau^2}{2rr_0} + 1\right)^2 - 1\right]^{-1/2}$$

(8.34)

We employ the Greek letter Σ in (8.29) as an abbreviated symbol to designate the sum of the four terms obtained by using the four possible combinations of signs:

$$(\pi + \theta + \theta_0), \quad (\pi + \theta - \theta_0), \quad (\pi - \theta + \theta_0), \quad \text{and} \quad (\pi - \theta - \theta_0).$$

The BT theory has been broadened substantially by Kinney *et al.* (1983), Chu (1990), and Davis and Scharstein (1997) to apply to wedge surfaces of different impedances. For example, for pressure-release situations, the zero pressure boundary condition is realized by reversing the signs of two of the terms so that $(\pi + \theta + \theta_0)$ becomes $-(\pi + \theta + \theta_0)$ and $(\pi - \theta - \theta_0)$ becomes $(-\pi + \theta + \theta_0)$. That form of (8.29) is called β_-.

The infinite wedge diffractions given by (8.27) to (8.34) may be supplemented by reflections as described previously and as determined by the geometry. For the reflection amplitude, the R in (8.24) is the total path length from source to reflecting facet to receiver.

8.5.2 Scattering at underwater ledges, mesas and escarpments

The theoretical prescription for scatter of a Dirac delta function pressure by an infinite rigid wedge has been effectively realized by several laboratory experiments designed to verify the predictions and to model scattering elements at sea. To represent the conditions of the theory, the following three experimental techniques are used.

(1) *Dirac delta function sound source*: In the time domain a short-duration, impulselike source is used, and the theory for a delta function source, Section 8.5.1, is convolved with the experimental source wave form to produce the predicted wave form for the scatter. This is then compared with the laboratory experimental wave form.

(2) *Infinite wedge*: The experiment is concluded before scatter is received from the ends of the finite wedge.

(3) *Rigid wedge*: The laboratory wedge material has a high acoustic impedance (ρc) relative to the surrounding medium. To maximize the contrast, usually the medium is air rather than water, and often the wedge material is plasterboard (also called sheetrock) which has a smooth surface, and which damps compressional and shear waves which are absent from the theory.

The right-angle wedge, an underwater ledge

Li and Clay (1988) studied reflected and diffracted components of forward scatter in the time domain by using a laboratory, right-angle, rigid wedge ($\theta_w = 270°$) as shown at the left of Fig. 8.10. The source and receiver were on a line 79 cm above the wedge, and the theoretical BT solution for a delta function source was calculated for ten different displacements at the constant source to receiver separation 81 cm (right side of Fig. 8.10). The direct signal and theoretical reflection and diffraction components are calculated from (8.27). The direct and reflected sound appear as sharply peaked curves which represent the Dirac δ function.

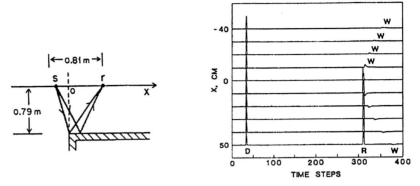

Fig. 8.10. *Left,* geometry of the theoretical and experimental test of the theory. *Right,* the theoretical wedge solution for 10 positions of a delta function source/receiver at fixed separation on a line over the wedge. The direct signal is D, reflected is R, and wedge diffracted is W. The half-amplitude diffraction occurs at $x = 0$, the midpoint of the source/receiver over the crest of the wedges. (From Li and Clay, 1988.)

Fig. 8.11. Free space reception of signal from a spark source. Receiver is a 0.6 cm electret microphone. Signal sampled at 10 μs intervals. (From Li and Clay, 1988.)

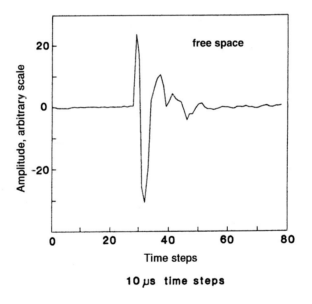

The much weaker diffraction signal is negative when the midpoint of the source/receiver is over the horizontal surface (positive x displacements) and positive when it is beyond the edge (negative x values). The diffraction is stronger when the midpoint between source and receiver is closer to over the crest of the wedge. When the reflection point is at the edge, $x = 0$, there is a half-amplitude "reflection" from the corner.

The laboratory experiment used a signal as shown in Fig. 8.11. The experimental result is at the left side of Fig. 8.12. The predicted scatter, at

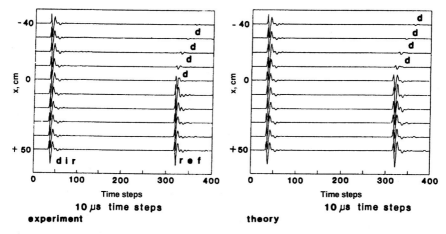

Fig. 8.12. Experimental and theoretical signals of direct, reflected, and diffracted sound from the wedge-scattering experiment sketched in Fig. 8.10. The direct and reflected (when it exists) signals occur at constant time delays, independent of source/receiver positions. The diffraction timing, amplitude, and phase depend on the position of the source/receiver with respect to the wedge ridge. (From Li and Clay, 1988.)

the right side of Fig. 8.12, is obtained by convolving the theoretical solution with the experimental signal shown in Fig. 8.11. Direct, reflected, and diffracted signals are all seen in Fig. 8.12. For positive values of x, there are reflections which dominate the total scattered signal; the delayed, negative diffraction is just barely evident. For negative values of x, the only scatter is the positive diffraction component.

The right-angle step ledge

Chambers and Berthelot (1994) have used the BTM method – (8.24) through (8.34) – to predict the contributions from the reflection and the two diffractions that occur in propagation over a step discontinuity (Fig. 8.13). Their extensive impulse measurements have confirmed that the theory provides accurate time-domain predictions for that geometry. They also show that the acoustic field scattered from the step is essentially unchanged if the step is smoothly curved with a radius less than the predominant wavelength of the sound. The experiment can be considered to be a representation of propagation over an escarpment at sea.

The geometry is at the left in Fig. 8.13 (not to scale). The source is a spark with peak-to-trough duration approximately 50 μs. In addition to the direct incident sound "i" (see Fig. 8.14), the step results in potential reflection from the top plate, "rl"; reflection from the bottom plate, "r2"; and diffraction from the top exterior corner at E, "dl,"; and component "d2" which is diffracted at the top corner and then reflected at the

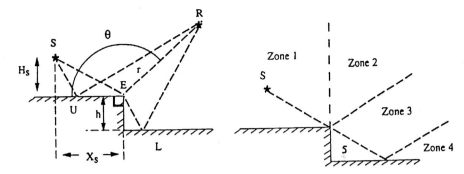

Fig. 8.13. *Left*, the geometry of the step discontinuity experiment, $H_s = 19.3$ cm, $X_s = 466$ cm, and $h = 1.8$ cm (the height in the figure has been exaggerated). *Right*, the five zones of potential interference between the four principal components of incidence, reflection, and diffraction. (From Chambers and Berthelot, 1994.)

Fig. 8.14. Experimental and theoretical reflections and diffractions within and near transition zones for a rigid step geometry. Receiver moves along arc 34.3 cm from corner E (see Fig. 8.13). (*a*) In zone 4 close to zone 3. (*b*) At edge of zone 3. (*c*) In middle of zone 3. (*d*) At edge of zone 3. (*e*) In zone 2, close to zone 3. (From Chambers and Berthelot, 1994.)

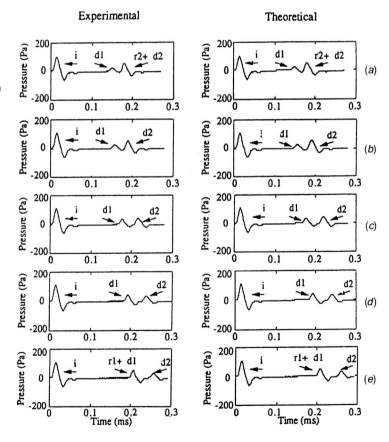

bottom of the step. Multiple diffractions, which would be of lesser strength, are not considered.

Two, three, or four of the five signal components can be received in the five different zones shown at the right of Fig. 8.13. A microphone in zone 1 receives i, r1, and d1; in zone 2, i, r1, d1, and d2; in zone 3, i, d1, and d2; in zone 4, i, r2, d1, and d2; in zone 5, d1 and d2. These expectations have been confirmed by moving a microphone at radius 34.3 cm over an arc that passes from zone 4 through zone 3 to zone 2. Note that in Fig. 8.14 the simple diffraction d2 moves in to replace the reflected diffraction r2 at the edge of zones 3 and 4, and the top plate reflection rl comes in as one moves into zone 2. The experimental and theoretical oscillograms are virtually identical.

The right-angle mound on a plane
The geometry of a 90° triangular rigid wedge protruding above a rigid plane (Fig. 8.15) has been examined by Li *et al.* (1994). In this more complicated case, in addition to reflections and "single" diffractions, there are significant secondary diffractions of the diffracted signals, which we call "double diffractions." The analysis of double diffraction is in Section 8.6.2.

8.5.3 Digital calculations for finite wedges

Equation (8.27) may be implemented for digital calculations by obtaining the average $p(t)$ over each of many small discrete intervals. This produces a series of impulses as shown in Fig. 8.16 from regions of the wedge as indicated in the inset of Fig. 8.16 and in Fig. 8.17. The integration has been performed by using the DCADRE routine (Digital

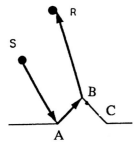
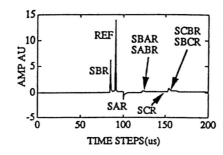

Fig. 8.15. Geometry (*left*) and reflection and diffraction amplitudes (*right*) for a 90° rigid wedge on a rigid plane. The paths are indicated by the combination of letters, for instance, SBR is a single diffraction at B; REF is a simple reflection; SABR is a double diffraction of the signal that starts at S, is diffracted first at A and then at B, and then goes to the receiver. In (x, y, z) cm; the source is at (20, −13, 0), receiver is at (20, +13,0) where y is the ridge direction. (From Li *et al.*, 1994.)

Fig. 8.16. Some of the discrete values of impulse pressure for diffracted backscatter from a rigid right-angle wedge ($\theta_w = 270°$) in air, calculated from (8.27). The impulse pressures are separated by $\Delta T = 1$ μs. The range is $r = r_0 = 50$ cm, and the angles of incidence and scatter are $\theta = \theta_0 = 146°$. (From Medwin et al., 1988.)

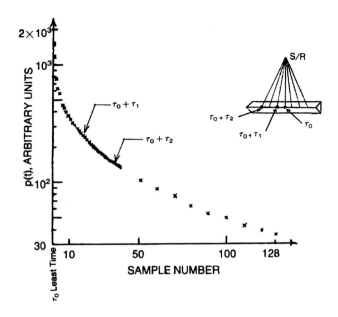

Fig. 8.17. Unfolded geometry showing discrete paths (solid lines) and their boundaries at $[n \pm (1/2)]\Delta T$ (dashed lines) for the assumption that $n = 0$ is the least-time path. The blackened regions on the wedge crest are Huygens sources for even-ordered time lags. (From Medwin et al., 1982.)

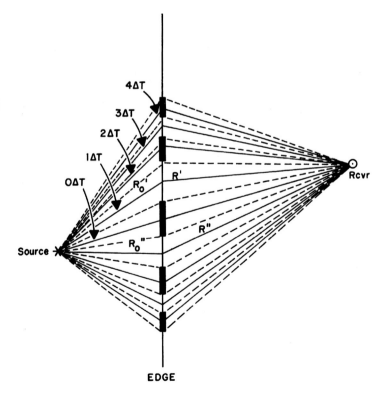

Cautious ADaptive Romberg Extrapolation), which is listed in IMSL Library Edition 9.2 (1984). In fact, away from the shadow boundary and the reflection direction, the pressure is generally changing slowly enough to permit the mid-time instantaneous value to represent the average with only a small error. The time increment ΔT is selected to produce the minimum bandwidth BW (Min) $= (2\Delta T)^{-1}$ for the desired frequency spectral description.

Where the pressure is rapidly changing as a function of travel time, the integration for the singular least-time point may be unstable. Keiffer *et al.* (1994) replaced the original integration range, as shown in Fig. 8.16, by a more robust integration with a different definition of the index, $n = 0$. In Keiffer's method, the separation between adjacent points is still ΔT, but the integration is centered on $\tau = (\Delta T)/2, 3(\Delta T)/2, 5(\Delta T)/2$, and so forth:

$$\left\langle p\left(\tau_0 + \frac{\Delta T}{2}\right)\right\rangle = \frac{1}{\Delta T} \int_{\tau_0}^{\tau_0+\Delta T/2} p(t)\, dt \tag{8.35}$$

In general,

$$\tau_n = \left(\frac{2n + 1}{2}\right) \Delta T \tag{8.36}$$

with $n = 0, 1, 2$, and so forth.

Once the diffracted pressure given by (8.27) and (8.29) has been digitized, the continuous diffracted signal may be interpreted as a series of contributions from Huygens wavelets that radiate sequentially from the wedge ridge, starting with the largest contributor from the intersection of the least-time path with the wedge. In this interpretation, the least-time impulse is the first and the strongest of the sequence; it is followed by pairs of wavelets of lesser strength from each side of the least-time intercept. Figure 8.17 shows this interpretation.

Simplifying approximations
The largest diffraction contributions occur for incremental times that are small compared to the least-time, $\tau \ll \tau_0$. Then certain analytical simplifications can be calculated,

$$\lim_{\tau \to \tau_0} (rr_0 \sinh Y)^{-1} \to (2\tau_0 \tau c^2 rr_0)^{-1/2} \tag{8.37}$$

and, except near the geometrical shadow boundary or reflection direction, we have the approximation

$$p(\tau) \simeq B\tau^{-1/2} \tag{8.38}$$

where

$$B = \left[\frac{\dot{V}\rho_A\beta}{4\pi\sqrt{2}(\tau_0 rr_0)^{1/2}\theta_w}\right] \tag{8.39}$$

and

$$\dot{V} = 4\pi P_0 R_0 (\Delta t)/\rho_A \qquad (8.40)$$

Under these conditions, except for the rapidly changing pressure at the least-time point, $n = 0$, the discrete pressures at $n\Delta T$ are given simply by

$$\langle p(n\Delta T)\rangle_{n\geq 1} = \frac{B}{\Delta T} \int_{(n-1/2)\Delta T}^{(n+1/2)\Delta T} \tau^{-1/2} d\tau \qquad (8.41)$$

$$= B\left(\frac{2}{\Delta T}\right)^{1/2} [(2n+1)^{1/2} - (2n-1)^{1/2}]$$

The effect of being close to the geometrical shadow boundary has been evaluated (Medwin *et al.*, 1982) by considering the small angular displacement, ε, away from the shadow boundary:

$$\theta = \theta_0 + \pi + \varepsilon \qquad (8.42)$$

It is found that (8.41) holds, provided that

$$\varepsilon \gg \left[\frac{c\tau_0}{(rr_0)^{1/2}}\right]\left(\frac{\tau}{\tau_0}\right)^{1/2} \qquad (8.43)$$

8.6 Diffraction by wedges and plates: frequency domain

8.6.1 Experimental verifications of predictions

In considering the spectral response implied by the BT impulse theory, one notes first that, sufficiently away from the shadow boundary or reflection direction, for times small compared with the least time, $\tau \ll \tau_0$, the Fourier transform of (8.38) gives the frequency spectrum,

$$P(f) = (B/2)(1+i)f^{-1/2} \qquad (8.44)$$

where B is given in (8.39). See Fig. 8. 18.

The $f^{-1/2}$ dependence is an analytical statement of the observed fact that high frequencies do not bend around corners as well as low frequencies. The behavior, which is documented in Fig. 8.18 is well known to noise control engineers.

A common reference pressure is the "white noise" spectral description of a Dirac delta function free-field point source,

$$P_\delta(f) = \frac{\dot{V}\rho_A}{4\pi R N \Delta T} = \frac{P_0 R_0}{R}\left(\frac{\Delta t}{N \Delta T}\right) \qquad (8.45)$$

where $N\Delta T$ is the total duration of the sequence of samples that has been Fourier-transformed, and $\dot{V}$ is the source strength, m^3/s. The time between samples, ΔT, is selected to yield the bandwidth, BW, that

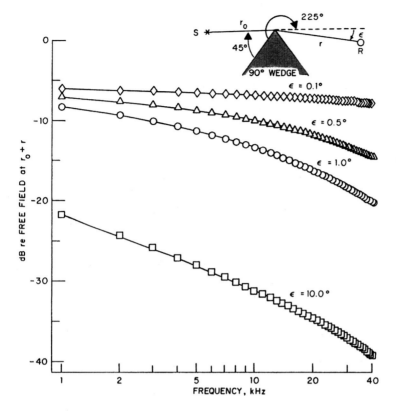

Fig. 8.18. Predicted frequency spectra of relative diffracted pressure as a function of angle of penetration, ε, into the geometrical shadow region when $r = r_0 = 100$ m, $\theta_0 = 45°$ for a right-angle wedge ($\theta_w = 270°$). Note the $f^{-1/2}$ behavior for low frequencies at $\varepsilon = 10°$. (From Medwin et al., 1982.)

is ultimately desired, where $BW_{\text{Min}} = (2\Delta T)^{-1}$. Practically, one sets $BW = 4BW_{\text{Min}}$. This achieves the smaller values of ΔT that are desirable for the rapidly changing strong early arrivals.

It is convenient to form the ratio of (8.44) to (8.45) and thereby obtain the relative diffracted pressure compared with the point source radiated pressure for a given frequency. This quantity is independent of the source strength $\dot{V}$ or the source reference pressure P_0 and is a popular way to show the relative strength of the diffracted signal. Such a description is presented in Fig. 8.18, where the spectrum is plotted as a function of penetration into the shadow region, ε. As ε approaches zero, the diffracted sound approaches one-half of the free-field pressure (-6 dB), independent of the frequency. For larger ε, the $f^{-1/2}$ asymptote is recognized as the slope for low frequencies, e.g., below 10 kHz at $\varepsilon = 10°$ in this graph.

Model of underwater ridge
Detailed predictions of scattering by underwater ridges (8.27) and its transform to the frequency domain (8.44) have been verified by

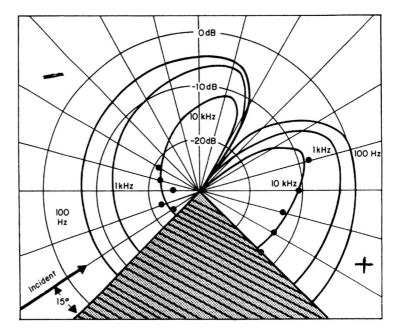

Fig. 8.19. Polar diagram of frequency domain diffraction loss compared with divergence loss for a free-field range $r + r_0$ for a right-angle rigid wedge, $\theta_w = 270°$ in air. The ranges are $r = r_0 = 25$ cm. The phase is positive for the right lobe and negative for the left lobe. Comparison with experimental data dots at 10 kHz is from Bremhorst (1978) and Bremhorst and Medwin (1978). (For details, see Medwin, 1981.)

several laboratory tests employing realistic rigid surfaces in air. In these validating experiments, solid wedges or plates are used in air because a surface is more nearly "rigid" in air than in water, due to the greater ρc mismatch. But also, since one must avoid extraneous scatter from the ends of a real wedge, the required size of the wedge can be one-fifth as great in air as in water.

The polar diagram of BT diffraction scatter for a right-angle wedge, Fig. 8.19, yields an instant comprehension of the relative amplitude and different phase of the scatter in different directions for three frequencies. This calculation models the backscatter and the forward diffraction into the "shadow" region of a ridge at sea. This figure, which shows diffraction loss (in dB relative to the level if the signal had diverged spherically from a point source to the total range $R = r + r_0$), when compared to experiment, was the first evidence of the effectiveness of the BT theory.

Model backscatter from the top of a wall

The predicted diffraction backscatter from the top of a wall, modeled by a rigid vertical plate, is shown for all angles at three frequencies in Fig. 8.20. The easily-measured experimental data for backscatter at angles 30° and less is another quantitative demonstration of the excellent predictive capability of the BT theory and the limited realm of agreement between experiment and HK theory.

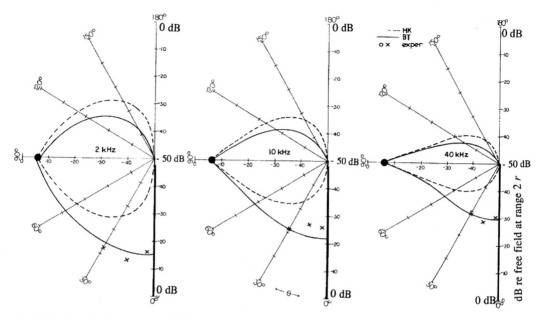

Fig. 8.20. Angular dependence of diffraction backscatter from a rigid plate
($\theta_w = 360°$) at 2, 10, and 40 kHz, incorrectly predicted by HK (dashed line) and
correctly by BT (solid line) theories, and measured at range 25 cm in air. (From
Jebsen and Medwin, 1982.)

Model backscatter from the corner of a mesa

When properly oriented, a right-angle rigid wedge is a model of the
corner at the top of the vertical face of an underwater mesa (Fig. 8.21).
Backscatter from the front corner of the mesa wall is modified by the flat
top (compare with Fig. 8.20). Figure 8.21 shows that BT theory agrees
with experiment at both angles tested. The HK predictions have an error
of approximately 8 dB for 30° incidence, 15 dB for 15° incidence, and
probably as much as 25 dB at near grazing incidence of 4°.

8.6.2 Double diffraction at a thick barrier

Huygens' interpretation of the action at a diffracting edge, e.g., Fig. 8.17,
would be that the original delta function source initiates reradiation from
"secondary sources" on the diffracting edge. The pressure signal received
at the field point is the accumulation of the radiation from delta function
secondary sources initiated at the proper times along that edge.

   In many problems (for example, shadowing by a seamount or *guyot*)
there is a second diffracting edge, between the first one and the receiver.
Figure 8.22 is an idealization of the geometry.

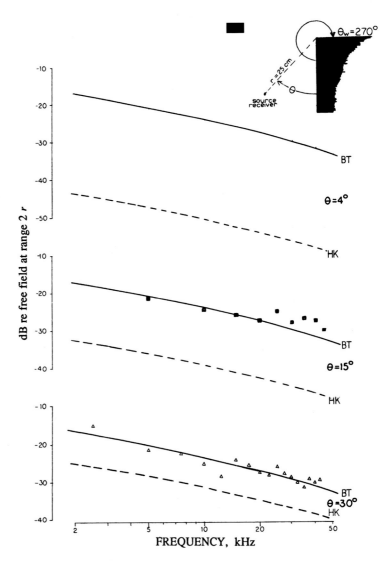

Fig. 8.21. Backscatter diffraction from a right-angle wedge, a model of an ocean escarpment, measured in air at range $r = r_0 = 25$ cm. Comparison between experiment, Biot–Tolstoy theory (BT) and Helmholtz–Kirchhoff (HK) theory at three different angles with respect to the wedge face. (Data from Bremhorst, 1978; theory from Jebsen and Medwin, 1982.)

From the Huygens point of view, there will be an infinite number of sources on the first edge, each of which spawns an infinite number of sources on the second edge. Practically, because they are far stronger than the later arrivals, only the earliest arrivals need to be considered.

The "secondary source" contributions come in pairs from points on either side of the least-time crestal source, which is located at the

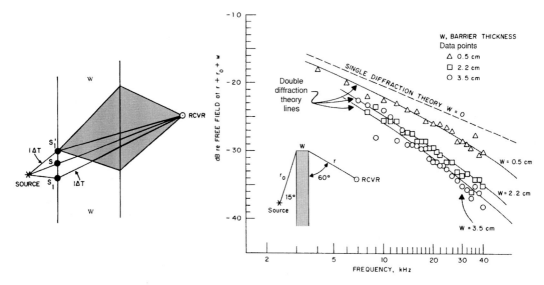

Fig. 8.22. *Left,* Double diffraction geometry for a barrier of thickness W. The general case is shown in the unfolded geometry with the least time contributor from S and the $n=1$ contributors, S_1 and S'_1, at $1 \Delta T$. One of the $1 \Delta T$ contributors is shown as the source of a second line of diffractors from the second edge. *Right,* relative spectral loss calculated for three thick-plate barriers at range $r = r_0 = 25$ cm. The double-diffraction laboratory data points compare well with the double-diffraction theory; the single-diffraction theory, dashed line, $w = 0$, does not describe the diffraction adequately. (From Medwin *et al.*, 1982.)

point of intersection of the source–receiver axis and the first crestline. These secondary sources will have the source strengths $\dot{V}_{ssn}$, which are necessary to produce the expected field after traveling the range R from their crest positions to the field position. That is, for a symmetrical geometry, the nth secondary source has the strength

$$\dot{V}_{ssn} = \left(\frac{1}{2}\right) \frac{\langle p(n\Delta T)\rangle}{p_\delta} \dot{V} \qquad (8.46)$$

where

$$p_\delta = \frac{\dot{V}\rho_A}{4\pi R\Delta T} \qquad (8.47)$$

The calculation of the mean value $< >$ follows the prescription of (8.35). Calculations for non-symmetrical contributions are considered in Medwin *et al.* (1982).

8.6.3 Forward diffraction at a seamount

Submerged mountains, "seamounts," can interrupt sound propagation and create a shadow zone behind them. This was conclusively

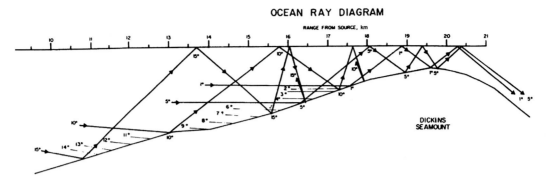

Fig. 8.23. Reconstruction of sound wave interaction at upslope of Dickins
seamount in experiments by Ebbeson and Turner (1983) and Chapman and
Ebbeson (1983). The sound source is about 20 km to the left of the crest. The
incoming rays are identified by their angles with the horizontal at the source.
The 15° ray reflects on itself after two sea surface reflections and three
interactions with the seamount; it does not go over the seamount. The 5° ray
forward reflects from the seamount three times and from the ocean surface three
times before it crosses over the seamount. The forward scattered wave at the
rough seamount surface diffracts over the crest and arrives at the hydrophone
sooner than the multiply reflected rays. However, it is weakened by scattering
from the rough ocean surface, as well as from the seamount. (From Medwin
et al., 1984a.)

demonstrated by Canadian experiments at Dickins seamount off the
coast of British Columbia (Fig. 8.23). (The seamount was named for
a respected scientist of the US Coast and Geodetic Services, not the
famous English author, Charles Dickens.)

In the experiment at Dickins seamount, a CW shadowing loss of
15 dB at 230 Hz was first observed by Ebbeson and Turner (1983).
Later, using an explosive source, a splitting of the signal was found by
Chapman and Ebbeson (1983). The earlier arrival showed a frequency-
dependent shadowing loss proportional to $f^{1/2}$; it was the only signal
received during high sea states.

The experimental results at sea have been explained by using a laboratory model to consider the two means by which sound may reach into
the shadow of the seamount. The later arrival, revealed by an impulse
source, is due to multiple reflections between the insonified surface of the
seamount and the ocean surface. This contribution, which repeatedly involves the ocean surface, cannot be found during high sea states because
of losses by scattering at the rough ocean surface. The rough surface
was modeled by a gravel-covered plate suspended over the laboratory
seamount. The laboratory ratio of (wavelength)/(surface rms height) was
the same as the ocean ratio λ/h for a Pierson–Moskovitz sea during the
35 knot wind.

At this point we are more interested in the earlier arrival, which the model proved to be due to multiple forward scatter along the insonified rough seamount face followed by diffraction over the crest of the seamount. Both at sea and in scale model experiments by Spaulding (1979) and Jordan (1981), this component was found to have the $f^{-1/2}$ dependence predicted by the simplified spectral version of the BT theory (8.44). The work is summarized in Medwin *et al.* (1984a).

The diffraction component can be roughly estimated by assuming that the seamount is a simple exterior rigid wedge of upslope and downslope 14° with the horizontal ($\theta_w = 280°$), where we ignore all the bumps along the way. A more accurate prediction is obtained by using the double-diffraction technique described in the previous section. Equations (8.39) and (8.44) shows that there is a range dependence $r^{-1/2}$ and a frequency dependence $f^{-1/2}$ (or $\lambda^{+1/2}$). Therefore, in order to have the laboratory scaling independent of range and frequency, the magnitude of the *diffraction strength, DS,* is defined in terms of the diffracted spectral pressure, P_D,

$$DS = 20 \log \left(\frac{P_D r_0}{P_0 R_0} \right) \left(\frac{r}{\lambda} \right)^{1/2} \qquad (8.48)$$

where $P_D = P_D(\theta, \theta_0, \theta_w, r, r_0, z)$ is the diffracted pressure given by (8.44) and P_0 is the reference pressure at the reference distance $R_0 (=1 \text{ m})$.

This definition allows us to compare air laboratory experimental data at frequencies of tens of kilohertz and ranges of tens of centimeters on the same graph as ocean data at frequencies of the order of 100 Hz and at ranges of several kilometers. The definition permits theoretical wedge diffraction calculations to also be presented as a function of r/λ, as shown in Fig. 8.24.

8.6.4 Finite wedge approximations to the infinite wedge

The Biot–Tolstoy solution is for an infinite wedge. The crest length is infinite, and the two facets that comprise the wedge are semi-infinite planes joined at the crest line. To use the BT equations in the real world, one truncates the infinite time series of the diffracted pulse. The truncation is done either (1) when the remaining terms can offer no significant correction to the total diffraction amplitude; or (2) at the time when the incident sound is beyond the end of the wedge; or (3) in order to obtain a desired frequency resolution (= 1/duration of signal) after performing the Fourier transform.

The scatter from a *finite* wedge, insonified by perpendicular incidence to the wedge crest at its midpoint, will lack the infinite length tail

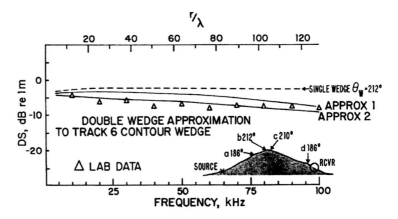

Fig. 8.24. Comparison of laboratory diffraction strengths (data triangles Δ) plotted versus r/λ (upper abscissa) with a theoretical calculation for a symmetrical single wedge of slope 14° and with two different freehand-drawn two-dimensional double-diffraction scale models of Dickins seamount shown in silhouette. The lower abscissa in kHz are the frequencies of the sound in the laboratory model where the range was several centimeters. In model approximation 1 it is assumed that the two wedges are $\theta_w = 186°$ and $\theta_w = 210°$ at crests a and c, respectively; in the second model it is assumed that the wedges have angles $\theta_w = 212°$ and $\theta_w = 186°$ at positions b and d. The wedges are assumed to have crest lines perpendicular to the sound track. The double-diffraction calculation for model approximation 2 agrees with experiment with less than 1 dB rms error. (From Mcdwin *et al.*, 1984a.)

of the impulse response shown for the "infinite" wedge in Fig. 8.16. The spectral diffraction from such a real wedge is found by transforming the truncated BT impulse response to the frequency domain. It will be a function of the sound frequency, the wedge angle θ_w, the angles θ_0 and θ, and the ranges r_0 and r.

Great interest is in the case of backscatter, $\theta_0 = \theta$ and $r_0 = r$. For backscatter at a given range, shorter wedges cause shorter duration responses – that is, the impulse response of Fig. 8.16 becomes more truncated. For backscatter from a given length of the wedge crest, greater ranges cause the wedge impulse response to be of shorter duration; again the response is truncated.

Finite wedge backscatter for a point source shows the effect of a Fresnel zone interference that depends on the length of the wedge and the sound frequency (recall Section 1.7.5 for point-source sound scattered by a finite plane surface). Also, the spectral diffracted pressure can be expected to decrease rapidly when the wedge crest length W is less than $(\lambda r)^{1/2}$, where λ is the wavelength in the transformed signal, and r is the range; the effect is reminiscent of Rayleigh scatter of a plane wave by a sphere that is small compared with the wavelength (recall Fig. 6.7 for plane wave incidence).

Problems

In doing the calculations for sound pressures, determine the relative amplitudes and ignore the source power.

8.1 Compare the exact values and the large κr approximation (8.10) for $H_0^{(2)}(\kappa r)$. Test the range of $0.2 < \kappa r < 5$.

8.2 Compute and plot the wave number components γ_m and κ_m for a waveguide that has free upper and lower interfaces. The waveguide is 10 m thick.

8.3 Plot the first three modes as functions of frequency. If cut-off exists, explain why.

8.4 Use the free top and free bottom water waveguide for simplicity. The waveguide is 20 m thick, and the source and the receiving array are at 15 m depth. The source has the frequency of 200 Hz and is at 10 000 m range. Horizontal arrays can be steered to locate the direction to a source. The line array is 750 m long and has 10 elements. The array processing includes time delays for digitally steering the beam to "look" in different directions. The steering directions are from $0°$ at normal to the line array to $\pm 90°$. (*a*) How many modes are excited? (*b*) Compute the array output *for each mode* as a function of steering angle for the source directions of $0°$ and end fire, $90°$ directions. (*c*) Repeat (*b*), except that all modes are received. For simplicity, ignore the mode interference terms.

8.5 Discuss the effects of water waves on sound transmissions. Read Clay, Wang, and Shang (1985). Compare their conclusions to those given in Tolstoy and Clay (1987, pp. 232–6).

8.6 How would you expect the time dependence of a surface to affect the transmission in a waveguide? Scrimger (1961) describes an elegant experiment: "Signal amplitude and phase fluctuations induced by surface waves in ducted sound propagation," (*J. Acoust. Soc. Am.* 33, 239–74, 1961). What are the most important regions for causing fluctuations of the sound transmissions?

8.7 Write computer programs for (8.24) through (8.34) so that you can solve the following problems.

8.8 Use your computer program of the BT theory to verify the graph of Fig. 8.19, which describes diffraction of 100 Hz, 1 kHz, and 10 kHz sound by a rigid wedge in air when the source and receiver ranges are 25 cm and $\theta_0 = \theta = 15°$.

8.9 Repeat the calculation of Problem 8.8 when the source and receiver are in air at a range of 100 m from the crest.

8.10 Repeat the calculation of Problem 8.8 when the source and receiver are at a range of 100 m from a right-angle ridge in the ocean. Repeat the calculation when the source and receiver are at a range of 1 km in the ocean (assume $c = 1500$ m/s).

8.11 Repeat the calculation of Problem 8.10 when the source is at range 1 km; angle $\theta_0 = 30°$, $45°$, $60°$, and $85°$; and receiver is at range 100 m, $\theta = 15°$.

8.12 Use your computer program of the BT theory to verify the predictions for backscatter diffraction by a rigid plate in air as shown in Fig. 8.20 for frequencies 2 kHz, 10 kHz, and 40 kHz at angles $0°$, $30°$, and $60°$.

8.13 Calculate the forward diffraction loss (compared with spherical divergence over the same distance) for a range of frequencies for a symmetrical, wedgelike seamount of slope $14°$ on each side of the wedge ($\theta_w = 208°$). Assume that the water is isothermal and ignore the effect of the water surface above the seamount. Assume that the ranges are $r = r_0 = 2$ km, that the source is on the upslope of the seamount, and that the receiver is on the downslope.

8.14 How would your answer to the previous problem change if, instead of a downslope, there is a horizontal mesa ($\theta_w = 194°$) on the right side of the wedge?

8.15 How would your answer to Problem 8.14 change if there is a horizontal mesa that extends for 1 km at the top of the seamount (a *guyot*)?

Further reading

See Chapters 17–24 for very recent, long-range, acoustical revelations of the physical characteristics of the ocean volume and ocean bottom. Extensive references are given.

Biot, M. A. and Tolstoy, I. (1957). Formulation of wave propagation in infinite media by normal coordinates with an application to diffraction. *J. Acoust. Soc. Am.*, **29**, 381–91.

A "block-buster" publication that was in the literature for 20 years before there was recognition of its vast applicability to a broad range of geometries from highway noise barriers to ocean escarpments and rough surfaces.

Parvulescu, A. (1995). Matched-signal (MESS) processing by the ocean. *J. Acoust. Soc. Am.*, **98**, 943–60.

A long-delayed description of his landmark concept (patented by Parvulescu in 1962, described publicly in 1969, and finally the basis for much ocean research beginning in the 1990s) wherein the multi-path propagation through the complex, time-varying ocean is interpreted advantageously by using time-reversal. See Chapter 19.

Medwin, H. and Clay, C. S. (1998). *Fundamentals of Acoustical Oceanography.* Academic Press.

See Chapter 11, "Waveguides: Plane Layers and Wedges." Derivations of the waveguide description of sound propagation and its signal processing. A development of the Biot–Tolstoy theory for wedge or channel geometry, including images. See also Section 12.4.3, Synthetic seismic profiles.

Chapter 9
Scatter and transmission at ocean surfaces

Summary

Real ocean surfaces and real ocean bottoms are rarely smooth, and they are usually not stable. However, a sufficient knowledge of, and wise assumptions about, the spatial and temporal statistics of an ocean surface allows one to predict the statistics of the scattered and transmitted sound. Dynamic laboratory-scale models have shown how this can be done.

The inverse problem is of particular interest to the physical oceanographer. In this chapter we learn how one can use the reflection and transmission of sound to characterize a remote, unseen ocean surface or rough bottom in terms of the statistics of its surface height and slope, or from the point of view of its spatial and temporal correlation functions and displacement spectrum.

Contents

* This section contains some advanced analytical material.

© H. Medwin 2005

9.1 Statistical descriptions of ocean surfaces and dynamic laboratory models*

9.1.1 The wind-blown surface

The basic, analytical descriptions of ocean waves were developed four decades ago in seminal publications by Cox and Munk (1954), Longuet-Higgins (1963), Pierson and Moscowitz (1964), Kinsman (1965), Phillips (1977) and others. These works describing the sea have been followed by dynamic laboratory scale models designed to accurately mimic and quantitatively study the complex ocean/acoustic behavior.

Oceanographers commonly describe the ocean surface in terms of the probability density function (PDF) of the displacements of the rough surface, the frequency spectrum of the displacements, and/or the directional wave spectrum. As we see in the following sections, not only those quantities but also the PDF of the slopes of the surface and the correlation function of the surface displacements are closely related to the scattering of sound from the surface. We introduce these quantities briefly here, and present examples of data from real seas as well as laboratory "seas." Properly designed dynamic or static laboratory scale models are effective devices for controlled studies of sound scatter.

The PDF of displacements of a wind-blown sea is close to Gaussian, as seen, for example, in Fig. 9.1. The Gaussian PDF is given by,

$$w(\zeta) = \frac{1}{h\sqrt{2\pi}} \exp\left[-\frac{1}{2}\left(\frac{\zeta}{h}\right)^2 \right] \tag{9.1}$$

where $w(\zeta)$ is the probability of displacement ζ from the mean surface and h is the rms height of the displacements.

The first two "moments" are

$$m_1 = \langle \zeta \rangle = 0 \tag{9.2}$$

and

$$m_2 = \langle \zeta^2 \rangle = h^2 \tag{9.3}$$

Modifications of the simple Gaussian are necessary for a more nearly correct specification of the ocean PDF. These are components of a Gram–Charlier series described by Longuet-Higgins (1962) in terms of the moments.

Two of the higher-order moments are the skewness,

$$m_3 = \langle \zeta^3 \rangle / (2\langle \zeta^2 \rangle^{3/2}) \tag{9.4}$$

and the kurtosis (or "peakedness"),

$$m_4 = (\langle \zeta^4 \rangle - 3\langle \zeta^2 \rangle^2)/(2\langle \zeta^2 \rangle^2) \tag{9.5}$$

* This section contains some advanced analytical material.

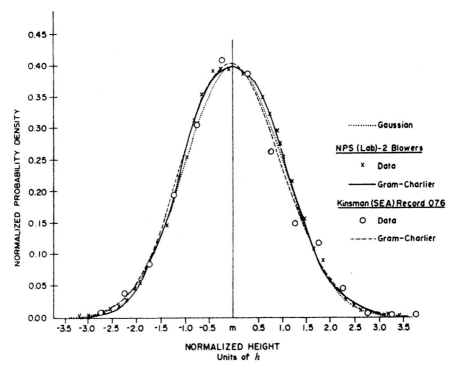

Fig. 9.1. Normalized PDF of water surface displacements. Ocean data are circles: dashed line is a fitted curve using first four moments of the Gram–Charlier expansion with $h = 2.72$ cm, $m_3 = 0.092$, and $m_4 = 0.031$ for wind 6.75 m/s, measured 12.5 m above surface at sea, from Kinsman (1960). Crosses are data for a carefully designed laboratory "sea" with $h = 0.119$ cm, $m_3 = 0.086$, and $m_4 = 0.073$; solid line is laboratory four-moment fit. The Gaussian is the simple dotted curve. (Medwin and Clay, 1970.)

Extensive studies of scattering from an oceanlike surface have been performed in laboratory scale models. The laboratory "sea" is generated by a paddle, or fans in a water–wind tunnel. The scaling is accomplished by using higher laboratory sound frequencies, in order to reproduce the ratio h/λ of an acoustical scattering trial at sea. The higher-order moments m_3 and m_4 should be similar for the scaling to be appropriate for acoustical scattering studies. In Fig. 9.1 the laboratory m_3 and m_4 are within the range of sea values for wind speeds 4 to 9 m/s.

The PDF of sea slopes is also important in sound scatter from the sea surface. Figure 9.2 (a) presents the crosswind (subscript c) and windward (subscript w) behavior for a 10 m/s wind at sea (Cox and Munk, 1954). The latter shows the skewness expected in the windward direction; both

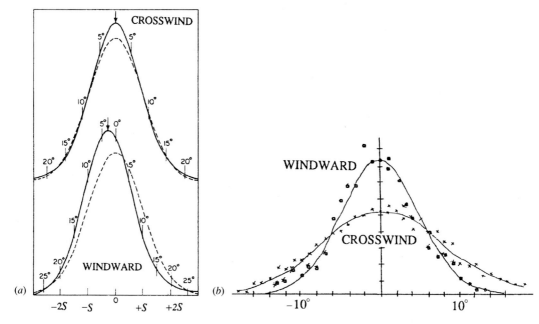

Fig. 9.2. (a) PDF of slopes measured optically during 10 m/s wind at sea. The solid lines are observed normalized distributions crosswind and windward; dashed lines are simple Gaussians of the same rms slope. (Cox and Munk, 1954.) (b) slope PDF for NPS laboratory 3 blower "sea" obtained by two-minute integrations of photocell output using a collocated point light source for various angles windward and crosswind. The lines are simple Gaussians of the same rms slope (windward, circles, $s_w = 7.8°$; crosswind, crosses $s_c = 4.9°$). (Ball and Carlson, 1967.)

measurements show peakedness. The mean squared slopes for clean water were empirically given as

$$s_w^2 = 0.000 + 3.16 \times 10^{-3} W \pm 0.004$$
$$s_c^2 = 0.003 + 1.92 \times 10^{-3} W \pm 0.002 \qquad (9.6)$$
$$s^2 = s_w^2 + s_c^2 = 0.003 + 5.12 \times 10^{-3} W \pm 0.004$$

where wind speed W is in m/s, measured at the traditional height 12.5 m above the surface at sea; s is the rms slope assumed to be independent of direction. Oily water has lesser rms slopes. Slopes of a dynamic laboratory model are shown in Fig. 9.2(b).

The temporal autocorrelation of heights $\zeta(t)$ is the temporal autocovariance divided by the mean squared height,

$$C(\tau) = (< \zeta(t)\zeta(t + \tau) >)/h^2 \qquad (9.7)$$

where τ is the time lag.

A "typical" temporal correlation at sea is shown at the top of Fig. 9.3. It is typical in that it shows a periodicity that corresponds to the frequency of the most prominent wave component, and a correlation magnitude that decreases as the time lag increases. A very similar scaled version of the temporal correlation at sea can be achieved by the use of a programmed plunger or fans operating in a sufficiently long water tank with the result as shown at the bottom of Fig. 9.3.

To manipulate analytical expressions for the correlation of acoustical scatter as a function of the correlation of ocean surface displacement, a first step is to approximate the temporal surface correlation – for example, by writing it empirically as

$$C(\tau) = \exp[-(\tau/T)^2 \cos \Omega_m \tau] \tag{9.8}$$

where T is the time lag for $C = e^{-1}$ and Ω_m is close to the angular frequency of the peak of the surface wave displacement spectrum (Fig. 9.3).

The general expression for the temporal autocovariance of the surface displacements is related to its spectral density, $\Phi(\Omega)$, (m^2/Hz) by the Wiener–Khinchine theorem, which gives the transforms. To compare with his graph, we follow the normalization of the transform relation as in Phillips (1977), but with our symbols:

$$\Phi(\Omega) = \frac{1}{2\pi} \int_{-\infty}^{\infty} h^2 C(\tau) e^{+i\Omega t} d\tau$$
$$C(\tau) = h^{-2} \int_{-\infty}^{\infty} \Phi(\Omega) e^{-i\Omega \tau} d\Omega \tag{9.9}$$

where $\Omega = 2\pi F$ is the angular frequency of a component $F(s^{-1})$ of the surface waves.

The spectral density is the part of the total mean-squared height in a 1 hertz bandwidth. Therefore the mean-squared height, h^2, is the integral over all frequencies:

$$h^2 = \int_0^{\infty} \Phi(\Omega) d\Omega \tag{9.10}$$

Similarly, the mean-squared slope, s^2, is obtained from the K^2-weighted spectrum, where $K = 2\pi/\Lambda$ is the wave number of the surface wave component of wavelength Λ. Using the expression for a gravity wave system, $\Omega^2 = Kg$, where g is the acceleration of gravity, 9.8 m/s^2, Cox and Munk (1954) show that the mean-squared slope is

$$s^2 = g^{-2} \int_0^{\infty} \Omega^4 \Phi(\Omega) d\Omega \tag{9.11}$$

The frequency spectrum of ocean displacements has been measured many times. Figure 9.4 shows several of these from Phillips (1977).

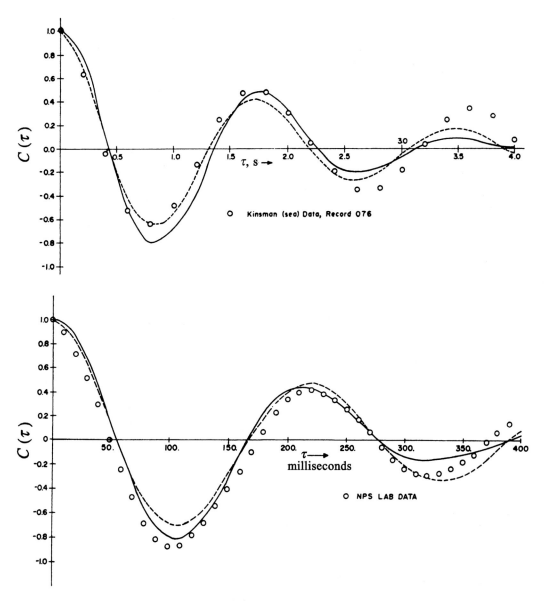

Fig. 9.3. *Above*, a temporal correlation function at sea for a low wind speed case from Kinsman (1960). Circles are data points. Solid line is a fit to (9.8) with $\Omega_m = (2\pi/1.78)\ s^{-1}$ and $T = 2.0$ s. *Below*, a temporal correlation function in a water wind tunnel at the Naval Postgraduate School; the solid line is fitted to (9.8). The NPS correlation constants are $\Omega_m = 9\pi\ s^{-1}$ and $T = 0.24$ s. The scaling is $(\Omega_m)_{Lab}/(\Omega_m)_{Sea} = 8.1$ and $(T)_{Sea}/(T)_{Lab} = 8.3$. Dashed lines are damped exponential fittings. (From Medwin and Clay, 1970.)

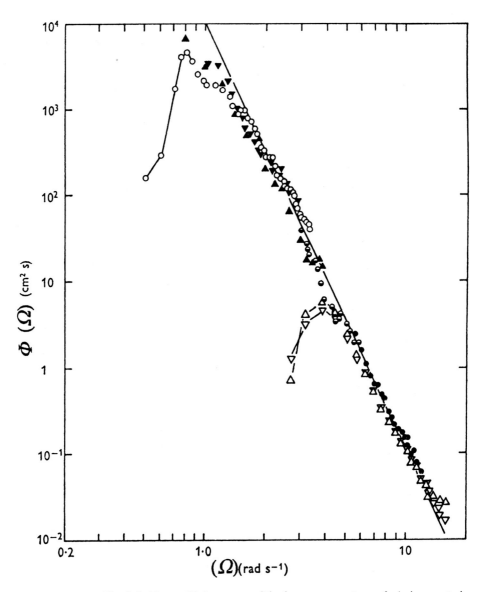

Fig. 9.4. The equilibrium range of the frequency spectrum of wind-generated waves from seven experiments, including 52 spectra. The shape of the spectral peak is shown in only three cases. The straight line has the slope −5, as in (9.12). (From Phillips, 1977.)

The Pierson–Moskowitz (1964) frequency spectrum is an empirical expression for the spectral density of ocean displacements due to wind-blown waves acting over an infinite fetch:

$$\Phi(\Omega, W) = \alpha g^2 \Omega^{-5} \exp[-\beta(\Omega_0/\Omega)^4] \quad \text{(m}^2/\text{Hz)} \qquad (9.12)$$

where Ω, radians/s, is the angular frequency of an ocean surface wave component; $\Omega_0 = g/W = $ nominal angular frequency of the peak of the spectrum; $W = $ wind speed (m/s) at 19.5 m above sea surface; $\alpha = 8.1 \times 10^3$; $\beta = 0.74$; $g = 9.8$ m/s^2.

The Pierson–Moskowitz spectrum is integrated over all frequencies to give the rms height in meters of such a surface, as a function of wind speed in m/s.

$$h = \int_0^\infty \Phi(\Omega) \, d\Omega = \left(\frac{\alpha}{\beta}\right)^{1/2} \left(\frac{W^2}{2g}\right) = 0.0053 W^2 \qquad (9.13)$$

In general, there is no reason to assume that the ocean surface is isotropic. Consequently, sound scatter from the ocean surface is usually dependent on the direction of the sound beam with respect to the axial direction of the surface wave system. To take account of this anisotropic surface, the *directional* wave spectrum must be specified. Kuperman (1975) does this in a general way. One very simple approximate form employed by Fortuin and de Boer (1971) and Novarini *et al.* (1992) assumes that there is a $\cos^2 \theta$ dependence of the spectral amplitude, where θ is the angle with respect to the wind direction. This gives simply

$$\Phi(\Omega, W, \theta) = \alpha g^2 \Omega^{-5} \exp[-\beta(\Omega_0/\Omega)^4] \cos^2\theta \qquad (9.14)$$

The character of the two-dimensional spatial correlation functions at sea can be seen from the ocean examples plotted at the top of Fig. 9.5, from the SWOP (Stereo Wave Observation Project) report (see Kinsman, 1965). A scaled version of the ocean correlation functions, at the bottom of Fig. 9.5, was obtained by the use of a set of fans in a water wind tunnel designed to study sound scatter from an oceanlike surface.

A useful view of the windward surface spatial correlation function is approximated by the simple empirical expression

$$C(\xi) = \exp[-(\xi/L_x)^2] \cos K_{xm}\xi \quad \text{Windward} \qquad (9.15)$$

where L_x is the correlation length in the x direction and K_{xm} is the wave number in spatial correlation space that corresponds to the most prominent surface wavelength in the x direction. Sometimes acoustician theorists set K_{xm} equal to zero and assume that the Gaussian form $\exp[-(\xi/L_x)^2]$ is an acceptable approximation.

The crosswind spatial correlation function does not oscillate with the spatial lag as does the windward. The function is sometimes approximated by either an exponential or a Gaussian form, depending on the particular sea. The Gaussian form is written as

$$C(\eta) = \exp[-(\eta/L_y)^2] \quad \text{Crosswind} \qquad (9.16)$$

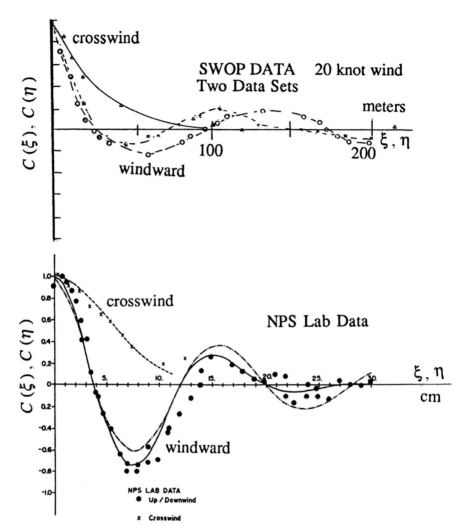

Fig. 9.5. *Above*, spatial correlation functions measured at sea by the Stereo Wave Observation Project (SWOP) during 10 m/s winds. Triangles and solid line are crosswind; circles, crosses, and dashed lines are windward. (From Cole *et al.*, 1960 and Kinsman, 1960.) *Below*, windward and crosswind spatial correlation functions generated by fans in a water wind tunnel at the Naval Postgraduate School. When fitted to (9.15) and (9.16) the constants for the laboratory sea are $L = 14.0$ cm, $K_{xm} = \pi/8$ cm^{-1}, and $L_y = 7.35$ cm. (From Medwin and Clay, 1970.)

where L_y is the correlation length in the y direction. For long-crested waves, $L_y \gg L_x$.

Figure 9.6 is a two-dimensional presentation of the surface correlation of displacements generated by fans in a laboratory "sea" to study the dependence of scattered sound correlation on surface displacement

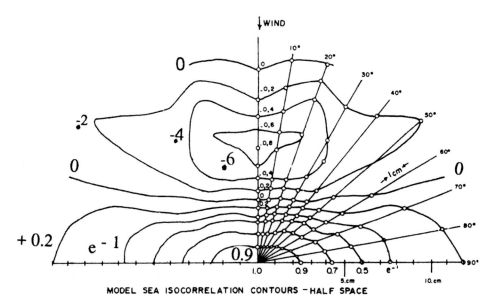

MODEL SEA ISOCORRELATION CONTOURS – HALF SPACE

Fig. 9.6. Two-dimensional spatial correlation of displacements in a laboratory sea generated by fans. The correlation lengths are spatial lags for the e^{-1} values. The data were collected by time averages for spatial lags in a fixed region, which is assumed by ergodicity to equal the data for those lags over many regions of a large surface area at an instant of time. The windward and crosswind components were shown in Fig. 9.5. (From Medwin and Clay, 1970.)

correlation. The windward and crosswind components are shown in Fig. 9.5.

Although (9.8) is a useful representation of the surface temporal correlation function, and (9.15) and (9.16) represent the components of the spatial correlation functions, laboratory research (Medwin *et al.*, 1970) has made it clear that the space–time correlation is *not* separable into its spatial and temporal components.

In the laboratory "sea" described by Fig. 9.7, the envelope of the crests moves at the group velocity of the waves,

$$u = L_x/T \tag{9.17}$$

while the individual crests move at the phase velocity,

$$v = \Omega_m/K_{xm} \tag{9.18}$$

An equation describing this "traveling correlation function" that reduces to the separate temporal and spatial correlation functions, (9.8), (9.15), and (9.16) – is

$$C(\xi, \eta, \tau) = \exp[-(\xi - u\tau)^2/L_x^2]\exp[-(\eta/L_y)^2]\cos(K_{xm}\xi - \Omega_m\tau) \tag{9.19}$$

Fig. 9.7. Map of the
correlation function $C(\xi, \tau)$ of
a wind-blown laboratory
water surface. The circles are
ξ, τ values that yield zero
correlation. These are
connected by lines whose
slope gives the phase velocity
of the correlation v, (9.18).
The shaded areas are
correlations greater than 0.5.
The correlation group velocity,
u, (9.17), is the slope of the
maxima of the correlation.
(From Medwin *et al.*, 1970.)

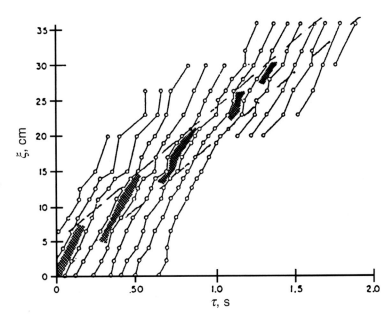

Fig. 9.7. Map of the correlation function $C(\xi, \tau)$ of a wind-blown laboratory water surface. The circles are ξ, τ values that yield zero correlation. These are connected by lines whose slope gives the phase velocity of the correlation v, (9.18). The shaded areas are correlations greater than 0.5. The correlation group velocity, u, (9.17), is the slope of the maxima of the correlation. (From Medwin *et al.*, 1970.)

A three-dimensional surface correlation function such as (9.19) has been used to predict the correlation of sound scattered from the sea surface (Clay and Medwin, 1970).

9.1.2 The rough ocean bottom

PDF of displacements, and rms roughness
The ocean bottom roughness is sometimes defined in terms of its PDF of displacements above and below the mean, as in Fig. 9.8, or, more often, in terms of the rms deviation from the mean, h, which is called the "bottom roughness," as shown in Table 9.1.

Spatial frequency spectrum
A different presentation exploits the spatial transform relations of the previous section. In analogy to the Pierson–Moskovitz *temporal* frequency spectrum for water waves, one specifies a *spatial* frequency spectrum in terms of the wave number, K (radians per wavelength) or, quite commonly, in terms of the cycles per unit distance, $\mathcal{K}$. A vast variability of spectra of ocean bottoms has been observed.

The framework adopted is to assume that the spatial spectral density of the roughness follows the empirical law

$$\Phi(\mathcal{K}) \sim \mathcal{K}^{-b} \qquad (9.20)$$

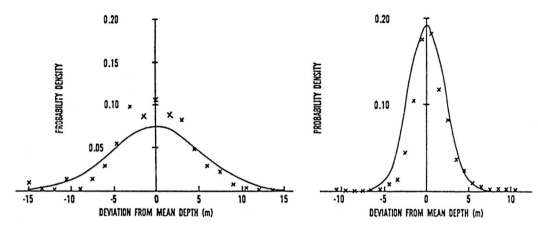

Fig. 9.8. Two examples of the PDF of the ocean bottom displacements, at scales of the order of meters. The solid lines are the Gaussian PDF after filtering through a high-pass spatial filter of wave number = 0.003 m^{-1}. On the left a Norwegian sea basin; on the right a Norwegian sea marginal plateau. (From Berkson and Mathews, 1983.)

where Φ is the "power"[1] spectral density in units of m^2/(cycle/distance), $\mathcal{K}$ is the spatial cycle frequency (e.g., cycles/m or cycles/km), and b is a constant that is fitted for the "province" of the bottom that is being described. The parameter b is often assumed to be 3, but values from less than 2 to 5 have been found, as shown in the following Table 9.1 from Berkson and Mathews (1983).

Sometimes the amplitude spectrum is specified rather than the "power" spectrum. Then, instead of (9.20), the empirical equation is

$$\mathcal{A}(\mathcal{K}) \sim \mathcal{K}^{-b/2} \qquad (9.21)$$

The appropriate spectral description of a backscattering patch of ocean bottom depends on the beam pattern "footprint" (that is, the insonified area) and the frequency of the sound. The description of the province is generally determined from echo soundings made from a ship on the ocean surface. Then the footprint is very large, and the returns are averaged, so that the scale of that spectral description is many meters or a fraction of a kilometer. On the other hand, a remotely operated undersea vehicle operating at higher frequencies at closer ranges will obtain backscatter from a much smaller footprint. The spectral description and

---

[1] The name "power spectral density" is sometimes used but the word "power" is an unfortunate, dimensionally incorrect, carryover from its similar use in electrical circuit theory.

Table 9.1 *Roughnesses of various sea bottoms*

Physiographic province	Ocean	Band-limited h, rms (meters)	b
Rise	Atlantic	3.7	3.2
Continental Slope	Atlantic	6.4	2.2
Seamount	Atlantic	3.6	2.1
Abyssal Plain	Atlantic	<1.3	—
Abyssal Plain	Atlantic	<1.5	—
Rise	Norwegian Sea	<1.1	—
Abyssal Hills	Pacific	3.4	4.9
Continental Shelf	Norwegian Sea	2.5	2.0
Marginal Plateau	Norwegian Sea	1.9	1.9
Abyssal Hills	Pacific	2.5	4.2
Continental Rise	Mediterranean	<1.4	—
Continental Rise	Norwegian Sea	1.2	—
Marginal Plateau	Norwegian Sea	2.1	1.5
Abyssal Hills	Pacific	2.3	2.2
Continental Rise	Mediterranean	<1.0	—
Basin	Norwegian Sea	5.4	1.8
Basaltic Interface*	Atlantic (av. of 50)	259 ±74	1.8 ±0.4
Basaltic Interface*	Pacific (av. of 50)	99±36	1.6 ±0.4

Rms displacement (roughness), h, and spectral slope parameter, b, for various band-limited topographies as defined in (9.20). Spatial wave number bandpass is 0.003 to 0.03 m^{-1} for all cases except the basaltic interfaces (*), where the band-pass is from 0.00006 to 0.003 m^{-1}. In cases where the rms roughness is less than the resolution of the measuring system, the upper limit of rms is given, and b is not estimated. Source: Berkson and Mathews (1983).

the rms roughness appropriate to such a usage is very different, and the predictions of backscatter will require a different value of b in (9.20). At intermediate ranges, the backscattering is determined by the tilt of the surface as well as the local patch descriptors. Quantitative methods for performing such an analysis have been described and large-scale backscatter data can be inverted to provide a statistical description of the sea-floor morphology.

The power law relations and spectral regions of applicability depend on physiographic provinces. Over limited dimensions on the sea floor and a limited spectral range, one can compute the root-mean-square roughness and spatial correlation functions from measurements of the displacements with respect to the mean level.

Circularly symmetric isotropic correlation function

A simple two-parameter isotropic surface correlation function can be written

$$C(r) = \exp\left[-\left|\frac{r}{L}\right|^n\right]$$ (9.22)

where L is the omnidirectional "correlation distance" and parameter n is a positive number. Examples of correlation functions and their spectra are shown in Fig. 9.9.

The function in (9.22) is the exponential when $n = 1$ and the Gaussian when $n = 2$. The power of n in the correlation function is an indicator of the shape of the surface. When $n > 1$, the correlation function decays gradually for $r/L < 1$; this corresponds to a surface having large radii of curvature. When $n < 1$, the correlation decays very rapidly in the region $r/L < 1$; this corresponds to a surface having sharp angular features. The corresponding spectra and correlation functions for the exponents $0.5 \leq n \leq 2.0$ are shown in Fig. 9.9. The Fourier transformations were evaluated numerically.

Laboratory model of a rough ocean bottom

A laboratory scale model of an "isotropic" rough surface was constructed by hand scattering a visually "uniform" distribution of #2 aquarium gravel (consisting of stones roughly 1/4", 6 mm in size) in order to scale-model the coupling between modes of propagation in a rough-surfaced waveguide (Kasputis and Hill, 1984). The results of very careful measurements of the statistics of the "uniform," "isotropic" surface are a cautionary demonstration of how unlikely it is to find an isotropic correlation function in the real world. (Furthermore, turbidity currents in the ocean can cause significant anisotropy of the correlation function.) The rms height was 1.5 mm; the correlations were different in different directions (Fig. 9.10), with correlation lengths that ranged from 1.2 to approximately 2.0 mm (depending on the direction), with an average value of 1.48 mm. Comparing with Fig. 9.9, we clearly see that the different values of n were all somewhat greater than unity.

9.2 Ocean surface forward scatter in the specular (mirror) direction[*]

Before we apply the general theory to scatter from a rough surface, it is helpful to consider the important case of surface scatter in the mirror (specular) direction under simplifying assumptions. The results of this section provide guidance in the interpretation of the more general

---

[*] This section contains some advanced analytical material.

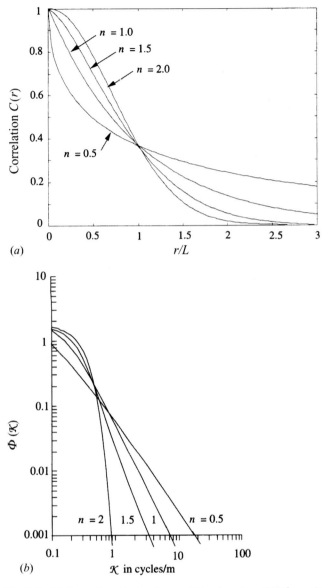

Fig. 9.9. (a) Examples of isotropic correlation functions $C(r)$ for various values of n, and (b) their spatial "power" spectra, $\phi(\mathcal{K})$, which are Fourier transforms of the spatial correlation functions. The Gaussian is the case $n = 2$.

problem. "Specular scatter" is defined as scatter in the direction of mirror reflection (i.e., the angle of scatter equals the angle of incidence and is in the same plane of incidence). Our important conclusion will be that specular scatter consists of a coherent component, which has a fixed-phase relation with respect to the incident sound, and an incoherent component. The coherent component will be found to depend on the rms

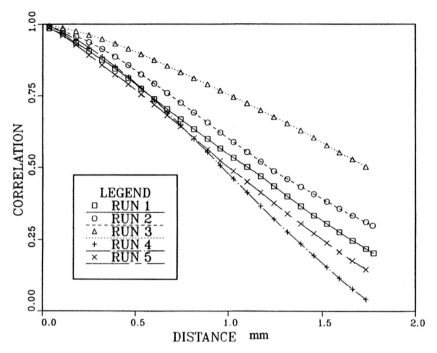

Fig. 9.10. Variation of correlation function for different directions in a laboratory scale model that used "uniformly spread" nominal 1/4", (6 mm), gravel. The Kasputis-designed micrometer measured heights in increments of 20 microns of spacing of 35 microns. (From Kasputis and Hill, 1984.)

height and the PDF of the surface roughness. The incoherent sound is a function of the two-dimensional statistics of the surface and the geometry of the experiment.

9.2.1 Mean coherent pressure, "acoustical roughness"

Consider Fig. 9.11, in which a spherical wave front at long range is approximated by a CW plane wave segment at (AA). It is incident on a rough surface that has homogeneous, stationary statistics as described in Section 9.1. Assume that all facets of the interface have a reflection coefficient $\mathbb{R}_{12}$ due to the impedance change at the interface, and that $ka \gg 1$ where a is the local radius of curvature and k is the acoustic wave number. With this "Kirchhoff Assumption," each ray experiences a phase shift relative to the phase of reflection from the mean surface. The phase shift depends on k, the surface displacement, ζ, and the angle of incidence, θ. The specular scatter contributions come from insonified horizontal facets. For a facet with an average displacement ζ, the path difference to the plane wave position at BB is $2\zeta \cos \theta$ and the spatial

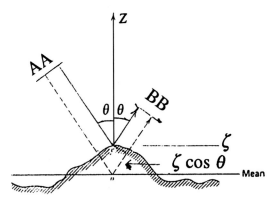

Fig. 9.11. Path differences for specular scatter from a typical horizontal facet at displacement ζ of a rough surface. In general, many such horizontal facets which are insonified by an incident beam are responsible for the radiation in the specular direction.

phase difference equals

$$k\,(2\zeta\,\cos\theta) \tag{9.23}$$

In terms of the perfect, mirror-reflected signal from the mid-surface, at any instant the real value of the pressure reflected at a horizontal facet of the rough surface will be given in terms of the mirror reflected pressure, p_{mir}

$$p_{rough} = \mathcal{R}_{12}\,p_{mir}\,\cos(2k\zeta\,\cos\theta) \tag{9.24}$$

We assume that during each ping the rough surface is "frozen" in time. The many horizontal facets of different elevations within an insonified surface result in many contributions to the total specularly scattered pressure at any instant of time. For a given surface configuration (a "realization"), the interfering sum of the contributing pressures produces a coherent pressure as represented by p_{rough}. Similarly, other surface realizations of the same statistics produce specularly scattered, phase-shifted signals.

The average pressure from these interfering reflections at a particular time depends on the statistics of the surface. For simplicity, assume that the PDF of the surface displacements is Gaussian. To find the average relative pressure, calculate the product of the relative amplitude of each component $-\,\mathcal{R}_{12}\cos(2k\zeta\,\cos\theta)$ times the Gaussian PDF from (9.1) – and integrate over all ζ. The average of the ratio of the coherent, specularly scattered pressure to the mirror-reflected pressure from the mean surface is

$$\left\langle \frac{p_{rough}}{p_{mir}} \right\rangle = \mathcal{R}_{12} \int_{-\infty}^{\infty} \{\cos(2k\zeta\,\cos\theta)\} \left\{ \frac{1}{h\sqrt{2\pi}} \exp\left[-\frac{1}{2}\left(\frac{\zeta}{h}\right)^2 \right] \right\} d\zeta \tag{9.25}$$

where h is the rms displacement of the surface. The integration yields the coherent reflection coefficient for a Gaussian rough surface,

$$\mathcal{R}_{coh} = \left\langle \frac{p_{rough}}{p_{mir}} \right\rangle = \mathcal{R}_{12} e^{-(2kh\cos\theta)^2/2} \qquad (9.26)$$

Note that, for CW, the stacking average over instantaneous values is equivalent to an average over the peak values $\mathcal{R}_{coh} = \langle p_{rough}/p_{mir} \rangle$.

The rough surface coherent pressure reflection coefficient, $\mathcal{R}_{coh}$, which is often called $\langle \mathcal{R} \rangle$ in the literature, is the product of the coefficient due to a change of medium, $\mathcal{R}_{12}$, and that fraction due to mutual phase cancellations of the scatter contributions, $\exp[-(2kh\cos\theta)^2/2]$. When the rough surface is symmetrical – that is, when the skewness moment m_3 (9.4) of the displacements is zero – the contributing phase-shifts from below and above the mean level cancel each other. Then, although the amplitude is reduced, the phase of the coherent mean pressure is the same as would occur for a mirror-reflected pressure from the mean surface. When there is peakedness or kurtosis (9.5), more of the horizontal facets are at small or zero displacements from the mean surface, and the coherent component is enhanced.

The specularly scattered coherent intensity relative to the mirror-reflected intensity is

$$\mathcal{R}_{coh}^2 = \mathcal{R}_{12}^2 e^{-g_R} \qquad (9.27)$$

where

$$g_R = \text{"acoustical roughness"} = (4k^2 h^2 \cos^2\theta) \qquad (9.28)$$

Sometimes the quantity $(g_R)^{1/2}$ is called the "Rayleigh Roughness Parameter," to honor the discoverer of the effect. Sometimes (9.28) is written in terms of the grazing angle, in which case the sine of the grazing angle replaces the cosine of the angle with the normal to the surface. An "acoustically smooth" surface, $g_R \ll 1$, has stronger specularly scattered energy because it has an rms height much less than the acoustic wavelength or is nearer grazing, $\theta \to 90°$.

The predictions of (9.27) have been carefully tested many times in laboratory wind-blown wave systems with a near-Gaussian PDF of heights. Mayo (1969) measured the incident and scattered signals in the specular direction at 50 kHz and 100 kHz, at seven angles of incidence and several surface rms heights, 0.1 cm $< h < 0.45$ cm (Fig. 9.12). The simple conclusion (9.27) is confirmed up to $g_R \cong 1$. Data from these and additional experiments have been analyzed by Clay, Medwin, and Wright (1973), who found that small deviations from the Gaussian PDF, and shadowing, profoundly change the coherent scatter at large g_R. An early determination of bottom rms roughness by studying coherent specular scatter from the deep sea floor was that of Clay (1966a, b).

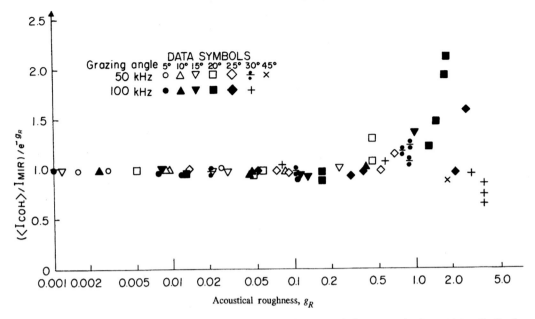

Fig. 9.12. Ratio of rough-surface, specularly scattered coherent intensity (I_{coh}) to smooth-surface reflection, I_{mir}, relative to the simple theoretical prediction e^{-g_R}, for various values of roughness parameter, g_R. (From Mayo, 1969.)

9.2.2 Acoustical determination of the rms height

For a general PDF of heights, $w(\zeta)$, (9.25) would be written

$$\left\langle \frac{p_{rough}}{p_{mir}} \right\rangle = \mathcal{R}_{12} \int_{-\infty}^{\infty} w(\zeta)\{e^{-2ik\zeta \cos\theta}\}\, d\zeta \qquad (9.29)$$

Note that the integral in (9.29) has the form of the Fourier integral transformation, with ω replaced by $2k\cos\theta$. Therefore one can calculate its inverse, which is the acoustical estimate, w_a, of the PDF of the surface displacements,

$$w_a(\zeta) = \frac{1}{\pi \mathcal{R}_{12}} \int_{-\infty}^{\infty} \mathcal{R}_{coh} \exp(2ik\zeta \cos\theta)d(k\cos\theta) \qquad (9.30)$$

The good agreement between the observed specular scatter and that predicted from the wave height as well as the result of inversion are shown in Fig. 9.13 for a laboratory experiment of Spindel and Schultheiss (1972).

9.2.3 Statistics of scatter in the specular direction

Assume that we have a CW beam specularly scattered from the ocean surface or bottom. As the ocean surface moves by a fixed sonar, or as a

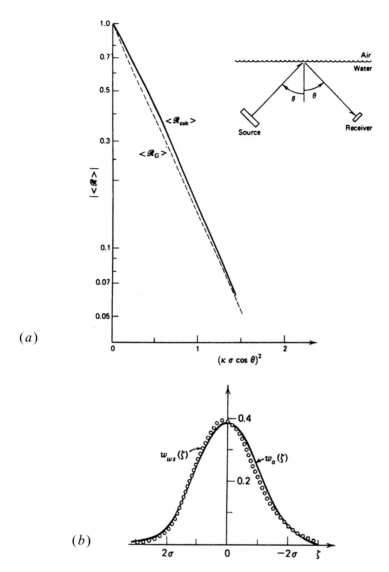

Fig. 9.13. (*a*) Coherent reflection coefficient for specular scatter, (R_{coh}), measured in a laboratory, wind-driven sea compared with the value predicted from a Gaussian PDF of surface displacements (R_G). (*b*) The PDF inverted from acoustical measurements, w_a, is compared with the wave staff measurements of the PDF, w_{ws}. (From Spindel and Schultheiss, 1972.)

sound beam on a ship moves past a rough bottom surface, the scattered pressure amplitude $|P|$ will fluctuate. From statistics theory, it is known that the mean-squared value of a randomly varying quantity is equal to the square of its mean value plus the variance of the quantity. Therefore the total specularly scattered intensity, which is proportional to $\langle |P^2| \rangle$, can be determined from the sum of the coherent component described in the previous section, $\langle|P|\rangle^2$, and an incoherent component, the variance of $|P|$,

$$\langle|P^2|\rangle = \langle|P|\rangle^2 + \text{Var}|P| \tag{9.31}$$

Var $|P|$, a measure of the random phase and amplitude of the scattered sound that depends on the correlation of surface displacements, will be considered in the next section. The similarity between variability of the envelope of scattered sound and the statistics of noise added to a sinusoid in an electrical circuit allows the results developed by S. O. Rice (1954) to be adapted to sound scatter. "Ricean statistics" for sound scatter can be defined in terms of the relative coherence,

$$\gamma = (\text{coherent intensity})/(\text{incoherent intensity}) = \langle|P|\rangle^2/\text{Var}|P| \tag{9.32}$$

For acoustically smooth surfaces, $g_R < 1, \gamma > 1$, there is a relatively large coherent component, and the specularly scattered pressure distribution is nearly Gaussian (9.1). At the other extreme, for acoustically rough surfaces, $g_R > 1, \gamma < 1$, the probability density function approaches the Rayleigh PDF for the sum of a large number of randomly phased, equal amplitude components,

$$w_{Rayl}(|P|) = \frac{\langle|P|\rangle^2}{h^2} \exp\left(-\frac{|P|^2}{2h^2}\right) \tag{9.33}$$

The distribution described by (9.33) was derived by Lord Rayleigh (1894, 1896, 1945) for the specific problem of determining the sum of a large number of randomly phased, equal-amplitude wave components. It can be shown (e.g., Beckmann and Spizzichino, 1963) that of all possible distributions of the amplitude of a field scattered by a symmetrically distributed rough surface, the Rayleigh distribution has the greatest variance; in terms of acoustic pressure during sound scatter, that value is

$$\text{Variance} = 0.212\langle(|P_{scat}|/|P_{mir}|)^2\rangle \tag{9.34}$$

where $|P_{mir}|$ is the amplitude of the pressure that would be reflected from a mirror-smooth surface.

Electrical engineers sometimes represent the magnitude and phase of a time-varying component by a "phasor." A phasor is a line drawn with a length proportional to the amplitude and at an angle that represents the relative phase of the component. Phasors add like vectors. From the phasor point of view, a near-Gaussian distribution is caused by the addition of a strong constant phasor (representing the mean value of the scattered sound) to much smaller, randomly distributed, time-variable phasors. On the other hand, the Rayleigh distribution is approached for

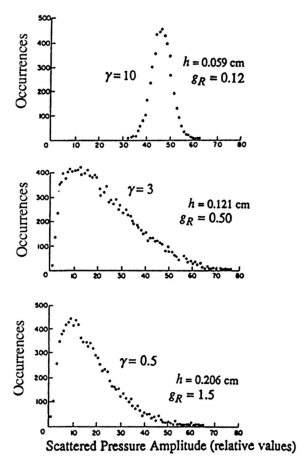

Fig. 9.14. PDF of specularly scattered, normal-incidence sound amplitudes
for three laboratory Gaussian water surfaces of rms heights, $h = 0.059$ cm,
0.121 cm, and 0.206 cm (*top to bottom*). The PDF of the pressure amplitude
goes from Gaussian to Rayleigh as the acoustical roughness increases (*top to
bottom*) from $g_R = 0.12$ to 0.50 to 1.5 and the relative coherence of the
specularly scattered sound intensity, $\gamma = \langle |P|^2 \rangle / \mathrm{var}\, P$, decreases from 10 to 3 to
0.5. The sound frequency was 70 kHz. (From Ball and Carlson, 1967. See also
Stephens, 1970.)

sound scatter when the constant phasor component, which represents
the coherent signal, is very small compared with the variable, randomly
phased contributions.

The change of the PDF of the scatter for surfaces of different acoustical roughnesses has been demonstrated by using a laboratory wind-
driven sea. In the top panel of Fig. 9.14, at $g_R = 0.12$, $\gamma = 10$, the co-
herent intensity $\langle |P^2| \rangle$ is 10 times the incoherent Var $|P|$, and the PDF
of the amplitude of the specular scatter is virtually Gaussian, with a
large mean value of $|P|$. When $g_R = 1.5$, $\gamma = 0.5$ (bottom panel), and

the coherent intensity is half the incoherent, the peakscattered amplitude has been greatly reduced and the PDF is nearly "Rayleigh".

9.2.4 Sound intensity of scatter in the specular direction: ocean scale-models

The coherent component of intensity in the specular direction decreases with increasing acoustical roughness (Fig. 9.13(a)). Extensive theoretical research has shown that at large acoustical roughness $g_R \gg 1$, the total scattered intensity will be dominated by the incoherent component. For plane-wave incidence at a surface with a Gaussian PDF and a Gaussian correlation function, the forms of the contributing components are shown in Fig. 9.15.

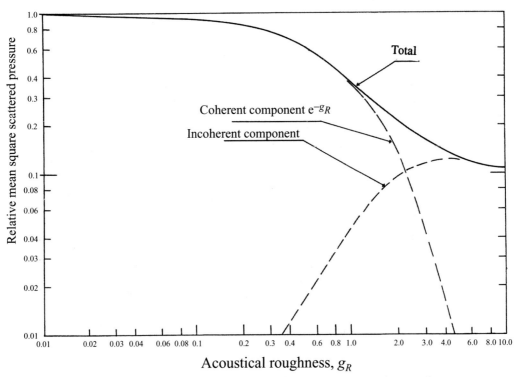

Typical behavior of coherent and incoherent components in specular scatter

Fig. 9.15. Specularly scattered, mean-squared pressure relative to mirror reflection as a function of the surface acoustical roughness, g_R, showing the dominating coherent component at low roughness and the controlling incoherent components at larger acoustical roughness that depends on the surface correlation function and the area insonified. To get these theoretical results, an incident plane wave of cross-section area A is assumed, as well as a Gaussian PDF of heights and a Gaussian spatial correlation function.

The incoherent scattering theory line of Fig. 9.15 has been calculated by assuming that the incident sound is a plane wave of cross-section area A, that the surface roughness has a Gaussian PDF of heights and an isotropic Gaussian spatial correlation function of displacements. The asymptotic high-frequency value is then found to be proportional to the insonified area and inversely proportional to the mean-square slope, $\langle s^2 \rangle$. Eckart (1953), Beckmann and Spizzichino (1963), and Tolstoy and Clay (1966, 1987) also show that, for such a Gaussian surface, the rms slope, s, is simply related to the rms height, h, and the isotropic spatial correlation length, L.

The general relation is

$$\langle s^2 \rangle = -h^2 \frac{\partial^2 [C(\xi)]}{\partial \xi^2}\bigg|_{\xi=0} \tag{9.35}$$

From (9.16), for an isotropic Gaussian spatial correlation function of correlation length L,

$$C(\xi) = \exp\left[-(\xi/L)^2\right] \tag{9.36}$$

After differentiating (9.36) to insert into (9.35),

$$s = \sqrt{2}h/L \tag{9.37}$$

where s is the rms value of the isotropic slope. However, note that real surfaces such as in Figs. 9.6 and 9.11 generally do not have an isotropic Gaussian spatial correlation function.

Laboratory experiments by Ball and Carlson (1967) and Medwin (1967), observing multifrequency, normal-incidence scatter in a wind wave tunnel, have confirmed the form of Fig. 9.15. At the top of Fig. 9.16 is a sketch of one experimental setup to measure the specular scatter at normal incidence. The PDF of the slopes was determined by photographing and laboriously analyzing ten cases of the glitter deflection at various positions below a string grid as shown typically at the bottom of Fig. 9.16. The rms slope s and ratio of the windward to the cross-wind slopes, s_w/s_c, were similar to Fig. 9.2 and were found from (9.6) to be equivalent to a wind 1.4 m/s at 12.5 m over an ocean surface. The rms slope has also been obtained by integration of the output of a photocell, using a co-located point source (see Fig. (9.16)) in this early experiment.

To determine both h and s by specular scatter, eight sequential pings of frequencies from 21 to 194 kHz, total duration 2.2 ms, were radiated perpendicular to the surface, from a low Q Mylar transducer. The PDF of the specular scatter at 21 kHz was nearly Gaussian; therefore it was assumed that, at that frequency, the scatter was coherent, and (9.27) was used to calculate the rms height, h. This allowed the specification of the

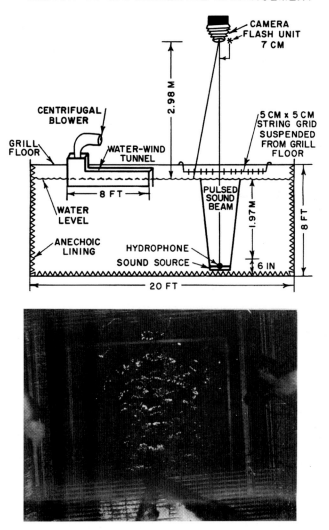

Fig. 9.16. *Above,* sketch of experimental arrangements. The floating water wind tunnel was constructed of wood in the form of an inverted "U" in cross-section. The 2.2 ms duration pulsed sound beam was a sequence of 12 to 29 cycles of each of eight frequencies from 21 to 194 kHz insonifying the surface in the transducer's near field. *Below,* a typical glitter pattern used to determine the PDF of slopes by the light deflection. (From Medwin, 1967.)

acoustical roughnesses $0.3 < g_R < 25$ for the range of frequencies of the experiment.

At $g_R = 10, 14, 19, 25$, the statistics of the scatter approached the Rayleigh PDF. Therefore, the specular scatter was assumed to be totally incoherent for these four asymptotic points (Fig. 9.17). The rms slope

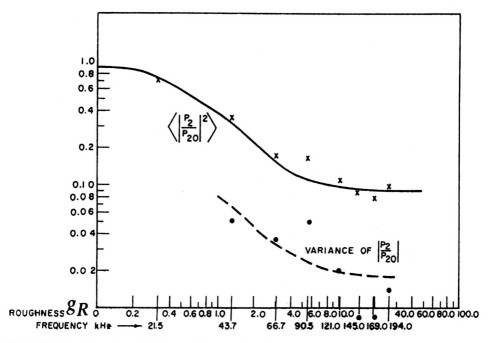

Fig. 9.17. Experimental results showing the normal-incidence, mean-squared, specularly-scattered pressure relative to mirror reflection for eight frequencies of data (symbol x) using the wind wave laboratory arrangement shown in Fig. 9.16. The nearly plane-wave insonification of the surface was over an area approximately equal to the area of the source, a 24 cm diameter piston. For $g_R < 1$, the relative intensity goes as $\exp(-g_R)$. For $g_R > 10$, the relative intensity is independent of frequency and is proportional to the insonified area and inversely proportional to the mean-square slope. The theoretical variance for a Rayleigh PDF (9.33) is shown by dashed line from data (symbol •) for the mostly incoherent specular scatter at $g_R > 1$. (From Medwin, 1967.)

s was calculated from acoustics theory for $g_R \gg 1$ and found to be in satisfactory agreement with the optical value.

In summary, the value of the rms slope, obtained optically, or from formulae such as (9.6) for the ocean surface, can be used to estimate the magnitude of the asymptotic (large g_R) incoherent component of specular scatter for a surface assumed to have a Gaussian PDF and a Gaussian correlation function. Inversely, the rms slope, s, can be estimated from measurements of specular scatter for large g_R values if the surface is Gaussian in height and correlation function.

A much more extensive set of laboratory measurements (Thorne and Pace, 1984) employed a parametric source (Section 4.3) to get narrow beams with a wide range of frequencies. An underwater pressure-release surface was constructed of low-density polyurethane with Gaussian PDF of displacements and a simple "isotropic" Gaussian correlation function to represent an ocean surface. The phase-sensitive scattering solution of Section A10.5 of Clay and Medwin (1977) was used to predict the incoherent normal incidence surface scatter. The inverse proportionality to the mean-square slope, s^2, was confirmed (for this Gaussian surface) to be independent of sound frequency for $g_R \geq 10$. The interrelation of $s, h,$ and L (9.37) was then applied to calculate s.

Ocean experiments

Clay (1966) showed that the bottom roughness is characteristic of sea bottom processes, and suggested ways to measure it. The methods were then used by Leong (1973) and Clay and Leong (1974) to consider a sea-floor area southwest of Spain. The 12 kHz echo-sounding profiles, the 3.5 kHz subbottom profiles, and the seismic profiles (25 to 100 Hz) were analyzed to estimate the sea-floor roughness in the four spatial wavelength ranges: less than 1 km, 0.6 to 3 km, 2 to 6 km, and 6 to 12 km. The histograms of the spectral estimates mapped consistently into small "provinces" on the sea floor.

Stanton (1984) employed a technique similar to the laboratory experiment in Fig. 9.17 to obtain the rms height h of ocean bottom "micro-roughness" at small values of g_R. He then supplemented this by obtaining bottom correlation information from the statistics of the specularly scattered sound. A ship-mounted 3.5 kHz sonar was used at water depths of 20–30 m to identify ripples, beds of rocks, and nodules.

9.2.5 Other surface scattering effects

Non-Gaussian surfaces

The several derivations that lead to the conclusions that rms height h can be obtained from the low roughness ($g_R \ll 1$) specular scatter, and rms slope s can be determined from the very high roughness data ($g_R \gg 1$), are based on surfaces defined by a Gaussian PDF of heights and Gaussian correlation function (simply, "Gaussian surfaces"). When the surface is not Gaussian, one may find several arbitrary surface realizations that produce scattered pressures that deviate from the above conclusions (Kinney and Clay, 1985).

Short-term signal enhancements

The theoretical predictions of specular scatter, and the measurements that lead to statistical data such as displayed in Fig. 9.14, are based on long-term observations for a Gaussian surface. But a Gaussian surface will produce very different scatter for shorter periods of time. It may even cause acoustic pressures greater than for plane surface reflection, for shorter periods of time (Medwin, 1967). For example, Perkins (1974) and Shields (1977) have used a laboratory wind-blown Gaussian surface to demonstrate signal enhancement (compared with average values) of the order of 3 to 5 dB for surfaces of roughness of $0.5 < g_R < 4$ during experiments that lasted for five "ocean" surface wave periods. At sea, this would correspond to specular scattering enhancements over many seconds to minutes for certain sound frequencies.

Long-range reverberation

Theoretically, it should be possible to use long-range reverberation to extract information about prominent underwater topographic features. Good progress has been made, so that it is now clear that major echoes in reverberation come from the faces of escarpments that are large enough and flat enough (compared to the acoustic wavelength) and properly oriented to coherently reflect the incident sound in a deterministic manner. A positive ocean-basin geomorphological identification has been made by use of 268 Hz sound traveling beyond two convergence zones over ranges of 100 km even in the presence of refraction, Lloyd's mirror interferences, and the cacophony caused by smaller diffractors at the western Mid-Atlantic Ridge (Makris *et al.*, 1995).

9.3 Rough surface scatter

The original theoretical research on scattering from rough surfaces dealt with incident plane waves and dates from the 1950s (Eckart, 1953; Isakovitch, 1952); the first book was by Beckmann and Spizzichino (1963).[2] A more recent, excellent book gives a detailed treatment of surface scattering using incident plane waves with the Helmholtz–Kirchhoff integral to compute scattering at rough interfaces (Ogilvy, 1992).

Direct applications to ocean acoustics requires the introduction of Fresnel corrections. Horton and Melton (1970) pointed out that the incident plane-wave assumption is equivalent to having a large planar transducer radiating to the scattering surface in the near field. In the incident plane-wave assumption, the scattering surface is small compared with the first Fresnel zone. (This was the condition in the experiment described in Fig. 9.16.) However, common underwater field experiments have geometries with small transducers far from the scattering surface. Then, the incident wave fields are curved, and Fresnel effects and corrections are important. We call the theory that includes Fresnel effects the HKF theory.

---

[2] Most of this electromagnetics reference deals with scalar waves that may be interpreted as acoustic waves.

The conditions for validity of the Kirchhoff and small-slope assumptions made in the original derivations is a field of active research. In general, one compares exact solutions for specific geometries with the various approximate theoretical approaches and looks for higher-speed methods of calculation. Thorsos (1988) points out that predictions based on the Kirchhoff approximation are generally acccurate for surfaces with a Gaussian PDF of heights and slopes, provided that the correlation length is greater than the sound wavelength $L > \lambda$. Kaczkowski and Thorsos (1994) use the parameter khs (where k is the acoustic wave number, h is the Gaussian rms height, and s is the Gaussian rms slope) to evaluate an efficient "operator expansion method" of solution. Thorsos and Broschat (1995) have extended the Voronovitch (1985) small-slope approximation method to speed the accurate calculation of scattering from rough surfaces.

9.3.1 Surface scatter of a spherical wave: backscattering strength

In ($M\&C$) we formally derived the scattering of a CW signal as a function of the sonar characteristics and the surface statistics.

Since the scatter is a function of the geometry, bistatic experiments are needed to characterize the 2D surface correlation function. On the other hand, in the specular direction, the rms height h can be determined directly as discussed in Section 9.2.

Commonly, for surface-scattered sound, the mean squared pressure is written in the simple form

$$\langle pp^* \rangle = p_0^2 R_0^2 \frac{A}{R_1^2 R_2^2} \mathcal{S} \tag{9.38}$$

Where the scattering coefficient, $\mathcal{S}$, includes all of the parameters that describe the experiment and the surface.

Therefore, one would expect to be able to characterize a rough surface acoustically by performing an experiment to get the surface-scattering coefficient

$$\mathcal{S} = \frac{\langle pp^* \rangle}{p_0^2 R_0^2} \frac{R_1^2 R_2^2}{A} \tag{9.39}$$

It is common to use the 10 times logarithm of $\mathcal{S}$, to obtain the *scattering strength of the surface* $\mathcal{SS}$, which is expressed in decibels:

$$\mathcal{SS} = 10 \log_{10} \mathcal{S} \text{ (dB)} \tag{9.40}$$

9.3.2 Apparent reflection coefficient and "bottom loss"

Some researchers consider the specularly scattered sound (mirror direction) as an apparent reflection, that is, a "reflection with reduced amplitude." This permits the characterization of an ocean bottom surface interaction in terms of the simple concept of "bottom loss per bounce." To do this, the squared "apparent reflection coefficient," $\mathcal{R}_a^2$, is written in terms of a point source at range R_1 from the surface, and the receiver at range R_2 from the surface, and is defined in terms of the received intensity, assuming spherical divergence.

$$\langle pp^* \rangle = p_0^2 R_0^2 \frac{\mathcal{R}_a^2}{(R_1 + R_2)^2} \qquad (9.41)$$

In fact, one needs to look at the concept critically to determine the implications of (9.41).

This is done in (M&C) where one finds that (unfortunately) this squared apparent reflection coefficient not only depends on the ranges R_1 and R_2 but also is a function of the frequency, the geometry, the beam pattern, and the correlation function of the surface. Equation (9.41) is therefore a superficially simple representation of the complex specular scattering process. Some of the real consequences of specular scatter were described from the point of view of laboratory experiments in Section 9.2.5. (See Section 9.4.2 later.)

When the apparent reflection coefficient is used, one expresses the bottom loss as

$$BL \text{ (dB)} = 10 \log_{10}(\mathcal{R}_a^2) \qquad (9.42)$$

When the bottom is smooth, the problems due to incoherent scattering disappear, and the bottom loss simply depends on the reflection coefficient $\mathcal{R}_{12}$

$$BL \text{ (dB)} = 10 \log_{10}(\mathcal{R}_{12}^2) \qquad (9.43)$$

9.4 Scale model and computer studies of ocean surface scatter

This section contains cautionary studies to clarify the use and misuse of simple engineering concepts that are in the literature of underwater acoustics.

9.4.1 Scale model studies

Dependence on angle of incidence
The dependence on angle of incidence of a spherical wave is revealed in an experiment described in Fig. 9.18. Seven cycles of frequency

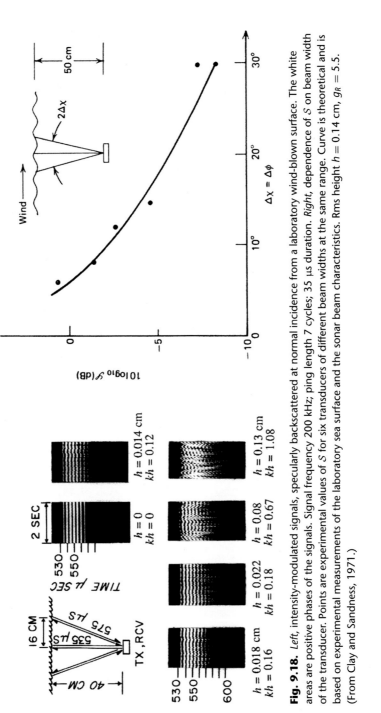

Fig. 9.18. *Left*, intensity-modulated signals, specularly backscattered at normal incidence from a laboratory wind-blown surface. The white areas are positive phases of the signals. Signal frequency 200 kHz; ping length 7 cycles; 35 μs duration. *Right*, dependence of S on beam width of the transducer. Points are experimental values of S for six transducers of different beam widths at the same range. Curve is theoretical and is based on experimental measurements of the laboratory sea surface and the sonar beam characteristics. Rms height $h = 0.14$ cm, $g_R = 5.5$. (From Clay and Sandness, 1971.)

200 kHz were incident at a wind-blown surface, and the display was intensity-modulated to show the fluctuations of the sound amplitude. When $g_R < 1$, the first return (from directly over the source) was virtually unaffected by the rough surface, but contributions from farther off-axis were increasingly incoherent. When $g_R > 1$, not only the off-axis scatter but also the normal scatter varied strongly with time. There was a clear dependence on the beam pattern.

Dependence on beam width

To show the dependence of the surface-scattering coefficient S on the beam width of an incident spherical wave, Clay and Sandness (1971) evaluated the theoretical expression for S and used their measurements of the surface correlation function and the sound beam geometry to obtain the predicted curve of S shown in Fig. 9.18. It agreed with the experimental values of S, obtained with six transducers of different beam width. Both determinations confirmed that, for a given rough surface, S *depends on the beam width.* Therefore, experimental determinations of S are system and geometry dependent, rather than solely rough-surface dependent!

Dependence on range

Thorne and Pace (1984) and Pace *et al.* (1985) have specialized to large acoustical roughness ($g_R \geq 10$) at normal incidence. This is the realm dominated by incoherent scatter, as indicated in Figs. 9.15 and 9.17. To simplify the calculation and the experiment, they evaluated the specular scatter of a Gaussian beam incident at right angles to a rough surface of pressure reflection coefficient $\mathcal{R}_{12}$ that had a Gaussian PDF of displacements and a Gaussian correlation function. The purpose of this model was to determine how S depends on range, and therefore to check on the region of applicability of the commonly used (9.39).

The experiment showed that the surface-scattering coefficient increases from low values in "near-field scattering" to stable values that can be used to define a rough surface in "far-field scattering." The increase in S for an increasing range, and the transition from near-field to far-field incoherent scatter, was verified by extensive laboratory experiments using manufactured polyurethane ("pressure-release") and gravel ("rigid") Gaussian surfaces. One of their figures for a pressure-release surface is Fig. 9.19.

It is shown that under these special conditions, within the far-field region, the scattering coefficient of the acoustically very rough interface,

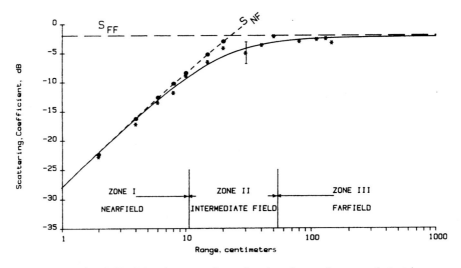

Fig. 9.19. Scattering strength as a function of range for a manufactured Gaussian pressure-release surface insonified at normal incidence by frequency 250 kHz ($g_R = 15$). The dots are experimental values for a smooth surface. The asterisks are experimental and the curved line is theoretical for a rough surface. S_{FF} is the far-field asymptote that occurs at $\theta_0 \leq s$, where θ_0 is the beam angle and s is the surface rms slope. (From Pace *et al.*, 1985.)

$g_R \geq 10$, has the constant value, independent of frequency,

$$S = R_{12}^2/(16\pi s^2) \tag{9.44}$$

where R_{12} is the pressure reflection coefficient at the interface and s is the rms slope.

The rms slope could be determined from (9.44) and the isotropic Gaussian correlation function could be calculated from (9.37).

9.4.2 Computer models of backscatter

Generally a surface is studied by "monostatic" backscatter (in which source and receiver are at the same point). The concept of "surface backscattering strength," BSS, has been used to characterize the backscatter from a surface that is assumed to consist of point scatterers. From (9.39),

$$BSS = 10\log_{10}\left(\frac{I_{BS}}{I_0}\frac{R^4}{R_0^2 A}\right) \tag{9.45}$$

where I_{BS} = backscattered intensity measured at the source/receiver; I_0 = source intensity at 1 m: R = range (meters) fom source/receiver to scattering surface: R_0 = reference range at source = 1 m; A = scattering area (m^2).

The simple derivation that led to the concept of a scattering coefficient S in (9.38) assumes that the point scatterers will cause the far-field

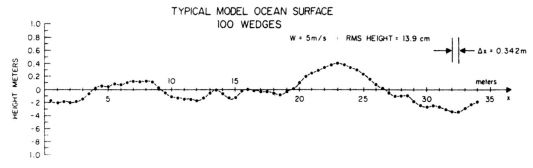

Fig. 9.20. Computer-generated model ocean surface with 100 wedges, appropriate to a wind speed of 5 m/s. Spacing between wedges, 0.342 m; rms height, 0.139 m; spatial correlation length, 2.7 m; surface rms slope, 6.3°. (From Medwin and Novarini, 1981.)

backscattering intensity range dependence to vary as R^{-4} from the surface. But scattering from the ocean surface is not point scattering. One notes that, at normal incidence, if the ocean surface is smooth, for a point source the relative backscattered (reflected) intensity varies as R^{-3} not R^{-4}. If the ocean waves are wedgelike, the relative backscattered intensity varies as R^{-3}. In either case, it is clear that the definition of BSS is critically dependent on the type of scatterer at the ocean surface.

The wedge is an element that can accommodate the various behaviors of a real rough surface. The flexibility derives from the fact that a wedge is composed of reflecting facets and a diffracting edge. Furthermore, at long range, a finite wedge looks like a point scatterer, particularly if the incident sound has a wavelength that is large compared with the wedge extent.

The theory for single wedges is given in Sections 8.5 and 8.6. A computer model based on single wedge scattering was exercised by Medwin and Novarini (1981) to determine the backscatter from an ocean surface made up of finite wedge elements. First a model ocean surface was generated, as in Fig. 9.20. The surface had a Gaussian PDF of displacements selected to produce an rms height appropriate to the wind (9.13) and wedge spacing in the windward direction to produce an appropriate surface-displacement correlation function (see Section 9.1.1). The extent of the wedge was set by the correlation length in the crosswind direction. For a 5 m/s wind speed, the model rms wave height was Gaussian, ranging from $h = 0.139$ m to 0.15 m compared with a theoretical value of 0.14 m. The windward correlation length was 2.7 m compared with the theoretical value 3.3 m. The rms slope was 6.3°, compared with $7.1° \pm 0.9°$ for a Cox and Munk type of calculation from (9.6).

Fifteen surfaces were defined, each containing nine wedges (ten facets). The equations for scattering from a pressure-release wedge were used to calculate the impulse response. Figure 9.21, which is a typical response curve, shows nine wedge diffractions and one case of a

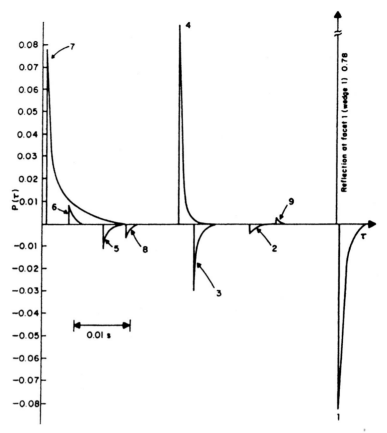

Fig. 9.21. Impulse response for one nine-wedge surface. The order of the wedge numbering was sequential along the surface; it differs from the time of arrival of the reflection and diffractions because of the range differences. For this case, wind speed is 20 m/s; range, 500 m. (From Medwin and Novarini, 1981.)

reflection from a facet. The magnitudes and phases of the wedge diffractions depend on the angle of incidence as noted for single wedges, in Chapter 8.

Some of the results of the computer simulation are shown in Fig. 9.22. The left figure demonstrates that the grazing angle (30°) backscatter may fall between R^{-3} and R^{-4}, depending on the frequency. This grazing angle was selected because it is more than three times the rms slope; therefore, reflections are virtually absent, and we are looking at only the diffraction effect. The range dependence is R^{-4} (point scattering) for the low frequencies only, that is when the acoustic wavelength is comparable to, or larger than, the surface correlation lengths. For 8000 and 16 000 Hz, however, the acoustic wavelengths are less than one-tenth the correlation lengths. Then the scatterers are wedgelike, and the

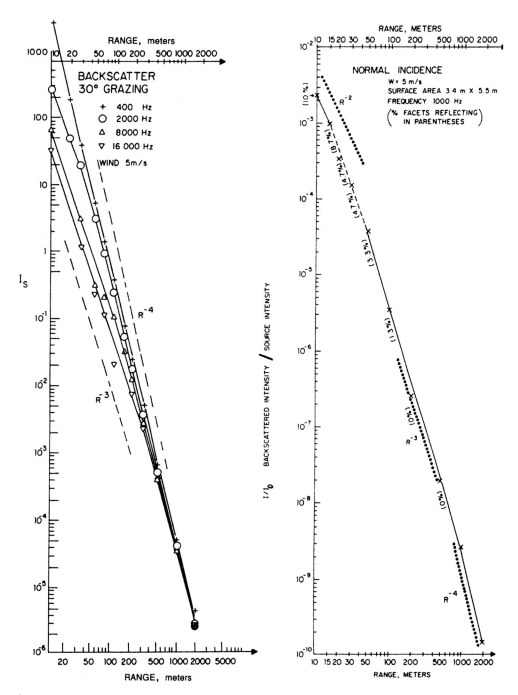

Fig. 9.22. *Left*, backscattered intensity as a function of range for a fixed area of surface defined by one spatial correlation length in each direction (3.4 m upwind × 5.5 m crosswind). Four frequencies shown for 5 m/s wind, 30° grazing angle. The dashed lines bounding the calculated scattered intensities are theoretical slopes for wedge scattering (R^{-3}) and point scattering (R^{-4}). *Right*, relative backscattered intensity for normal-incidence backscatter at frequency 1000 Hz. The dotted lines show the idealized slopes for facet reflection, wedge scatter, and point scatter. (From Medwin and Novarini, 1981.)

backscattered intensity follows the R^{-3} law to about 200 m. At greater ranges, interference between the wedge scatters causes the transition to the R^{-4} behavior.

The backscatter is large when a facet *reflects*; for example, it is large near vertical incidence because horizontal facets are then very common (see Fig. 9.2). Backscatter decreases to very small values for incidence angles nearer to grazing because there are no vertical facets. The *diffracted* sound is the *only* significant component of intensity during backscatter at angles of incidence very much greater than the rms slope. For example, from Fig. 9.2 and tables of areas under the Gaussian PDF, at an angle of incidence greater than twice the rms slope, only 5 percent of the facets will be perpendicular to the incoming rays and thereby reflect. In that case the major part of the backscatter will come from diffraction and will depend sensitively on the correlation function of the surface.

The graph at the right of Fig. 9.22 shows the effect of horizontal facet reflections that occur with a high probability at normal incidence and that will be found more often at shorter ranges than at long ranges. Within about 10 m of the surface, about 9 percent of the facets are effectively reflecting; from 10 to 30 m, about 4 percent of the facets produce reflections, and these dominate and give the appearance of a reflecting surface, rather than a scattering surface, and determine the R^{-2} behavior. Beyond 50 m, less than 1 percent of the facets are oriented to cause reflection and, because of the dominating interference of the diffractions from the wedges, the net result appears like point scattering and varies as R^{-4}.

This exercise has shown us that scattering from the ocean surface is *not* always point scattering. This suspicion had been growing for two decades (e.g., Mikeska and McKinney, 1978).

In all cases, it would be prudent to verify that the range dependence is R^{-4} before one assumes that the measurement is of far-field point scattering from ocean surfaces or bottoms, and before one applies (9.39) or (9.40) to characterize the surface scatter.

9.5 Backscatter from bubbles below the ocean surface

From Chapter 6 it is clear that both the sound attenuation and speed are functions of bubbles under the surface, and that the bubble density is a non-uniform function of depth and location. There are several consequences:

(1) Significant energy loss can occur as underwater sound approaches the surface; this excess attenuation is a function of the sound frequency and the local bubble distribution.

(2) Sound approaching the surface is refracted due to the generally greater bubble density nearer the surface; for low-frequency sound (<10 kHz), the sound speed is less than the bubble-free speed, and the angle of ray incidence becomes more nearly normal at the rough surface.

(3) Bubble resonance frequencies fluctuate as a function of the fluctuating ambient pressure at depth, (6.67) and Fig. 6.21. Therefore, sound that interacts with the bubbly region fluctuates in amplitude and phase as a function of time and position (Medwin *et al.*, 1975).

The now historical Fig. 9.23 shows the first quantitative explanation of the potential effect of adding the omnidirectional scatter of postulated bubble densities to directional rough surface backscatter calculated from the Beckmann and Spizzichino theory (1963). The paper was written in the days before there were any data of ambient bubble densities at sea. The combination of the two scattering processes produces large backscatter that is strongly dependent on angle of incidence near the normal and less, essentially constant, backscatter for grazing angles $\phi_g <$ 60°, due to unresolvable bubbles below the surface. The calculation was

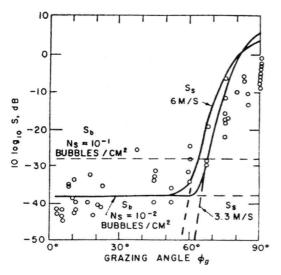

Fig. 9.23. Backscatter from a rough surface for two wind speeds and backscatter from two postulated below-surface unresolved column-resonant bubble densities (horizontal lines). The circles are experimental data of Urick and Hoover (1956) at 60 kHz during 4–5 m/s winds. (From Clay and Medwin, 1964.)

done for the frequency of 60 kHz to compare with the acoustical data available.

9.6 Point-source transmission through the air–sea interface

9.6.1 Smooth interface

Plane wave
Reviewing from Section 1.7, a plane wave going from a fluid medium $\rho_1 c_1$ through a smooth interface into a medium $\rho_2 c_2$ is described by the pressure transmission coefficient,

$$\tau_{12} = \frac{2\rho_2 c_2 \cos \theta_1}{\rho_2 c_2 \cos \theta_1 + \rho_1 c_1 \cos \theta_2} \qquad (1.77)$$

The ray direction is given by Snell's Law,

$$\theta_2 = \arcsin\left(\frac{c_2}{c_1} \sin \theta_1\right) \qquad (1.78)$$

where θ_1 is the angle of incidence measured to the normal of the interface, and θ_2 is the angle of transmission measured to the normal.

For sound going from air to water, $\rho_2 c_2 \cong (1000 \ \text{kg/m}^3)$ $(1500 \ \text{m/s}) \gg \rho_1 c_1 \cong (1/\text{kg}\,\text{m}^3) (330 \ \text{m/s})$; therefore for a plane wave $\tau_{12} \cong 2$, the pressure is doubled at the surface.

For a plane wave from air into water, there is a critical angle of incidence

$$\theta_1 = \theta_c = \arcsin(c_1/c_2) \cong \arcsin(330/1500) \cong 13° \qquad (9.46)$$

Plane-wave theory shows that for $\theta_1 > 13°$, there is an evanescent wave in the water (Section 1.7).

Point source
Invoking the Kirchhoff assumption one can argue that, for a point source, the plane-wave transmission coefficient given by (1.77) may be used at all incident points on the surface. Consequently, only a cone of energy with incident angles less than 13° would penetrate from the air into the smooth sea. To determine the dependence on source height, H, and receiver depth, D, we calculate the effect of a ray spreading at a smooth interface (Fig. 9.24). This is done by comparing the incident intensity passing through a conical shell at angle θ_1 and incremental width $d\theta_1$ with the transmitted intensity at angle θ_2 and incremental width $d\theta_2$ at the surface.

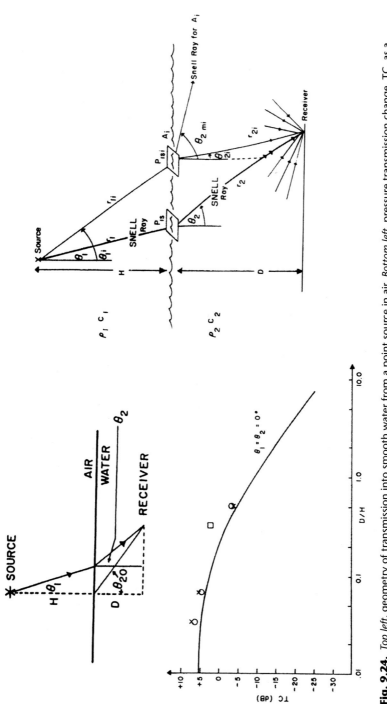

Fig. 9.24. *Top left,* geometry of transmission into smooth water from a point source in air. *Bottom left,* pressure transmission change, TC, as a function of the depth to height ratio, D/H, for sound at normal incidence through a low-roughness ocean surface. Solid line is the theoretical prediction; data are low-roughness experimental values for $0.04 < g_t < 0.15$; circles for a helicopter sound source, crosses and squares for flyovers by a P3 aircraft. *Right,* geometry for a rough ocean surface. (From Medwin, Helbig, and Hagy, 1973.)

At the surface, where the surface pressures in media 1 and 2 are p_{1s} and p_{2s}, the incident and transmitted powers through a ring of area A are

$$\Pi_1 = \left(\frac{p_{1s}^2 A \cos \theta_1}{\rho_1 c_1} \right) \tag{9.47}$$

$$\Pi_2 = \left(\frac{p_{2s}^2 A \cos \theta_2}{\rho_2 c_2} \right) \tag{9.48}$$

For a source at height H (see Fig. 9.24), the incident intensity at the surface is

$$I_{1s} = \frac{\Pi_1 \cos^2 \theta_1}{2\pi H^2 \sin \theta_1 d\theta_1} \tag{9.49}$$

The transmitted intensity received at depth D is

$$I_2 = \frac{\Pi_2}{2\pi (H \tan \theta_1 + D \tan \theta_2)} \left(\frac{H \tan \theta_1}{\sin \theta_2} + \frac{D}{\cos \theta_2} \right) d\theta_2 \tag{9.50}$$

Finally, after significant algebraic manipulation, Hagy (1970), Medwin et al. (1973) show that the transmission change in sound pressure level for a mirrorlike surface is given by

$$TC = 20 \log \frac{P_2}{P_{1s}} = -20 \log \left[\left(1 + \frac{c_2 D \cos \theta_1}{c_1 H \cos \theta_2} \right) (2 \cos \theta_1 \cos \theta_2)^{-1} \right] \tag{9.51}$$

The transmission change in dB to "ground zero," immediately below the source, is

$$TC_z = TC - 20 \log_{10}(\cos \theta_1) \tag{9.52}$$

where TC is the transmission change in dB across the smooth interface for the geometry shown in Fig. 9.24. At normal incidence, when $H \gg D$, the pressure in the water will be twice as high as the incident pressure ($TC = +6$ dB), whereas when $D \gg H$, the pressure in the water decreases proportional to the depth – that is, 6 dB per double distance. The smooth-surface theory (9.51) has been verified in an anechoic tank and then at sea (Fig. 9.24). In the ocean experiment, the sound source was a hovering helicopter or fly-by aircraft. The "almost smooth surface" for the ocean analysis was in the frequency band 100 to 200 Hz, so that, with the sea rms height $h = 14.5$ cm, the acoustical roughness for transmission was $0.04 < g_t < 0.15$.

9.6.2 Rough interface

The acoustical roughness parameter for transmission, g_t, depends on the in-water and in-air phase shifts relative to the interface. It is different from that for reflection, g_R by virtue of the different sound speeds and different angle of refraction into the second medium. It was derived by Hagy (1970) and may be calculated from

$$(g_t)^{1/2} = h(k_1 \cos\theta_1 - k_2 \cos\theta_2) \qquad (9.53)$$

or

$$(g_t)^{1/2} = k_1 h[\cos\theta_1 - (c_1/c_2)\cos\theta_2] \qquad (9.54)$$

where h is the rms height of the surface; θ_1 is the angle of incidence in the first medium, where the sound speed is c_1; and θ_2 and c_2, are for the second medium.

For transmission from air to water, $c_2 > c_1$ and $\theta_2 > \theta_1$ so that the first term dominates.

For large acoustical roughness of the surface, energy that would not have penetrated beyond the critical angle for a smooth surface *is* transmitted through the angled facets of the rough surface. These incoherent components may be calculated by applying the theory of the previous section to transmission.

The ratio (P_2/P_{1s}) was measured in extensive at-sea experiments using the Floating Instrument Platform FLIP of Scripps Institution of Oceanography as a base of operations. The 100 to 1000 Hz band noise of hovering US Navy helicopters or fly-by P3C aircraft were the sound sources at height 180 m. The sound receivers were sonobuoy hydrophones at 6 m and 90 m depth.

When the source was over the shallow hydrophone $(\theta_1 = \theta_2 = 0°)$ and $D/H = 0.03$, the transmission change was $+6 \pm 1$ dB not only at low roughness but also at higher frequencies, where $g_t = 4.2$. (See Fig. 9.25.) This was due to the incoherent contributions that came in from other positions beyond the normal-incidence direction. For the hydrophone at depth 90 m, where $D/H = 0.5$, there is a similar effect. Off-axis transmission contributions through a rough surface provide a *TC* that shows more energy than the theoretical loss 4.1 dB as predicted for smooth surfaces by (9.51).

When the 180 m high-sound source is offset from the hydrophones, so that the Snell angles in the water are 85° for the 6 m depth hydrophone, and 40° for the 90 m hydrophone, there are incoherent contributions from insonified regions beyond the transmission cone. Significantly, much more sound is received at larger surface roughness, especially by the shallow hydrophone, in general agreement with the theoretical solution (Fig. 9.26).

The significant difference between the results for smooth surface, Fig. 9.24, and those for rough surfaces, Figs. 9.25 and 9.26, should be noted.

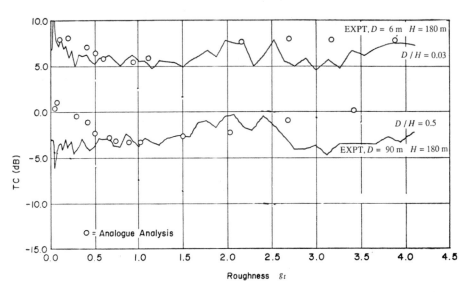

Fig. 9.25. Normal acoustic pressure transmission change as a function of surface acoustical roughness for transmission, g_t, for a point source at $H = 180$ m above the sea. The receivers were at depth $D = 6$ m and 90 m, directly under the source. The surface rms height was $h = 13$ cm; correlation length was $L \simeq 150$ cm. The source was a hovering helicopter. Solid line is digital analysis with 2 Hz frequency resolution for the noise in the 100 Hz to 1000 Hz band. The circles are spectral levels derived from analog band filtering. (From Helbig, 1970. See Medwin, Helbig, and Hagy, 1973.)

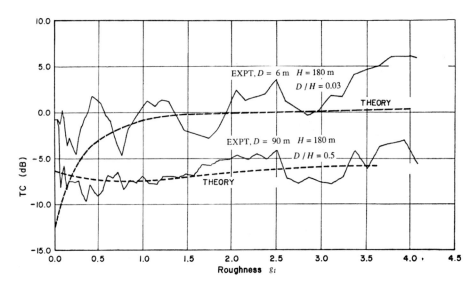

Fig. 9.26. Offset acoustic pressure transmission change as a function of surface acoustical roughness for transmission, g_t, for a point source at $H = 180$ m above a sea of rms height $h = 13$ cm. The receivers at depth $D = 6$ m and 90 m are offset by 75 m from the ground-zero point under the source. The offset angles are 85° for the 6 m hydrophone and 40° for the 90 m hydrophone. Solid line is digital analysis from 100 Hz to 1000 Hz with 2 Hz frequency resolution. (From Helbig, 1970. Dashed line is HK theoretical solution by Hagy, 1970. For theory, see also Medwin, 1972; for experimental details, see Medwin, Helbig, and Hagy, 1973.)

Problems

9.1 Integrate the Pierson–Moskowitz spectrum over all frequencies to obtain the rms height, h.

9.2 Use the two graphs in Fig. 9.5 to give analytical equations – such as (9.15) and (9.16) – for the SWOP correlations. Plot those equations and compare with actual data in Fig. 9.5.

9.3 Derive the PDF of a swell wave, $\zeta = a \sin K_x x$. Answer $w(\zeta) = \pi^{-1}(a^2 - \zeta^2)^{-1/2}$ for $\zeta < a$.

9.4 A 38 kHz sound beam is specularly scattered from a sea surface of rms height $h = 2$ cm. Determine the angles at which the relative coherent intensity $\mathcal{R}_{coh} > 0.7$.

9.5 Use the data on coastal bubble densities, resonant at 60 kHz to calculate the proper position of the horizontal bubble backscattering line in Fig. 9.7.

9.6 Use "reasonable" values to describe a rough sea surface and calculate the backscatter of a "reasonable" sonar as a function of angle of incidence.

9.7 Repeat the calculation of the previous problem to determine forward scatter as a function of the grazing angle of incidence.

9.8 Derive (9.50) for the transmitted intensity across a smooth interface between two fluid media.

9.9 Verify (9.51) for transmission change in dB across a smooth interface.

Further reading

See Chapters 10, 11, 22–24 for descriptions of recent and current acoustical measurements of the physical and biological characteristics of the near ocean surface and ocean bottom. Extensive specific references are given.

Farmer, D. "Acoustical studies of the upper ocean boundary layer"
Chapter 10 of this book.

Medwin, H. and Clay, C. S. (1998). *Fundamentals of Acoustical Oceanography*. Academic Press.
See Section 13.3 Helmholtz–Kirchhoff–Fresnel solution for surface scatter of a spherically diverging wave. Also, bottom scattering described in Chapter 14, "Mapping the Sea Floor: Side-Scanning, Swath Mapping, Spatial Spectra and Correlations."

Part II

Studies of the near-surface ocean

Chapter 10
Acoustical studies of the upper ocean boundary layer

DAVID FARMER
University of Rhode Island

Summary

This survey of acoustical oceanography of the upper ocean is far from comprehensive, and omits, among other topics, the important new developments in acoustical studies of sea-ice, the application of acoustical techniques for biological research, and the use of acoustical imaging for the study of internal waves and topographically induced effects. Nevertheless, given the widespread application and development of new techniques for upper ocean research, it does give some indication of the possibilities that exist. The ocean provides a natural acoustical signal, rich in frequency diversity, temporal variability, and directionality, that can be exploited to learn about the air–sea interface, the weather, and the details of wave breaking which play so important a role in air–sea transfer of momentum and other properties.

Active acoustical measurement can determine the bubble scattering and velocity field, the waves properties, and organized motion of the upper ocean boundary layer. The competing influences of buoyancy, dissolution, turbulent resuspension, and advection, together with the intermittent injection of bubbles by wave breaking, generate a constantly evolving bubble size distribution, which can now be measured with *in situ* instrumentation. While bubbles are important to many aspects of air–sea interaction including gas exchange, acoustical and optical effects, nuclei for biological and chemical effects, among others, their role as tracers appears to offer particular promise. Bubble size distributions in particular offer the opportunity for testing models of bubble formation, near-surface turbulence, and circulation. The fact that size distributions change with time provides constraints on the age of bubble clouds. To cite just one example: bubbles collecting in Langmuir convergence zones can be expected to be "older" than bubbles recently injected by breaking waves. Their greater age will have allowed more buoyancy sorting and dissolution thus modifying their size spectrum. As measurements of turbulence, advection, wave breaking, gas saturation, and near-surface circulation are integrated in comprehensive observational programs, the extent to which we can reconcile measured bubble size distributions with models of the fluid dynamical environment will serve as a critical test of our understanding of this dynamic environment.

Contents

10.1 Introduction

The upper ocean boundary layer extends from the air–sea interface down through the actively mixed near-surface waters where the stratification is weak. Heat, gas, moisture, and momentum are transferred across the air–sea interface; active mixing beneath this interface connects these exchanges to the ocean interior. Vertical exchange of properties plays a central role in the stabilization of the earth's climate and accurate parameterization is critical to the satisfactory modeling of the coupled atmosphere–ocean system and climate prediction. The upper ocean boundary layer also overlaps the euphotic zone where sunlight sustains photosynthesis, creating the first step in biological production. Acoustical methods are helping elucidate the complex interplay between atmosphere and ocean and the physical and biological processes of this energetic and climatically sensitive portion of the water column.

In the absence of wind, the ocean surface can be a benign environment, but higher wind speed leads to enhanced exchanges across the air–sea interface, along with increased near-surface mixing. Accurate measurements of fluxes and small-scale processes associated with the wind driven surface layer become difficult at higher sea states, and a variety of acoustical methods offer significant advantages for observing the upper ocean in these conditions. Passive detection makes use of the background sound field caused by naturally occurring noise sources such as wave breaking, precipitation, or the fracturing of sea-ice. Nature provides the signal. Our challenge, as acoustical oceanographers, is to learn how to interpret it. Active methods, on the other hand, involve the artificial generation of sound together with detection of absorption, scattering or other effects on the signal due to oceanographic phenomena. Both approaches can involve remote sensing in that the measurement volume may be well removed from the sensor, a particularly valuable characteristic in the challenging upper ocean environment, where instruments are vulnerable. On the other hand, some observations cannot be acquired in this way and *in situ* acoustical sensors are proving helpful in acquiring measurements that are inaccessible to the remote methods. This brief overview discusses a representative range of acoustical approaches that have proved useful in this context. This is an evolving field in which new techniques are continuously being developed.

10.2 Passive detection of breaking waves

The relationship between wind speed and ambient noise has a long history and is well described by the Knudsen curves (Fig. 5.11). Even casual observation of the sea surface would suggest that the primary noise source must be associated with breaking waves. An early application involved the inversion of measured noise to recover wind speed (Shaw et al., 1978); surprisingly, the relationship between wind speed and noise spectrum level is quite robust (Vagle et al., 1990). A great advantage of this approach is that the noise field can be measured from well beneath the sea surface, even from the sea floor, thereby providing long records of ocean surface weather. This application led naturally to an exploration of the detailed mechanisms responsible for noise generation along with an interest in using the finer scale features of the signal to learn more about the properties and distribution of wave breaking. The motivation here is as much oceanographic as acoustic and includes the determination of wave breaking frequency and breaking characteristics, air entrainment, energy dissipation and consequent mixing.

Much can be learned from a single hydrophone placed beneath the ocean surface. Consider the contribution to the signal received by a hydrophone at depth h, due to a surface sound source of intensity I_0 at radius r. An individual source close to the surface (Fig. 10.1) tends to radiate sound as a dipole so that the source strength varies as $I_0 \sin^2 \theta$ where θ is the transmission angle measured with respect to the sea surface.

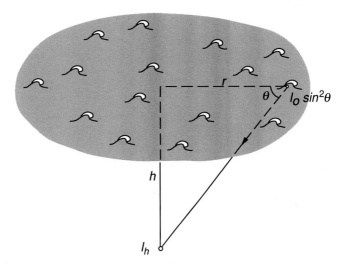

Fig. 10.1. Sketch showing location of a hydrophone relative to a field of breaking surface waves. The hydrophone is at depth h and detects a signal of intensity I_h (from Farmer & Vagle, 1989).

The signal is attenuated as it passes through the water column. In the absence of dense bubble clouds the attenuation is due to shear viscosity and molecular relaxation effects and depends on the path length $(h^2 + r^2)^{1/2}$. Neglecting refraction effects, the signal strength I_h at the hydrophone is then (Farmer and Lemon, 1984):

$$I_h = I_0 \sin^2 \theta (h^2 + r^2)^{1/2} \exp[-\alpha(h^2 + r^2)^{1/2}] \qquad (10.1)$$

where α is the attenuation coefficient applicable to the frequency of interest. If the data are integrated over time, so that there is on average a spatially dense and uniform distribution of sources, the integrated signal level I_r at the hydrophone can be expressed as

$$\int_0^\infty 2\pi r I_0 \sin^2 \theta (h^2 + r^2)^{1/2} \exp[-\alpha(h^2 + r^2)^{1/2}] dr$$

$$= 2\pi I_0 E_3(\alpha h + \beta) \qquad (10.2)$$

where E_3 is the exponential integral function of third order. Refraction may also affect this result for deeply placed hydrophones and, as discussed later, near-surface bubble fields are also important. However, we can see the relative contribution of a given source by integration of (10.2) about a circle of unit width:

$$F(r) = 2\pi r I_0 \sin^2 \theta (h^2 + r^2)^{-1} \exp[-\alpha(h^2 + r^2)^{1/2}] \qquad (10.3)$$

The sensitivity of the hydrophone to sources at different ranges depends both on hydrophone depth and upon the attenuation, which is frequency

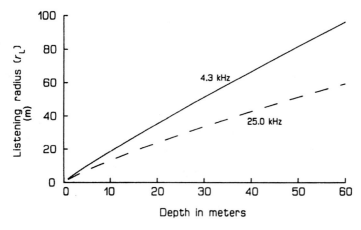

Fig. 10.2. Effective radius of ocean surface (see (10.4)) detected at two different frequencies as a function of depth, assuming the sound sources are dipole (from Farmer & Vagle, 1989).

dependent. This sensitivity is conveniently expressed (Farmer & Vagle, 1989) in terms of a listening radius r_L, equal to one half of the second moment of F:

$$r_L = \tfrac{1}{2}h[\{E_1(\alpha h) - E_3(\alpha h)\}/E_3(\alpha h)]^{1/2} \qquad (10.4)$$

Figure 10.2 shows the listening radius calculated for two different frequencies as a function of depth. Differences in measured signals with respect to frequency arise from the frequency dependence of acoustic absorption.

Bubbles are efficient at scattering and absorbing sound if the acoustic frequency matches the bubble's resonance frequency. Bubble attenuation can be incorporated by including a second exponential term $-\beta/\sin\theta$ to (10.1) and following expressions (the argument of the exponential integral in (10.4) becomes $(\alpha h + \beta)$. The effect of bubble attenuation can be seen vividly at higher wind speeds by displaying the measured sound spectrum level at one frequency as a function of sound spectrum level at a different frequency (Fig. 10.3). At wind speeds above about $15\ \mathrm{m\,s^{-1}}$ bubble clouds start to attenuate the signal significantly, with the greatest attenuation occurring at around 30 kHz, consistent with the greater volume scaled density of smaller bubbles resonant at approximately this frequency. Above approximately 14 kHz the ocean actually becomes quieter with increasing wind speed due to the insulating effect of the bubble clouds (Farmer and Lemon, 1984). This unexpected result is not generally included in the Knudsen curves but it does have interesting consequences. Under assumptions about the consistent spectral shape of the noise source at different wind speeds, the inferred attenuation provides a basis for inverting the sound spectrum level to recover the

Fig. 10.3. Ambient noise spectral levels (NSL) simultaneously detected at frequencies of 4.3 kHz and 25 kHz. In general, high winds produce a greater signal, but the higher frequencies can be attenuated at high wind speed due to a layer of bubbles causing the NSL relationship to "roll over," i.e., the 25 kHz NSL decreases as the 4.3 kHz NSL increases, with increasing wind speed. Additional high frequencies are contributed by precipitation noise, additional lower frequencies by ship noise.

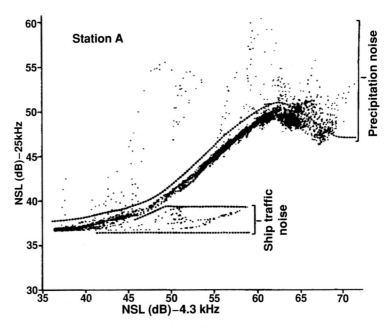

bubble size distribution. Bubble attenuation also presents a complicating factor in attempts to invert precipitation induced noise at higher wind speeds, so as to infer rainfall rate (Nystuen and Farmer, 1989).

It is clear from Fig. 10.2 that as the hydrophone depth *decreases* or the observed frequency *increases*, the effective measurement area becomes smaller. As the measurement area decreases, the signal strength will fluctuate more because there will be fewer breaking events within the effective listening radius. If our interest is in the properties of individual wave breaking characteristics, it is necessary to move our hydrophone closer to the surface and replace time averaged signals with continuous measurement. For a near-surface hydrophone, temporal variability of the signal strength can be interpreted in terms of individual breaking events. By taking into account the range and attenuation effects described above, signal variability can be used to infer the number density or mean spacing of the breakers, the tendency for breaking to repeat within the wave group, and properties of the wave group itself.

10.2.1 Measurements with hydrophone arrays

If we want to go much further than this in our study of breaking waves, we need to know exactly where the wave is with respect to the sensor. One approach is to deploy an array of hydrophones such as shown in Fig. 10.4 (Farmer and Ding, 1992). If the sound source is sufficiently

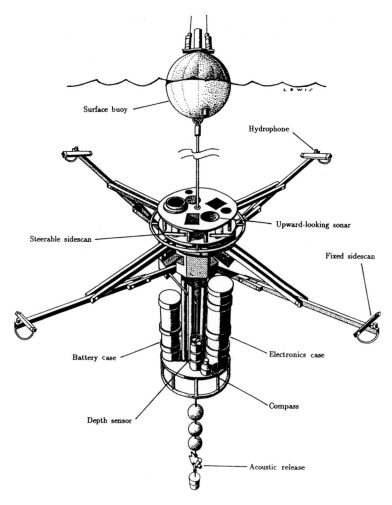

Surface buoy

Hydrophone

Upward-looking sonar

Steerable sidescan

Fixed sidescan

Battery case

Electronics case

Compass

Depth sensor

Acoustic release

Fig. 10.4. An example of a freely drifting instrument for both passive and active acoustic measurement. The instrument is suspended from a surface float at a depth of 20–30 m by a rubber cord, which effectively decouples it from surface motion. Hydrophones at the ends of each arm provide an aperture with which to detect the angle of arrival of broadband signals. Sidescan sonars are pointed at an oblique angle to the surface. Vertical sonars of different frequency are used to track the vertical structure of the bubble field. The instrument records data internally (from Farmer & Ding, 1992).

distant that it appears as a point source, cross-correlation of the signals detected at each hydrophone identify the time delay between its arrival first at one hydrophone and then at the other (Ding and Farmer, 1992). We can display the correlation intensity to show the correlation peak associated with the source as it moves across the sea surface above the hydrophones (Fig. 10.5). Similar time delays can be calculated for arrivals at orthogonal hydrophone pairs. Knowing the time delays, the instrument depth and its orientation, and the location of the breaking wave can be found. From continuous measurements it is possible to monitor the positions of breaking waves and their speed and direction of travel within the effective listening area of the instrument (Fig.10.6). Much useful information can be derived from these measurements. From the dispersion relationship, in which the speed of a wave is related to its

Fig. 10.5. Correlation intensity due to the sound of individual wave breaking events arriving at different hydrophone pairs (see Fig.10.4) as a function of time delay. Correlation signals from the hydrophone pairs may be combined so as to track the motion of a given breaking event as it moves across the sea surface (from Ding & Farmer, 1992).

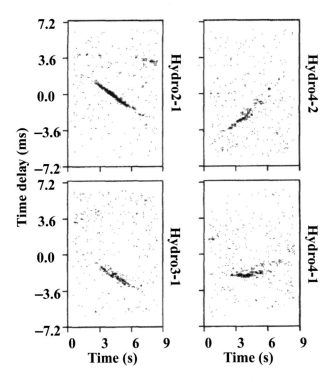

Fig. 10.6. Tracks of individual breaking waves shown relative to the instrument's hydrophone array (see Figs. 10.4, 10.5).

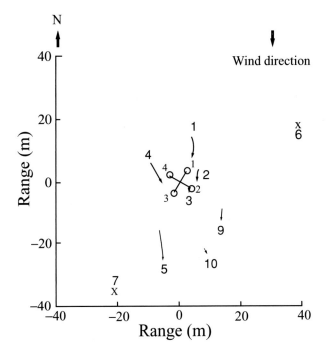

length, we can derive the scale of individual breaking waves (Ding and Farmer, 1994a). The observations can therefore tell us the way in which wave breaking extracts energy from the wave field, in particular the part of the wave spectrum at which dissipation due to breaking is occurring. It turns out that shorter waves, significantly shorter than the dominant wavelength in the spectrum, contribute most to breaking. Measurements of this type provide guidance on the proper incorporation of dissipation in models of wave fields.

From an oceanographic point of view, it would be highly desirable to learn the relationship between radiated sound and the energy dissipated. Laboratory studies have shown that a significant fraction of the energy dissipated by a wave is expended in air entrainment and the radiated acoustic energy is related to the energy lost by breaking (Lamarre and Melville, 1991). Other studies have shown that dissipation increases with wave speed, at least for steady waves (Duncan, 1981). Since the measurement approach described above can be used to determine the position and velocity of the breaking event, the acoustic source level of individual breaking events may also be determined as a function of velocity. Figure 10.7 shows an example of such a comparison for breaking waves in the open ocean and illustrates that a much tighter relationship exists between the source level and the speed with which it travels across the ocean surface, than the wind speed (Ding and Farmer, 1994b). From a remote sensing perspective it would be valuable to search for a relationship between acoustic power flux radiated by breaking waves and energy dissipation. The frequency and distribution of breaking events can be determined from the spatial mapping of waves discussed above. Under certain assumptions, the energy dissipation can be inferred for the observed wave field, allowing a comparison between the acoustic power flux radiated by breaking waves and the wave energy dissipation per unit area (Fig. 10.8). The acoustic energy is of order 10^{-8} of the dissipation rate. While observations such as these should be considered preliminary, they do point towards an important remote sensing capability of acoustical measurement of the rate of wave energy dissipation due to breaking.

This discussion has not touched on the details of sound generation within a breaking wave. From an acoustical point of view the challenge is to explain the sound spectrum radiated by the wave in terms of the size distribution of bubbles created. Since bubble creation in a turbulent environment is likely to occur in a cascade as bubbles are successively broken into smaller and smaller sizes, the interpretation is unlikely to be straightforward. This is a topic of active study, especially with respect to the measurement of bubble source distributions within breaking events, and depends on detection of bubble break-up with high speed photography.

Fig. 10.7. The mean dipole source level of breaking waves expressed as a function of (*a*) wind speed, and (*b*) the speed of the breaking wave ("event speed"). Much better correlations are found with the former (from Ding & Farmer, 1994b).

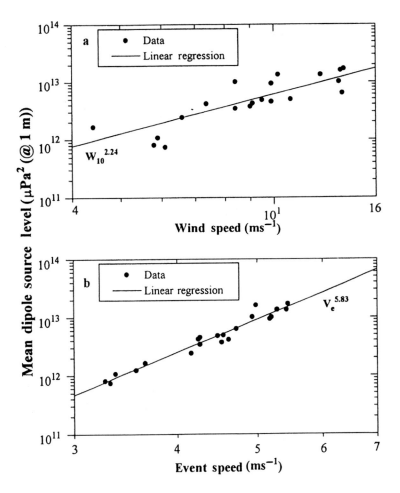

Fig. 10.8. The acoustic power flux (watts per square meter) radiated by breaking waves as a function of the estimated wave dissipation per unit area (from Ding & Farmer, 1994b).

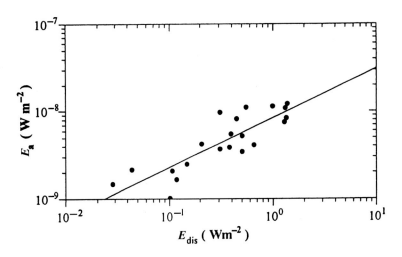

Additional issues are raised by the possibility that the dense plume of bubbles created by a breaking wave is subject to collective oscillation, giving rise to much lower frequency radiation than achieved by breathing mode oscillation of small bubbles. There is evidence from laboratory studies that such lower frequency signals are indeed generated, however the link between dense bubble plumes observed at sea and corresponding collective oscillations has yet to be demonstrated unambiguously.

The discussion thus far has been concerned with the sound generated by breaking waves detected in the far field of moving point sources. The actual shape of the sound source can also be inferred using a near-field array of hydrophones, although the signal processing for this case becomes quite challenging. Figure 10.9 shows a reconstruction of the

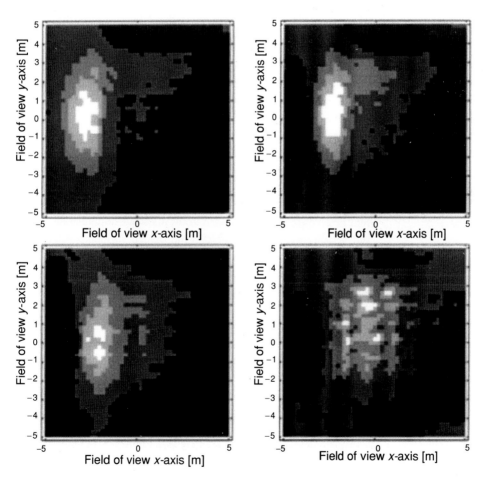

Fig. 10.9. Successive images of the sound source due to an individual breaking wave, synthesized from ambient sound detected with a small subsurface hydrophone array (from Andrew *et al.*, 2001).

spatial and temporal evolution of the sound source due to a passing breaking wave (Andrew et al., 2001).

10.3 Sonar measurement of directional wave fields and organized circulation

A particularly insightful application of sonars in the study of the upper ocean boundary layer is the use of slant beam deployments directed at a shallow angle towards the ocean surface (i.e., Pinkel and Smith, 1987). The transmitted pulse grazes the surface and is scattered by bubbles and surface roughness. A standard sidescan transducer generating a "fan beam" is oriented such that the broad axis of the beam cross-section is in elevation and the narrow axis in azimuth. The sonar is directed at a shallow angle to the surface. A small correction is required to transform the time delay of the detected signal into horizontal range (see Section 3.9). Bubbles tend to be dominant scatterers at wind speeds above a few $m\,s^{-1}$ and can occur several meters beneath the sea surface. At any given range the transmitted pulse will be scattered by the surface and by all bubbles at that range lying within the beam. In this sense the scattering is distributed in depth although typically dominated by the near-surface, since the bubbles tend to have an exponential depth dependence with an e-folding scale of order a meter or less. In typical applications, broadband pulses are used (Pinkel and Smith, 1992; Trevorrow and Farmer, 1992) and the signal may be Doppler processed to recover velocity components. An important development of this concept involves the use of a phased array, allowing simultaneous intensity and velocity measurements over a pie-shaped segment (Smith, 2001). Variations of this configuration include mechanical rotation in azimuth, similar to that used in radar systems, so as to recover a two-dimensional field of intensity or velocity.

The potential of this technique is best understood by considering the behavior of bubbles injected by breaking waves. Air entrained by breaking waves is progressively broken into smaller and smaller bubbles, a mechanism bounded by the Hinze scale, at which turbulent pressure fluctuations at the bubble surface are just balanced by the restoring force due to surface tension. The result is a spectrum of bubble radii distributed vertically and horizontally in the neighborhood of the breaking event. Their subsequent behavior depends upon buoyancy which progressively removes the larger bubbles, dissolution which tends to remove the smaller bubbles, turbulence which redistributes the bubbles locally and can keep the bubbles in suspension, and advection which organizes the surviving bubbles over longer periods. Most practical narrow beam and Doppler sonars tend to operate at frequencies of 100 kHz or more, and are therefore sensitive to bubbles of radius 30 μm or less. These bubbles are "small" in the sense that their buoyant rise rate is less than typical near-surface rms turbulent velocities and they are not quickly lost. These bubbles therefore tend to collect in convergence zones where they descend until finally going into solution. Small bubbles are essentially passive with respect to wave orbital motions and therefore serve as excellent tracers of surface waves.

Doppler analysis (Section 2.6) of slant beam sonar signals allows calculation of the resolved component of the surface or near-surface velocity as a function of range, along the axis of the sonar beam. The measured velocity field consists of motion due to surface gravity waves, near-surface circulation such as convection and Langmuir circulation, together with residual currents and instrument motion. The wave components may then be inverted to recover the resolved component of the wave field. Since the direction of wave propagation cannot be resolved unambiguously with a single sonar orientation, measurements from two or more directions are required. The directional wave spectrum is recovered using the dispersion relation together with an optimization of the data from the different sonar beams (Fig. 10.10). Some care is required in the interpretation of such measurements. Images acquired with slant beam sonars tell us nothing about the depth dependence of the bubbles contributing to the scattered signal. Wave orbital motions decay as $\exp(-kz)$, where k is the wave number of the surface wave and z is the depth. A spectral correction based on vertical sonar measurements of the 1D wave spectrum, derived from surface backscatter rather than bubble scatter, may be required.

Whereas the Doppler processed signal reveals wave and other near-surface motions, intensity measurements reveal the distribution of bubbles. With low pass filtering to remove wave effects, the scattering intensity typically shows bubbles organized into elongated structures approximately aligned with the wind. The bubble clouds are organized by the subsurface circulation, especially pairs of counter rotating vortices referred to as Langmuir circulation which draw the bubbles into convergence zones. While larger bubbles escape to the surface, the smaller ones descend 10 m or more before dissolving completely. Narrow beam sonars can be mechanically rotated like a scanning radar, to obtain a 360 degree image of the scattering field (Fig.10.11). The resulting bubble distributions tend to be aligned with the wind, but are far from strictly linear structures and often reveal Y-junctions and other patterns that challenge our understanding of near-surface circulation. Low pass filtered Doppler measurements provide the corresponding velocity field associated with these features, although the separation of wave motions (of the order m s^{-1}) and Langmuir circulations (of the order cm s^{-1}) can be demanding, especially from a drifting platform that is itself subject

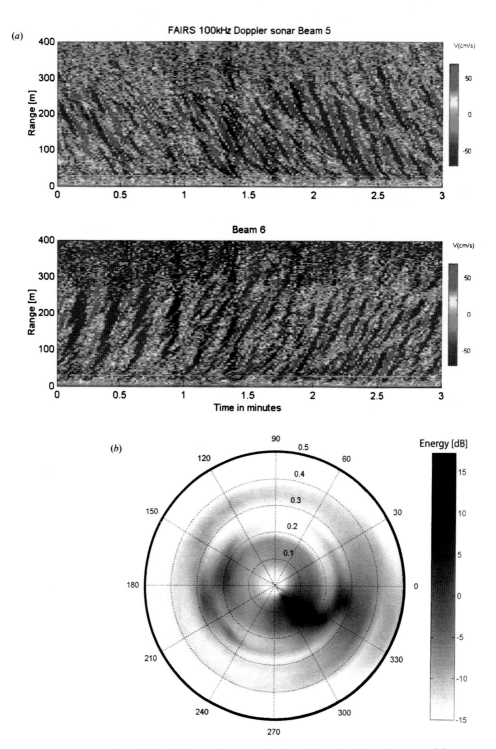

Fig. 10.10. (a) Time series example of two Doppler sidescan images of the ocean surface acquired with two orthogonally directed 100 kHz narrow beam sonars. The time series shows the orbital velocities of waves propagating away from (upper image) and towards (lower image) the instrument. (b) Polar display of the directional wave spectrum calculated from the Doppler sonar time series shown above (from Farmer *et al.*, 2002). (See colour plate section.)

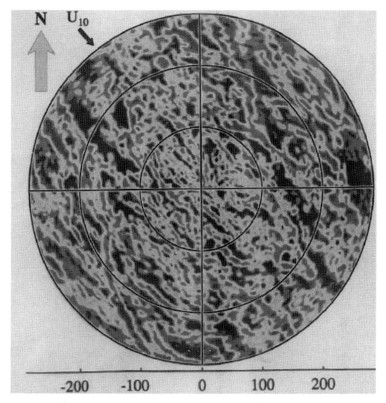

Fig. 10.11. A polar view of backscatter from the sea surface acquired with a combination of four rotating sonars. The image is acquired in about 30 s. Wind direction is from the northwest (U_{10}). (See colour plate section.) Red indicates higher backscatter intensity corresponding to denser bubble distributions.

to wave motion. If the wind picks up after relatively calm conditions and there is no well-established near-surface bubble field to mask the breaking events, these may become clearly identifiable.

Slant beam measurements can be usefully combined with vertical sonar observations which detect the vertical structure of bubble clouds; measurements of this type formed the basis for initial studies of bubble clouds and their role in air–sea gas flux. Bubble penetration is typically associated with enhanced scatter from the slant beam sonars, consistent with an interpretation in terms of subduction by convective rolls or Langmuir circulation.

10.4 Velocity profiling into wave crests

The remote sensing characteristics of sonar measurement are especially useful in extreme sea states. For example, there are both practical and

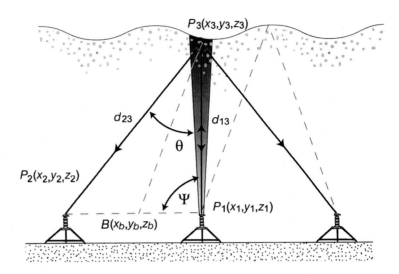

Fig. 10.12. Deployment configuration for a bottom mounted bistatic sonar for Doppler measurement of three components of velocity as a function of elevation. The vertically oriented narrow beam sonar transmits a signal which is then scattered by bubbles and other inhomogeneities to the three fan-beam hydrophones (only two are shown) as well as to the vertical sonar on the sea floor (from Farmer *et al.*, 2002).

scientific reasons for wanting to know the velocity field in the crests of large, steep waves which can be hazardous to offshore engineering structures. The relevant "overturning moment" acting on an offshore platform is that produced by the largest steep wave encountered, depending primarily on the horizontal flow speed in the wave crest. Extrapolation of classical perturbation solutions to these extreme states, especially where there is incipient breaking, may not be justified. The application of a bottom mounted bistatic Doppler sonar provides an example of the way in which acoustical methods can contribute to the resolution of such problems, as well as highlighting some of the challenges (Farmer *et al.*, 2002). Figure 10.12 illustrates a deployment on the floor of the North Sea. A single narrow beam sonar transmits a broadband pulse up towards the sea surface. The pulse is scattered back to three fan-beam hydrophones, and also to the transmitting sonar which switches into receive mode. Scattering occurs from bubble clouds and from the surface itself. The vertical component of velocity can be determined as a function of elevation from the vertically oriented Doppler sonar. However, reception of the scattered signal by the remaining hydrophones on the sea floor allows the horizontal components to be resolved.

In the bistatic setup shown in Fig.10.12, P_1 is the transmitter, P_2 a receiver and P_3 a moving target that could be a bubble cloud or the sea surface. A typical ray path from P_1 is scattered isotropically by P_3 and detected at P_2, where the angle $\angle P_1 P_3 P_2 = \theta$. For a transmitted signal of wavelength λ_0, the Doppler frequency shift from the moving target is

$$\Delta f = -\frac{1}{\lambda_0} \mathbf{v} \cdot (\mathbf{r}_1 + \mathbf{r}_2) \qquad (10.5)$$

where $\mathbf{v}$ is the velocity vector of the moving target and $\mathbf{r}_1$, $\mathbf{r}_2$ are the unit vectors from transmitter and receiver to the target. Since $|\mathbf{r}_1| = |\mathbf{r}_2| = 1$,

$$\Delta f = -\frac{2}{\lambda_0} \cos\left(\frac{\theta}{2}\right) v_b, \qquad (10.6)$$

where v_b is the velocity component along P_3, the bisector of θ. If Δf can be measured, the velocity component along the bisector can be determined. From the four receivers, the three components of velocity as a function of elevation can then be derived.

This technique proves successful when there is a strong wind and well developed wave field (the conditions of interest!), but breaks down at low wind speeds. In order to understand this limitation, consider again the deployment configuration in Fig. 10.12. A narrow beam (3.5°) is used for transmission. However, for scatterers close to the sea surface, the signal detected at a receiver will arrive at the same time as specular scatter from the sea surface. Although the transmission angle in this case, indicated by the dashed line intersecting the surface on the right-hand side of Fig. 10.12, lies outside the main lobe, the target strength of the surface relative to the bubbles can be great enough to exceed the side lobe transmission intensity, thus contaminating the signal of interest. This possibility can be analyzed using standard acoustic propagation and scattering theory and predicts that at lower wind speeds such contamination is indeed a problem, but at higher winds, above 15 m s^{-1}, the bubble density and surface scattering losses are such that bubble scattering dominates and we obtain a useful measurement. Since we have a total of four receivers on the sea floor to calculate three components of velocity at each depth, the inversion is over-determined and we can use this fact to confirm the validity of the acoustic model. Figure 10.13 shows a short time series of the wave elevation and horizontal velocity at the ocean surface, and also illustrates the way in which the linear surface gravity model makes reasonable predictions for small waves, but greatly underestimates the speed in the crest of large steep waves.

Surface elevation time series can provide a valuable record of the one-dimensional wave spectrum. At lower sea states a vertically oriented sonar can be used to identify the sea surface elevation from the step-increase in observed target strength, so that the surface may be automatically measured using a signal threshold detection. At higher sea states, the presence of dense bubble clouds will reduce the magnitude of

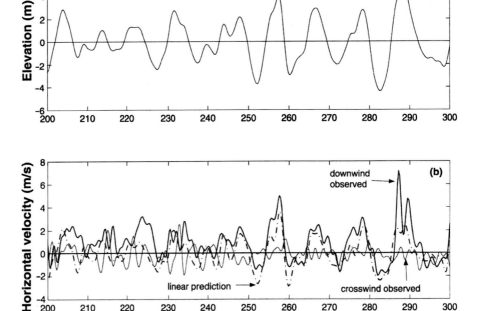

Fig. 10.13. A short time series acquired from the bistatic sonar of Fig. 10.12 showing (*a*) ocean surface elevation, (*b*) downwind and crosswind velocity components at the sea surface corresponding to the waves measured above. Also shown (dash–dot–dash) is a linear prediction of the downwind surface velocity, which severely underpredicts for the large, steep wave at 283–290 s (from Farmer *et al.*, 2002).

this step and make it much harder to identify. The result is that surface elevation time series are subject to drop-outs at times when the threshold detection fails, as shown in the upper time series of Fig.10.14. However, Doppler processing allows us to find a way out of this problem, since the vertical elevation is related to the vertical velocity. We may integrate the vertical velocity at the expected location of the sea surface, so as to obtain a refined estimate of its location. The first-order derivative of surface elevation should be close to the vertical velocity. The calculation can be repeated iteratively and converges rapidly. The lower time series in Fig.10.14 shows the surface elevation corrected in this way. These examples are presented in order to illustrate the way in which different types of acoustical data, such as multiple receivers, intensity and Doppler, can be combined to overcome some of the limitations encountered with simpler systems.

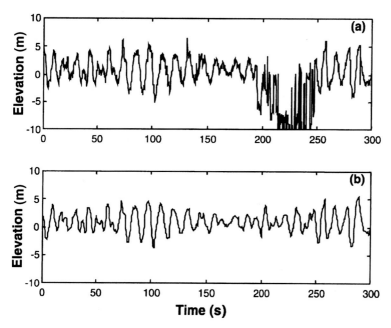

Fig. 10.14. Time series of surface elevation derived from a vertically oriented
sonar (*a*) using threshold detection – note "drop-out" due to dense bubble
cloud which obscures surface – and (*b*) corrected surface elevation using
Doppler signal to estimate surface with an iterative calculation.

10.5 Bubble injection, bubble clouds, and bubble size distributions

Bubbles created by breaking waves play a role in gas transfer and serve as
useful tracers of upper ocean turbulence and circulation. The pioneering
work on gas transport by bubbles was carried out with a single frequency
sonar (Thorpe, 1982). Bubbles are effective acoustic scatterers owing to
their high quality factor at resonance. For bubble sizes typically encountered in the upper ocean the acoustic cross-section at resonance is of
order 1000 times the geometrical cross-section. Consequently, the size
distribution of bubbles is central to an interpretation of the upper ocean
acoustical environment and it is worth discussing the physical processes
at play in determining bubble sizes, as well as the acoustical methods
for detecting them.

10.5.1 Measuring bubble size distributions

The high acoustical cross-section of bubbles greatly facilitates their detection. Several acoustical techniques have been developed for bubble

size measurement, most of which depend on linear acoustics and on the modification of the bulk acoustical properties of a fluid due to the presence of bubbles. Since the resonance frequency of bubbles depends upon their radius, acoustical measurement of the size distribution of an ensemble of bubbles using this approach requires a suitable range of frequencies. The bulk approach has been used in three different ways: with multiple frequency sonars, with broadband propagation loss and time delay over a short path, and with resonators.

Multiple frequency sonars may be used to measure target strength as a function of range and frequency which may be inverted to recover bubble size distributions subject to certain assumptions (Szczucka, 1989; Vagle and Farmer, 1992). In practice this approach is difficult to accomplish with confidence. Unless special precautions are taken, sonars of different frequencies will normally have different beam widths and will not insonify identical volumes. The results will be sensitive to beam properties and sonar sensitivities. It must be assumed that the bubble field is uniform across the sampled volume. Bubble induced attenuation along the sonar beam must be allowed for in the inversion, although the bubble population contributing to this attenuation is not known *a priori*. Other challenges include the need to avoid near-field effects, the tendency for a sonar to collect microbubbles on the transducer face in supersaturated water leading to calibration shift, and the size and power requirements of lower frequency sonars needed to span the primary range of bubble sizes.

Short range propagation systems use a broadband source and a separate hydrophone typically separated by a few tens of cm (Medwin, 1965; Lamarre and Melville, 1994; Vagle and Farmer, 1998). The source produces a broadband pulse which is selectively attenuated by bubbles along the path. The frequency dependent attenuation is then inverted to recover the bubble size distribution. There is also a corresponding frequency dependent travel time that can be measured and inverted to get the bubble distribution. This technique is relatively simple to implement, but there are some subtleties. As with any system that uses a short broadband pulse, the temporal response of the bubbles needs to be considered; in other words, if the pulse duration T is comparable to or less than $Q\tau$, where Q is the quality factor at resonance and τ is the acoustic period, steady state solutions are no longer applicable (see discussion in Vagle and Farmer, 1998). This effect will be greater at lower frequencies and even when properly accounted for has the effect of reducing the sensitivity. Use of longer pulses is of course possible, but decreases the distance the measurement can be made from a reflector, such as the sea surface, without contaminating the measurement. An additional complication is that strong acoustical sources, required in noisy environments

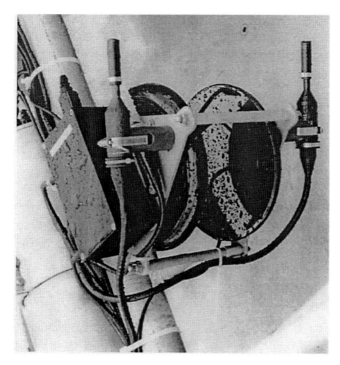

Fig. 10.15. Acoustical resonator consisting of two plates with broadband acoustical excitation. Attenuation of the harmonics is used to estimate bubble population (from Farmer, Vagle & Booth, 1998).

or with higher bubble densities, invoke a nonlinear response that must be incorporated in the inversion.

One way of increasing the measurement sensitivity and avoiding the temporal acoustical response of bubbles in short pulse measurements, is to use a resonator, first used by Medwin and Breitz (1989) and further developed and analyzed by Farmer *et al.* (1998) (Fig. 10.15). White noise injected between two reflecting plates excites many harmonics spanning a wide frequency range. The harmonics are modulated by the presence of bubbles. The resonator response is detected with a hydrophone, which may be in the form of an acoustically sensitive coating on one of the plates, and the resulting signal is inverted to recover the bubble population. Internal consistency of the measurement may be checked by separately using both the attenuation of each harmonic and its frequency shift, which are related through the Karmer–Kronig relationship. In addition to its high sensitivity, the resonator measurement is confined to the fixed volume between the plates. It may be operated close to reflecting boundaries such as the sea surface, and can use long or even continuous excitation, thus avoiding short pulse effects discussed above. However, care is required to avoid, or correct for, an effect on bubbles that is peculiar to resonant cavities. Leighton has drawn attention to the fact that a bubble's resonant response in a standing wave can be modified by the resonant pressure field. This effect has been numerically examined for the case of the resonator with random bubble distributions and shown to lead to narrow-band peaks and dips in the resonator sensitivity. Careful choice of instrument geometry can minimize coincidence of these peaks and dips with the resonator harmonics.

Non-linear techniques for the measurement of bubble sizes have been pioneered by Leighton and his colleagues (i.e., Leighton et al., 1991; Leighton, 1992; Leighton et al., 1997). Non-linear techniques have many interesting features and constitute a wide subject which lies beyond the scope of this report. We note that non-linear acoustic measurement of bubble sizes is discussed elsewhere in this volume (Chapters 4 and 21).

10.5.2 Bubble size distributions

Air is trapped in the overturn of the wave crest and successively broken into smaller and smaller bubbles. The creation process is very brief and results in a size distribution that depends in some way upon the rate of air entrainment and the intensity of the turbulence that breaks the bubbles. However this initial bubble size distribution quickly changes under the influence of gas dissolution, buoyancy, and the redistributing effects of turbulence. Successive break-up of bubbles has been studied in the laboratory. Ultimately the bubbles have a radius small enough that surface tension is sufficient to resist the turbulent pressure fluctuations and the cascade to smaller radii ceases. The bubbles descend in a turbulent jet and may be associated with vortical structures created by the moving breaker. Turbulence decays rather quickly, as $t^{-5/4}$ for isotropic turbulence in unstratified flow, faster if bubble buoyancy effects are significant. The suspension and transport of bubbles by turbulence is therefore a rapidly changing function of the fluid dynamical environment just beneath the surface.

Scaling arguments, in which a steady rate of air entrainment leads to bubble break-up under the influence of turbulence, imply a bubble spectrum within the active breaker having a power law radius dependence of $N(a) = N_0 a^{-n}$ where $a = -10/3$ (Garrett et al., 2000). Recent observations by Deane and Stokes have confirmed this relationship. There are some subtleties to this interpretation. For example, a steady-state rate of formation and break-up in bubbles would lead to a spectral pile-up at small radii, due to the fact that the smallest bubbles able to resist turbulent pressure fluctuations could not dissolve as rapidly as they are produced. No such spectral pile-up is observed, suggesting that fluctuations in turbulent intensity smooth out the process leading to the relatively smooth power law distributions observed.

The arguments leading to an initial power law size distribution depend only on the supply of air and the effects of turbulence. Once the bubble distribution has emerged, however, it quickly evolves under the influence of buoyancy, gas dissolution, and the residual turbulence. Over the radius range $a < 500$ μm the rise speed is $dz/dt \approx Aa^2$, which has the effect of rapidly removing the larger bubbles. For small bubbles whose surface is quickly stabilized by an organic coating, the rise speed is close to the Stokes Law for a solid sphere. The dependence of gas transfer on bubble radius is weak in this range so, neglecting pressure variations, the effect of dissolution on radius may be approximated as $da/dt = -D$; after time t,

$$N(a, t) = N_0(a + Dt)^{-n} \qquad (10.7)$$

If the bubbles are uniformly distributed over depth H, with injection periodicity T, the time averaged size distribution is found by integrating in radius space and time:

$$N(a, z) = \frac{N_0}{(1-n)DT} \left\{ \left[a^3 + \frac{3D(H+z)}{A} \right]^{\frac{1-n}{3}} - a^{1-n} \right\} \qquad (10.8)$$

Figure 10.16 shows this result for nominal values of $H = 2$m, $T = 60$ s, $n = 2$, $A = 1.7 \times 10^6$ m^{-1} s^{-1} and $D = 10^{-6}$ m s^{-1}. The bubble spectrum is multiplied by a^3, and is therefore volume scaled. The general shape and peak volume scaled bubble radius a_{max} is quite similar to

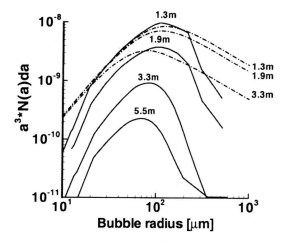

Fig. 10.16. Volume scaled bubble density as a function of bubble radius. Solid lines: time averaged bubble densities observed during a storm at different depths. Dashed lines: predicted bubble densities that use a model described in the text. Failure to predict deeper bubbles follows from the model's failure to account for Langmuir circulation.

our averaged data near the surface, but this simple model misses several features and overestimates the number of larger bubbles. First, bubbles are carried well below the injection depth by Langmuir circulation. This accounts for measurable-size distributions at 5.5 m in Fig.10.16, which is much greater than typical injection depths. Second, the redistribution effects of turbulence are not included. Third, this is a time averaged model and is sensitive to the breaking periodicity used, as well as the other parameters. Not withstanding these simplifications, it is a remarkable fact that time averaged bubble size distributions measured in many ocean environments have a volume scaled peak close to the prediction.

Bubbles with rise speeds significantly greater than the rms turbulence velocity of the background flow will rise to the surface and disappear. As the turbulence decays, the radius separating freely rising bubbles and those in turbulent suspension decreases. The bubble spectrum is eroded in this way from the large radius side. Small bubbles are more rapidly affected by dissolution (the details depend on saturation levels of dissolved gases in the water and on bubble depth). Thus a given bubble spectrum is eroded at both small and large radii and quite quickly tends towards a dominant radius. We refer here to the volume scaled bubble distribution spectrum; there will always be more small bubbles than large bubbles, but for many purposes, including acoustic behavior, number density tends to be less important than volume. The balance between dissolution and buoyancy effects leads to a volume scaled peak in the spectrum at a radius of order 100 μm, but subject to variation depending on pressure, gas saturation level, the time since bubble injection, and other factors. This, in turn, results in a frequency range of 25–35 kHz at which scattering and attenuation due to bubbles in the upper ocean is greatest.

10.6 Turbulence measurements with coherent sonar

Coherent Doppler sonars have proved useful for the measurement of small scale velocity structures, especially in a turbulent environment. The principles behind this technique are well established (Section 2.6) and will not be repeated here, except to note that in general the transmissions must be repeated sufficiently rapidly that the scatterers in the water column remain coherent between any pulse pair. Other constraints relate to ambiguities in the speed measurement and the maximum speed that can be unambiguously observed. Careful design of the transmission procedure can greatly reduce the ambiguity problem, but decorrelation of the targets will generally limit the useful range–velocity performance at high turbulence intensities.

In the upper ocean boundary layer it is helpful to be able to measure turbulence just beneath the surface. The rapid evolution of the turbulence also motivates measurements in wave number space rather than lengthy time series, requiring the assumption of some homogeneity along the short acoustical path. Initial attempts have been made to carry out such observations, linking them to the effects of wave breaking using a coherent Doppler profiler measuring velocities along a path slightly less than 1 m. Sharp peaks in the time series of turbulence dissipation appear to be directly associated with breaking (Gemmrich and Farmer, 2004), and provide a basis for examining energy losses due to breaking and the suspension of bubbles. A particularly difficult aspect of such measurements is arranging to suspend the instruments at a fixed depth beneath the surface without contaminating the turbulence measurement.

In one successful approach a freely drifting self-contained instrument includes a light-weight vertical sensor support fixed to the main instrument housing with hinged arms and drogued so as to point into the wind. This arrangement allows the sensors to follow the surface closely, yet remain upwind of the primary instrument housing so as to minimize wake effects. The precise instantaneous sensor depth is measured with a wire gauge. An alternative arrangement uses a vertical sensor array supported beneath four small floats which are tethered from the Floating Instrument Platform (FLIP); simultaneous photography allows the subsurface measurements to be referenced to instantaneous records of breaking waves.

10.7 Acoustical tracking of neutrally buoyant floats

Upper ocean measurements have also benefited from the application of neutrally buoyant float technology (D'Asaro et al., 1996). Floats have been fitted with various sensors and permit a Lagrangian observation that leads to measurement of vertical fluxes of heat and momentum which are fundamental to our knowledge of air–sea interaction and upper ocean physics. While acoustical instruments are, or will become, part of the sensor load of such floats, acoustical positioning offers unique possibilities for Lagrangian measurement of three-dimensional fluid motions. Vertical displacement is easily determined through observations of the hydrostatic pressure on the float. Horizontal motion presents a greater challenge in the open ocean, but can be achieved through acoustical tracking with drifting surface floats that are themselves tracked by GPS (Fig. 10.17). Since the surface floats themselves tend to drift with the water mass containing the neutrally buoyant float, dispersion of the surface and subsurface floats is reduced. This technology has been demonstrated, but its full potential has yet to be realized.

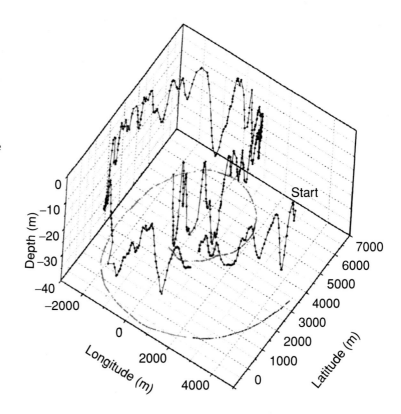

Fig. 10.17. Three-dimensional track of a neutrally buoyant float deployed in the open ocean. The float is acoustically tracked with three surface floats which have GPS positioning and tend to drift with the subsurface float. The float generally exhibits an inertial response on which is superimposed its vertical motions within the actively mixing surface layer.

10.8 The future

One of the most challenging aspects of upper ocean studies at higher wind speeds is that so many different processes are interacting. The wind drives waves, wave breaking deposits momentum into the surface layer; wave breaking entrains air and generates intense turbulence, and the wave and current shear interact to drive Langmuir circulation, which in turn organizes the bubble field and drives the smaller bubbles to greater depth. Bubbles contribute to the exchange of gases, especially the weakly soluble species, while dense bubble populations near the surface may suppress the turbulence. Increasingly, researchers will need to measure more variables in their search for a deeper understanding of these inter-dependent processes and the role they play in vertical exchange. This brief and somewhat selective overview provides examples of ways in which acoustical methods can contribute to this new and exciting field of research.

Innovative application of acoustical methods will play a key role in the development of our understanding of upper ocean physics.

Chapter 11

Using underwater sound to measure raindrop size distribution

JEFFREY A. NYSTUEN
University of Washington

Summary

The underwater sound of rain is loud and distinctive. It can be used as a signal to detect and measure oceanic rainfall. These measurements are needed to support climatological studies of the distribution and intensity of global rainfall patterns. Individual raindrops produce sound underwater by their impacts onto the ocean surface and, more importantly, by sound radiation from any bubbles trapped underwater during their splashes. Because different raindrop sizes produce distinctive sounds, the underwater sound can be inverted to quantitatively measure drop size distribution in the rain. Acoustical Rain Gauges (ARGs) are being deployed on oceanic moorings to make long-term measurements of rainfall using this acoustical technique.

Contents

11.1 Why listen to raindrops?

Rain is one of the most important components of climate. Knowledge of its distribution and intensity is important not only to farmers and flood control planners but also to meteorologists, oceanographers, and climatologists. This is because the formation of a raindrop in the air is accompanied by latent heat release. This heat release is one of the primary sources of energy driving atmospheric circulation. Thus, understanding the global patterns of distribution and intensity of rainfall is needed to improve weather and climate forecasting. Furthermore, layers of relatively fresh water due to rain at the ocean surface are now thought to significantly affect oceanic circulation (Anderson *et al.*, 1996), another component of global climate. Unfortunately, rainfall is also very difficult to measure, especially over the ocean where few people live and where rain gauges commonly used on land don't work. But we all know that rain falling onto a tin roof makes a lot of noise, and so does rain falling onto water. In fact, rain falling onto water is one of the loudest sources of underwater sound. So maybe we can measure oceanic rain by listening to it from below the ocean surface.

11.2 How do raindrops make sound underwater?

There are actually two components to the sound generated by a raindrop splash. These are the splat (impact) of the drop onto the water surface and then the subsequent formation of a bubble underwater during the splash. The relative importance of these two components of sound depends on the raindrop size. Surprisingly, for most raindrops, it is the bubble that is, by far, the loudest sound source. Bubbles are one of the most important components of underwater sound. They have two stages during their lifetimes: screaming infant bubbles and quiet adult bubbles. As a bubble is created, in general it is not in equilibrium with its environment. It radiates sound (screams) to reach equilibrium. The frequency of the sound is well defined (Minnaert, 1933):

$$f_r = \frac{1}{2\pi a}\sqrt{\frac{3\gamma P_0}{\rho_0}}$$

and depends on bubble radius, a, local pressure, P_0, local water density, ρ_0, and a geophysical constant, $\gamma = 1.4$ (see Section 6.5.2). The important observation is that the size of the bubble is inversely proportional to its resonance (ringing) frequency. Larger bubbles ring at lower frequencies. The sound radiated is often loud and narrowly tuned in frequency (a pure tone). But quickly, after just tens of milliseconds, a bubble in water becomes a quiet adult bubble and changes roles. It absorbs sound, and is especially efficient at absorbing sound at its resonance frequency.

Table 11.1 *Acoustics of raindrop sizes. The raindrop sizes are identified by different physical mechanisms associated with the drop splashes.*

Drop size	Diameter	Sound source	Frequency range	Splash character
Tiny	<0.8 mm	Silent		Gentle
Small	0.8–1.2 mm	Loud bubble	13–25 kHz	Gentle, with bubble every splash
Medium	1.2–2.0 mm	Weak impact	1–30 kHz	Gentle, no bubbles
Large	2.0–3.5 mm	Impact	1–35 kHz	Turbulent, irregular
		Loud bubbles	2–35 kHz	bubble entrainment
Very large	>3.5 mm	Loud impact	1–50 kHz	Turbulent, irregular
		Loud bubbles	1–50 kHz	bubble entrainment, penetrating jet

Naturally occurring raindrops range in size from about 300 microns diameter (a drizzle droplet) to over 5 mm diameter (often at the beginning of a heavy downpour). As the drop size changes, the shape of the splash changes and so does the subsequent sound production. Laboratory and field studies (Medwin *et al.*, 1992; Nystuen, 2001) have been used to identify five acoustic raindrop sizes (Table 11.1). For tiny drops (diameter < 0.8 mm), the splash is gentle, and no sound is detected. On the other hand, small raindrops (0.8–1.2 mm diameter) are remarkably loud. The impact component of their splash is still very quiet, but the geometry of the splash is such that a bubble is generated by every splash in a very predictable manner (Pumphrey *et al.*, 1989). These bubbles are relatively uniform in size, and therefore frequency, and are very loud underwater. Small raindrops are present in almost all types of rainfall, including light drizzle, and are therefore responsible for the remarkably loud and unique underwater "sound of drizzle" heard between 13 and 25 kHz, the resonance frequency for these bubbles.

Interestingly, the splash of the next larger raindrop size, medium (1.2–2.0 mm diameter), does not trap bubbles underwater, and consequently medium raindrops are relatively quiet, much quieter than the small raindrops. The only acoustic signal from these drops is a weak impact sound spread over a wide frequency band. For large (2.0–3.5 mm diameter) and very large (>3.5 mm) raindrops, the splash becomes energetic enough that a wide range of bubble sizes are trapped underwater during the splash, producing a loud sound that includes relatively low frequencies (1–10 kHz) from the larger bubbles. For very large raindrops, the splat of the impact is also very loud with the sound spread over a wide frequency range (1–50 kHz). Thus, each drop size produces sound underwater with unique spectral features that can be used to acoustically identify the presence of that drop size within the rain.

11.3 Inverting the sound field to measure drop size distribution

The sound intensity, I_0, at the surface is related to the drop size distribution in the rain by:

$$I_0(f) = \int A(D, f)V_T(D)N(D)\,dD \qquad (11.1)$$

where f is frequency, $A(D, f)$ is the transfer function describing the radiated sound as a function of frequency for a given drop size, D, V_T is the terminal velocity of the drop, and $N(D)$ is the drop size distribution in the rain. In a measurement situation, this equation is discrete and given by:

$$\mathbf{I}_0(f) = \mathbf{A}(D, f) \bullet \mathbf{DRD}(D) \qquad (11.2)$$

where $\mathbf{DRD}$ is the drop rate density, $\mathbf{DRD} = V_T \bullet N(D)$. If the inversion matrix, $\mathbf{A}$, is known, then, using singular value decomposition, (11.2) can be inverted:

$$\mathbf{A} = \mathbf{U} \Lambda\, \mathbf{V}^T$$
$$\mathbf{DRD} = \mathbf{V}[\Lambda^{-1}(\mathbf{U}^T\, \mathbf{I}_0)] \qquad (11.3)$$

to make a measurement of the drop size distribution in the rain. The transfer function $A(D, f)$ has been determined by empirical decomposition of field observations (Nystuen, 2001) and is shown in Fig. 11.1.

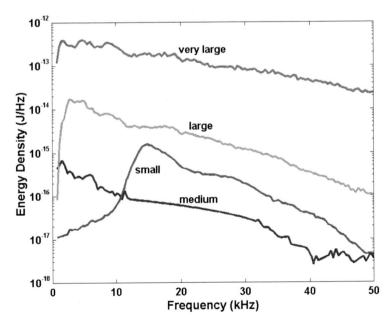

Fig. 11.1. Radiated acoustic energy densities for very large, medium, and small raindrop sizes. This forms the mathematical basis for the inversion of the sound field to obtain drop size distribution (Nystuen, 2001).

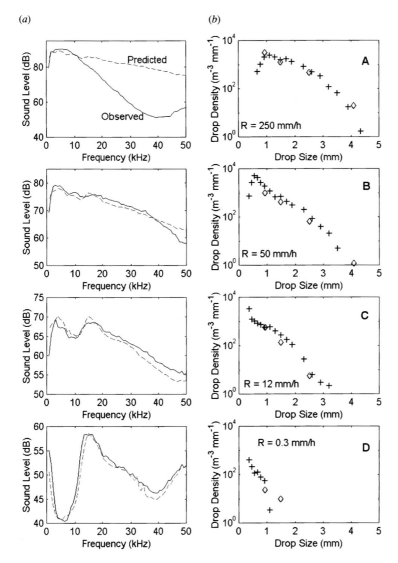

Fig. 11.2. Four examples of the inversion of the sound field to obtain drop size distribution (DSD). A mechanical device, known as a disdrometer, provides comparison data. The disdrometer relates the momentum of individual raindrop strikes to drop size. (right) The disdrometer data are shown, "+," with the acoustical inversion results "◇." (left) The observed sound field (solid line) is compared to the sound field predicted with the disdrometer data (the "forward problem").

In order for this inversion to work, unique acoustical signatures are needed for each drop size. Only four unique acoustical signatures are identified. Furthermore, because the signal from the very large drop category is several magnitudes louder than the medium drop category, independent detection of very large drops is needed to obtain realistic drop size distributions. This can be done acoustically based on field experience (Nystuen, 2001). Figure 11.2 shows some examples of the inversion, both the forward problem, that is predicting the sound field given the drop size distribution, and the inversion, measuring the drop size distribution given the sound field. In general, the predictions match the observations, although there is one situation where consistent disagreements are present.

During extremely heavy rainfall, the forward problem overestimates the observed sound levels above 10 kHz, often by many decibels (Fig. 11.2(a)). In fact, the sound levels above 30 kHz can often become less than those observed at lighter, but still very high, rainfall rates (compare Figs. 11.2(a) and (b)). Since heavier rain has more sound sources present, the best explanation for this observation is attenuation by small bubbles below the water surface. These bubbles are presumably mixed downward by the turbulence of the very large raindrop splashes, forming a layer through which newly generated sound (at the surface) must pass. The smallest bubbles will remain in the water longest, due to buoyancy, and thus the attenuation is greatest at the higher frequencies. Since attenuation by air bubbles is strongly dependent on bubble size (M & C, Chapter 8), the size of bubbles present in this layer can be identified. For this example, the bubbles range in size from about 200 μm radius (resonance at 15 kHz) down to less than 60 μm radius (resonance at 50 kHz).

11.4 Making acoustical measurements of rainfall at sea

How can we use our knowledge of the sound generated by raindrops to make measurements of rainfall at sea? Because of its inherent temporal and spatial variability, rainfall is a relatively difficult geophysical quantity to measure. Useful measurements often require both spatial and temporal averaging. We need instruments that detect and record the underwater sound signal from the rain and that can be deployed over a long period of time. Since we understand how raindrops make sound underwater, we should be able to recognize the rain signal when we hear it, and then be able to quantify it.

11.5 Acoustical rain gauges

Acoustical Rain Gauges (ARGs) have been designed and built at the Applied Physics Laboratory, University of Washington for autonomous deployment on ocean surface moorings. An ARG contains an ITC-8263 hydrophone, signal pre-amplifiers, and a recording computer (Tattletale-8). The nominal sensitivity of these instruments is −160 dB relative to 1 V/μPa and the equivalent oceanic background noise level of the pre-amplifier system is about 28 dB relative to 1 μPa^2 Hz^{-1}. Bandpass filters are present to reduce saturation from low frequency sound (high pass at 300 Hz) and aliasing from above 50 kHz (low pass at 40 kHz). The ITC-8263 hydrophone sensitivity also decreases above its resonance frequency, about 40 kHz. A data collection sequence consists of four 1024 point time series collected at 100 kHz (10.24 ms each) separated by 5 seconds. Each time series is fast Fourier transformed (FFT) to obtain a 512-point (0–50 kHz) power spectrum. These four spectra were averaged together and spectrally compressed to 64 frequency bins, with frequency resolution of 200 Hz from 100–3000 Hz and 1 kHz from 3–50 kHz. The shapes of these spectra are evaluated individually to detect the acoustic signature of rainfall and then are recorded internally.

The overall temporal sampling strategy is designed to allow the instrument to record data for up to one year without servicing and yet detect the relatively short rainfall events present in the tropics (Nystuen, 1998). In order to achieve this, the ARG is designed to enter a low power mode "sleep mode" between each data sample. For these deployments, the ARGs "sleep" for 9 minutes and then sample the sound field. If "rain" is detected, the sampling rate changes to 1 minute (or 3 minutes if "drizzle" is detected) and stays at the higher sampling rate until rain is no longer detected. Some "noise" will trigger the high sampling mode and must be removed from the data.

A sound source at a free surface, the ocean surface, is an acoustic dipole, radiating sound energy downward in a $\cos^2\theta$ pattern where θ is the zenith angle. This allows the intensity of surface generated sound at some depth, h, below the surface to be given by:

$$I(h) = \int (I_0 \cos^2 \theta) \quad atten(p) \, dA \qquad (11.4)$$

where I_0 is the sound intensity at the surface and $atten(p)$ describes the attenuation due to geometric spreading and absorption along the acoustic path, p. If the sound source is uniform at the surface and absorption and refraction are neglected, the measurement should be independent of depth. For any particular deployment, the attenuation along the acoustic path can be complicated, but has only resulted in minor corrections in other studies (Vagle et al., 1990). The ARGs have been deployed at 38 m depth on the mooring lines (wire cable). The depth was chosen to be above the thermocline, lessening the effects of acoustic refraction, and to maximize sampling area, so that the buoy itself does not occupy a significant portion of the effective listening area. Equation (11.4) can be used to estimate the effective sampling area at the surface. Neglecting refraction and absorption and assuming that the receiving hydrophone is omnidirectional, 90% of the signal is arriving from a sampling area equal to:

$$sampling\ area \cong \pi(3h)^2 \qquad (11.5)$$

where h is the depth of the ARG. The integrating area of the hydrophone is important for two reasons. First, rainfall is inhomogeneous on all scales, but rainfall measurements are needed on large temporal or spatial scales. An instrument with a large inherent sampling area should produce a better "mean" rainfall statistic. Second, the large spatial sampling allows the short temporal sampling periods being used for each data sample to include many individual raindrop splashes.

Sub-surface instruments have several other advantages over surface instruments. They are not subject to harsh environmental conditions that destroy surface instruments including, for example, storm waves and sea-ice. Fouling by marine animals is reduced. Surprisingly, vandalism and piracy are big problems, even in the middle of the ocean. Sub-surface instruments are less likely to be detected by pirates and vandals. It is a remote sensing measurement, which means that the measurement sensor does not interfere with the quantity being measured. This technology is passive, introducing no acoustic disturbance into the environment and thus poses no potential harm to marine mammals or other forms of life in the ocean. Furthermore, sub-surface moorings are an inexpensive alternative to surface moorings. Currently, NOAA surface moorings cost about $100 000, depending on instrumentation, and require larger ships to deploy and maintain them. Sub-surface moorings cost roughly one-third as much, depending on instrumentation, and are smaller and more easily deployed.

11.6 Acoustical weather measurements

In the frequency range from 200–50 000 hertz, naturally generated sound at the sea surface is predominately produced by wind-driven breaking waves and precipitation. In turn, these physical processes generate sound principally through the production of bubbles during splashing at the ocean surface. And on the scale of individual bubbles, the sound is the resonant ring of newly formed individual bubbles within the splashes (Medwin and Beaky, 1989; Medwin *et al.*, 1992). Because wind-driven breaking waves and raindrop splashes generate different distributions of bubbles sizes, the sound from breaking waves can be distinguished from the sound of precipitation. This allows each sound source to be identified and then quantitatively measured (Nystuen and Selsor, 1997). Bubbles can also absorb sound. Smaller bubbles are mixed downward into the ocean surface by turbulence to form clouds, plumes, and layers. Sound newly generated at the surface must pass through these bubble structures into order to reach the measurement hydrophone. Since ambient bubbles absorb sound principally at their resonant frequency, which is inversely proportional to size, the distortion of the measured sound spectrum is largest at higher frequency.

Some examples of sound produced by wind and rain are shown in Fig. 11.3. The sound generated by wind has a distinctive shape and is relatively quiet when compared to the sound generated by rain. Drizzle

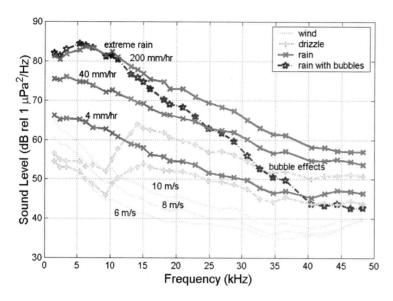

Fig. 11.3. Examples of the spectral signal for different geophysical sound sources. These sound spectra were recorded on an ocean surface mooring in the South China Sea (Nystuen et al., 2000). The surface instrumentation was stolen by pirates during the experiment, however the sub-surface acoustic rain gauges were not detected and provided data throughout the experiment.

has a distinctive peak in the spectrum from 13–25 kHz, due to a unique signal produced by small bubbles formed during the splashing of small raindrops (Pumphrey et al., 1989), and the sound of heavy rain is very, very loud. The rain-generated spectra also have relatively more high frequency energy from 5–15 kHz than the wind-generated spectra. In fact, rainfall is the predominant source of sound when it is present. It can be detected and quantified even in the presence of high winds (Nystuen and Farmer, 1989) or when other noises are present.

In the ocean, there are sometimes other underwater sounds that can interfere with acoustical weather measurements. In order to make acoustical weather measurements, it is first necessary to identify the sound source. Figure 11.4 shows time series of oceanic underwater sound at three different frequencies. The different sources of sound are identified by comparing spectral intensity levels, spectral shapes, and temporal variances of sound intensities. Sound spectra not consistent with known geophysical signals (wind, rain, and drizzle) are assumed to be "noise" and are removed from the data record. In general, noise due to biological activity or shipping are of short duration or are localized and do not interfere with the geophysical interpretation of the sound signal.

Once classification is obtained, there are several algorithms available to quantify wind speed (Vagle et al., 1990) and precipitation (Nystuen et al., 1993). Figure 11.5 shows the quantitative acoustic interpretation of the time series shown in Fig. 11.4. Wind speed agreement is excellent for winds 3–15 m/s. The absolute difference between the ARG wind speed estimate and the R. M. Young anemometer is 0.5 ± 0.4 m/s for several months of comparisons (Nystuen and McPhaden, 2001). For wind speeds

Fig. 11.4. An example of a time series of oceanic underwater sound at three different frequencies (3, 8.5, and 21 kHz) recorded from an Acoustical Rain Gauge (ARG) mounted on a deep ocean mooring at 40 meters depth. The mooring is part of the NOAA TAO array and is located at 10° N, 95° W.

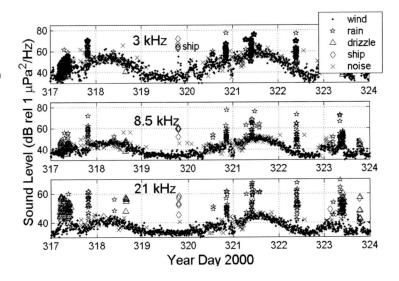

Fig. 11.5. Geophysical interpretation of the underwater sound time series shown in Fig. 11.4. Quantitative comparisons of the acoustic measurement to a surface-mounted R. M. Young (RMY) anemometer and a RMY rain gauge are shown. The acoustic wind speed measurement uses the algorithm from Vagle *et al.* (1990).

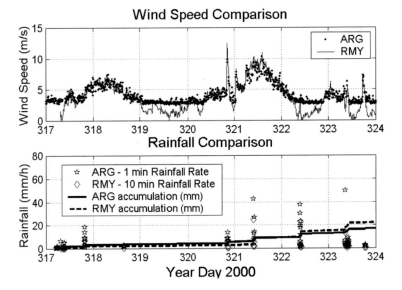

below 3 m/s, there is no wave breaking at the ocean surface, and thus there is no acoustic signal with which to measure the wind speed. Above 15 m/s there are relatively little comparison data available.

In order to validate the acoustic rainfall measurement long-term time measurements have been made at several deep ocean surface Tropical Atmosphere Ocean (TAO) moorings in the tropical Pacific Ocean. Rainfall accumulations for the Year 2001 at the TAO moorings at 10° and 12° N, 95° W are shown in Fig. 11.6. A comparison is made with a collocated Tropical Rain Measuring Mission (TRMM) satellite rainfall products

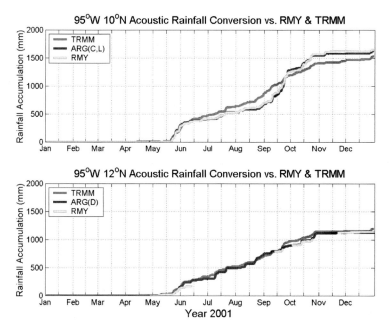

Fig. 11.6. Comparison of rainfall accumulation measurements from ARGs, collocated R. M. Young (RMY) rain gauges, and satellite estimates (TRMM) of precipitation at two deep ocean moorings. Fouling and other problems caused some loss of data for the R. M. Young rain gauge at 12° N. The accumulation data for this instrument are offset to match the ARG accumulation total after periods of non-performance.

(Product 3B42) and collocated R. M. Young rain gauges mounted on the TAO moorings. This part of the ocean has a distinctive rainy season beginning in May and lasting into October. When both instruments are working, the agreement between the ARG measurement and the surface-mounted R. M. Young (RMY) collection-type rain gauge is excellent. However, fouling and piracy affect the RMY gauge measurement at times, while mooring self-noise (acoustic) sometimes affects the ARG measurement. Exact agreement between the TRMM estimate and the other two measurements should not be expected. The sampling strategies are very different (spatial averaging versus high temporal resolution). However, seasonal agreement should be expected. And except for a few "events," this agreement is observed.

11.7 Conclusions

The underwater sound of rain is a loud and distinctive signal that can be used to detect and measure rain at sea. Individual raindrops make sound underwater by two distinct mechanisms: the impact of the raindrop onto the ocean surface and sound radiation from any bubbles trapped underwater during the splash. For most raindrops, the sound radiation by bubbles is, by far, the louder sound source. Because the geometry of their splashes regularly trap a bubble of uniform size, small raindrops

(0.8–1.2 mm diameter) are unexpectedly loud underwater. These drops are responsible for the remarkably loud "sound of drizzle" heard between 13–25 kHz. Medium raindrops (1.2–2.0 mm diameter) are relatively quiet, while large (2.0–3.5 mm diameter) and very large (>3.5 mm) raindrops have energetic splashes which can trap larger bubbles. These bubbles radiate sound at frequencies as low as 1 kHz. Because the different raindrop sizes produce sound with distinctive features, the sound field can be "inverted" to measure the raindrop size distribution within the rain. This is a good measure of rainfall rate, or other interesting features of rainfall.

Although there are sometimes man-made or biological noises that are loud and could potentially interfere with the acoustical measurement of rain, these noises are generally intermittent or geographically localized. When rain is present, the sound from rain dominates the underwater sound field. There are two features of rain and drizzle generated sound that allow detection of rain at sea. These are the relative level (very loud) and the relatively higher sound levels at higher frequency (over 10 kHz) when compared to wind. By monitoring for these distinctive spectral features, it is possible to detect and then quantify rainfall at sea. Acoustical Rain Gauges (ARGs) have been built and deployed on ocean surface moorings to demonstrate the potential for this acoustic technology to provide long-term measurements of rainfall.

11.8 The future

The passive acoustical measurement of sea surface processes represents a new tool for the oceanographic science community. The instrumentation is reliable and robust and can be deployed from a wide variety of ocean instrument platforms, including bottom-mounted and sub-surface moorings, drifters, profilers, and autonomous underwater vehicles (AUVs). Furthermore, by acoustically monitoring rainfall, and other surface processes including wind speed, from below the surface, many of the fouling, physical damage, and vandalism problems that affect surface instruments will be avoided. This acoustic technology will provide long-term measurements of wind speed and rainfall, and contribute to understanding the forcing of wind stress and fresh water hydrology on the ocean ecosystem.

Part III
Bioacoustical studies

Chapter 12

Active acoustical assessment
of plankton and micronekton

D. V. HOLLIDAY
BAE Systems

T. K. STANTON
Woods Hole Oceanographic Institution

Summary

Particles and bubbles in the sea are responsible for much of the scattering and reverberation one observes when one makes a noise underwater. Many, perhaps even most, of the particles in the ocean are either living or have a biotic origin (e.g., detritus). Sound is routinely used by biological oceanographers to detect life that ranges in size from protists, only a few tens of microns long, up to tens of meters for the great baleen whales.

Contents

12.1 Introduction

Zooplankton are an important part of the aquatic food web. Adult zooplankters generally fall in the size range from tens of microns to tens of millimeters in length, but it is worth remembering that every adult started life as a much smaller organism, e.g. as a tiny egg. The word "plankton" refers to communities of organisms that either swim not at all, or are weak swimmers with respect to the horizontal currents they encounter. In the sea, these organisms are a primary source of food for

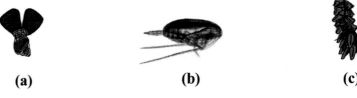

(a)   (b)   (c)

Fig. 12.1. Zooplankters have many diverse sizes and shapes. Different species and even parts of individual organisms may have different physical properties (e.g., density and compressibility). These differences modify the ability of an organism to reflect sound. The shelled pteropod (*a*) has both fluid and elastic properties in different parts of its body. The calanoid copepod illustrated in (*b*) might be modeled as fluid surrounded by a very thin elastic (chiton) shell or as simply a fluid with different characteristics than the surrounding water. The siphonophore (*c*) is actually a colony of organisms in which different members take on different functions (e.g., prey capture, digestion, etc.). In the illustration shown, the member of the colony at the top has become a carbon monoxide-filled float, providing buoyancy for the rest of the colony. When gas inclusions are in an organism those features often dominate in the echo formation process. The genera illustrated here are not to scale.

the great baleen whales, fish, and seabirds. If an aquatic animal doesn't eat zooplankton or micronekton directly, it generally eats something that does subsist on these trophic levels in the food web. Zooplankton and micronekton are ubiquitous in their geographic distribution, occur in both fresh and salt water environments, and are sometimes present in very large numbers, e.g., tens of thousands to millions per cubic meter. There are numerous species of zooplankton and micronekton, and they take on many physical shapes (Figs. 12.1 and 12.2).

Micronekton, such as euphausiids, mysids, and larval fish are distinguished from both fish and zooplankton primarily by their swimming ability. These constituents of the food web are able swimmers, both in the horizontal and vertical directions. Many species of micronekton participate in diel vertical migrations of several hundred meters. They are also capable, in some cases, of maintaining aggregations in the presence of substantial horizontal currents. There is no consistent definition in terms of size, but for our purposes we will consider micronekton to range from about a centimeter to a few centimeters in length. It should be remembered that most adult species of micronekton and even large fish start their lives as eggs, growing through a larval stage. For a while, at least, they are a part of the assemblage of plankton one finds in aquatic environments. For simplicity, we refer to both zooplankton and micronekton as "zooplankton" in this chapter unless it is necessary to distinguish them for a particular reason.

Because of the vast expanse of the oceans and the ubiquity of the zooplankton and micronekton, it is not practical to assess either group

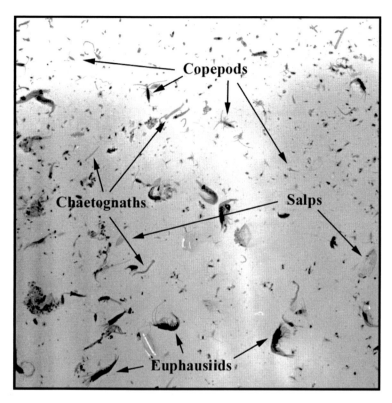

Fig. 12.2. Assorted zooplankton and micronekton from a MOCNESS net haul in the Gulf Stream. In terms of diversity of organism shapes, this collection is typical of the epipelagic water column in temperate waters. It contains several species and life stages of copepods, several euphausiids, chaetognaths, salps, a variety of crustacean nauplii, and at least one pelagic annelid marine worm, and even a radial colony of a large phytoplankter, *Trichodesmium*, a colonial marine cyanobacteria. Similar diversity in size and shape is quite typical of most marine plankton communities.

purely through means of nets, pumps, traps, or cameras. Such approaches have been consistently shown to give inaccurate estimates of both the biomass and spatial distribution of small living organisms in aquatic environments. Two primary factors contribute to making sampling with conventional biological tools problematic in defining distribution and biomass for these organisms. One is the patchy nature of the distribution of life in the sea. In fact, in the absence of such patchiness, a predator can be shown to spend more energy getting from "bite to bite" than it gets from each bite. This would, for long time scales, bode very badly for life in the sea, and patchiness is often invoked to explain how animals survive in the ocean. The other reason is active avoidance of nets, pumps, and traps by the animals. The ability to sense an approaching net and avoid it probably stems from a need to avoid being eaten by predators. The sea has few hiding places, so honing one's escape and evasion strategy is also a necessary and important activity if you are potentially prey. Active acoustical methods provide means by which the populations can be more thoroughly sampled than has proven possible with other more direct methods. However, acoustical methods offer many interesting challenges when interpreting the data one gathers with currently available sensors. The best available current methods consist of a combination of advanced hardware, acoustical scattering models, and inversion algorithms.

12.2 Acoustical scattering by zooplankton

Since the material properties of zooplankton generally differ from those of the surrounding water, acoustical signals from an active acoustical system will be scattered by the presence of the organisms. While this scattering occurs in all directions, bio-acousticians usually work with energy that is scattered back towards the sound source. The pressure, p_{bs}, of the backscattered signal (echo) is described in the following general equation:

$$p_{bs} = p_0 e^{ikr} r^{-1} f_{bs} \tag{12.1}$$

where p_0 is the amplitude of the incident pressure signal before scattering, and k is the acoustical wave number, defined as $k = 2\pi/\lambda$, with λ defined as the acoustical wavelength. The symbol r represents the distance between the sonar or echosounder and the scatterer, and f_{bs} is the fish backscattering amplitude in meters. The scattering is fully described in f_{bs} and is sometimes described in logarithmic terms as the target strength (TS) where the reference is 1 m^2.

$$\begin{aligned} TS &= 10\log_{10}[|f_{bs}|^2/1 \text{ m}^2] \\ &= 10\log_{10}[(\sigma_{bs})/1 \text{ m}^2]. \end{aligned} \tag{12.2}$$

The unit of target strength is decibels (dB) and is expressed relative to a reference area of 1 m^2. The target strength is given both in terms of the square of the magnitude of the backscattering amplitude as well as the backscattering cross-section σ_{bs}.

When multiple targets are present, the volume scattering strength (S_v) and volume scattering coefficient (s_v) are used. These common measures of acoustical scattering can be expressed in terms of the volume scattering strength and the numerical density of organisms as:

$$S_v = 10\log_{10}[(s_v)/s_v, \text{ref}] \tag{12.3}$$

and

$$s_v = n\overline{\sigma_{bs}} \tag{12.4}$$

where the overbar on σ_{bs} indicates a mean value, as averaged over the assemblage of scatterers in the volume of water being examined. Also see Section 7.3. This is an important, but subtle point. Any single summation of echoes from a group of randomly located scatterers, such as fish or zooplankton, is itself random. The Fourier transform, or power spectrum of a single realization of such a set of echoes is also random. After all, it is only a mathematical transformation performed on the random sum of small pressure waves as they arrive over a period of time at the sensor. It is *only* by averaging a number of the Fourier transformations of individual time records from a set of echo ranging cycles from an active acoustical sensor that one gets a useful measure of the power spectrum (energy content vs. frequency) of the collection of echoes that make up volume reverberation.

While the scattering amplitude appears in a simple compact form in (12.1), its evaluation, when based on the real physical shape and properties of an organism is generally quite complex. The scattering by zooplankton has a complex dependence on the organism's size (relative to wavelength of sound), shape, orientation, and its material properties. These quantities vary, sometimes significantly, across the many species one finds in an aquatic environment. The challenge is to adequately characterize these organisms acoustically to allow echoes from zooplankton assemblages to be accurately interpreted. One successful approach has been to group the organisms into gross anatomical categories: (*a*) elastic shelled, (*b*) quasi-fluid, and (*c*) gas-bearing. Each of these categories has associated with them distinctly different acoustic scattering characteristics as displayed in Fig. 12.3.

Mathematical scattering models have been developed in order to account for these differences when interpreting data from acoustical surveys. There are numerous models of varying complexity per anatomical

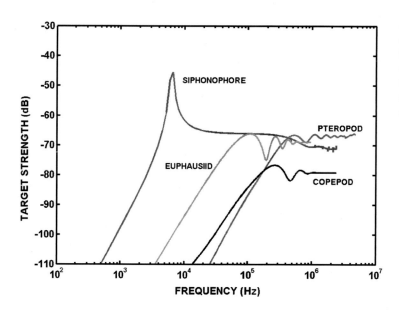

Fig. 12.3. Target strength as a function of frequency for individuals of specific distinct sizes for three anatomical groups of zooplankton and a euphausiid.

group. Some models are quite complex and require a computer for evaluation. These models are general accurate over a wide range of conditions. Other models are much simpler and can be expressed in analytical form. Generally, these latter models are valid over a narrower range of conditions.

The choice of models and the complexity needed for a particular problem is usually dictated by how much one knows about the animals that are present in one's echo sounder or sonar beam. If not much is known, as is often the case, some of the simpler models will often suffice. Sometimes one has, either from *a priori*, or *in situ*, collections, some idea of what subset of all possible organisms might be present at a particular place and time. If that is the case, a more complex model may allow one to assess the numbers (or biomass) of organisms present with more precision, if not accuracy. In some cases researchers hope that eventually, elaborate models of scattering may reveal subtle differences in the acoustical signature of an organism, leading to its identification at a species level or classification into a larger group of species or genera.

The acoustical scattering as calculated by one of the more complex models is illustrated in Fig. 12.4. The method used, in this case for a small copepod, involved a distorted wave Born approximation and the 2D geometric representation of the animal.

The small adult crustacean, *Corycaeus* spp., illustrated in Fig. 12.4 had an overall length of about 3.5 mm. The image was used as a template and was scaled in the DWBA-based calculations of target strength (Fig. 12.5) to represent the range of sizes over which this morphology is reasonable. Naupliar (post-egg/pre-juvenile) sizes for marine organisms usually take on shapes that differ substantially from the adults. In fact, even a trained scientist will usually admit that recognizing them as a particular species from microscopic images is something of an art. This particular crustacean has sufficient axial symmetry that a 2D image is an adequate descriptor of its geometry. Its shape was approximated by a series of cylinders, and was entered into a complex mathematical model

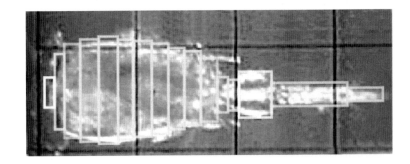

Fig. 12.4. A series of cylinders was used to approximate the shape of a small marine copepod, *Corycaeus* spp. for use in modeling the scattering of sound from this specimen.

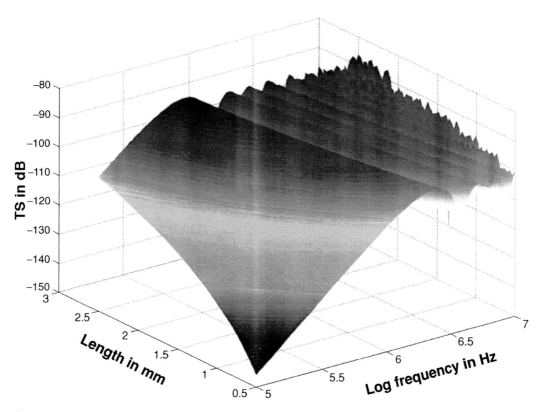

Fig. 12.5. The acoustical reflectivity of a small marine copepod, *Corycaeus* spp., as approximated by the series of contiguous cylinders shown in Fig. 12.4 is illustrated here as a function of acoustic frequency for a range of animal sizes.

and used to estimate the amount of sound scattered when the incident angle was normal to the page, as shown.

For a particular length animal the complex frequency dependence illustrated in Fig. 12.5 is a "fingerprint" which embeds physical information about the scatterer. The details of the scattering at high frequencies, i.e., the interference patterns in the geometric scattering region, are the result of the summation of all of the pressure waves that have interacted with the animal and returned to the receiver as an echo. This complex "fingerprint" is the result of both positive reinforcements and negative pressure wave interferences. These interactions arise as the scattered signals from different parts of the animal are phase shifted and amplitude modified when the sound interacts with the animal's internal and external parts. Thus, information about the internal and external morphology of the animal is encoded in this "fingerprint." The challenge is to extract it in a meaningful way.

It is important to note that the predicted scattering has a quite complex dependence on the length of the zooplankter, even for smaller products of the acoustical wave number and the animal's size. At higher frequencies, e.g., where the product ka is close to or greater than one and the scattering is relatively strong, the reflectivity is both non-linear and non-monotonic. Here, a is a measure of the particle size, often the radius of a sphere that would include the same volume as the actual animal, or its length, while k is the acoustical wave number. For a single frequency the target's reflectivity is not a simple, nor, without *ex situ* information, a reliable indicator of animal size. In the geometric scattering zone ($ka \gg 1$), a larger animal can actually scatter a particular frequency of sound at levels less than one that is smaller. This scattering behavior makes it difficult to interpret data collected at a single acoustical frequency. Under some conditions, however, it allows one to use volume scattering strengths collected from an assemblage of scatterers at several frequencies to reliably estimate the size–abundance or size–biomass spectra of a zooplankton assemblage.

While it is a bit more complex in practice, the simple description of the "inverse" calculations one uses to extract information about an animal's abundance and size can be described as follows. Presume the target strength calculations illustrated in Fig. 12.5 are perfectly accurate and noise free, which of course is not usually a great assumption in most real-world situations. If one were to measure the volume scattering strengths in a small volume of sea water at several frequencies (e.g., in the band between 100 kHz to 10 MHz), a single curve would result, defining a spectral "fingerprint" for the inhabitants of that volume of water. If there were only one individual *Corycaeus* in the sample volume, then that curve would exactly match one of the target strength versus frequency curves from which the 3D surface in Fig. 12.5 is constructed. Picking out that curve would tell one the size of the animal that had been insonified. If there were 100 animals in the volume sampled acoustically, then the measured volume scattering strength curve would have the same shape, with frequency, as for it did for one animal, but the scattering levels would be 100 times (or 20 dB) higher. Since the measured volume scattering coefficient (s_v), scales linearly with the number of animals in the measurement volume, dividing the volume scattering coefficient by the backscattering cross-section for one animal, as estimated from the mathematical model for target strength, will reveal the number of animals present. In other words, with a measurement of s_v at several frequencies, one can solve (12.4) for n.

If there were two animals of different sizes, but the same shapes in the water insonified, then one would have need to find two frequency-dependent curves in Fig. 12.5, which when combined matched the measured volume scattering strength dependence on frequency. If there are multiple sizes, and each size includes a different number of animals, then one must resort to setting up a series of equations which express the dependence of the measured volume scattering strength at each frequency as the sum of the scattering cross-sections at each size present, multiplied by the number of animals at those sizes. Mathematically, this can be expressed as shown in (12.5). One can then solve for n_j, the number of scatterers at each size j. In this formulation, the index i represents the ith frequency. The volume scattering strength at the ith frequency is s_{vi}. As long as there are at least as many measurements of s_v as there are unknowns in this equation, and there is no "noise" in either the measurement process or the modeling process, solving for the abundance at each size is straightforward. Given the abundance, and the sizes, one can easily calculate biomass distributions with size as a variable, or total biomass.

$$s_{vi} = \sum_j \sigma_{ij} n_j \tag{12.5}$$

This basic process has also been extended to more than one kind of scatterer (e.g., see Figs. 12.1 and 12.3). In that case, one extends the form of (12.5) to read:

$$s_{vi} = \sum_j \sigma'_{ij} n_j + \sum_j \sigma''_{ij} n_j + \sum_j \sigma'''_{ij} n_j + \cdots \tag{12.6}$$

or in more general form,

$$s_{vi} = \sum_{k=1}^{K} \sum_{j=1}^{N_k} \sigma_{ij}^k n_j^k \tag{12.7}$$

Here, K is the number of different models needed to describe all of the acoustically different organisms present in one's sample volume. Note that, since *every* animal present scatters sound at every frequency, if one does not include all of the animals and models needed to describe the problem, the answer one gets will reflect the incomplete question being asked! In practice, usually only a few animals will dominate the acoustical scattering, but if one doesn't formulate the question correctly, one will get a "noisy" or incorrect answer. With acoustical methods, as with other methods such as nets, you collect what is there (if your sampling gear works well), not just what you are interested in observing. With acoustics, as with nets, however, it is important to note that each sensor or system has limits. With nets, it might be mesh size or avoidance. With sound, it might be the sample volume examined, the patchiness of the organisms, or the number and placement of the frequencies used. Every sampling system has its limits. For some problems one configuration or method may be better than another. The challenge is to understand the limits and benefits of each and use what is appropriate.

12.3 Ocean instrumentation for use in studying zooplankton with acoustics

Sound scattering by zooplankton and micronekton strongly depends on the acoustical frequency used (e.g., Figs. 12.3 and 12.5). This frequency-dependence can be exploited for interpretation of acoustical data and is very useful for minimizing ambiguities in the interpretation of measured scattering profiles. Generally, it is best to make measurements of volume scattering at several frequencies, if not many, to aid in the interpretation of volume scattering strength measurements from aquatic environments. The amount of information one can extract increases with the number of independent measurements one makes. Although research into other variables has been done, the most common approach is to measure the frequency dependence of volume reverberation or discrete echoes from marine organisms. The question of how far apart a set of frequencies must be in order to be independent is beyond the scope of this chapter. It will suffice to mention that instruments have been built with up to 21 frequencies spanning a range from 100 kHz to 10 MHz for the purpose of studying assemblages of small zooplankton. A typical instrument in the first decade of the twenty-first century employs six to eight frequencies, usually falling between about 250 kHz and 3 MHz. This is a trade-off between the information, e.g., size resolution and accuracy in the estimate of the biomass desired, and the cost of implementing each acoustical channel.

The means by which the sensors are deployed can be almost as important as which sensors are used. Since zooplankton and micronekton are generally centimeters in length and smaller, the acoustic frequencies typically used are in the 100s of kHz and greater, and sometimes well into the MHz region. Since the signals at these higher frequencies cannot travel far before they are attenuated to levels below ambient sea and electronic receiver noise backgrounds, the sensor platform must be positioned reasonably close to the organisms. This can be done in a variety of ways. Several implementations of acoustical zooplankton sensors are illustrated in Fig. 12.6.

A twenty-one frequency "Multifrequency Acoustical Profiling System" (MAPS) is shown configured for cast operation Fig. 12.6(a) and for use in a to-yo'ed mode of operation in Fig. 12.6(b). The MAPS technology employed a frequency band between 100 kHz and 10 MHz. It was used to about 200 m depths and data were usually collected in depth

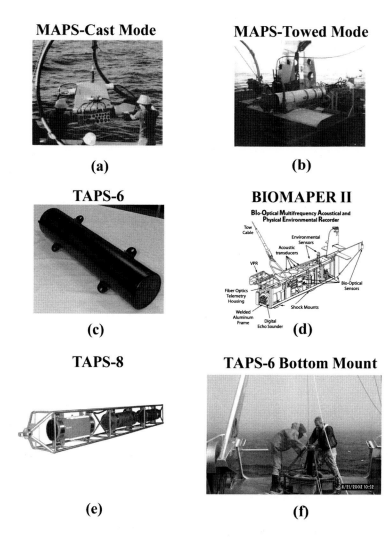

MAPS-Cast Mode

(a)

MAPS-Towed Mode

(b)

TAPS-6

(c)

BIOMAPER II

Bio-Optical Multifrequency Acoustical and
Physical Environmental Recorder

(d)

TAPS-8

(e)

TAPS-6 Bottom Mount

(f)

Fig. 12.6. Bio-acousticians have developed a variety of sensors and modes of deployment for studying zooplankton and micronekton. The sensors illustrated are the MAPS (*a*), (*b*); the TAPS-6 (*c*); the BIOMAPER II (*d*); the TAPS-8 (*e*); and a TAPS-6 in a bottom-mounted configuration (*f*). Each sensor, and its deployment configuration is discussed in the text.

bins of about 2 m. Binning at a finer depth interval was difficult because the motions of the ship often coupled to the MAPS package, smearing the data in the vertical dimension. No longer in use, the MAPS was built in the late 1960s and employed a mix of analog components, discrete transistors, and the integrated circuits of that day. Its main function was as a high-resolution research tool that allowed scientists to understand how sound scattered from assemblages of plankton. Successive families of sensors have been built on the knowledge thus gained, taking advantage of major advances in electronics and computers that were driven by general consumer demands, not science or engineering forces. The predecessors to the MAPS were built with vacuum tube technology and required much of an entire laboratory space on a research vessel with several racks of electronic gear. Those same capabilities are now available in a couple of cubic feet of instrumentation, require much less power, and are considerably less expensive.

A more modern, more compact cousin of the MAPS, called a TAPS-6, employs only six frequencies, and has been used since the early 1990s. It is routinely deployed in a cast mode or in a bottom-mounted frame (Fig. 12.6(f)). In that configuration it is used as an inverted six frequency echo sounder over distances that are usually determined by the absorption of sound at the higher frequencies and by the amount of plankton in the water column. This sensor has also been towed on a fishing trawl and used to look ahead of a MOCNESS zooplankton net system, allowing one to direct the net sampling to interesting places in the water column.

The BIOMAPER (Fig. 12.6(d)) is a five frequency towed device with simultaneous up and down-looking transducers. It is optimized for studying micronekton, as the sample volumes are necessarily larger than the MAPS or the TAPS, which are optimized for looking at zooplankton. It has been extensively used to assess zooplankton and micronekton populations on Georges Bank in the GLOBEC program, and has also seen use in difficult Antarctic environments.

A TAPS-8 is also shown (Fig. 12.6(e)), mounted in an in-line mooring frame, lying horizontally in this picture. The TAPS is the unit closest to the reader. The two pressure cases most distant from the reader enclose two batteries for mooring the sensor in the Coastal Gulf of Alaska for several months at a time. This system includes two frequencies, 102 and 265 kHz, in addition to the ones used in the TAPS-6. Its beams "look" horizontally when moored in-line with a surface or subsurface buoy to hold it up in the water column. The four lowest frequencies are used in an interleaved "ping" sequence to examine larger volumes for micronekton to ranges of about 32 m. All eight frequencies are used to examine smaller (liter-size) volumes from about 1.5 to 3 m from the sensor. The operating modes are "firmware" driven, but can be modified in the lab by changing an EEPROM before deployment to optimize the data collection in a particular science study. For example, comparisons were made of the scattering from volumes inside a meter distance from the mooring cage and those beyond a meter. The results suggested that algal growth on the instrument cage itself might be a food source for small crustaceans, thereby modifying the local trophic environment and causing aggregation. Volumes examined beyond a meter from the cage seemed to indicate minimal, if any, change with range from the cage. This was important, as one really wants to know the density of organisms as they naturally occur, not as they are changed by the instrument used to sense their presence.

With three batteries for power and ballast, a TAPS-6 system mounted on a frame and placed on the bottom can "look" vertically (Fig. 12.6(f)). Volume scattering strengths are then collected at 265, 420, 700, 1100, 1850, and 3000 kHz. The vertical resolution for this sensor and mode is 12.5 cm at intervals through a 25–30 m water column. The data are collected as multi-ping averages (programmable, usually 8–32 echo ranging cycles) and are normally reported every minute for three to four weeks via a small surface telemetry buoy located nearby. In this mode the TAPS-6 has also been used for longer deployment intervals (months) with cables being used for shore power, and for transfer of data and system control signals. This mode of operation was developed to study zooplankton distributions at ecologically critical sub-meter vertical scales. As much as 86% of the total zooplankton biomass has been found in sub-meter thick layers in a 50 m deep water column. The use of the TAPS-6 in this particular deployment mode allows one to study how these thin layers of zooplankton are distributed in relation to other trophic levels (phytoplankton, micronekton, and fish) as well as the fine scale physics (turbulence, mixing, stratification) and the chemistry of the water column. Biofouling is the usual limit on deployment times for this system in coastal waters, and deployment times have varied with season and local biological production rates for fouling organisms such as barnacles. Direct sampling sub-meter thick layers of zooplankton and micronekton with conventional nets, pumps and traps is currently a major challenge for biological oceanographers. Acoustics is being used to direct attempts at conventional sampling of these layers. At the same time, it also allows one to extract size and abundance information directly from these critical-scale distributions.

Co-registration of acoustical, optical, and chemical data along with the local fine-scale physical oceanographic data is necessary so that scientists can understand the costs and benefits zooplankton get from this newly discovered small scale behavior. Populations often depend on large-scale phenomena and forcing, but individual zooplankters live in a world with ambits of centimeters to tens of meters, and seconds to months. Understanding their world on their scales is important if we are ever to be able to model biophysical interactions at a level which will accurately replicate behavior, including reproduction and foraging. Some scientists think understanding how an individual animal lives will help in predicting recruitment success of higher trophic levels, such as commercially important fish. The bioacoustics community is responding to some of these concerns by developing appropriate tools to supplement and direct more conventional sampling strategies. In some cases such as thin layer ecology, for now, acoustical methods appear to be the only option at present for studying the zooplankton components of these structures. Along a parallel path, remote optical methods are currently being developed to examine phytoplankton layers.

12.4 A few examples of data collected with acoustical tools

12.4.1 Towing an array of sensors along a transect

The power of acoustical data is often first realized through its display, which illustrates patches and layers of biological sound scatterers. Since the acoustical signal typically travels such a great distance through the water, a continuous record of scattering can be obtained for a given direction of sound propagation. Since multiple pings are collected from a moving platform, then the array of data can be assembled into a two- or three-dimensional image of scattering. Such an image (Fig. 12.7) collected with a predecessor of BIOMAPER II on Georges Bank, just east of Cape Cod, MA, reveals that the scattering in the water column is quite complex. One of the strengths of the BIOMAPER program has been the simultaneous use of a plethora of ancillary physical oceanographic and imaging optical sensors during the to-yo process. This, and some state-of-the-art display technology aboard the ship has allowed scientists to observe and interpret the data stream while at sea, rather than later in a laboratory ashore when there is no chance to adapt one's sampling strategy to what is being observed *in situ*, and in real time. This kind of strategy and capability has significant implications for both data quality, and for improved efficiency in using large, costly ships to accomplish research in difficult, sometimes even dangerous environments.

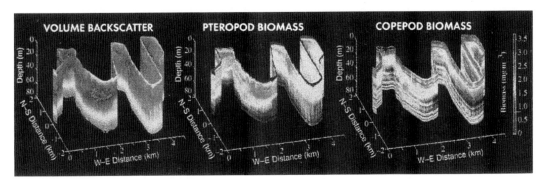

Fig. 12.7. Volume scattering measurements at 420 kHz were made with a predecessor of BIOMAPER II and used to estimate the spatial distributions of pteropod and copepod biomass along a transect on Georges Bank, off the eastern coast of the State of Massachusetts, USA. Echogram of volume scattering strength (left) used in combination with *in situ* video images of the organisms to estimate distribution of two types of zooplankton (middle and right). From Wiebe *et al.* (1997). (See colour plate section.)

12.4.2 Interpreting acoustical data

If one uses an acoustical sensor, even a simple one, to make measurements of sound scattering in the sea (or in a lake or river), interpreting the resulting patterns of scattering or displayed image remains a challenge. In the simple case in which all organisms are the same length and type and exhibit the same behavior (that is, orientation distribution), then a larger echo will generally correspond to the presence of more organisms. However, once the assemblage consists of a mix of sizes and types of organisms, the interpretation becomes far more complex, especially when the proportions of sizes and types vary in time and space. In this latter, more complex case, it is possible that a larger echo may be due to a different type of zooplankton that might be a more efficient scatterer than the ones in the surrounding waters.

These challenges have led bioacousticians to work on what they call the "inverse" problem. Initially, underwater acousticians were principally interested in catching the organisms in the sea and then, with their abundances and physical shapes and characteristics, predicting the levels and characteristics of sound scattering that would occur from that assemblage. This is called the "forward" problem and the emphasis was on predicting the acoustical environment. At the end of the twentieth century, it was recognized that acoustics could be used by scientists whose primary interests was in the ocean, its boundaries, and what resided in its volume. Using acoustical measurements to examine and understand life in the sea formed the basis of what is now called the "inverse" problem. These activities are an important part of the science of acoustical oceanography.

Below, we offer some information and data that has resulted from applying acoustics to a specific problem in marine ecosystems science. Our example focuses on how oceanographers are trying to understand the role of fine-scale distributions of several parameters that partially describe some of the ocean's physical features (temperature, salinity, shear, and mixing). The questions being asked involve why extremely intense sub-meter thick, vertical distributions of phytoplankton, zooplankton, and fish are commonly present in several coastal marine environments. Using high-frequency sound up to nearly 90% of the zooplankton biomass have been observed in very thin (e.g., 20 cm) layers when the total water depth was 50 to 100 m. Comparable amounts of the phytoplankton biomass have been observed in similar layers by using a variety of underwater optical sensors (e.g., fluorometers, multi-spectral sensors, and transmissometers). With the entire water column habitat available, an obvious question is "What advantage do the zooplankters find in distributing themselves in this way?". Another, is "Do they have a choice?". Since zooplankton often use phytoplankton as a food resource, one might think that these layers would be overlapping, if not perfectly co-registered in the vertical and horizontal. Indeed, sometimes this is the case, but many times, it is not. In fact, there appears to sometimes be active avoidance of a phytoplankton layer by zooplankton, which are undertaking diel migrations through the water column. These questions, and questions involving the role, if any, of the ocean physics in creating these thin layers of organisms, simply cannot be addressed with the conventional tools that oceanographers of the twentieth century have used so effectively.

A large part of the problem involves just sampling from a ship that is heaving up and down by several feet every few seconds in the waves. When this motion is coupled to an instrument hanging over the side of the ship, even with the best possible measurement of the sensor depth, the sensor data are smeared and the resolution is usually insufficient to study layers of only 10–30 cm thickness. Another complication is that these structures tend to move somewhat coherently in depth with the local internal wave field. The best net systems, when towed across an internal wave field, can't currently be controlled in depth with sufficient accuracy to reliably sample within, and on each side of such a thin biological structure as it moves up and down. In several studies, this problem has been addressed by tethering one of the TAPS-6 instruments discussed earlier some 10–15 m below the surface at the top of a large float. The sensor then "looks" up through the water column and sends data to shore over a cable. In another configuration, a TAPS is mounted, as is illustrated in Fig. 12.6, in a frame, and placed on the seabed. A cable to shore is needed for lengthy deployments, but for a few weeks, a small surface buoy can be used to support an antenna for two-way communications with the TAPS.

A five hour record of acoustical data collected in the northeastern part of Monterey Bay, CA in August 2002 revealed some complex temporal changes in volume scattering strength in the water column over the acoustical sensor (Fig. 12.8(a)). A 420 kHz record from a TAPS-6, mounted on the bottom as in Fig. 12.6(f) is used to illustrate how acoustics was employed at this site to examine the distribution of biomass in the water column. Twenty-four echo ranging cycles were collected in an interleaved sequence during contiguous 80-second intervals. These were averaged into 12.5 cm depth bins and telemetered to a laptop on shore from a small surface buoy that was anchored about 30 m away from the position of the TAPS. The shore station was on a hill, about 10 km north of the study site.

Each "pixel" of the data set (80 s by 12.5 cm by 6 frequencies) was then processed using an NNLS-based inverse code that simultaneously sorted the data into optimal fits to two zooplankton scattering models, allowing for multiple sizes for each shape that we assumed was present.

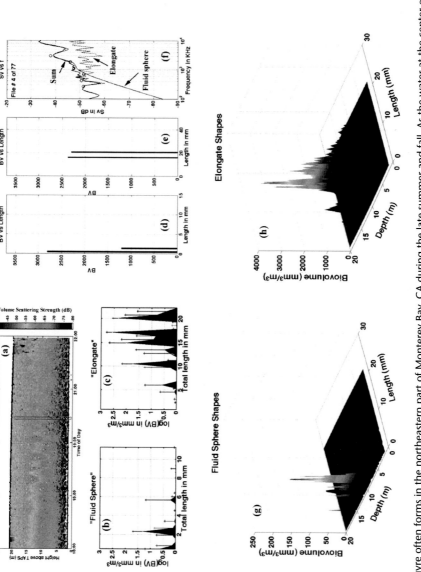

Fig. 12.8. A small gyre often forms in the northeastern part of Monterey Bay, CA during the late summer and fall. As the water at the center of this gyre warms, the water column stratifies, phytoplankton grows within the gyre, and zooplankton responds to an increased availability of food. The 420 kHz data displayed in panel (*a*) were collected at 80-second intervals with an upward-looking, bottom-mounted TAPS. The sensor was positioned within this gyre with the specific intent of examining the zooplankton response to changes in phytoplankton abundance in the upper mixed layer and near the pycnocline. A layer of scatterers was observed during a 5 h period near the depth of the thermocline. The pycnocline supported a packet of internal waves at the boundary between the warm water in the upper mixed layer and the underlying water mass. During this particular observation period the thermocline deepened and the layer became more diffuse as the tide receded and evening approached. Panels (*b*)–(*h*) illustrate the results of inverse processing for the time interval and depths included in the red box of panel (*a*). These subplots and the analyses that led to each are discussed in the text. (See colour plate section.)

One of the animal shapes assumed was modeled as fluid sphere. This is often used to approximate the dependence of scattering on length and acoustical frequency for small crustaceans such as calanoid copepods. It is often used if limited information is available on what species, genera or taxa of the nearly omnipresent small crustaceans might be dominating the acoustical scattering in a specific place and at a specific time. The second shape was an elongate scatterer. We computed its expected scattering "fingerprint" with the DWBA-based method discussed earlier and employed the shape of a mysid. This is a shrimp-like animal that sometimes burrows into loose sediments during the day and comes out to feed in the water column at night. Such animals were expected to be active at this site, as were euphausiids, a plankter that has a similar shape.

The analysis of a single pixel from the acoustical data during this period revealed that in order to explain the scattering we observed from this pixel, two sizes of animals were needed for each of the two shapes we assumed were present (Fig. 12.8(d), (e)). The contribution to the acoustical scattering from the aggregate abundance for each shape of animal is displayed versus frequency, as is the sum of their contributions (Fig. 12.8(f)). The small circles are the measured scattering at each frequency. With the constraint that no negative abundances are allowed, it is the difference between the total calculated scattering and the measured scattering levels which is minimized in arriving at the numerical abundances for each shape of animal.

Several hundred pixels, encompassing a range of depths and five of the 80 second intervals were selected from about 20:00 PST for processing (see the red box in Fig. 12.8(a)), one pixel at a time. The size–$\log_{10}$ (biomass) distributions calculated from each pixel were averaged over each of the selected times and depths. The result of this processing (Fig. 12.8(b)) reveals the presence of three sizes of scatterers in the fluid sphere-like shapes (e.g., copepods), but at relatively low abundances. Experience has taught us, based on net and pump collections done when this kind of data have been collected alongside the acoustics, that these "fluid sphere-like" scatterers are usually an assemblage of small copepods, nauplii and other small crustaceans (e.g., see Fig. 12.2). At 20:00 PST the dominant small crustacean at this location was slightly over 2 mm long and if one filtered a cubic meter of water, one would have found a biovolume (analogous to a displacement volume and a measure of biomass) about 10 mm^3 of animals at that size in the collection. The bar that extends above the shaded area (which defines the mean biovolume) in Fig. 12.8(b) represents the mean biovolume at the indicated size, plus the standard deviation, measured over all the pixels in the "red box." A similar display is provided for the distribution of "elongate" scatterers (Fig. 12.8(c)). Animals at about six sizes and abundances that ranged up to about 100 mm^3/m^3 were evident for the elongate shape assumed. Without samples, we cannot tell whether these discrete sizes represented distinct species or taxa, or perhaps were different stages of the same crustacean. Mysids tend to grow in "spurts," maintaining their size until they molt, and grow again. In other locations where we have collected specimens, it has been common to find several stages of the same species present at the same time and place.

In addition to averaging over all of the inverse results for each pixel in a selected time and depth interval, we also find it useful to display biomass as profiles of biovolume for each scatterer shape (Figs. 12.8(*g*), (*h*)). It is clear from this visualization mode that the layer that is being changed in depth was inhabited by animals of very different sizes and shapes. This layer was located near a shallow pycnocline that supported an internal wave packet (a group of waves at a density interface caused by warm, relatively fresh sea water floating on top of colder, more saline water). Examination of the distributions of biomass and the sizes measured acoustically with the TAPS, reveals that the larger, elongate animals were living mostly in the surface layer and their numbers decreased as they approached the cold water boundary. A distinct layer of smaller, copepod-shaped animals was located near the pycnocline and its depth was changed as the internal wave progressed shoreward over the TAPS. From some advanced optical sensors that were deployed on a winch near the TAPS, we know that a thin layer of phytoplankton was also associated with these structures. Sometimes the zooplankton layers are coincident in depth with the phytoplankton layers, a food resource, but they have also been observed apparently avoiding the phytoplankton. When this has happened, toxic species have been observed in the layers of algae. These processes, and the underlying physics (e.g., shear and turbulent mixing) that is associated with these layers is poorly understood at present, but is the subject of current work and planned future research.

12.5 The future

It seems clear that acoustical sensors will find many future applications in defining the response of zooplankton and micronekton to fine-scale ocean structure. These new research areas are opening now, largely because we have the means for the first time to observe animals and their behaviors on temporal and spatial scales that impact individuals. It is an exciting time to be a bioacoustical oceanographer!

Chapter 13

Models, measures, and visualizations of fish backscatter

JOHN K. HORNE
University of Washington

J. MICHAEL JECH
NOAA Northeast Fisheries Science Center

Summary

Fish are complex acoustical targets due to the anatomical structures and behavioral traits. The size, shape, and composition of an animal will influence the amount of sound it reflects. Relevant acoustical behaviors include locomotory and aggregation behaviors that influence fish orientations, densities, and distributions within the water column. Acoustical backscatter models are quantitative tools used to investigate how sound is reflected from individual and groups of aquatic animals. Models are best used iteratively with field or laboratory measures to validate model predictions, to help identify processes that influence backscatter intensities, and to identify potential research areas. Since humans are visually oriented, acoustic visualizations are used to summarize complex data sets, to aid in interpretation of results, and to translate findings to non-specialists. This chapter examines fish as acoustic targets, the development and use of backscatter models, how biological factors influence backscatter, and what future developments may occur in fisheries acoustics research.

Contents

13.1 Introduction

Two events in the early twentieth century focused the use of sound as a sensing tool in aquatic environments: the sinking of the Titanic and the need to locate enemy submarines during World War I. In response to these events, the British initiated the Anti Submarine Division (ASDIC) program and launched the application of sound as a sensing tool. The possibility of using sound to detect fish was introduced in 1927 by Rallier du Batty who attributed "false" signals on his echosounder to cod on the Grand Banks. The ability of fish to reflect sound was confirmed by Kimura in 1929 when the motion of fish in a tank interrupted a beam of sound between a transmitter and a receiver. Correlation between marks in the water column and trawl catches provided the first documented observations of fish echoes.

Fish are complicated scatterers of sound due to their size, shape, composition, and behavior. Our current understanding of backscattered sound from mobile animals is not complete. The development and use of models that predict acoustic backscatter provide tools that quantify factors influencing the intensity and variability of backscattered sound. Combining models with field and laboratory measurements provides an optimal approach to increase the understanding of backscatter from fish. Information acquired during acoustic surveys (i.e., *in situ*) or through experimental (i.e., *ex situ*) measures is used to highlight knowledge gaps. The iterative use of backscatter models and measures validates model predictions, quantifies the relative importance of processes that influence backscatter intensities, and identifies potential research areas. As confidence in the ability of a model to accurately predict backscatter increases, the potential range of applications for predictions that cannot be experimentally determined also increases. The goal of all fisheries acoustics research is to convert returned acoustic energy to fish lengths, spatial and temporal density distributions, abundances, and species-specific biomass estimates that can be used by resource managers and ecologists.

A common result of all acoustic measurements is large data volumes. It is easy to collect gigabytes of data during a survey but the quality and effective use of the data remains the responsibility of the user. Each frequency in a broadband or multifrequency system, and each beam in a multibeam sonar results in at least one unique data stream. Integration and interpretation of multiple data streams has lagged our ability to collect data. Visualization of acoustic data provides a mechanism to integrate multiple data sets and aids in the interpretation of organism distribution patterns in a format that is understood by non-specialists. We use visualizations throughout this chapter to illustrate how fish influence attempts to acoustically size, map, and count populations.

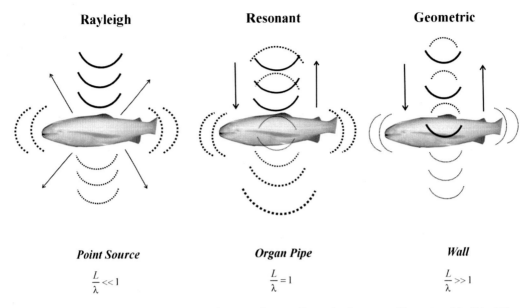

Fig. 13.1. Schematic diagram illustrating how sound is scattered by fish. If the fish length L is much less than the acoustic wavelength λ, then the fish scatters sound as a point source and the scatter is in the Rayleigh scattering region. When the length of the fish equals the acoustic wavelength sound resonates in the swimbladder like an organ pipe and is termed resonant scattering. When fish length exceeds the acoustic wavelength, the fish reflects sound similar to a planar surface (i.e., a wall) and is termed geometric scattering.

13.2 Fish and sound

Sound scattering by fish can be characterized three different ways depending on the length of the fish relative to the wavelength of the acoustic carrier frequency (Fig. 13.1). If the ratio of fish length (L) to acoustic wavelength (λ) is much less than unity, fish will reflect sound as an omnidirectional point source. Sound intensity in this Rayleigh scattering region is proportional to the fourth power of L/λ. As fish lengths increase or acoustic wavelengths decrease and the value of the L/λ ratio approaches 1, sound resonates within the swimbladder and the sound intensity increases. When fish length exceeds the insonifying acoustic wavelength (i.e., $L/\lambda > 1$) sound reflection is specular (i.e., like a planar surface). This region is referred to as the "interference" or geometrical scattering region.

Insonifying carrier frequencies in the resonance and geometric scattering regions are used to acoustically detect fish. For most swimbladder bearing fish, the resonance region typically extends from hundreds of hertz (Hz) to a few kilohertz (kHz). Lower frequencies can travel great distances, but have coarse spatial resolutions. As frequency increases, spatial resolution and ability to detect smaller organisms increases, but the detection range and the sampling volume decrease. These inter-dependencies illustrate the need for a range of frequencies to acoustically study aquatic organisms.

13.3 Fish as acoustical targets

From an acoustic perspective, biological backscattering properties of a fish can be examined in four major categories: morphological, ontogenetic, physiological, and behavioral. Morphological factors include fish anatomy and material properties. Ontogenetic factors are growth, development of the body, and the onset of sexual maturity. Feeding and the annual production of reproductive material (i.e., gonad) are two physiological factors that influence backscatter. Behavioral factors can be divided into individual and group behaviors. Individual behaviors include orientation (i.e., tilt, roll, yaw) of the animal relative to the incident wave front and avoidance reactions that influence activity and orientation. Group behaviors include shoaling, schooling, and avoidance reactions to predators or vessel noise.

Fish, also known as nekton, are vertebrates composed of muscle, a skull, a spinal column with ribs, fin rays, and are all covered by skin and scales or bony plates on the exterior. The bones of most fish are not as calcified as human bones, but still provide structural support to the organism. Locomotion occurs through movements of the fins and undulations of the body. The presence or absence of a swimbladder is the single anatomical attribute that has the greatest influence on sound scattering. A swimbladder is typically a gas-filled membrane, shaped like an irregular, prolate spheroid (Fig. 13.2). The dorsal (i.e., back or upper) surface tends to be relatively straight and is constrained by the vertebral column and musculature. The ventral (i.e., lower or stomach) surface is not reinforced by skeletal elements or musculature. Swimbladders typically comprise 3–5% of the fish body volume, have a variety of shapes, and are single or multi-chambered (see Whitehead and Blaxter, 1964). Deepwater fish (i.e., meso- and bathypelagic species) may have a lipid-filled or a minimized swimbladder. In addition to maintaining buoyancy, swimbladders are used to produce sound on spawning grounds and to enhance sound reception through connections to the Weberian ossicles (Popper *et al.*, 2002).

Physostomous fish contain a swimbladder with an esophageal or anal duct connecting the swimbladder to the water. These species inflate the swimbladder by swallowing air at the surface and do not have organs that inflate the swimbladder at depth although this is under dispute. Deflation of the swimbladder is achieved by expelling air via the esophageal or

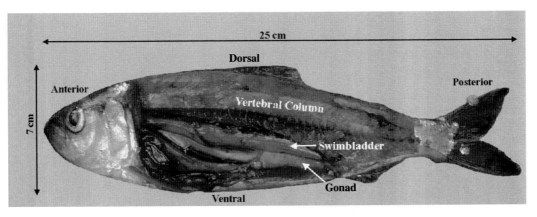

Fig. 13.2. Internal structure of an alewife (*Alosa pseudoharengus*) highlighting the vertebral column, swimbladder, and gonad. General terms used to describe viewer perspectives (dorsal = back or upper, ventral = stomach or lower, anterior = front, posterior = rear) are indicated on the alewife photograph.

anal duct. Physoclistous fish have a closed swimbladder without a duct. A vascular organ called a *rete mirable* is used to pump gas into the swimbladder. A gas gland is used to absorb excess gas. Adjusting to ambient pressure is slow among physoclist species but access to air is not necessary to increase gas volume at depth. In general, physoclist species tend to remain at a fairly constant depth, as rapid ascents to the surface will rupture the swimbladder. Physostomes can vertically migrate over a wider depth range.

All fish reflect sound, but those with a gas-filled swimbladder will scatter more sound than an identical fish without a swimbladder. Reflected sound from the swimbladder comprises 90% or more of the energy backscattered by a fish (Foote, 1980a). Any anatomical structure with an acoustic impedance different from that of water will reflect sound. The acoustic impedance depends on the density (ρ) of an object and the speed of sound (c) through that object. The proportion of sound reflected at the interface of two objects (e.g., swimbladder and flesh) depends on the difference between the acoustic impedances of the two objects. The higher the difference, the greater proportion of reflected sound.

Ontogeny (i.e., development) and material properties of the fish body and swimbladder directly influence the density, sound speed, and resulting acoustic impedance of any structure. In addition to the obvious density differences among tissues (e.g., muscle, bone), densities of body parts differ among life history stages. Radiographs show that relative densities of skeletal structures are significantly different among young-of-the-year, juvenile, and adult walleye pollock (*Theragra*

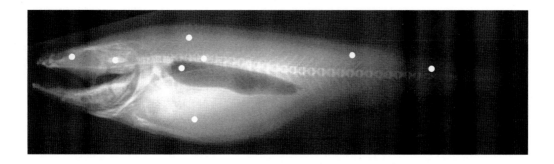

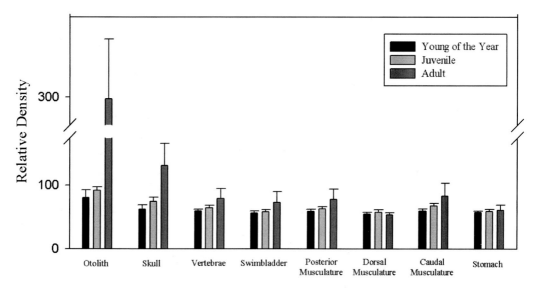

Fig. 13.3. Walleye pollock (*Theragra chalcogramma*) lateral radiograph (upper figure) showing locations of relative density measurements (white dots). Average relative densities (mean + standard deviation, $n = 10$), measured using standardized grey scales for transmittance, of young-of-the-year, juvenile, and adult walleye pollock at eight locations on the fish body (lower figure).

Chalcogramma) (Fig. 13.3). Density ratio, g and sound speed ratio, h are used to characterize acoustic reflective properties at the interface of any two surfaces see Table 13.1. Higher g and h values result in larger amounts of reflected sound energy. Material properties of body components in non-gas-bearing aquatic organisms influence the magnitude and variability of reflected sound greater than swimbladdered fish due to the relative contributions of body parts to total reflected energy. Table 13.1 summarizes g and h values for fish species with and without gas-filled swimbladders.

Ontogeny of the fish body and the swimbladder will alter the intensity of reflected sound throughout the life history of the animal. First

Table 13.1 *Density (g) and sound speed (h) values for swimbladders (sb) and fish bodies (fb) from a variety of fish species (see (13.1))*

Species	g_{sb}	g_{fb}	h_{sb}	h_{fb}
Anchovy (*Engraulis mordax*)	0.001267	1.023		
Atlantic cod (*Gadus morhua*)	0.001204	1.039	0.2315	1.054
Orange roughy (*Hoplostethus atlanticus*)		1.09		1.09
Orange roughy (*Hoplostethus atlanticus*)	0.875	1.017	1.028	1.043
Black and smooth oreo (*Allocytus niger, Pseudocyttus maculatus*)	0.122	1.027	0.2257	1.043
Fish flesh		1.03–1.06		
Fish flesh				1.03–1.08
Fish bone		2.04		3.75
A. cod eggs (*Gadus morhua*)		1.005		1.004
A. cod larvae (*Gadus morhua*)		0.9751		0.9997

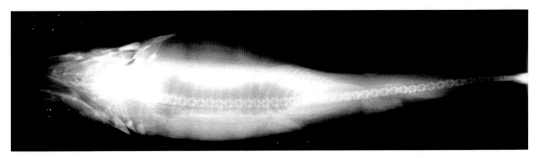

Fig. 13.4. Dorsal radiograph of walleye pollock (*Theragra chalcogramma*) showing anterior lobes of swimbladder.

inflation and subsequent growth of the swimbladder will dominate echo intensity at all life-history stages. The amount of proportionate or allo-metric (i.e., non-linear) growth of swimbladders is not well documented. Allometric changes in swimbladder volumes have been observed among teleost species and among individuals within a species. One example is the development of lateral lobes at the anterior of a walleye pollock swimbladder as the fish matures (Fig. 13.4).

Gut fullness and gonad development are two physiological factors that influence the amount of sound reflected from a fish. Any deformation of the swimbladder volume due to gut cavity contents or changes in surface area due to stretching of the stomach will alter the scattering characteristics of the fish body and swimbladder. Ona (1990) showed that mean swimbladder volume could be reduced by up to 90% due to gut content and up to 35% due to gonad production. Swimbladder volume reduction occurs primarily in the dorso–ventral plane, with one or both ends of the swimbladder collapsing before the middle. Gut contents and gonad can also contribute directly to the total backscatter from a fish. Prey species in the stomach, such as swimbladdered fish or zooplankton that contain oil sacs, will increase the amount of backscattered sound. Gonad, especially eggs, can be rich in lipids that scatter more sound than an equivalent amount of flesh or water.

Depth and orientation are the two primary behavioral factors influencing the amount and direction of reflected sound. As fish move up and down in the water column, pressure-induced changes in swimbladder volume (important for resonance scattering) and changes in swimbladder surface areas (important for geometric scattering) will influence the amount of backscattered sound. Boyle's law (volume inversely proportional to pressure at constant temperature) has been used to predict swimbladder volumes as a function of depth. But swimbladders are not bubbles. The ability to actively adjust the amount of air among physoclists and the presence of muscle or connective tissue attached to the swimbladder restricts the isometric contraction or expansion of the swimbladder. Active biological control of swimbladder shape restricts the application of depth–swimbladder volume or depth–swimbladder surface area relationships for many fish species.

Within the geometric scattering region, the orientation of the animal relative to the incident wave front influences the amount of sound reflected by fish (Foote, 1980b; Horne and Jech, 1999). As the value of fish length to acoustic wavelength ratio (L/λ) increases, sensitivity of echo intensity to animal orientation (i.e., pitch, roll, and yaw) also increases. Fish tilt angles are variable and range from negative while descending, to horizontal while maintaining depth, and increasingly positive while ascending. Swimbladders are typically angled posterior downward relative to the medial axis of the fish (see Figs. 13.1 and 13.2). Maximum backscatter will occur when the swimbladder surface closest to the sound source is orthogonal to the incident wave front.

Group behaviors influencing backscatter amplitude include coalescence and avoidance reactions. Aggregative behaviors include shoaling and schooling. Shoaling is distinguished from schooling by the lack of coordinated, polarized movement. Echo intensities of schooling fish are less variable than for shoaling fish due to similar lengths and orientations among individuals within schools. Avoidance reactions by fish to potential predators or vessels will affect acoustical measurements by altering orientations of individual fish, packing densities, or trajectories of fish away from the acoustic beam.

Table 13.2 *Summary of acoustic scattering models and measures of aquatic organisms or anatomical structures*

Organism or structure	Geometric form or measurement
Fish body	Gas-filled sphere
	Arrays of point scatters
	Fluid-filled cylinder
Fish swimbladder	Gas-filled spherical bubbles
	Gas-filled spheroid bubbles
	Gas-filled cylinders
	Gas-filled mesh
Whole fish	Gas-filled swimbladder
	Gas and fluid-filled cylinders
Empirical models	Literature review
	Caged
	Tethered
	In situ
	Statistical

13.3.1 Backscatter model history

Anderson (1950) summarized the challenge when predicting acoustic backscatter by marine organisms: "The exact theoretical study of the acoustic effects of the individual scatterers is prohibitive in its complexity since no simple geometric form can be attributed to marine life, and further, the material of which the scatterers are composed is in general not homogeneous." Acoustic scattering theory has advanced a great deal since this time but backscatter models still approximate the complex shapes and compositions of aquatic organisms (Table 13.2) and continue to evolve.

13.3.2 Fish as a sphere

The simplest shape for known acoustic scattering is the sphere. Because most fisheries acoustics applications examine sound scattered back to the transducer, we will consider only backscattered sound in this chapter. Anderson (1950) provides an approximation of backscatter by a fluid sphere. The reflectivity factor (proportional to the acoustic backscattering cross-sectional area) is dependent on the acoustic frequency, sphere radius, the ratio (g) of the density of the sphere (ρ') to the density of the environment (ρ), and the ratio (h) of the sound speed in the sphere (c') to the sound speed in the environment (c):

$$g = \frac{\rho'}{\rho} \quad h = \frac{c'}{c} \tag{13.1}$$

Backscatter from a fluid sphere depends on g, h, and the acoustic wavelength to sphere radius ratio (ka) (Fig. 13.5). The material property values of curve "C" in Fig. 13.5 were set at: $c_{water} = 1460$ m s^{-1}, $c_{sphere} = 1570$ m s^{-1}, $\rho_{water} = 1000$ kg m^{-3}, and $\rho_{sphere} = 1080$ kg m^{-3}, to represent a typical fish body. A common feature of all plots is the complex pattern of backscatter response in the geometric scattering region. Non-monotonic backscatter curves highlight the difficulty when converting echo intensities to organism lengths. Spherical shapes have been used to model fish swimbladders for many decades and have led to many insights on spatial and temporal distributions of aquatic organisms. These models are most appropriate when modeling backscatter at or near resonant frequencies.

13.3.3 Resonant backscatter modeling

Most derivations of resonance swimbladder models begin with Andreeva's (1964) formulation for scattering by swimbladdered fish. She assumed a spherical swimbladder encased in an "infinite" fish body. This approach reasonably approximated the acoustic cross-section and resonance frequency of swimbladdered fish, but differences between model predictions and *in situ* backscatter measurements motivated efforts to improve accuracy of backscatter estimates. Love (1978) modeled the swimbladder as a viscous elastic shell and incorporated a depth and species-specific correction for a prolate spheroid that was derived by Weston (1967).

The resonance frequency and shape of the peak in the backscatter curve is dependent on the swimbladder size and shape. The shape of the resonance peak is defined by the "Q" value, which is the ratio of the peak amplitude to peak width. To illustrate the dependence of resonance frequency and Q on the size of the swimbladder we calculated the resonance frequency for a swimbladder approximated as a sphere, as a prolate spheroid (i.e., adding Weston's correction), and the backscattering cross-sectional area as a function of insonifying frequency. Digitized radiographs of 12 Namibian pilchard (*Sardinops ocellatus*) were used to image swimbladder morphometry (Jech and Horne, 2002). The mean standard length (*SL*) was 21 cm, the mean swimbladder length was 9.4 cm (major axis), and the mean swimbladder width was 1.2 cm (minor axis). The resulting mean swimbladder elongation index (i.e., ratio of major to minor axis) of the pilchard was 7.83. The resonance frequency of the mean fish length using a spherical swimbladder was 1887 Hz. The

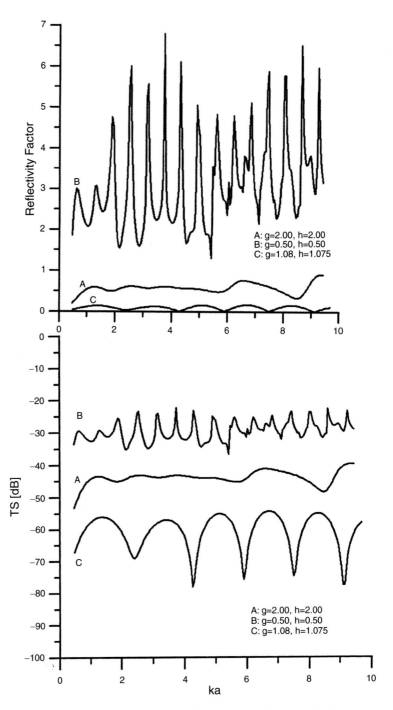

Fig. 13.5. Reflectivity of a 22.5 mm radius sphere as a function of *ka* for various values of *g* and *h* (upper graph). Reflectivity of the same sphere expressed as target strength ($TS = 10 \log_{10}(a^2 R^2 / 4)$) (lower graph).

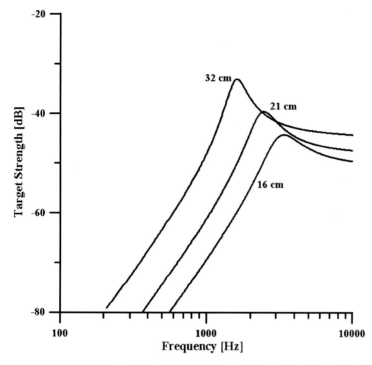

Fig. 13.6. Predicted target strengths (dB) of three different Namibian pilchard (*Sardinops ocellatus*) lengths in the resonance region as a function of frequency.

predicted resonance frequency incorporating Weston's correction was 2395 Hz for a fish at 50 m depth. If we isometrically scale (i.e., length, width, and depth change in equal proportions) the swimbladder as a percentage of fish length, the resonance frequency changes as a function of fish length (Fig. 13.6). The Weston corrected resonance frequencies for 16 cm, 21 cm, and 32 cm pilchard are 3194 Hz, 2395 Hz, and 1597 Hz. As fish length increases, the amplitude of the resonance peak increases and sharpens (i.e., Q increases).

These model predictions demonstrate that features of backscattering response curves in the resonance region may be used to index target sizes. If fish species differ by size, then species classification and discrimination may be possible using backscatter measurements in the resonance region (Holliday, 1972). Two other advantageous features of using resonance backscattering to index fish size or type are the single high amplitude peak and the low influence of fish orientation on echo amplitude.

13.3.4 Geometrical backscatter

The majority of fisheries echosounders and sonar systems operate in the 12 kHz to 420 kHz range. While there is no defined boundary that separates the resonance from geometric scattering regions, Love (1971) suggested that the geometric scattering region for fish is nominally bounded by $0.7 \leq L/\lambda \leq 200$ (where L is fish length and λ is the acoustic wavelength). Within this interval, fish lengths potentially span 3 cm to almost 8 m at 38 kHz ($\lambda \approx 4$ cm at $c = 1500$ m s^{-1}). Depending on the combination of acoustic carrier frequency and organism length, nearly all fisheries acoustics measurements are made within the geometric scattering region.

Realistic models of backscatter at geometric scattering frequencies must incorporate the anatomy and morphology of all major scattering components. Due to the complexity in developing analytical or numerical scattering models for irregular shapes, most models approximate fish bodies and swimbladders as objects that have analytical or numerical solutions. Three geometries have exact analytical solutions: a sphere, an infinite cylinder, and an infinite rectangular slab (Partridge and Smith, 1995). None of these shapes accurately represent a fish's shape, so approximate solutions for finite, elastic objects have been developed.

An important step in the development of geometric scattering models was made by Foote (1985) who combined a Kirchhoff–Helmholtz approximation with swimbladder morphology to predict fish backscatter. This model has recently been expanded with the incorporation of diffraction in boundary-element models (Foote and Francis, 2002). Another important development was the approximation for deformed, finite, elastic cylinders (Stanton, 1989). Clay (1991) modified this bent cylinder model and derived a ray-mode solution using a combination of gas- and fluid-filled cylinders (Clay, 1992).

A significant application of the finite cylinder model to fish backscatter was made by Clay and Horne (1994). They modeled acoustic backscatter of Atlantic cod (*Gadus morhua*) using a Kirchhoff-ray mode (KRM) model combined with a low mode cylinder solution. Model predictions closely matched empirical target strength measures. Utilizing digitized images of the fish body and swimbladder, the KRM model predicts backscatter as a function of fish length, orientation (tilt and roll), and acoustic frequency. Morphometries of the fish body and swimbladder are obtained from dorsal and lateral radiograph images, dissection, or computerized tomography (CT) scans, and then digitized. The digital morphometry data are used to construct a series of contiguous finite cylinders along the sagittal axis of the fish (Fig. 13.7). Complex scattering lengths $\mathcal{L}$ are computed as a function of frequency (f) for each cylinder, and then summed over the swimbladder ($\mathcal{L}_{sb}(f)$) and fish body ($\mathcal{L}_{fb}(f)$). Backscatter from the fish body is calculated separately from the swimbladder. Swimbladder backscatter calculations incorporate orientation and position of the swimbladder relative to the fish body (Horne

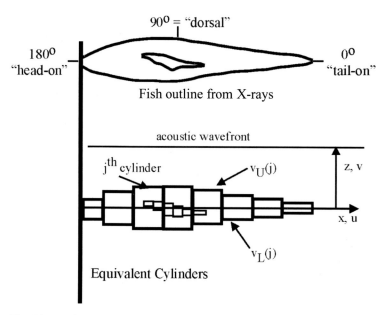

90⁰ = "dorsal"

180⁰
"head-on"

0⁰
"tail-on"

Fish outline from X-rays

acoustic wavefront

j^{th} cylinder

$v_U(j)$

z, v

x, u

$v_L(j)$

Equivalent Cylinders

Fig. 13.7. Schematic representation of traced radiograph fish body and swimbladder silhouettes (upper) and the corresponding equivalent cylinders used in the KRM model (lower). Incident scattering angles are designated as: 0° is "tail-on," 90° is "dorsal," and 180° is "head-on" incidence.

and Clay, 1998). Fish body and swimbladder scattering lengths are added coherently to estimate backscatter from the whole fish ($\mathcal{L}_{wf}(f)$).

The backscattering intensity (scattering length $\mathcal{L}$) can be normalized by the fish length (L_{fish}) to calculate reduced scattering length (*RSL*):

$$RSL = \frac{|\mathcal{L}(f)|}{L_{fish}} \qquad (13.2)$$

where $\mathcal{L}$ denotes the swimbladder, fish body, or whole fish. The reduced backscattering cross-section (σ_{bs}) is computed from the complex scattering lengths by:

$$reduced \; \sigma_{bs} = \frac{|\mathcal{L}(f)|^2}{L_{fish}^2} \qquad (13.3)$$

The reduced target strength (*TS*) is defined as:

$$reduced \; TS = 20 \log_{10} \left[\frac{\mathcal{L}(f)}{L_{fish}} \right] \qquad (13.4)$$

The non-normalized σ_{bs} is obtained by multiplying reduced σ_{bs} by L_{fish}^2. Target strengths are obtained by adding $20 \log_{10}(L_{fish})$. Jech *et al.* (1995) demonstrated that the KRM model is not species specific, and that it provides realistic predictions of backscatter from the fish body and swimbladder.

13.3.5 Empirical backscatter models

Analytic or numeric backscatter models are used to investigate mechanisms that influence acoustic scattering from regular and irregular shaped objects. Empirical models are statistical relationships between observed and measured variables. In fisheries acoustics, the most common relationship, target strength to length, is used to convert acoustic backscattering cross-sections to fish lengths.

Experiments used to parameterize these regressions are based on *in situ* or *ex situ* measurements. *In situ* experiments assume that fish are behaving "normally" and that backscatter measurements include natural variability. Challenges associated with *in situ* measurements include: finding suitable densities and length distributions of the species of interest; the ability to observe a fish's orientation without affecting behavior; minimizing backscatter from other organisms; obtaining backscatter measurements under all conditions; and obtaining representative length samples of fish that are insonified. *Ex situ* experiments include laboratory or controlled field measurements of tethered fish or fish in cages. These experimental data can provide direct comparisons of model predictions to measurements under specified conditions. For example, comparison of KRM model predictions to broadband data from live alewife (*Alosa pseudoharengus*) illustrates similarities and differences between empirical measures and KRM model predictions (Fig. 13.8). The agreement between model predictions and measures decreases as incident angles approach "end-on" aspects. An additional caveat is that we don't know how measurements from *ex situ* experiments represent scattering in the natural environment. In the alewife experiment, individual fish were restrained in a monofilament sock to minimize swimming motions.

13.4 Visualizing acoustic backscatter variability

One of the first things observed when acoustically measuring fish is that backscatter intensities vary. A set of target strength measurements from a tethered but free-swimming lavnun (*Mirogrex terraesanctae*) in Lake Kinneret, also known as the Biblical fish from the Sea of Galilee, illustrates the point (Fig. 13.9). Two distinct activity patterns are present. When first released, the fish swam vigorously and target strengths varied by up to 25 dB between successive pings. When the fish calmed, backscatter amplitudes decreased but still varied by as much as 5 dB between successive pings. Variability in backscatter intensity is influenced by physical and biological factors. Physical factors relevant to fish as acoustic targets are the speed of sound through water and the insonifying acoustic wavelength. Backscatter models can be used to quantify and illustrate the relative influence of biological factors.

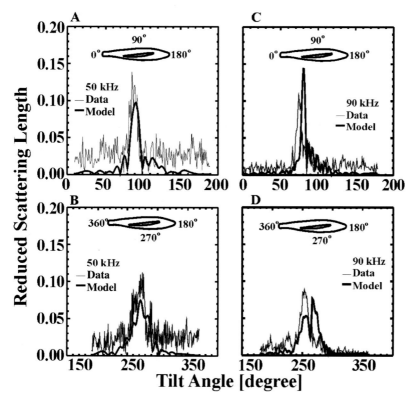

Fig. 13.8. Comparison of measured (thin line) to KRM modeled (thick line) reduced scattering length (*RSL*) as a function of tilt angle for a 256 mm alewife (*Alosa pseudoharengus*) at 50 kHz (left panels A and B) and 90 kHz (right panels C and D). The fish was rotated 360° to insonify all dorsal and ventral tilt angles. See Reeder *et al.* (2004) for details.

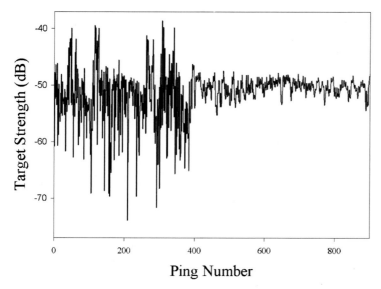

Fig. 13.9. Target strength plotted as a function of ping number (i.e., elapsed time) for a tethered lavnun (*Mirogrex terraesanctae*). The fish swam vigorously until approximately ping 400. See Horne *et al.* (2000) for details.

Differences in body and swimbladder morphology influence intensity and variability of backscattered sound. Potential variability in backscattered intensities is illustrated in the backscatter response curves of individual Namibian pilchard (Fig. 13.10). In model calculations, all pilchard were isometrically scaled to the mean standard length of the group (210 mm), over a frequency range of 10 kHz to 200 kHz (L/λ range: 1.4 to 28), and at a tilt angle of 90° (i.e., dorsal incidence). Sound speed (c) and density (ρ) parameter values were set at: $c_{water} = 1505$ m s^{-1}, $c_{fb} = 1575$ m s^{-1}, $c_{sb} = 345$ m s^{-1}, $\rho_{water} = 1026$ kg m^{-3}, $\rho_{fb} = 1070$ kg m^{-3}, and $\rho_{sb} = 1.24$ kg m^{-3}.

The dominant characteristic of all backscatter plots is the undulating pattern of peaks and nulls. These peaks and nulls result from constructive and destructive interference between the incident wave front and backscatter from the anatomical interfaces of the fish (e.g., fish body–swimbladder interface). Fish body backscatter curves contain many more peaks and nulls than those found in swimbladder backscatter curves. Maximum backscatter amplitudes occur at L/λ values less than five. Backscatter amplitudes from the fish body are generally lower than those from the swimbladder. A common feature in fish body and swimbladder backscatter curves is large amplitude ranges of 40 dB or more, which is a factor of 10 000. Whole fish curves closely resemble swimbladder curves as reflected sound from the swimbladder dominates backscatter from the fish.

Intraspecific (i.e., within species) backscatter variability of Namibian pilchard is apparent in the range of amplitudes and in the location of amplitude peaks and nulls. Variability decreases among fish with decreasing L/λ values. At L/λ values less than about five, KRM model predictions are less sensitive to the orientation of the swimbladder. Reduced sensitivity of the model to swimbladder orientation continues through the resonance region. The mean backscatter response for the fish group is calculated from RSL or σ_{bs} values and then converted to mean TS (i.e., the mean TS is not the mean of the logarithmic target strengths). The mean curve characterizes the backscatter response of the group over the length and frequency range modeled but does not characterize amplitude variability.

Changes in the orientation of the swimbladder relative to the incident wave front will modify backscatter intensities. From an acoustic perspective, swimbladder orientation is a combination of ontogeny and behavior. Similar to the beam pattern of a transducer, backscatter from a fish is directional. The beam pattern of a fish depends on the insonifying frequency, the shape of the fish body, and the shape and orientation of the swimbladder (Fig. 13.11). As frequency increases, the main lobe of the mean backscatter beam pattern decreases in angular width, the peak

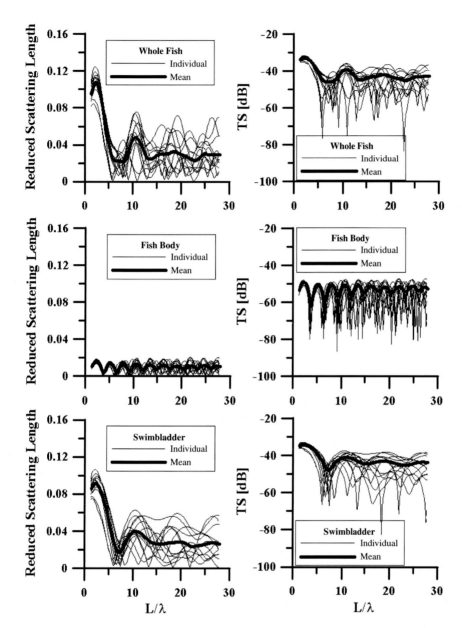

Fig. 13.10. KRM predicted reduced scattering length (*RSL*) (left side) and target strength (right side) plotted as a function of fish length to acoustic wavelength ratio (L/λ) for 12 Namibian pilchard (*Sardinops ocellatus*). Fish were scaled to a length of 210 mm and backscatter was calculated from 10 kHz to 200 kHz (L/λ range: 1.4 to 27.9) at a tilt angle of 90° (dorsal incidence). Light lines represent predicted backscatter for each individual, and the dark line represents the mean.

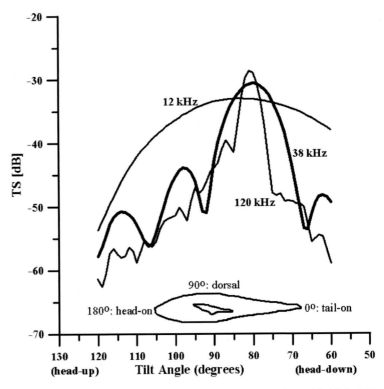

Fig. 13.11. Mean target strength (combined fish body and swimbladder) of 12 Namibian pilchard (*Sardinops ocellatus*) plotted as a function of tilt angle at 12 kHz, 38 kHz, and 120 kHz. The fish body and swimbladder traces are presented for angular perspective.

amplitude increases, and the number of side-lobes (smaller peaks off the main lobe) increases. Another characteristic of backscattered beam patterns is that the tilt angle of the peak amplitude corresponds to the aspect angle of the swimbladder relative to the sagittal axis of the fish. The maximum echo amplitude occurs when the swimbladder is perpendicular to the incident wave front and not necessarily when the fish body is horizontal. The peak amplitude of the Namibian pilchard occurs at approximately 80°, which corresponds to an average 10° posterior downward tilt of the swimbladder.

KRM model predictions can be used to characterize backscattered beam patterns for an individual or group of fish across a range of fish lengths, frequencies, and tilt angles. In the mean backscatter response surface of the 12 Namibian pilchard (Fig. 13.12), the effect of tilt angle on backscatter amplitude is low at low L/λ values and increases as L/λ values increase. Maximum backscatter amplitudes form a ridge along the L/λ axis where the swimbladder is perpendicular to the incident wave

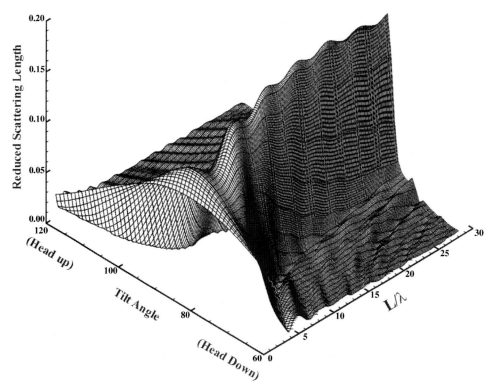

Fig. 13.12. Mean backscattering response surface of 12 Namibian pilchard (*Sardinops ocellatus*) scaled to a standard length of 210 mm. Reduced scattering length (*RSL*) is plotted as a function of tilt angle and the ratio of fish length (*L*) to acoustic wavelength (λ).

front. Backscatter amplitudes generally increase with increasing L/λ. Values drop as the tilt angle deviates from 85°.

Our definition of morphology includes material properties as well as anatomy. Slight changes in density and sound speed influence the amount of sound reflected at an interface. KRM model backscatter predictions from a 42 cm (fork length) walleye pollock at 120 kHz for the body, swimbladder, and whole fish differ depending on the density (g) and sound speed (h) ratio values (Fig. 13.13). Fish body target strengths differed by as much as 10 dB from the reference ($g_o = 1.039$, $h_o = 1.068$) as g values increased and h values deviated from h_o (Fig. 13.13(a)). Deviations among swimbladder predicted target strengths were minimal (<0.0004 dB) over the entire range of g and h values (Fig. 13.13(b)). Swimbladder target strength differences were more sensitive to sound speed contrasts than density contrasts. Target strengths differed by a maximum of -2 dB from the reference values for the whole fish (Fig. 13.13(c)). Deviation patterns observed for the whole fish mimicked

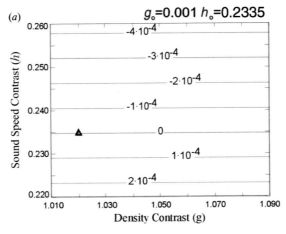

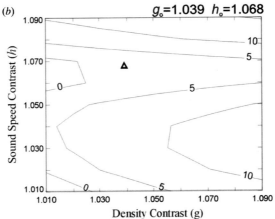

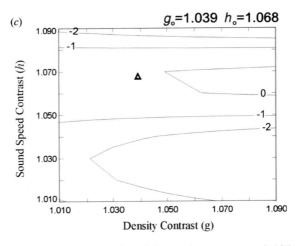

Fig. 13.13. Contour plots of changes in target strength (dB) of a walleye pollock (*Theragra chalcogramma*) as a function of sound speed (h) and density (g) ratios for the (*a*) swimbladder, (*b*) fish body, and (*c*) whole fish. Triangles indicate the reference sound speed (h_o) and density (g_o) contrast values.

those observed for the fish body. Larger whole fish target strength deviations are expected for species lacking gas-filled swimbladders.

Changes in fish orientation due to behavior will significantly influence backscatter intensity and variability. Variability in intensity directly affects accuracy of acoustic size to fish length conversions and the resulting density, abundance, or biomass estimates. For example at 38 kHz, a horizontally swimming pilchard would have a target strength of approximately -43 dB. If that same fish swims toward the bottom at a $10°$ angle, then the target strength would increase to -31 dB. Using Love's (1971) target strength to length equation, the estimated length of the horizontally swimming fish would be 12 cm while the downward swimming fish would be 48 cm. Assuming fish mass is proportional to the cube of the length, this 400% change in estimated length potentially alters biomass estimates by a factor of 64.

Incorporating behavior in acoustic-based estimates of fish length can improve accuracy of abundance and biomass estimates. One method includes a probability distribution of tilt angles in the target strength to length regression (Foote, 1980c). Valid acoustic to biological conversions require knowledge of the population tilt angle distribution. Orientation measurements of fish are difficult to obtain. One approach uses *in situ* tracked targets to determine fish aspect angles. This technique assumes that fish tilts are aligned with the observed track. Observations of fish swimming naturally or in cages are quantified using underwater video or photographic images. The orientation of the camera must be known to accurately measure fish tilt and roll. Visual ranges are limited under water, which means that optical instruments must be close to the fish and that supplementary light sources may be required. The presence of a tow body or frame, camera, and artificial light at depth may cause avoidance reactions or alter "natural" tilt angles. Techniques used to minimize behavioral responses include the use of low-light cameras and choosing wavelengths of supplementary lights that are not visible to fish.

13.5 The future

Fisheries acoustics continues to evolve at a rapid pace. Availability of new, digital hardware and backscatter model development has catalyzed research in many areas. Integration of modeling efforts with experimental measures is becoming a standard approach and will continue to guide the use and interpretation of fisheries acoustic technologies. The current analytic trend is to increase the amount of information used in target classification and identification. This approach originated in plankton acoustics and is being repeated in fisheries investigations. Information is increased through wider frequency bandwidth or a larger transducer swath from an increased number of angular perspectives.

Wider frequency bandwidth is achieved using multifrequency or broadband sonar systems. Traditional fisheries echosounders transmit narrowband signals in the 12 kHz to 420 kHz range. Multiple or multifrequency echosounders increase the number of discrete frequencies but this approach requires additional hardware and strategic deployment of transducers. Broadband (bandwidth > 10% of the center frequency) echosounders use a single or reduced number of transducers over a continuous frequency range. Increased bandwidth improves the ability to resolve targets using pulse compression techniques and potentially improves species classification and identification. Broadband systems are constrained by reduced signal-to-noise ratios at increased ranges, and by the dependence of insonified volume (i.e., beam pattern) on frequency. Since beam width decreases with increasing frequency, comparisons of volume scattering measurements among frequencies are problematic.

Multibeam and sector-scanning sonars use a single frequency and increase the number of beams transmitted and received by a single transducer. Beam-forming techniques are used to separate the swath into multiple, narrow beams with beam widths of a few degrees. When combined, individual beams can create a swath up to 180° or a rotating swath of 10° to 20°. These swaths are wider than traditional fisheries echosounder beams ($\leq$ 15° total angular width as measured at the half-power points) and provide a greater visual representation of three-dimensional spatial dynamics of schooling fish. Multibeam sonar technology has recently been transferred from military and commercial applications to fisheries acousticians. Sector-scanning sonars were developed to search for antisubmarine mines while multibeam sonars were initially used to inspect pipes and map bathymetry. Calibration procedures, volume backscatter processing algorithms, and target detection software are under development to extend these technologies beyond qualitative descriptions of fish schools.

Backscatter models will help facilitate the application of new sonar technologies to quantitative assessments and ecological investigations. Multibeam and sector-scanning sonars provide wider acoustic swaths but insonify targets at many different angles. Existing target strength to length regressions based on dorsal incidence measurements are insufficient when converting acoustic sizes to fish lengths using multibeam data. Three-dimensional scattering models that accurately represent fish anatomy and material properties are needed to convert echo amplitudes to organism lengths at non-normal insonifying angles. Our current analytic and numeric backscatter models provide robust estimates over frequency and orientation ranges commonly encountered during acoustic surveys. Greater use of alternate imaging technologies combined with modifications to existing or development of new backscatter models, should extend the frequency and orientation range of backscatter model estimates.

Over the next decade we predict further integration and coalescence of acoustic technologies. Environmental sensors are routinely added to tow bodies or fixed monitoring sites and the resulting data streams are spatially or temporally indexed in common databases. Integration of biological and environmental sensors will advance the development of multispecies and ecosystems approach in resource conservation management. The amount of information collected by acoustic sensors will also continue to increase. Combining broadband and multibeam technologies in a single instrument should provide adequate bandwidth and beam width to acoustically survey, discriminate, and track all organisms in the water column and to classify substrate types. A study of vessel noise production relative to fish hearing and avoidance has motivated the research community to require quieter sampling platforms. Research vessels are being constructed to match minimum noise requirements and alternate platforms such as autonomous underwater vehicles (AUVs) are being used to supplement acoustical surveys (e.g., Fernandes *et al.*, 2000). Entire fish distribution and abundance surveys conducted by autonomous vehicles using multibeam–broadband sonars may not be that far off in the future.

Chapter 14

Bioacoustic absorption spectroscopy: a new approach to monitoring the number and lengths of fish in the ocean

OREST DIACHOK
Naval Research Laboratory, Washington, DC

Summary

This chapter provides an introduction to bioacoustic absorption spectroscopy, a tomographic method for estimating number densities of fish with swimbladders as a function of length and depth. This method requires measurements of transmission loss (*TL*) over a broad band of frequencies between a fixed source and a fixed hydrophone, identification of absorption lines, which are related to the resonance frequencies of fish with swimbladders, classification of absorption lines by matching measured and calculated resonance frequencies of species and year classes, and estimation of number densities by matching measured and calculated absorption spectra. The experimental procedures, benefits, and potential for practical applications of this method will be reviewed.

Contents

14.1 Introduction

Bioacoustic absorption spectroscopy (BAS) is a tomographic method for estimating number densities of fish with swimbladders (Diachok, 1999; Ching and Weston, 1971). It requires measurements of transmission loss (*TL*) over a broad band of frequencies between a fixed source and a fixed hydrophone. *TL* is defined as the difference in signal level between 1 meter and other ranges from a source. *TL* is controlled by the sound speed profile in the water, the physical properties of the bottom and surface, and by absorption coefficients of biological layers. This method employs sources with low source levels (170 dB re 1 μPa at 1 meter), is non-invasive, and is not sensitive to unknowable details of swimbladder morphology and tilt distributions. Classification of measured absorption lines is based on matching of measured and calculated resonance frequencies, f_0, which are a function of the effective radius, eccentricity, and depth. Number densities are derived from measured absorption coefficients of biological layers and calculated extinction cross-sections. The latter may be derived from first principles provided that average separations between fish, $s > 1/4\lambda_0$ (where λ_0 is the wavelength at f_0), which is generally valid except for schooling fish (Diachok, 1999).

14.2 Resonance frequency of individual fish

Most species of fish, which occur in the ocean in large numbers, have air filled swimbladders. Swimbladders provide buoyancy and enhance hearing. Examples of the typical shapes of swimbladders found in sardines and anchovies (O'Connell, 1955) are shown in Fig. 14.1. The sardine swimbladder has a single chamber, and is attached by ducts to the gut (near the center), the ear (on the left), and the "*pore*" (on the right). The

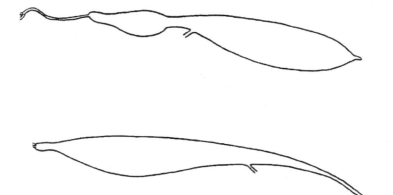

Fig. 14.1. Typical cross-sections of the swimbladders of two physostomes, anchovies (top) and sardines (bottom).

anchovy swimbladder has two connected chambers and is attached by ducts to the gut and the ear.

The resonance frequency, f_0, of a spherical bubble may be described with the equation:

$$f_0 = (1/2\pi \, r)(3\gamma)^{1/2} p^{1/2}/\rho_0^{1/2} \qquad (14.1)$$

where r is the radius of the bubble, p is the pressure, ρ_0 is the density of sea water, and γ is the ratio of specific heats of the gas in the swimbladder.

To calculate the resonance frequency of an elongated bubble, r needs to be replaced by the "effective" radius of the swimbladder, defined as the radius of a sphere, which has the same volume as the swimbladder (Holliday, 1972); and the magnitude of f_0 needs to be corrected for the eccentricity of the bubble. The resonance frequency of an elongated bubble is significantly greater than the volume resonance frequency of a spherical bubble with the same volume (Feuillade and Werby, 1994; Weston, 1967). To account for this effect, (14.1) should be modified to:

$$f_0 = e(1/2\pi r)(3\gamma)^{1/2} p^{1/2}/\rho_0^{1/2} \qquad (14.2)$$

where e is a non-dimensional correction, which is a function of the eccentricity (the ratio of the major to minor axes), ε, of swimbladders. The geometrical shapes of swimbladders may be approximated by prolate spheroids or cylinders with end caps (Feuillade and Werby, 1994). According to calculations based on the prolate spheroid and cylindrical models, a value of ε equal to 5 corresponds to e equal to approximately 1.15.

The size of the swimbladder increases as fish grow. The relationship between r_0, the effective radius at the surface, and L, the length of anchovies (Holliday, 1978) is shown in Fig. 14.2. Other species exhibit a similar behavior (Lovik and Hovem, 1979; Blaxter and Batty, 1990).

Fig. 14.2. Effective radius of swimbladder vs. length of anchovies, and fourth-order polynomial fit to Holliday's (1978) data.

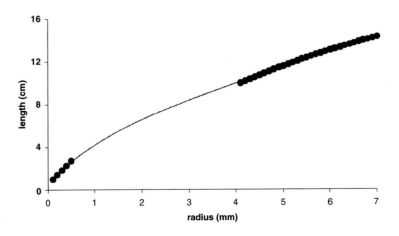

Measurements of the depth dependence of the resonance frequency indicate that the majority of fish in the ocean fall into one of two categories, physoclists and physostomes. The latter have a pneumatic duct between the gut and the swimbladder, permitting passage of air swallowed at the surface. The volume of the swimbladder of physostomes varies inversely with pressure, in accord with Boyle's law. Physoclists have a gland that secretes gas into the swimbladder, which they employ to maintain a constant volume of gas in the swimbladder, independent of depth (Blaxter and Batty, 1990).

The amount of gas in adult physostomes' swimbladders is only approximately constant. Their swimbladders are connected to the stomach by a narrow duct, to the auditory "bulla" system by very fine ducts, and in some species (e.g., herring, sardines, and sprat) to the *pore* via the anal duct, as illustrated in Fig. 14.1. Other species' swimbladders (e.g., anchovies) are only connected to the stomach and the auditory "bulla" system.

As a physostome moves downward, a small amount of gas passes from the swimbladder to the auditory system, and in the reverse direction during ascents (Blaxter *et al.*, 1979). This phenomenon results in a small, depth dependent reduction in the magnitude of r_0. In addition, physostomes lose gas very slowly (over periods of tens of days) by diffusion through the swimbladder membrane, and can replenish gas by occasionally swallowing air at the surface.

Despite these complexities, measured resonance frequencies of species, which have swimbladders that are attached to ducts at two and three points are in good agreement with the assumption that swimbladders are "free." Figure 14.3 shows laboratory measurements and

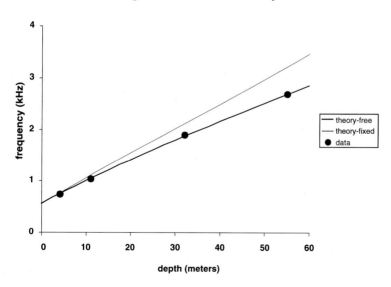

Fig. 14.3. Measured and calculated resonance frequencies of herring. Calculations assume that the swimbladder is "free" and fixed at both ends.

calculations of resonance frequencies vs. depth of herring (Lovik and Hovem, 1979), and calculations of f_0, which assume that the swimbladder is "free" and ε is independent of depth; and fixed at both ends. In the latter case the length of the swimbladder is independent of depth and the diameter and ε decrease with depth in accord with Boyle's law. These calculations assume a value of r_0 of 0.76 cm. This value is consistent with the value of r_0, 0.78 cm, which was derived from Lovik and Hovem's measurements of the maximum diameter and maximum length of herring swimbladders and the assumption that the shape of the herring swimbladder may be approximated with a pair of cones. Assumptions that the shape of the herring swimbladder may be approximated with prolate spheroids or cylinders with end caps resulted in values of r_0, which were too large. Measurements of f_0 were in closer agreement with calculations which assumed that the swimbladder was "free," than fixed at two ends.

Similar calculations were performed for comparison with Lovik and Hovem's measurements of sprat (not shown). Again, measurements were in closer agreement with calculations, which assumed that the swimbladder was "free," than fixed at two ends.

Measurements of f_0 of anchovies provide additional support for the "free" model of the swimbladders of physostomes. Figure 14.4 shows a comparison of measurements and calculations of resonance frequencies of anchovies at depths between 6 and 25 m. These data were derived from laboratory measurements (Baltzer and Pickwell, 1970), and backscattering (Holliday, 1972) and bio-absorptivity (Diachok, 2001) experiments at sea. Both sets of field measurements were made at night, when most anchovies are dispersed. Measurements are in accord with calculations which assumed that the swimbladder was "free."

The good agreement between the measured and calculated depth dependence of f_0 of these physostomes (herring, sprat, and anchovies) may be related to the lateral dependence of the physical properties of

Fig. 14.4. Measured and calculated resonance frequencies of individual anchovies in the laboratory (▲), and dispersed anchovies at night in the ocean from bio-absorptivity (●) and backscattering measurements (■).

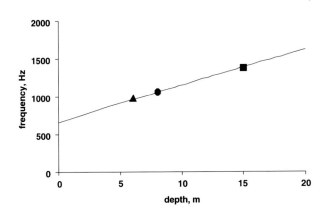

the swimbladder, which were not considered in the idealized models described above. The good agreement may also, to some extent, be fortuitous, since other phenomena (described below) can affect the depth dependence of f_0.

Discrepancies between measured and calculated values of f_0 may be expected due to swimbladder compression as a result of an enlarged gonad or a full stomach. Ona's (1990) laboratory measurements of the volumes of cod (physoclist) swimbladders suggest that enlarged gonads can reduce r_0 by a factor of 0.85 (compared to r_0 of a "normal" fish) and significantly increase the magnitude of ε. The increase in f_0 due to enlarged gonads may be expected to have the most pronounced effect on f_0 during a few months before and during the peak spawning month. His measurements also indicate that a nearly (3/4) full stomach can reduce r_0 by a factor of 0.79 (compared to r_0 of a fish with an empty stomach) and significantly increase the magnitude of ε. Theoretically, decreases in r_0 and increases in ε lead to higher values of f_0. The effects of a full stomach on r_0 and ε of physostomes may affect near-surface measurements of f_0, but are hypothetically less significant at greater depths, since the volumes of their swimbladders and their compression by other organs decrease with depth. Furthermore, lipids (oils) in fish flesh may also cause swimbladder compression. Note that r_0 (at the surface) varies inversely with the percentage of lipids; 30% lipids can reduce r_0 by as much as 20% (Blaxter and Batty, 1990).

In view of these uncertainties, theoretical calculations of f_0 should be considered approximate, and classifications of absorption lines at sea should be confirmed with concurrent trawling data.

14.3 Observations of temporal changes in resonance frequencies at sea

In their natural habitat physostomes, such as sardines and anchovies, are generally dispersed and occupy near-surface layers at night. Limited evidence suggests that at dawn they approach the surface to "gulp" air, a phenomenon known as "pre-dawn rise" (Woodhead, 1966). This phenomenon apparently occurs just prior to their descent to the bottom at sunrise, where they form schools. The structure of schools resembles comets; they consist of a "nucleus," where average separations are approximately one fish length and "fuzz," a region surrounding the nucleus, where the average separation between fish is large. At dusk, the schools generally return to the surface where they disperse.

An experiment, which was designed to determine the effects of bioacoustic absorptivity on transmission loss in the Gulf of Lion in the

Fig. 14.5. Length
distributions of sardines and
anchovies in the Gulf of Lion,
September 1995.

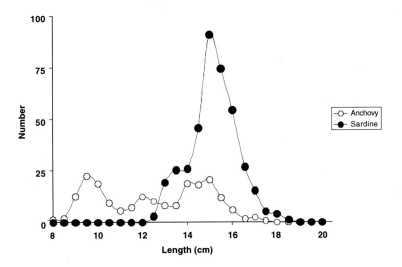

Mediterranean Sea, revealed absorption lines which were consistent with
this diurnal migration pattern (Diachok, 1999). The measurements were
made with a broadband source, which was deployed 5 m above the bottom
in 68 m of water and a vertical array of hydrophones, which spanned
most of the water column and was deployed at a range of 12 km from the
source. *TL* measurements were made over 7-hour periods, which spanned
sunrise and sunset. The observations of the resonance frequencies will
be discussed in this section; observations of absorption coefficients will
be considered in Section 14.4.

Nearly concurrent trawling data indicated that sardines and anchovies constituted approximately 60% and 30%, respectively of the
fish at this site. In the Gulf of Lion sardines and anchovies spawn
in January and July respectively. The experiment was conducted in
September. Length distributions of sardines and anchovies are shown in
Fig. 14.5. Sardine distributions exhibit peaks at lengths of 15.5, 13.5,
and 9.5 cm, which correspond to 3.7 +, 2.7, and 1.7 year olds; whereas
anchovy distributions exhibit peaks at 14.5, 12, and 9.5 cm, which correspond to 3.2 +, 2.2, and 1.2 year olds. Smaller (juvenile) fish could not
be adequately sampled because the mesh size (1 cm) of the net was too
large. These data clearly indicate that the dominant year class/species at
this site was 3.7+ year old (15.5 cm long) sardines.

The acoustic data shown in Fig. 14.6 show the frequencies at which
the transmission loss data exhibited the highest maxima as a function
of time. The highest losses at night occurred at a frequency of about
1.4 kHz. This is approximately equal to the resonance frequency of
15.5 cm sardines at 25 meters. A systematic change in resonance frequency between 1.4 and 3.0 kHz occurred during twilight. These data do

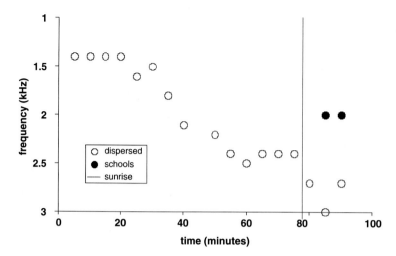

Fig. 14.6. Resonance frequencies of 15.5 cm long sardines as a function of time during dawn.

not provide evidence of "pre-dawn rise," possibly because the depths of the source (63 m) and hydrophones (25 m) were not optimized to detect the effects of bio-absorption layers near the surface. The observed time-frequency "signature" is consistent with the twilight descent of 15.5 cm long sardines from 25 m to 65 m. The descent starts approximately 85 minutes before sunrise (at astronomical twilight), and is completed at approximately 5 minutes after sunrise, where the resonance frequency is approximately 3 kHz.

Approximately ten minutes after sunrise, absorption losses abruptly diminish and the absorption line at the apparent resonance frequency bifurcates. During daytime two absorption minima were evident in the data: one at 3 kHz, which corresponds to dispersed fish and another at approximately 2 kHz, consistent with the calculated resonance frequencies of schools (Diachok, 1999). The latter are lower than f_0, when the separation between fish in schools, s, is less than λ_0, the wavelength at f_0 (Diachok, 1999). This condition is satisfied in schools of 15.5 cm sardines, when s is typically equal to about one fish length and λ_0 equals 1.1 m. The magnitude of the downshift is a function of the average separation and the total number of fish per school. This two-step process during sunrise (downward descent followed by school formation) is consistent with concurrent echo sounder data, shown in Fig. 14.7. This sequence of events occurs in reverse at sunset (Diachok, 1999).

Figure 14.8 provides a comparison of measured and calculated resonance frequencies of dispersed sardines as a function of depth (Diachok, 1999) at night just before twilight and during daytime just after sunrise. The good agreement between calculations and measurements supports the "free" model of sardine swimbladders.

Fig. 14.7. Echo sounder data at 38 kHz showing downward descent of 15.5 cm long sardines during dawn.

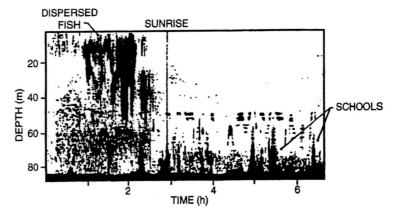

Fig. 14.8. Measurements and calculations of resonance frequencies of dispersed sardines.

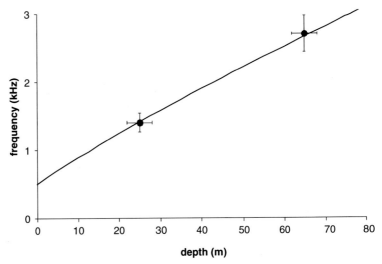

14.4 Extinction cross-section

The attenuation coefficient, α, nepers/m due to an ensemble of identical spherical bubbles may be calculated from the following expressions (Weston, 1967):

$$\alpha = 1/2(n\,\sigma_E) \qquad (14.3)$$

$$\alpha = [n\lambda_0 r/Q_0][(1 - f_0^2/f^2)^2 + 1/Q_0^2] \qquad (14.4)$$

where r is the effective radius (in meters), which is a function of depth, σ is the extinction cross-section, n is the number density (number/m^3), λ_0 is the wavelength (in meters) at f_0, and Q_0, which is defined as $f_0/\Delta f_0$, is the inherent Q of the bubble. At the resonance frequency the attenuation is

$$\alpha = n\,r\,Q_0\lambda_0 \qquad (14.5)$$

A more accurate estimate of n may be derived by matching measured absorption spectra with (14.4).

14.5 Derivations of number densities

Experimental measurements of absorption coefficients, which were derived from transmission loss measurements in the Gulf of Lion, are shown in Fig. 14.9 (Diachok, 1999). The most prominent absorption line at approximately 1.3 kHz is consistent with calculations, which were based on (14.4); the measured values of depth d (meters); trawling data which indicated that the dominant fish during this experiment were 15.5 cm long sardines; and values of r_0 and ε_0, which were extrapolated from Jech and Horne's (2002) X-ray measurements of the swimbladder dimensions of 24 cm long pilchard sardines (*Sardinops ocellatus*). The extrapolated values of r_0 and e of 15.5 cm long sardines were 0.80 cm and 10, respectively. The relatively weak absorption line at 0.9 kHz is correlated with the presence of adult 14.5 cm long anchovies at 6 meters (Holliday, 1978). Absorption lines at higher frequencies are correlated with younger year classes of these species.

Absorption spectra, which were derived from broadband transmission loss measurements, were compared with calculated absorption

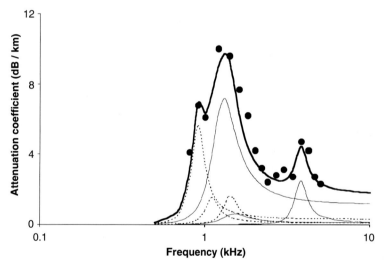

Fig. 14.9. Absorption coefficients vs. frequency at night derived from transmission loss measurements, which were made in the Gulf of Lion (•) (Diachok *et al.*, 2004); and (*a*) calculated absorption coefficients due to 15.5, 13.2, and 6.5 cm long (3.7-, 2.7-, and 0.7-year-old) sardines, which resonate at frequencies 1.3, 1.5, and 3.9 kHz respectively at a measured depth of 25 meters (three solid curves from left to right, respectively), (*b*) calculated absorption coefficients due to 14.5, 12.0, and 9.5 cm long (3.2-, 2.2-, and 1.2-year-old) anchovies, which resonate at frequencies of 0.9, 1.1, and 1.4 kHz respectively at an assumed depth of 6 meters (three dashed curves from left to right respectively), and the sum of (*a*) and (*b*) (heavy black line). The ratio of the number of sardines to anchovies was assumed to be equal to 2.0, in accord with trawling data.

spectra (based on (14.4)) due to ensembles of identical fish with swim-bladders over the frequency range 0.5–5.0 kHz, and near-coincident trawling data. Only night-time data, when the separation between these fishes is large (compared to wavelength), were considered. Measured distributions of the lengths of sardines and anchovies, which are shown in Fig. 14.5, were separated into groups, which were of nearly uniform length, i.e., year classes. Matching of measured and calculated absorption spectra, which are shown in Fig. 14.8, yielded estimates of the number densities and Qs of species and year classes. The spatial number densities, n_A (which equals the product of n and layer thickness), of adult sardines, which were derived from absorptivity measurements $(1.1/m^2)$ and nearly coincident echo sounder data $(0.6–1.6/m^2)$, were consistent. The inferred values of Qs of sardines and anchovies were 2.3 and 4.2, respectively. The latter is comparable to Diachok's inference of Q of year classes of anchovies, 4.4, which were derived from Holliday's (1972) backscattering measurements in the Southern California Bight. Inferred Qs were also consistent with laboratory measurements of Q_0, which is described in the next section.

It is instructive to compare the magnitudes of attenuation coefficients due to fish and chemical relaxation in sea water. The loss in signal level due to bio-absorptivity at 1.3 kHz at 12 km was 18 dB, which translates into an effective attenuation coefficient, averaged over the water column, of 1.5 dB/km (Fig. 14.9 provides attenuation coefficients associated with a bio-absorbing layer, which is 10 m thick). For comparison, the attenuation coefficient due to chemical relaxation at 1.3 kHz is much smaller, viz., 0.1 dB/km.

14.6 The Q of absorption lines

The Q of absorption lines associated with dispersed fish may be approximated with the following equation, which is a modified form of an equation due to Weston (1967):

$$1/Q^2 = 1/Q_z^2 + 1/Q_L^2 + 1/Q_0^2 \tag{14.6}$$

where Q_z and Q_L are associated with the distribution of fish in depth and length, and Q_0 is the inherent Q of a single fish. Other possible causes of spectral spreading are probably small.

Experimental measurements indicate that physical contact between the swimbladder and fish flesh, and/or attachment to other organs has a large effect on Q_0. Laboratory measurements by Batzler and Pickwell (1970) reveal that Q_0 of swimbladders extracted from anchovies was 21 at a depth of 6 m, whereas Q_0 of swimbladders within anchovies at the same depth was 4.5. Figure 14.10 provides a summary of previously

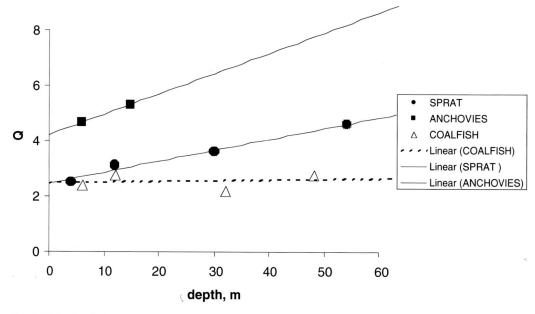

Fig. 14.10. Q_0 of physostomes, anchovies (■) and sprat (•), and an "adapted" physoclist, coalfish (△), as a function of depth.

reported measurements of Q_0 of physostomes, anchovies, sprat, and the physoclist, coalfish. These data show that Q_0 of physostomes increases with depth. The measurements of sprat were made by Lovik and Hovem (1979). The measurements of anchovies at 6 m were made by Baltzer and Pickwell (1970) in the laboratory; the measurement of anchovies at 15 m represents a lower bound derived by Diachok (2001) from Holliday's (1972) backscattering experiment. The increase of Q_0 with depth of physostomes is qualitatively consistent with the projected decrease in the amount of contact between the swimbladder and internal organs as r decreases with depth. In addition, these data indicate that the value of Q_0 of anchovies, 4.5, is significantly larger than sprat (2.8) at the same depth (6 meters), possibly because anchovy swimbladders are attached to two ducts, whereas sprat swimbladders are attached to three ducts. These data also show that the value of Q_0 of "adapted" coalfish, a physoclist, is independent of depth. This may be related to the depth independence of r of physoclists.

14.7 Conclusions

Experiments have shown that bioacoustic absorptivity due to fish with swimbladders can have a large effect on transmission loss, and that this phenomenon can be exploited to provide estimates of number density of year classes. Concurrent trawling data is required, at present, to confirm classification of absorption lines, which are associated with the resonance frequencies of fish swimbladders. Number densities derived from nearly coincident absorptivity and echo sounder data have been shown to be comparable. Interpretation of absorption spectra are complicated by the presence of schools, which resonate at lower frequencies than the resonance frequencies of individuals.

14.8 The future

There are at present no cost-effective means for monitoring number densities of juvenile fish. As a result, fishers have no way of avoiding casting their nets on predominantly juvenile fish congregations. Bioacoustic absorption spectroscopy permits isolation of signals due to juvenile and adult fish in frequency space and, as a result, has the potential to revolutionize fisheries acoustics and commercial fishing operations. BAS measurements could be conducted with a broadband source deployed from a ship equipped with dynamic positioning, and a hydrophone deployed on a glider. Another possible application: fixed observing systems for monitoring the dynamics of migrating species.

Chapter 15

Passive acoustics as a key to the study of marine animals

DOUGLAS H. CATO
University of Sydney, Australia

MICHAEL J. NOAD
University of Queensland, Australia

ROBERT D. McCAULEY
Curtin University, Australia

Summary

The vastness of the oceans makes it difficult to study marine animals because they are so often out of sight, except in a few specialized environments such as the clear shallow waters of tropical reefs. Because sound travels so much further in the sea than light or other forms of electromagnetic waves, it is natural to turn to acoustics for finding and studying marine animals. Active sonar is used for this purpose but the sounds that the animals produce themselves may also be exploited because so many have high source levels and are detectable at great distances. These vocalizations can be quite spectacular, and are intriguing in terms of the biological function they perform. This chapter describes some methods used and recent research in the use of fish and marine mammal sounds to study behavior, distributions, and movements, as well as acoustical methods of estimating population sizes and rates of increase in numbers. Even a single hydrophone can provide useful information, while multiple hydrophones can provide detailed information about movements and behavior.

Contents

411

15.1 Introduction

Life began in the sea where it has developed to a level of diversity unparalleled on land. Yet we do not often see marine animals in the wild because of the poor penetration of light through water and it is difficult to study them in their natural environment. Acoustics provides the means of monitoring animals at much greater distances than visual observations allow. Marine animals have evolved an extensive range of capabilities to exploit underwater sound, particularly through their vocalizations, such as communication signals and breeding (courtship) displays, and in some species such as dolphins, echolocation (active sonar) and signals for individual recognition. Their sounds tend to have high source levels so are detectable at substantial distances. Quite simple listening systems are effective to distances of tens to hundreds of meters for fish and invertebrates and tens of kilometers for whales. In addition, vocalization is such an important component of marine animal behavior that passive acoustics is essential in behavioral studies.

15.2 Sounds of the marine animals

Animal sounds vary from a few microseconds duration at frequencies up to hundreds of kilohertz (dolphin clicks: Au, 1993; shrimp snaps: Everest et al., 1948; Cato and Bell, 1992) to 15 to 45 s duration at frequencies as low as 20 Hz (blue whales: Cummings and Thompson, 1971; McCauley et al., 2000). The sounds of most animals, however, are within the audio frequency range and durations lie between 0.1 to 5 s (Tavolga, 1964, 1967; Richardson et al., 1995).

Invertebrates produce sounds that are usually short, impulsive, and broadband, such as clicks or series of clicks (Tavolga, 1964). The best known is the snapping shrimp which abounds in shallow warm water where it provides a continual background from about 1 kHz to more than 300 kHz. Other invertebrate sounds are made by scraping or impact of hard parts, usually clicks or series of clicks.

Fish produce a wide range of sounds (Tavolga, 1964). Some are short, impulsive, and broadband, click-like sounds. Many fish, however, use the swimbladder, a gas filled sac, to enhance sound production. Because the flesh of fish has an acoustic impedance similar to water, the swimbladder is acoustically similar to an air bubble in water (Section 6.4), which is a very efficient source of sound. Many species of fish drum on the swimbladder with attached muscles, producing drumming and knocking sounds of varying rates. Frequencies are usually in the range 50 Hz to a few kilohertz (Tavolga, 1964, 1967). Some species contract the attached muscles at a rate close to the resonant frequency of the bladder, producing harmonic sounds with fundamentals usually in the range 10 Hz to 300 Hz. Source levels (mean square pressure levels) of fish sounds vary widely from 110 to 160 dB re 1 μPa at 1 m, although most are in the range 130 to 150 dB re 1 μPa at 1 m (Cato, 1980; Myrberg *et al.*, 1986; D'Spain *et al.*, 1997; McCauley, 2001).

Marine mammal sounds generally fall into categories according to the classification: baleen whales (mysticetes), toothed whales (odontocetes), and seals. Baleen whales produce a wide range of sounds, from the long, low-frequency tonal signals (15–100 Hz) of the blue whales, to the variety of ever-changing sounds of the humpback song, in the range 50 Hz–10 kHz (Richardson *et al.*, 1995). Odontocete sounds include whistles and impulsive clicks. Much is known about the way dolphins use clicks for echolocation (Au, 1993; Section 5.5.2). Whales produce the most intense sounds of all marine animals, with most source levels (mean square pressure levels) at one meter from 165–185 dB re 1 μPa (Richardson *et al.*, 1995).

15.3 Use of passive acoustics to determine direction and location

One hydrophone can indicate the presence of the animal and something of its behavior, although detection ranges vary substantially with variations in propagation loss and the ambient noise against which the signal must be detected. Two hydrophones provide substantially greater capability. Figure 15.1 shows two hydrophones on the x axis at positions $H_1(0, 0)$ and $H_2(s, 0)$ receiving sound from a source at $Q(x, y)$. Since the source is closer to hydrophone H_1, the signal from the source will

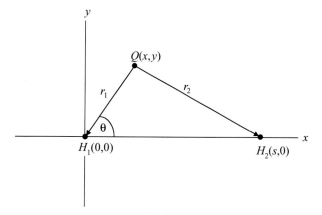

Fig. 15.1. Sound from a source at position $Q(x, y)$ is received by hydrophones at H_1 and H_2, Cato *et al.*, 1998.

arrive at H_1 before it arrives at H_2, and the received pressure will be higher at H_1 than at H_2. In a homogeneous, isotropic medium (constant speed of sound, no boundaries), the difference in the arrival times of the signals to the two hydrophones and the ratio of received pressures are enough information to determine the position of the source, though with an ambiguity (see Section 15.3.3). In the real ocean, travel time differences are affected by variations in the sound speed and interference from multiple propagation paths such as the reflections from the sea surface and the bottom. Received pressures vary substantially with variations in the propagation.

It turns out that the difference in arrival times is, however, quite robust and is little affected by sound speed variations, mainly because the paths to the two hydrophones experience similar variations. The ratio of received pressures is far more affected by environmental conditions, and is of more limited use, though it can be useful when the source distance is comparable to or less than the hydrophone separation.

15.3.1 Determination of the source direction and location using arrival time differences for distant sources

Figure 3.18, shows a wave front arriving at a line array of hydrophones where the distance of the source is much greater than the size of the array, so that the received signal is effectively a plane wave, see (3.57). The angle ϕ, the direction or bearing of the source from the array depends on τ, the arrival time difference between hydrophones, and the speed of sound c. Since $\sin\phi$ is proportional to τ, the accuracy of estimating ϕ decreases as ϕ increases and is highest broadside to the array (perpendicular to the array) and least at endfire (along the array) (see Section 3.9.5).

A second array located in another position would provide a second bearing and the source location could be determined by the intersection of the bearings. There is an uncertainty or ambiguity in determining the direction and position of the source. In Fig. 15.1, the signals from a source at position $Q'(x, -y)$ below the line through the hydrophones, would produce the same times of arrival as those from the source at Q, so we cannot determine if the source is at Q or Q'. Using two hydrophone arrays does not necessarily resolve this ambiguity. It depends on the source position and the relative orientation of the arrays, as in the example of Fig. 15.2. There is an ambiguity in both direction and location using the two arrays with similar orientation (the pairs H_1, H_2 and H_3, H_4), but not if either of these were used with the pair H_5, H_6 which has a different orientation. Sometimes external factors can resolve ambiguities.

Three hydrophones are enough to localize a source since this provides three pairs. In practice, extra hydrophones can provide more array

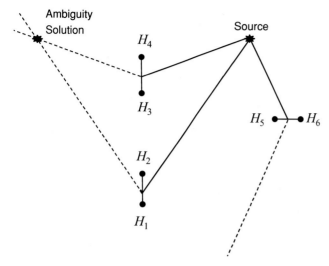

Fig. 15.2. The position of the source is determined by the intersection of the bearings from hydrophone pairs. There is an ambiguity using hydrophones pairs H_1, H_2 and H_3, H_4, but not if either of these is used with hydrophone pair H_5, H_6. This assumes that the distance of Q is much greater than the separation of hydrophones in any pair, so that the received wave front is close to plane.

orientations and greater accuracy, since it increases the chance that some arrays will have the source nearer broadside where accuracy is highest (six hydrophones provide 15 pairs in Fig. 15.2).

15.3.2 Determination of source location by arrival time differences for any source distance

In many cases, especially in the study of marine animal behavior, we are interested in sources that are much closer relative to the hydrophone separation, so that the plane wave front assumption is no longer valid. This section gives a more general solution for any source distance.

In Fig. 15.1, the time for sound to travel from the source at $Q(x, y)$ to the hydrophones at H_1 and H_2 is r_1/c and r_2/c respectively. The difference in the times of arrival τ is then given by

$$c_0\tau + r_1 = r_2 \tag{15.1}$$

where c_0 is the speed of sound which for this purpose can be assumed constant. Squaring both sides and substituting $r_1^2 = x^2 + y^2$ and $r_2^2 = (s - x)^2 + y^2$ gives

$$2c_0\tau \sqrt{x^2 + y^2} = s(s - 2x) - c_0^2\tau^2 \tag{15.2}$$

Squaring both sides again and collecting terms in x and y leads to

$$4x^2\left(c_0^2\tau^2 - s^2\right) - 4sx\left(c_0^2\tau^2 - s^2\right) + y^2\left(4c_0^2\tau^2\right) = \left(c_0^2\tau^2 - s^2\right)^2$$

Dividing by $(c_0^2\tau^2 - s^2)$, then by $c_0^2\tau^2$, gives

$$\frac{4x^2 - 4sx + s^2}{c_0^2\tau^2} + \frac{4y^2}{c_0^2\tau^2 - s^2} = 1$$

which can be written as

$$\frac{\hat{x}^2}{a^2} - \frac{y^2}{b^2} = 1 \tag{15.3}$$

where $a^2 = c_0^2\tau^2/4$, $b^2 = (s^2 - c_0^2\tau^2)/4$, and $\hat{x} = x - s/2$. Equation (15.3) is the equation of a hyperbola which is symmetrical about the x axis and has asymptotes crossing the x axis at $x = \hat{x}$, the point midway between the hydrophones. Thus for a particular arrival time difference, the source can lie anywhere on the hyperbola given by (15.3). The asymptotes are the lines of bearings for the plane wave case for distant sources. A second pair of hydrophones would give a second hyperbola for the particular source, so that the source position is given by the intersection of the two hyperbolas. An example with three hydrophones is shown in Fig. 15.3 for a source at position Q. The hyperbolas intersect at two points, giving an ambiguity in the determination of the source position, as in the case of distant sources discussed above. Three hydrophones are sufficient to localize, but more well placed hydrophones increase the accuracy and reduce the likelihood of ambiguity.

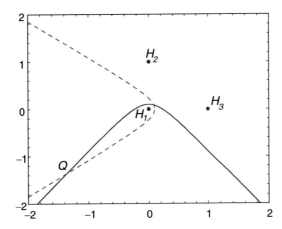

Fig. 15.3. The arrival time difference between hydrophones H_1 and H_2 of sound from a source at Q indicates that the source is somewhere on the hyperbola shown by the solid line. The time difference between hydrophones H_1 and H_3 indicates that the source is on the hyperbola indicated by the dashed line. The source position is thus given by the intersection of the two hyperbolas, but there is an ambiguity since there are two points of intersection.

Arrays of many hydrophones are often used in ocean acoustics because the capability to detect signals in noise increases as the number of hydrophones increases. In many marine animal applications, however, the signal levels are so high relative to the noise that the substantially greater cost and complexity of multi-hydrophone systems are not justified. Simple systems that have been used to monitor marine animals include those reported by Cummings *et al.* (1964), Watkins and Schevill (1972), Clarke (1980), Cummings and Holliday (1985), Freitag and Tyack (1993).

We have tacitly assumed that the source and hydrophones of Fig. 15.3 are in the horizontal plane, and treated this as a two dimensional problem. Rotating the plane containing the source and a pair of hydrophones about the line through the hydrophones does not change the result, so that the possible source position determined from two hydrophones is actually on the surface swept out by rotation of the hyperbola. The location of the source in three dimensions using two pairs of hydrophones is on the line forming the intersection of the surfaces swept out by rotation of the two hyperbolas. Often the geometry of the measurements is such that a two dimensional solution is adequate, for example, in shallow water for distances significantly greater than the water depth.

Measurement of time of arrival difference

Arrival time differences are usually very small, and accurate localization requires measurement accuracy of milliseconds or less. Signals from the hydrophones in an array must be recorded on the same multi-channel device (tape recorder or computer) or time synchronized to minimize the errors. The most effective method of measuring τ is by cross correlation of the signals from hydrophone pairs. The cross correlation as a function of time delay τ' is

$$R(\tau') = \frac{1}{NT} \int_{-T/2}^{T/2} f_1(t) f_2(t + \tau') dt \tag{15.4}$$

where $f_1(t)$ and $f_2(t)$ are the signals received at the two hydrophones and N is a normalizing function equal to the product of the rms values of $f_1(t)$ and $f_2(t)$ so that $R(\tau')$ lies between -1 and $+1$.

When $\tau = \tau'$, the cross correlation is maximum since the two received signals are in phase. Sinusoidal signals, however, are in phase for all values of τ' that are multiples of the sinusoidal period, so τ cannot be determined reliably for very narrow band signals. For transient signals common from marine animals, the period of integration T is usually chosen to enclose the duration of the signal as received on the two hydrophones. Software known as ISHMAEL (Mellinger, 2002) is available to perform cross correlations between pairs of hydrophones, calculate the hyperbolas, and determine the position of the source from three or more hydrophone pairs.

15.3.3 Determination of location using two hydrophones, by differences in arrival times and received levels

Using differences in arrival times and received levels, only two hydrophones are required for localization (with ambiguity), and the distance of the source can be determined without knowing the hydrophone separation. This method is more limited, however, because differences in received levels are much more variable than time differences, so errors tend to be larger. Acceptable accuracy requires the source to be significantly closer to one hydrophone than to the other and knowledge of the propagation conditions. The logistics are simpler and so it can be used where it is impractical to deploy multiple hydrophones with the high positioning precision required for localization using arrival time differences alone. For example, in estimation of the abundance of vocalizing animals during ship transits, the spatial density can be estimated by deploying sonobuoys at intervals and determining the distance to each source. Only a rough idea of the positions of the sonobuoys is needed.

In Fig. 15.1, let $k = r_2/r_1$ and $r_2 > r_1$. Let the received intensity at H_1 and H_2 from the source at $Q(x, y)$ be I_1 and I_2 respectively. We assume that propagation loss is proportional to ar^n, where a and n are constants, so that the received intensity at H_1 is

$$I_1 = I_0 / \left(ar_1^n \right) \tag{15.5}$$

where I_0 is the source strength (intensity at unit distance). The difference in received level DL at the two hydrophones is

$$DL = 10 \log_{10}(I_1/I_2) = 10n \log_{10}(r_2/r_1) = 10n \log_{10} k \tag{15.6}$$

When $a = 1$ and $n = 2$, propagation loss is according to spherical spreading. This would be applicable for example, for relatively short distances less than the water depth for vocalizations that are sufficiently broad band that the interference effects like Lloyd's Mirror (Section 1.5.6) are not significant. These circumstances are often the case in studying vocalizing fish. Fitting (15.5) to propagation loss as a function of distance is likely to be suitable for a wider range of conditions, again for sounds that are not very narrow band.

Substituting $r_2 = r_1 k$ in (15.1) gives

$$r_1 = c_0 \tau / (k - 1) \tag{15.7}$$

This simple expression for the distance of the source from the closer hydrophone is useful so long as the source is significantly closer to one hydrophone than the other ($k \gg 1$). This can be seen by noting that

$$\partial r_1/\partial \tau = c_0/(k-1) = r_1/\tau \quad \to \infty \quad \text{as} \quad \tau \to 0$$

and $\quad \partial r_1/\partial k = -c_0\tau/(k-1)^2 = -r_1/(k-1) \quad \to -\infty \quad \text{as} \quad k \to 1$

so that r_1 changes very rapidly as $\tau \to 0$ or $k \to 1$ and small errors in either τ or k would result in large errors in the estimate of r_1. The significance and magnitude of these errors, and ways of reducing them, are discussed by Cato (1998).

Evaluation of (15.7) does not require knowledge of the positions of the hydrophones or their separation. The accuracy of the result, however, will tend to improve as the hydrophone separation increases.

Determination of the source position

If the positions of the hydrophones are known, the position of the source can be estimated, with the usual ambiguity that arises when arrival time differences are used. From Fig. 15.1,

$$k^2 = \frac{r_2^2}{r_1^2} = \frac{(s-x)^2 + y^2}{x^2 + y^2} \tag{15.8}$$

and

$$\cos\theta = x/r_1 \tag{15.9}$$

where θ is the angle between the line $H_1 Q$ and the x axis (Fig. 15.1). Substituting $y^2 = r_1^2 - x^2$ into (15.8) and rearranging leads to

$$x = s/2 - r_1^2(k^2 - 1)/(2s) \tag{15.10}$$

Using the expression for r_1 from (15.7) with (15.9) and (15.10) gives

$$\cos\theta = \frac{s(k-1)}{2c_0\tau} - \frac{c_0\tau\,(k+1)}{2s} \tag{15.11}$$

Equations (15.7) and (15.11) provide the distance of the source from hydrophone H_1, and the angle of the source direction relative to the line through the hydrophones at the closer hydrophone, respectively. There is the usual ambiguity. Rotation of the plane containing Q, H_1, and H_2 about the x axis (Fig. 15.1) does not change the above results. Thus $Q(x, y, z)$ lies on the circumference of this circle, centered at $(x, 0, 0)$ with radius $\sqrt{y^2 + z^2}$. If the differences in depths of the source and the hydrophones are much less than their separations, the angle θ is approximately the azimuthal bearing of the source relative to the line through the hydrophones. It is not possible, however, to determine if θ in Fig. 15.1 is positive or negative (since $\cos\theta = \cos -\theta$) so that the source can be either side of the line through the hydrophones.

Manipulation of (15.10) gives

$$(k^2 - 1)x^2 + (k^2 - 1)y^2 - s^2 + 2sx = 0$$

which leads to $(x - x_0)^2 + y^2 - \rho^2 = 0$ (15.12)

where $x_0 = -s/(k^2 - 1)$ $\rho = sk/(k^2 - 1)$ (15.13)

Equation (15.12) is a circle with radius ρ, and center at $(x_0, 0)$. It is the locus of the point $Q(x, y)$ for fixed values of k and s, and thus DL. The circle lies in the plane containing the points. Since rotation of this plane containing Q, H_1, and H_2 about the x axis does not change the above result, the three dimensional locus of the source position, $Q(x, y, z)$, is the surface of the sphere swept out by rotation of the circle about the line H_1 and H_2. Thus from the difference in received level we determine that the source lies on the surface of a sphere of radius ρ centered at the point $(x_0, 0, 0)$ so that the position of the source is on the intersection of this sphere and the surface swept out by rotation of the hyperbola determined from the arrival time difference.

In the plane containing the source and the two hydrophones, the source must lie on the circle determined from difference in received levels and the hyperbola determined by the difference in the arrival times, its position being one of the two points of intersection of the circle and the hyperbola. Fig. 15.4 shows the circles for various values of DL (for square law propagation) and the hyperbolas for various values of τ.

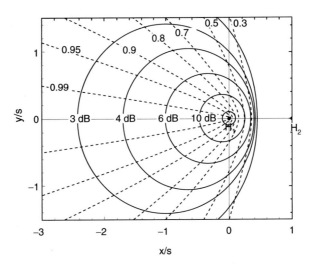

Fig. 15.4. Localization of a source as the intersection of the circle determined from the difference in levels and the hyperbola determined from the difference in the times of arrival between the two hydrophones at H_1 and H_2, separated by distance s. There are two points of intersection due to the ambiguity, Cato et al., (1998).

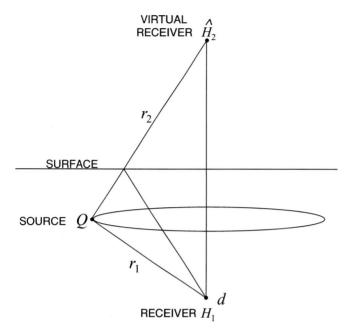

Fig. 15.5. Localization in the vertical plane using one hydrophone and a virtual hydrophone provided by the surface image of reflected sound from the source (Cato *et al.*, 1998).

Use of difference in level requires accurate measurement of the received levels so that the hydrophones need to be properly calibrated. The effectiveness depends on how well the propagation loss is known (see Cato (1998) for further discussion of the significance of these factors and the errors involved).

Use of the surface reflection to provide an additional virtual hydrophone
When the source to receiver distance is less than the water depth, the surface reflected path provides a surface image which may be used as a virtual hydrophone if the direct, surface reflected and bottom reflected arrivals are separated (Fig. 15.5). Two horizontally separated hydrophones and their surface images provide effectively four hydrophones and the arrival time differences to pairs can be used to localize a source.

Vertical localization is possible with one hydrophone, using the arrival time difference and the difference in received levels by the direct and surface paths. Reflection from the sea surface occurs with negligible loss for rms wave heights much less than an acoustic wave length (Section 9.2), a situation that occurs with many biological sounds in all but very rough weather. Hence (15.10) and (15.11) can be used with the geometry of Fig. 15.5 replacing that of Fig. 15.1. If the depth of the hydrophone is known, the source can be localized vertically to the horizontal circle shown in Fig. 15.5.

15.4 Application to tracking migrating humpback whales

Humpback whales are medium-sized mysticete whales found in all the oceans of the world. During summer, humpbacks feed in high latitude areas, but in autumn, start long migrations to tropical breeding grounds where they calve and mate over winter and spring. During the breeding season, male humpback whales produce long complex vocalizations known as "songs" because of their repetitive structure (Payne and McVay, 1971). Each song consists of several "themes," each theme being a string of similar repeated "phrases." Each phrase is a sequence of sounds or "units."

An individual male may sing continuously for several hours, producing a variety of sounds such as groans, moans, grunts, roars, trills, yaps, violin-like sounds, bellows, and squeaks. Most energy is in the frequency range 50–2500 Hz and source levels range from 175–188 dB re 1 µPa at 1 m (Richardson *et al.*, 1995). This continuous stream of high level sounds makes singers excellent subjects for passive acoustic tracking for ranges up to tens of kilometers. Frankel *et al.* (1995) used three moored buoys each with bottom-mounted hydrophones to track singing whales off Hawaii. Noad and Cato (2002) and Noad (2002) used a similar set up to track singing whales off Peregian Beach on the east coast of Australia. Both used the arrival time differences on pairs of hydrophones (Section 15.3.2). The acoustic tracking, in conjunction with simultaneous land-based visual tracking of singers and non-singers using a theodolite, allowed studies of behavior in relation to singing, calibration of acoustic counts against visual counts of whales for survey purposes, and estimation of swimming speeds of singers.

15.4.1 Example of tracking a singing humpback whale

This section describes the results of tracking a singer using arrival time differences on three hydrophone pairs, as in Section 15.3.2. Three hydrophone buoys were moored approximately 750 m apart and parallel to the coast in 20 m water depth off the east coast of Australia. Humpback whales pass close to this coast during the annual migrations between the Antarctic feeding grounds and the breeding grounds within the Great Barrier Reef. The hydrophone on each buoy was firmly fixed to a mooring which ensured it did not move while the buoy was free to swing around the mooring with the wind and currents. Each buoy transmitted the signals from the hydrophones by radio to a nearby shore station, where they were recorded on tape. For analysis, samples of the signals were captured by a stereo soundcard in a desktop computer from each of the three pairs of buoy hydrophones.

An example of a humpback whale sound recorded on one pair of hydrophones is shown in Fig. 15.6 as spectrograms and wave forms. The different arrival times of the sound at the two hydrophones are evident. The cross-correlation function of the signals at the two hydrophones, calculated using a routine written in MATLAB (Mathworks Inc.), is also shown as a function of the sample number. MATLAB calculates the cross correlation by shifting one sample record relative to the other in steps equal to the sample interval, starting with the last sample of the first record aligned with the first sample of the second record, and ending with the first sample of first record aligned with the last of the second record. Thus correlation of two records of 4000 samples each produces a result that is 8000 samples long, the two records being aligned at sample number 4000.

Cross correlation was performed for each buoy-pair to estimate the arrival time differences between the hydrophones, and the MATLAB routine then calculated the hyperbolas for the three pairs, and the inter-sections of the hyperbolas were found iteratively. In practice, the three hyperbolas usually intersected at three points rather than one and the position of the whale was taken as the geometric center of the triangle formed by the intersecting points. Ambiguity was usually not a problem in this experiment as most whales passed seaward of the hydrophones which were close to the coast, so the ambiguity usually resulted in a solution inshore – impossible for a swimming whale! On the other hand, having the three hydrophones parallel to the coast meant that for much of the time, the whales were broadside to the arrays where the localization had the greatest accuracy.

The track of one singer over 7 h, determined by this method is shown in Fig. 15.7. Also shown is the track of the same whale determined from the theodolite fixes from Emu Mt. The theodolite gives the bearing of the whale from the horizontal angle measured from a reference bearing, and the range is calculated from the vertical angle to the whale from the horizon. The visual positions were taken when the whale surfaced whereas the acoustic positions were taken when submerged (the received sound amplitude decreases as the source approaches the surface making it difficult to detect acoustically). Hence acoustic and visual positions do not coincide though they are expected to follow similar tracks, as in Fig. 15.7. Singers surface on average about once every 10 min, but are detectable from their sounds every few seconds, so acoustics provides many more opportunities for localization.

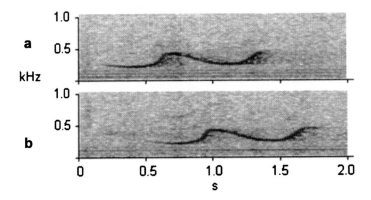

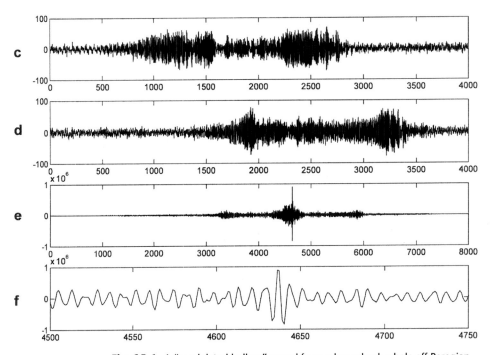

Fig. 15.6. A "modulated bellow" sound from a humpback whale off Peregian Beach. (*a*) and (*b*): spectrograms on hydrophones B and C (8 kHz sampling rate, 512 point FFT). (*c*) and (*d*): waveforms corresponding to (*a*) and (*b*). (*e*): full cross-correlation function, (*f*): expanded view of the cross-correlation function around the peak. For (*c*) – (*f*), the signals were down-sampled to 2 kHz, 8 bit (each sound signal 2 s in duration) to reduce computation time. The maximum cross-correlation value occurs at sample value 4636, which is 636 samples from the mid-point. The signal therefore arrived at hydrophone B 318 ms before hydrophone C.

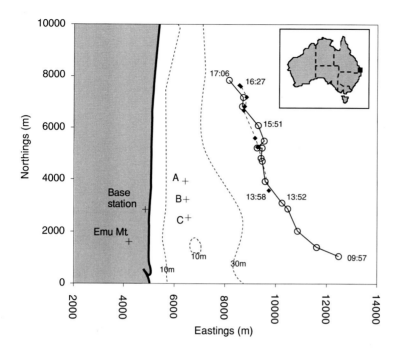

Fig. 15.7. Acoustical and visual tracks of a singing humpback whale during part of its northward migration along the east coast of Australia. Acoustical positions are circles while visual positions are diamonds. Acoustic data were received by hydrophones at A, B, and C and transmitted to the base station by radio, while the visual data were collected using a theodolite on Emu Mt. (73 m).

Accurate localization by arrival time differences requires accurate determination of the hydrophone positions and calibration of the system. For sources at distances significantly greater than the hydrophone separation, the asymptotes of the hyperbolas intersect at very small angles, so that small errors in bearing cause proportionally larger errors in estimate of distance. At Peregian, the position of each hydrophone was accurately measured using two theodolites on the beach, with a diver holding a surveyor's staff and prism vertically above the hydrophone. The accuracy of localization was checked by comparing visual and acoustics locations of a source, such as the boat or imploding light bulbs, with the position determined by GPS. The array at Peregian, with hydrophones spaced approximately 750 m apart, suffered mean range errors increasing from approximately 5% of range at 2 km to 10% at 10 km and 18% at 20 km.

15.5 Application of methods using a single hydrophone to movements of fish and whales

15.5.1 Application to fish

Many species of fish are vocal and their sounds are responsible for a substantial part of the ambient noise of the ocean. Fish use vocalizations for a number of purposes (Tavolga, 1964, 1967) providing many opportunities for acoustic localization. The following describes the results of

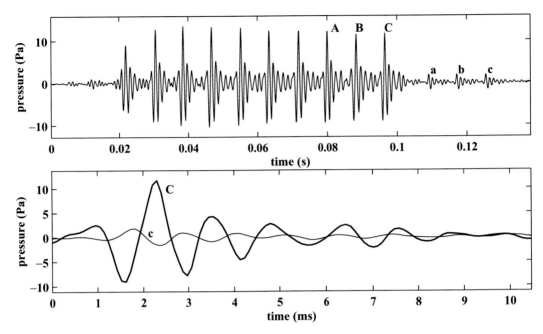

Fig. 15.8. Top: Waveform of the call at a bottomed receiver from a fish *Terapon theraps*. Surface reflections (a, b, c) of the last three direct pulses (A, B, C) are evident. Bottom: The last pulse of the wave form by direct path (C) and surface reflected path (c) with peaks aligned.

localization using one hydrophone and the virtual hydrophone formed by the surface image (Section 15.3.3), i.e., using the differences in arrival times and received levels between the direct and surface reflected paths.

Many recordings have been made of fish sources in Australian waters using single hydrophones located on the sea floor (McCauley and Cato, 2000). An example of a call of the species *Terapon theraps*, a very vocal fish known locally as flagtail trumpeter, is shown in Fig. 15.8. It was recorded in 28 m of water in the Gulf of Carpentaria, on the northern coast of Australia. The pulsed nature of fish calls is clearly evident on this plot. Each pulse represents a single "tug" of the swimbladder, in this fish, by specialized muscles attached to the anterior, dorsal, swimbladder end. The muscles rapidly expand the swimbladder, which is then allowed to oscillate, producing the damped decay seen for each pulse. The pulse rate in this instance was 121 Hz. Three trailing pulses can be seen following the primary pulses, considered to be surface reflections, as highlighted by the lettering of Fig. 15.8. On the lower plot of Fig. 15.8 the direct and appropriate surface reflection are overlaid.

With short impulsive sounds like those of Fig. 15.8, it is more appropriate to measure the integral of the pressure squared over the pulse

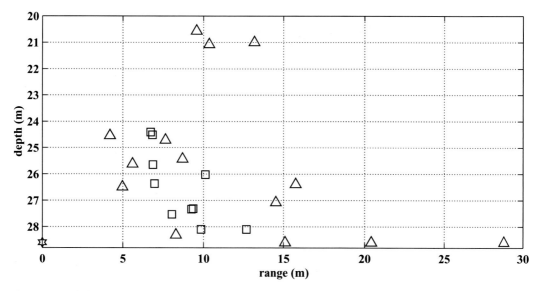

Fig. 15.9. Calculated locations of a calling fish *Terapon theraps* from a single bottomed hydrophone (indicated by the star at zero range). Two call types are differentiated by the different symbols (square and triangle).

duration (which is proportional to the energy flow through unit area) rather than the mean squared pressure, to determine the difference in level between the direct and surface reflected paths. The location of several fishes from this recording set, calculated using this technique is shown on Fig. 15.9 for two types of calls (different symbols) from *Terapon theraps*. It can be seen that calling was concentrated near to the bottom. Using these estimations of range and assuming spherical spreading, the source levels of the 24 calls were calculated to vary from 147–151 dB re 1 μPa (rms) at 1 m. Note that horizontal direction is not determined by this method.

15.5.2 Application to whales

A second example is presented for impulse signals made by blue whales recorded from a bottomed receiver in 450 m water depth off the Western Australian coast. An example of a series of pulses, with most energy between 20–40 Hz, can be seen on the top panel of Fig. 15.10. Since the hydrophone was on the bottom, all arrivals except the direct arrival reached the receiver via a surface reflection. For several pulses within this sequence it was possible to discriminate the first five arrivals and to make use of each reflected path as a virtual hydrophone. The time differences between the direct arrival and each of the first four reflected arrivals were used to determine the hyperbolas for each virtual hydrophone with the real hydrophone. The positions of a point on the direct pulse and

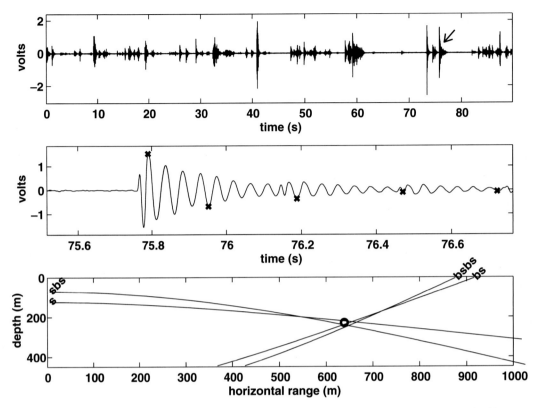

Fig. 15.10. (Top) Time series of several impulsive signals produced by blue whales off the Western Australian coast. (Center) Expanded waveform of the pulse highlighted by the arrow in the top panel. The crosses refer to the same part of the waveform for the following paths: (left to right) the direct, surface reflected, bottom–surface reflected, surface–bottom–surface reflected and the bottom–surface–bottom–surface reflected. (Bottom) The hyperbola of possible solutions for each of the arrival times relative to the direct arrival, with the estimated source position shown by the circle.

the same point on each of the first four reflected pulses are shown by the crosses in the middle panel of Fig. 15.10. The intersections of these hyperbolas then provide an estimate of the source location, as shown in the bottom panel of Fig. 15.10, at a range of 640 m and a depth of 240 m. There were a number of different sources evident in the full recording, with the source of the signal immediately preceding that shown by the arrow estimated to be at 1780 m range and 332 m depth and the pulse at 41 s at 970 m range and 212 m depth. It has been established that the study area from where these recordings were taken is a feeding area for whales and that during daylight, their food source is compacted into layers between 200–400 m depth, correlating with the source location of the impulse signals.

This technique relies on accurate discrimination between the various reflected paths. For long signals or very shallow sources (high in the water column) this discrimination may not be possible, thus the technique is limited to impulse signals which are preferably deep in the water column (hence having a reasonable time delay between their direct and surface arrival). The technique of using one hydrophone, and with virtual hydrophones provided by reflected paths, determines an estimate of range and depth of the source in the vertical plane, but not the direction or bearing of the source.

15.6 The future

More sophisticated techniques of acoustical oceanography are already being used to study behavior and movements of marine animals. Acoustical oceanography may also reveal how the animals themselves use acoustics to learn about their environment and their fellow animals. Environmental conditions cause sound signals to undergo a number of changes that add information about the environment and about the location and movement of the source. Acoustical oceanography is the study of this information and what it can tell us about the sources and their environment. Do marine animals also exploit this information? Probably they do, and probably they do it better than we do. Animals evolve to exploit any information they can sense that is useful to them. Through countless generations of trial and error, evolution would have favored the survival of those individuals that, by accident, turned out to have enhanced response to the acoustical information that gave them an advantage in finding food or mates. Their response is instinctive. Our study of acoustical oceanography can, however, show what is possible in the way that animals might exploit acoustical information in their environment. It will also provide the tools to test this experimentally.

Chapter 16

The acoustical causes of collisions between marine mammals and vessels

JOSEPH E. BLUE
EDMUND R. GERSTEIN
Leviathan Legacy, Inc., Florida

Summary

Whales and other marine mammals are vulnerable to boat, barge, and ship collisions. Though more commonly identified and reported in coastal areas, collisions are not restricted to shipping lanes or shallow water areas. A common denominator is that they all occur near the surface where the acoustical laws of reflection and propagation can significantly limit the ability of marine mammals to hear and locate the sounds of approaching vessels. Of major concern and motivation for this chapter are ship collisions with the North Atlantic right whales, as they may be on their way to extinction. There are only approximately 350 of them surviving. Knowlton *et al.* (2001) indicate that about 35% of right whale deaths are caused by large ships.

To address the problems of collisions between watercraft and marine mammals, one must understand their psycho-acoustical hearing characteristics, the acoustical characteristics of their habitats, watercraft noise, and near-surface propagation. This chapter covers possible acoustical reasons for collisions between watercraft and marine mammals and provides some guidance to researchers in the field who are not well schooled in acoustics.

Analysis of some acoustical causes of these collisions shows that the following factors contribute to whales' acoustical difficulties of hearing and locating approaching ships: (i) masking of ship noise by ambient noise including that from other ships emitting more intense noise, (ii) acoustical shadowing ahead of ships that have their propellers mounted above keel level, (iii) Lloyd's Mirror effect (see Section 1.5.6) that lessens ship noises heard at low frequencies near the surface, (iv) spreading loss from the noise dominant propellers to the bow of ships (Ross, 1976) and (v) downward refraction (Section 2.3). The confluence of these acoustical factors poses significant ecological challenges with respect to the whale's detection and localization of approaching ships. This chapter discusses these factors and illustrates their effects with data from controlled vessel passages recorded with vertical hydrophone arrays.

Contents

16.1 Introduction

With increased commerce and international shipping, vessel collisions with marine mammals have become a global concern. A variety of whales, primarily the mysticeti, baleen whales which include gray, sei, humpback, fin, blue, and right whales are hit by ships in the open sea and in coastal shipping lanes (Fig. 16.1).

Fig. 16.1. North Atlantic right whale killed by ship collision (Center for Coastal Studies, MA).

The underlying acoustical causes of collisions had not received the attention of conservation biologists and regulators until recently. While the cumulative effects of increased traffic and noise on the behavior and physiology of marine mammals is difficult to quantify, the direct effect of sound intensity on hearing is measurable and is documented for many vertebrate species (Fay, 1988). Auditory masking is one of the more thoroughly studied psycho-acoustical phenomena across taxa. Masking is a perceptual phenomenon that occurs when the audibility of one sound is decreased by the presence or occurrence of another sound. The presence of a competing background noise or masker measurably affects the hearing threshold for a sound or signal within the frequency band of the background masker. For instance, having a conversation in a quiet room where the ambient noise levels are low is quite different from trying to have the same conversation standing next to a busy freeway. The background noise from the freeway obscures or masks the words of the speaker necessitating the speaker to shout above the noise. The resulting increased energy under the masked condition can be charac-terized by the critical ratio (CR) (Fletcher & Munson, 1937). CR is the level of sound above a signal-to-noise ratio (SNR) of 0 dB required for an animal to hear a tone. It is derived by subtracting the masking noise in dB from the masked threshold in dB. For example, when a 1.6 kHz signal is presented against a 90 dB (re 1 µPa) one-third octave back-ground noise, a West Indian manatee requires a signal intensity of a least 114 dB to be able to detect the signal (Gerstein, 1997). The resulting CR at 1.6 kHz is 24 dB. The CR is conserved, so that at a higher ambient noise level of 100 dB re 1 µPa, the same signal would need to be 124 dB re 1 µPa before the manatee could hear it. This conserved relationship is important as high ambient conditions can conceivably push masked hearing thresh-olds above the received or even the actual source levels of approaching vessels (Gerstein, 2002). CRs have been measured for the West Indian manatees several pinnipeds, and the Atlantic bottlenose dolphin.

Although we have not directly measured the CR for any of the great whales, CR ranging from 10 to 60 dB across frequency bands for a diverse sample of mammals are reported in Fay (1988). A conserva-tive prediction that these whales require at least 13 dB above ambient conditions to hear ship noise is an arguable assumption. It is important to recognize that, whatever the CR values, it significantly affects the ability of whales to hear and locate the sounds of approaching ships. As underwater ambient conditions near the surface and in coastal areas are dramatically influenced by climatic, biological, and anthropogenic sources, these environments tend to be more dynamic and noisier than deep water conditions.

Determining perceptual hearing abilities, in particular masked detection and directional hearing abilities of the great whales, would be of great interest. However, even without this information, we can still argue that, if received sound pressure levels of ship noise fall near or below the prevailing ambient levels, whales would have difficulty detecting and locating these sounds. Direct physical measurements of ship noise spectra, intensity, and propagation can be used to estimate acoustical masking of biologically significant sounds, as well as the noise of approaching ships.

In the next sections basic acoustical theory on probable causes of ship collisions with whales is presented. The theory is not developed in rigorous detail, as much of it can be found elsewhere in this book or in basic textbooks on acoustics. As summarized above, independent of auditory or behavioral considerations, there are at least five measurable physical acoustical factors that directly affect any animal's abilities to detect and localize the sounds of approaching ships. In the following sections, we present brief descriptions of the physical factors that affect the acoustic intensity of sounds from a ship's propeller that make their detection by marine animals difficult. Noise measurements from watercraft are then presented to illustrate the effect of these factors.

16.2 Downward refraction

The student is assumed to have a basic knowledge of the laws of reflection and refraction of acoustical waves. These properties are addressed in most elementary texts on acoustics and optics, and are discussed in Chapters 1 and 2 of this book. One factor that can affect the sound pressure level near the surface is the downward refraction of sound rays caused by negative temperature gradients that become well defined in the seasonal thermocline during the summer and fall. With respect to ship noise, not all rays from a ship are horizontal. Wave radiation is generally spherical. While the horizontal ray is diffracted downward, the typical upward grazing ray at the surface will not be diffracted as deep. The refraction of the grazing ray is dependent upon the depth of the propellers, which determines the upper limit of spherically spreading rays. Unless the negative sound speed gradient is very steep, downward refraction is not a significant factor in ship strikes. As will soon become evident, other near surface effects severely reduce sound pressure levels at shorter distances before downward refraction is even a factor.

16.3 Spherical spreading from propellers

Laist *et al.* (2001) showed that most ships that hit whales are greater than 80 meters long. Ross (1976) showed that the ship's propellers are the primary noise sources on ships. At speeds of a few knots and above, the propellers become the acoustical center of ship noise. In propagating propeller noise from the stern to the bow of a 100 m long ship, noise level spreading results in a 40 dB loss. In noisy habitats and busy shipping lanes, losses from spreading might reduce propeller noise at the bow to levels below the CR necessary for detection. Spherical and cylindrical spreading are covered in Chapter 1 of this book, but, as we shall see, spreading is not the only, nor is it the most significant, contributor to the ship strike problem.

16.4 Acoustical shadowing (bow null effect)

Acoustical shadowing in front of ships occurs when propellers are located above keel depth. The majority of ships that kill whales are > 80 meters in length and have propellers located above their keel depths. The configuration helps drive these large vessels, as well as smaller tugs, in line with their centers of mass and protects propellers from damage in case a keel hits bottom. Ship noise, with wavelengths less than ship stern dimensions, are reflected from the stern. Frequencies with wavelengths larger than stern dimensions can diffract around the ship's hull. As discussed later in this chapter, the amplitude of the sounds at these low frequencies are lessened by the Lloyd's Mirror effect. Figure 16.2

Fig. 16.2. Simplified sketch of acoustic diffraction and shadowing of propellor noise around the hull. Dominant frequencies are indicated.

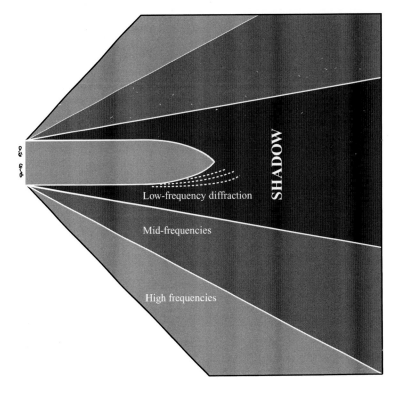

depicts acoustical shadowing caused by propellers being located above the ship's keel level.

The shadow zone is shown as being sharp, but it is really not. A suspended hydrophone recording sound at a few meters deep near the ship's path with the ship going past it will record low frequency sounds of the ship. The frequency content of the propeller noise will increase as the ship passes until, at some point, the geometrical diffraction limit is considered reached. The acoustical shadowing has been called the bow null effect by some researchers. Data illustrating acoustical shadowing is given later in this chapter.

16.5 Lloyd's Mirror effect

Lloyd's Mirror effect, interference between source and image, is covered in Section 1.5.6 and Fig. 1.10. The interference effects are strongest in the frequency range 30 to 3000 Hz. Above these frequencies, roughness of the sea surface often tends to wipe out the coherence between source and image. At very low frequencies, the nearly 180-degree phase shift between the direct and image sources nearly cancel each other leaving very little detectable low frequency sound near the surface. While near the surface, marine mammals exploit and may rely more on higher frequency hearing sensitivity to communicate and interrogate their environment. As marine mammals must return to the surface to breath, the majority of their lives are spent near the surface and Lloyd's Mirror effect may have been partially responsible for shaping their audiograms (Gerstein *et al.*, 1999). Lloyd's Mirror effect on ship noise propagation, and the resulting auditory detection challenges for marine mammals, is clearly shown or inferred in some of the data presented later in this chapter.

16.6 Ship noise and masking

A leading cause of collisions of marine mammals and watercraft is the masking of low noise vessels by the sounds from higher noise level vessels. Proposed regulations to mitigate collisions between marine mammals and watercraft intuitively focuses on reducing vessel speed and is based, in part, on the belief that marine mammals have more time to get out of the way of slow vessels than they do faster ones. That is true when detection is by visual detection of observers on board vessels. The opposite is shown here to be true when detection is made underwater. *A vessel traveling fast is so much noisier than the same vessel traveling slow that it can be detected farther away by marine mammals underwater.* The additional range of detection over that of the slower vessel is sufficient to allow more time for marine mammals to avoid the faster vessel than the slower one.

Noise from jets and propellers has been studied extensively for aircraft (Richards and Mead, 1968). Lighthill's theory of aerodynamic sound is the basis of many of these studies. The fact that these studies were for aircraft instead of for vessels in water makes no difference as both air and water are fluids. The derivation for the intensity of sound radiated by a propeller (assuming that propeller noise is omnidirectional) yields the result that intensity, I, is proportional to the tip velocity, U of the propeller to the sixth power.

$$I \propto U^6 \qquad (16.1)$$

For those who are more familiar with jet noise than propeller noise, jet noise intensity is proportional to the speed, V of the aircraft to the eighth power. The difference in jet and propeller noise arises because jet noise is generated largely by turbulence but propeller noise is generated by vortex shedding and propeller slap on the vortices.

Ship speed, V, and propeller tip speed, U, can be expressed as

$$V = U + \delta_1(U) \qquad (16.2)$$

where $\delta_1(U)$ is a term (not necessarily small) that accommodates for the increased requirement on tip velocity to account for the ship's load and its environmental parameters.

The additional load on the propeller increases with speed and makes the acoustic intensity nearer to the seventh power of speed. For our purposes, we shall deal with the tip velocity to show that marine mammals, swimming underwater, detect fast vessels at farther distances and longer times than identical slower vessels. The source level (SL)[*] of a ship's propeller is

$$SL = 10 \log I = 20 \log P \propto 60 \log U \qquad (16.3)$$

In order to show that the same vessel at fast speeds can be detected by the mammal at *longer* times in advance of arrival, the animal's CR is assumed to be satisfied at slow speed, at time T_S. The same vessel, moving at faster tip velocities, can be detected at time T_F. Since the SL of a propeller is proportional to its tip velocity to the sixth power, the difference in source levels for the same ship at two different tip velocities (U_F and U_S) is

$$\Delta SL = 10 \log I_F / I_S = 10 \log (U_F)^6 / (U_S)^6 = 60 \log (U_F / U_S) \qquad (16.4)$$

---

[*] See discussion of dB references in Section 2.5.

Considering the losses for spherical and cylindrical spreading, and leaving out the references, engineers write:

$$\text{SPL}_{\text{sph}} = \text{SL} - 20 \log R \tag{16.5}$$
$$\text{SPL}_{\text{cyl}} = \text{SL} - 10 \log R \tag{16.6}$$

where SL denotes source level and R denotes range. Spherical spreading is typically used for deep water and cylindrical for shallow. An understanding of the geometry of source and receiver depths and ranges is necessary in making the correct choice rather than just blindly applying these equations, because many multipath arrivals (>10) are necessary for cylindrical spreading. The sound pressure levels (SPLs) necessary for detection are assumed to be the same for slow and fast boats. The ratio of ranges where this occurs for spherical spreading can be calculated from:

$$\text{SL}_\text{S} - 20 \log R_\text{S} = \text{SL}_\text{F} - 20 \log R_\text{F} \tag{16.7}$$
$$\Delta\text{SL} = \text{SL}_\text{F} - \text{SL}_\text{S} = 20 \log R_\text{F}/R_\text{S} = 60 \log (U_\text{F}/U_\text{S}). \tag{16.8}$$

Since

$$R_\text{F}/R_\text{S} = (3/2)\,(U_\text{F}/U_\text{S}) \tag{16.9}$$

the time to collision:

$$T = R/V = R/(U + \delta(U)) \tag{16.10}$$

Let $T_\text{S} = R_\text{S}/U_\text{S}$ and define $\delta(U_\text{S}) = 0$ for the slow speed. Substituting $R_\text{S} = T_\text{S}U_\text{S}$ and $R_\text{F} = T_\text{F}U_\text{F}$ into (16.8) one obtains:

$$20 \log(T_\text{F}U_\text{F}/T_\text{S}U_\text{S}) \propto 60 \log U_\text{F}/U_\text{S} \tag{16.11}$$

Define $T_\text{F}/T_\text{S} = u$ (relative time to collision) and $U_\text{F}/U_\text{S} = w$ (ratio of tip velocities). Solving (16.11) for u:

$$u_{\text{sph}} = 10^{3 \log w} \tag{16.12}$$
$$u_{\text{cyl}} = 10^{6 \log w} \tag{16.13}$$

Equations (16.12) and (16.13) show that marine mammals have more time to avoid a fast vessel than the same vessel at a slower speed for the case when the animals can detect the vessel at both speeds.

These equations are plotted in Fig. 16.3. For these curves, the same vessel is assumed. From the spherical spreading curve, one finds that the same vessel going twice the speed allows a whale eight times the "time to collision". For cylindrical spreading, the time ratio is 64 for doubling the speed. The graph breaks down when the noise from the slower vessel falls below the ambient noise, but the slower vessel still cannot be detected because of masking. For vessels that emit omnidirectional sound, the masking area can be quite large as shown in Table 16.1.

Table 16.1 *Relative masking areas*

Relative speed (U_F/U_S)	Relative masking range R_F/R_S	Relative masking area $(R_F/R_S)^2$
1	1	1
2	64	4096
3	729	531 441

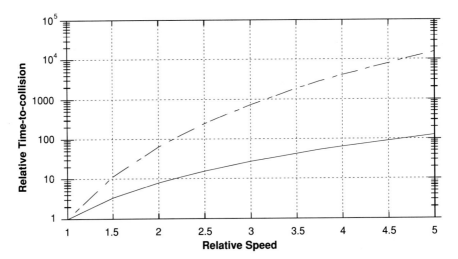

Fig. 16.3. Relative time to collision with a whale for a vessel at higher speeds compared to a slower vessel. Solid line for spherical spreading and dotted for cylindrical.

16.7 Data acquisition and analysis

The acoustical factors that may contribute to collisions between marine mammals and watercraft are not easily separable. However, their influences are observable in the time series obtained from hydrophone array recordings of watercraft passing near the suspended vertical array. A large cruise ship, the Fantasy, was the noise source. Figure 16.4 shows the results of spectral analyses of segments of the hydrophone outputs from a range of 173 m at 1.5 m and 7.5 m depths. The difference in the low frequency ends of the spectra at the two depths is clearly seen and is 30 to 40 dB below at 100 Hz. The effect of shadowing is seen as a rapid drop in spectral amplitude with increasing frequency as the ship's stern dimensions become significantly large compared with the wavelength of the propeller

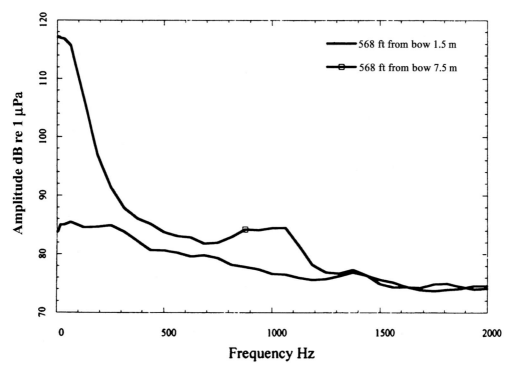

Fig. 16.4. Spectra from hydrophone recording taken 173 m ahead of the ship's bow at hydrophone depths of 1.5 and 7.5 m.

noise. The spreading loss from the propellers to the measurement point (434 m) is 52.7 dB. The confluence of all of these acoustical factors is quite large (> 100 dB), but the components cannot be measured independently.

Figure 16.5 shows the ship noise spectra at several distances as the ship approaches and the propeller comes even with the hydrophone that is suspended at 1.5 m depth. Note that the sound pressure levels do not rise significantly above ambient levels until the ship's propellers are even with the hydrophone. These data illustrate acoustical shadowing when the propellers of a large ship are above the ship's keel level. While the size and geometry of the shadows ahead of ships may vary, propeller noise is more intense aft of the stern and off the port and starboard sides. Observations of whales surfacing in front of ships and the high incidence of strikes on Northern Atlantic right whale females and calves suggests the whales, unaware of ships in their direct path, may seek refuge by actively swimming or surfacing into the quieter zones directly in front of ships. Once there, hydrodynamic forces can sweep adults and especially calves into the propellers. The high incidence of calf strikes and scarred

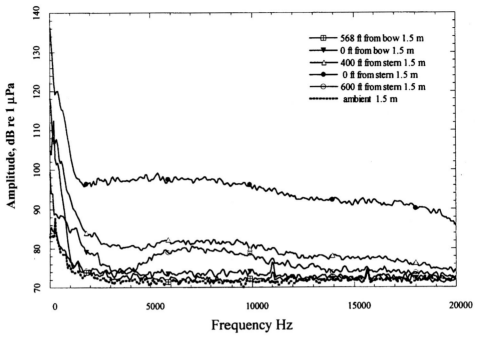

Fig. 16.5. Spectra from hydrophone recording taken 173 m at hydrophone depth of 1.5 m at various ranges, and locations.

females exceeds the normal probability of chance encounters and suggests that the whales may be active participants in the phenomenon seeking safety near the surface, in front of approaching ships. The data clearly demonstrates that acoustical shadows and near-surface propagation seriously challenge the whale's ability to detect and locate ships. A directional acoustical source, attached to the bow of a ship could selectively "fill in" the shadows to alert whales of an approaching ship.

Data from large ships at several speeds is difficult to get because of the expense of operating the vehicle (Laist *et al.*, 2001). The data on the Fantasy was obtained as it left port. Data taken during World War II (Fig. 5.15) shows the source spectral levels at various speeds. Ross (1976) states that propeller noise predominates after speeds of only a few knots as shown in Fig. 5.15. These data were taken with a bottom-mounted hydrophone instead of a near-surface one where Lloyd's Mirror effect would decrease the low-frequency components of the ship's noise.

To demonstrate the effect of watercraft speed on noise intensity and time-to-collision with marine mammals in their path, motorboat noise was used. Figure 16.6 shows the spectral results at 3 mph (1.34 m/s) and 24 mph (10.7 m/s). Also shown are the ambient noise spectrum, the manatee audiogram and the minimum critical ratios for broadband

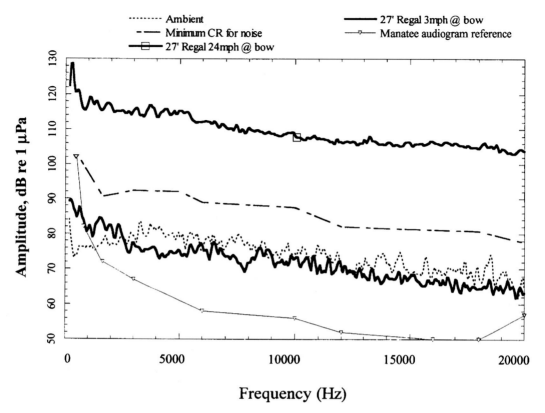

Fig. 16.6. Noise spectra from an 8.23 m "Regal" motorboat at speeds of 1.34 and 10.7 m/s (3 mph and 24 mph).

noise. From this plot, one sees that the boat moving at the slower speed is barely detectable while the faster boat is readily detectable.

The difference in spectral levels for the two speeds is ≥ 40 dB. The theoretical difference for sound pressure levels is 54.2 dB based on a tip velocity ratio of 8. The difference could be attributable to the extra tip velocity needed to bring the motorboat up on a plane where power requirements drop. Such a difference is not expected for vessels that do not "plane". The additional 40 dB sound pressure level from the faster speed results in an additional range of 165 m. At 10.7 m/s, the additional range gives a marine mammal 15.4 seconds to detect the boat and avoid it.

16.8 Discussion

Blanket reduction of vessel speeds without compensating for the associated acoustical consequences and challenges of near-surface propagation may be counter-productive to the protection of whales as well as for another marine mammal the West Indian manatee. In many instances slowing vessels can actually increase collision risks by reducing the audibility of the approaching vessels. This increases the probability of masking, while extending vessel transect times and the subsequent opportunities for collisions. After years of speed regulations to protect the West Indian manatee in Florida, the number of mortalities and non-lethal collisions has increased to record highs. Since propellers are the only sources that contribute substantially to vessel noise, reducing tip rotation and cavitation results in lower acoustic levels. In naturally occurring ambient conditions, the sounds of these quieter vessels are effectively masked and fall below the manatee's CR for detection. Masking of slower vessels by naturally occurring ambient noise, as well as by intrusive sounds from faster vessels in the area, presents a serious problem for manatees. This masking is responsible for the multiple scar patterns on the backs and tails (up to 50 different strikes) from repetitive encounters with different slow moving boats.

For whales the acoustical challenges are even more serious. For ships longer than 80 meters (the size implicated in most reported strikes) losses from spherical spreading can be greater than 38 dB at the bow. The slower tip rotation results in lower source levels and spectral components that are further attenuated by 20 to 50 dB by the Lloyd's Mirror effect. Add acoustical shadowing to the mix and detection is even more difficult or impossible for whales. Whales at sufficient depth may readily detect approaching ships but then they may seek quiet and/or refuge in the acoustical shadows.

Placing acoustical considerations aside, there may be a variety of behavioral factors which also effect a whale's perception, and subsequent reaction or non-reaction to ship sounds, ranging from preoccupation with feeding, mating, and states of arousal and general habituation to ship noises. Whales preoccupied with sex or feeding may not yield to ships regardless of acoustical conditions or their detection abilities. Such preoccupation and the possibility of habituation may occur when the sounds of ships are not associated with a direct or perceived threat to the animal's safety. The sounds of many distant ships may be audible to whales swimming at sufficient depths and these sounds may be perceived as part of a benign ambient landscape. If a whale has not had a negative experience with a ship, it may not be perceived as a threat. However, Richardson *et al.* (1995) report that whales do consistently exhibit some avoidance reactions to sounds of ships, so there may be some generalized avoidance responses to moving sound sources that are loud enough. The sounds of ships that would actually cause injury are most likely perceived by whales only at the moment of impact or immediately following the incident, so associative learning may not be possible for those animals which manage to survive collisions (Gerstein, 1944; Turhune, 1999). There are a plethora of different ship noise signatures. So even if a particular ship sound is recognized as a past threat, a whale may not generalize or associate the threat to a different ship's signature. Cummings (1985) suggests that playbacks of orca sounds recorded in the North Pacific were not perceived as a threat to southern right whales because the recorded calls differed from orcas in the region. The inconsistency of ship and boat signatures and the possible habituation of omnidirectional vessel noise are problems for both manatees and whales. To negate habituation, a consistent modulated sound could be projected directionally from the bows of vessels so that they are detected only by the animals in the path of an approaching ship. Such a device in conjunction with slow moving vessels could provide whales with directional cues. However, without such a device, whales near the surface have no consistent detectable acoustical reference with which to associate the threats of approaching ships.

16.9 The future

Collision mitigation needs to be applied as soon as possible for large ships in areas where North Atlantic right whales are found. As ship strikes account for up to 35 % of their known mortality and it is imperative to reduce this threat immediately. In particular, we need to protect females and calves which congregate in calving areas off the coast of Florida and southeast Georgia. At their current rate of decline the survival of just two additional females per year may be vital to the survival of the species.

The problem of acoustical shadowing needs prompt attention for many marine mammals, inclusive of right whales and manatees. Acoustical shadows pose deceptive and circumstances which may lure marine mammals into the paths of approaching vessels. Large ships that are killing whales, and slowly moving barges and tugs that kill manatees can be provided with acoustical arrays that selectively "fill in" the shadows so these marine mammals can detect vessels they now cannot hear. The arrays need to be highly directional to negate habituation and minimize the potential of additive noise problems. What sounds work best for the different marine mammals will require continued exploration and monitoring. However, since whales and manatees already associate the loud sounds of distant fast ships and boats with danger, as observed by their avoidance behavior, modulated ship noise projected forward from bows should be utilized as an immediate "stopgap" mitigation before more animals die. The association of modulated ship noise projections with approaching danger would be strengthened and resistant to habituation by using a consistent directional projection, detectable only to marine mammals in the direct path of approaching ships.

Marine mammal researchers have only begun to seriously consider the acoustical reasons for collisions between watercraft and marine mammals. There is a need for team efforts between marine mammal researchers and underwater acousticians to better address the collision problem. Much work remains to be done on the topic of masking, as that seems to be an area that can lead to a better understanding of the effect of vessel speed and noise intensity. Most of the data available on sound intensity, ship speed, and propellers is World War II vintage. New propeller designs have since been implemented that increase drive efficiency and lessen propeller noise. The effect of these designs on masking and overall acoustical detection by whales also needs to be evaluated. Greater understanding of acoustical factors is needed by regulators and managers who are responsible for the rules of waterways intended to mitigate collisions between vessels and marine mammals.

Chapter 17
Whale monitoring

CHING-SANG CHIU
CHRISTOPHER W. MILLER
Naval Postgraduate School, Monterey, CA

Summary

In this chapter, an acoustical technique to automatically detect, classify, and count the vocalizations of whales is described. The technique utilizes a matched filter to simultaneously detect and identify the call that is being censused as it scans through the data stream of a hydrophone. The technique is exemplified for the census of blue whale calls in the Monterey Bay National Marine Sanctuary (MBNMS), using time series measured by a bottom-mounted, cabled-to-shore receiver operated by the Ocean Acoustic Observatory of the Naval Postgraduate School. In order to achieve high-probability, concurrent detection and classification at a low false-alarm rate, the matched filter demands, for initialization, an accurate specification of the source signal, i.e., the actual waveform transmitted by the whale devoid of multipath interference. It also demands that the transmitted waveform possesses some exploitable characteristics that are somewhat unique and robust. Using blue whale vocalizations as an example, a deverberation scheme for the reconstruction and characterization of the source signals is illustrated.

Contents

17.1 Introduction

"There is inadequate knowledge of the ecology and population dynamics of such marine mammals and of the factors which bear upon their ability to reproduce themselves successfully…" The Marine Mammal Protection Act of 1973 acknowledged that the activities of man were severely impacting marine species, but also admitted that little was known about these populations and much would have to be learned if we were to make an impact in their recovery.

Monitoring whales has become important in recent years. Biologists continue to monitor population densities to determine if conservation efforts of threatened and endangered species have been effective. Increases in commercial ship traffic have caused more ship-strike incidents, in which whales are injured or killed as a result of impacts with vessels. Increased shipping and boat traffic has also resulted in an increase in the ambient noise levels in the ocean (Andrew *et al.,* 2002), which may affect whale communication. Monitoring is also important to determine where the whales aren't. Seismic exploration, construction, and naval operations are among the activities which transmit sounds into the water at levels that can affect the behavior and health of marine mammals, and it is important to determine if there are animals in the vicinity during these operations.

Whale monitoring has traditionally relied on aircraft and ship-based visual techniques. Drawbacks of these visual techniques include the high cost of ship time, limited visual coverage in both space and time, and poor accuracy. A factor that contributes to the latter is that the long-diving whales may not clearly present themselves during visual surveys and therefore are difficult to track. In recent years, towed hydrophone arrays have been added to the surveying ships, to assist in the sighting and tracking of vocalizing whales.

Acoustical recordings of whales have been studied for years (Walker, 1963). Throughout the cold war, the US Navy installed and maintained a vast network of hydrophone arrays for the purpose of detecting Soviet submarines. These SOund SUrveilance Systems (SOSUS) detected widespread low frequency (17 Hz) signals over much of the world's oceans. Initially this signal was thought to be a low frequency sonar system, however it was soon discovered that these were actually calls from blue (*Balenoperta musculus*) and fin (*Balenoperta physalus*) whales. With the end of the cold war, the Navy decommissioned many of these SOSUS arrays and their data was made available for scientific and educational use. One of these arrays, located at Point Sur, California, was turned over to the Naval Postgraduate School (NPS) to establish an Ocean Acoustic Observatory (OAO).

Shore-based hydrophone arrays offer unique advantages over visual techniques because the ocean is largely transparent to low-frequency sound. Low frequency vocalizations can be heard for long distances, and beamforming techniques can be used to track whales for hundreds of miles. Continuous monitoring of some of the baleen species has been accomplished on a basin scale. Acoustical data combined with visual data would greatly enhance regional-to-global population estimates. Some of the recent research of the subject area can be found in Watkins *et al.* (2000), Clark and Fristrup (1997), McDonald and Fox (1999), Moore *et al.* (1998), Stafford *et al.* (2001), and Chiu *et al.* (2002).

Acoustical monitoring does have limitations, however. Only whales that are vocalizing can be monitored. If vocalizations are periodically recorded on a hydrophone, is there some way to determine if this is one whale, or multiple whales traveling together? There is evidence that in both the fin whale and humpback whale populations, only the males are vocal and the reason for vocalizations is to attract a mate. If this were true, acoustical monitoring of whale populations would provide a conservative (under sampled) count of the whale populations, as the female population would not be represented. High ambient noise levels near the hydrophone (ship traffic, seismic events, etc.) also have the ability to mask (drown out) the whale calls, making them difficult if not impossible to detect. As long as we acknowledge these limitations to the technique, passive detection of whales still provides a remarkable opportunity to study these animals, their behavior, and perhaps some insight into their language.

17.2 Blue whale calls

Blue whales primarily produce low-frequency sounds, which are perhaps some of the best-studied whale sounds (Cummings and Thompson, 1971; Clark, 1994, among many others). The vocalizations of the Northeastern Pacific blue whales generally consist of a sequence of "A" and "B" calls. The A call can be characterized as a train of short, amplitude-modulated pulses with a fundamental carrier frequency at about 18 Hz and a strong fifth harmonic at 90 Hz. The B call is a long ($\sim$12 second) frequency-modulated moan with a fundamental at 17 Hz and a strong third 51 Hz harmonic (Fig. 17.1). The A and B call harmonics provide robust, clear signals that are spectrally separated from each other. For the purpose of this discussion it is these signal harmonics that will be referred to as A and B calls, and used for autodetection investigation and application. We choose not to exploit the fundamental-frequency components of the signals because they are not always present or clearly present in

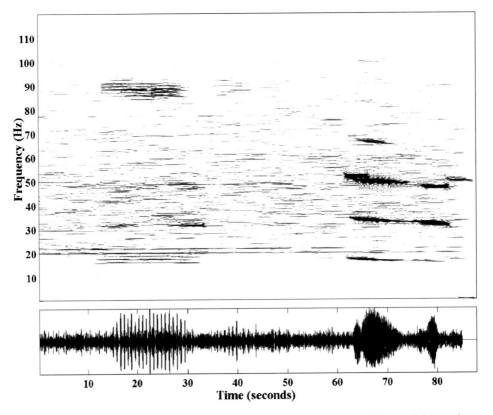

Fig. 17.1. Spectrogram (top) and time series (bottom) of blue whale A and B calls. Pulse train structure of the A call is clearly shown between 15 and 31 s, with the fundamental frequency at 18 Hz and strong harmonic at 90 Hz. The B call structure, between 64 and 82 s, shows the fundamental frequency at 17 Hz, with harmonics at 34 and 51 Hz.

the call data that we have collected. Such dropouts at the fundamental frequencies may be related to a low-frequency propagation cutoff imposed by the water depth and sediment properties where the calls are originated.

The Naval Postgraduate School (NPS) Ocean Acoustic Observatory (OAO) receiver is located at the southern portion of the Monterey Bay National Marine Sanctuary (MBNMS) on the sea floor (Fig. 17.2). The unique locale of this cabled-to-shore receiver, the continuous, real-time nature of the data stream, and the abundance of blue whale signals received, provide a unique opportunity to investigate autodetection and censusing techniques for blue whale vocalizations in the Monterey Bay National Marine Sanctuary. The ability to scan through a large amount of data to time-stamp and count the blue whale calls quickly and accurately is key to realizing long-term, continuous, real-time censusing.

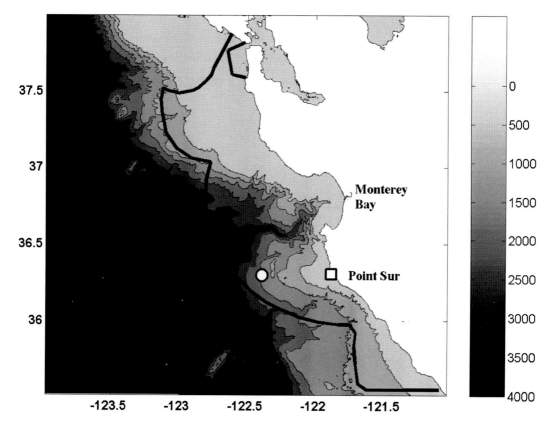

Fig. 17.2. The NPS Ocean Acoustic Observatory (square) located at Point Sur, California. The hydrophone receiver location is also shown (circle) in relation to the Monterey Bay National Marine Sanctuary boundaries (black line). The bar to the right of the figure provides the depth scale (in meters) for the bathymetric contours displayed.

17.3 Detection and classification

There are two elements to consider when designing an automated algorithm to census whale calls in the data stream: detecting the arrival of a signal at the hydrophone, and correctly classifying it (by type, species, etc.). Classification is an important extension of the detection process that provides the quantity of the searched signals and safeguards the quality (accuracy) of the counts. Detection theory has been widely studied: telecommunications, radio, radar and digital networking are among the many applications relying on accurate signal detection.

Detection is the process of making a decision about the presence, or lack, of a signal, $s(t)$, in the received data stream, $r(t)$, which always contains noise, $n(t)$. Figure 17.3 depicts the basic components in an automated detection process. It typically consists of an optimal detector

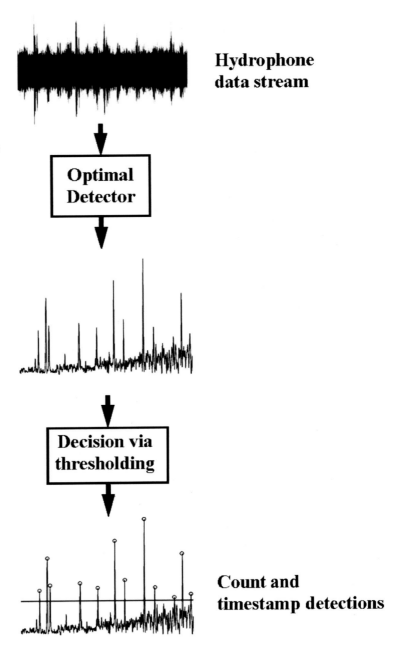

Fig. 17.3. Detection process flow diagram showing the steps from raw data (top), detection filtering (middle) and thresholding steps (bottom). "Detections" are found from picking the peaks of the detector output (circles), which are above the constant threshold line, in the time series.

Hydrophone data stream

Optimal Detector

Decision via thresholding

Count and timestamp detections

a filter that acts to reduce noise relative to a signal, to convert an input data stream into a simple, thresholdable time series. A detection is declared and recorded whenever the output of the detector exceeds a preset threshold value, the choice of which depends on the accuracy requirement. The four possible decisions of a simple detector are:

detecting the signal when the signal is present (correct detection or "hit"); not detecting a signal when the signal is present ("miss"); claiming a detection when no signal is present ("false alarm"); and not detecting a signal when no signal is present (correct null decision). False alarms and misses occur because the signal at the receiver can fluctuate randomly due to unpredicted ocean variability, because the noise at the receiver is also random, or due to poor threshold selection. Ocean variability can cause the received signal level to drop significantly; excessive noise input can cause the detector output to exceed the threshold value; a threshold value too high will cause excessive missed detections and a low false alarm rate, while a low threshold will provide a high detection rate but increased false alarms. When the signal is present; the total possible outcomes are a correct detection or miss. Stated in terms of probabilities, the probability of detection, p(D), plus the probability of a miss is equal to 100%. Similarly, the probability of false alarm, p(FA), plus the probability of a correct null decision is also equal to 100%. Due to the direct relationship of these parameters, the accuracy requirement for a given threshold value is well defined in terms of the minimum acceptable p(D) and the maximum acceptable p(FA). Common approaches to determining p(D) and p(FA) include controlled at-sea playback experiments and realistic computer simulations.

It is impossible to demand the highest possible p(D) and lowest possible p(FA), simultaneously, because both probabilities increase as the threshold is lowered. This can be seen from the bottom pane in Fig. 17.3. As the threshold is lowered, the noise will begin to exceed the threshold value causing false alarms at the detector output. To avoid over-counting the whale calls, i.e., to come up with conservative estimates, it is important to choose a reasonably low p(FA) by tolerating a moderate number of misses. The best combination of acceptable P(D) and P(FA), and hence the best threshold value, are of course application dependent. If you were trying to detect mines in a harbor you would accept many false alarms in order to guarantee that you would achieve a 100% detection rate.

The optimal detector for extracting a known source signal in random noise is a matched filter. The filter is designed to maximize the signal level and minimize the noise level at the output. It makes full use of the known signal characteristics by correlating the received time series, $r(t)$, with a normalized replica of the source signal being searched, $s(t)$. The operation can be expressed, mathematically, as

$$c(\tau) = \int r(t)s(t-\tau)dt \qquad (17.1)$$

where $c(\tau)$ is the output of the detector and τ the scanning (time shift) parameter. By using the signal (or a catalog of signals) of interest in the correlation, the matched filter is also performing signal classification during the detection process. Classification during detection makes this an ideal filter as long as one is working with a limited set of known signals, with robust characteristics. Any variation of the actual transmitted signal from the replica causes degradation in the detector output.

If the data stream were squared and the replica signal were replaced by a simple boxcar function with duration equal to a short integration time, then (17.1) would correspond to an energy detector. The energy detector is optimum for detecting signals of unknown or unspecified characteristics. Thus, the energy detector does not provide any classification of the signal(s) that it detects, but reveals that significant energy has been received. When the energy exceeds the preset threshold, an "event" is declared. This event can be used to: notify an operator that an event is in progress and needs to be reviewed; trigger a data acquisition system to collect the data for future analysis; or initiate a classification process. The classifier could be a simple "table lookup" approach, using a matched filter to compare the received data to a known database of signals; or more sophisticated algorithms based on wavelet analysis or non-linear techniques.

17.4 Source signal characterization

In order to utilize the matched filter for automated detection and/or classification, it is clear from (17.1) that the source signal must first be characterized. Accurate estimation of the source signal can be accomplished with multiple hydrophones (such as in a towed array or an array of distributed sonobuoys), deployed in close proximity to the vocalizing whales in dedicated at-sea experiments. It is important to recognize that the ocean is highly reverberant. The whale calls received at the hydrophones differ from the calls that the whale transmits. The received signal contains many overlapping and interfering copies of the original signal, each arriving from one of the many raypaths (multipaths) connecting the source and receiver. Therefore, deverberation of (i.e., removal of the multipath interferences from) the received signal is required in order to retrieve the characteristics of the actual source signal, such as source levels, duration, waveform structure, and other details of the vocalization including call-to-call variations (robustness).

It is well known that any time series, $g(t)$, can be represented by a weighted sum of sinusoids or Fourier components, $e^{i2\pi ft}$, of various frequencies. Mathematically, this representation is expressed as

$$g(t) = \int_{-\infty}^{+\infty} G(f)e^{i2\pi ft}df \qquad (17.2)$$

where the amplitude $G(f)$ of the sinusoids is called the spectral density function of $g(t)$, and $G(f)$ is calculated from the Fourier transform,

$$G(f) = \int_{-\infty}^{+\infty} g(t)e^{-i2\pi ft}dt \qquad (17.3)$$

A deverberation method applicable to towed-array data is described next utilizing this frequency-domain representation of signal and noise.

The spectral density function $R_p(f)$ of the data time series $r_p(t)$ measured by hydrophone number p is related to the spectral density function $S(f)$ of the source signal $s(t)$ weighted by the source-to-receiver transfer function, i.e., the multipath propagation model, $H(f; \vec{x}_w, \vec{x}_p)$, and contaminated by additive noise $N(f)$:

$$R_p(f) = H(f; \vec{x}_w, \vec{x}_p) S(f) + N(f) \qquad (17.4)$$

where $\vec{x}_w$ and $\vec{x}_p$ are the positions of the whale and the hydrophone, respectively.

In the presence of J eigenrays, $H(f; \vec{x}_w, \vec{x}_p)$ can be modeled, following the development in Chapter 2, as

$$H(f; \vec{x}_w, \vec{x}_p) = \sum_{j=1}^{J} A_j(f)e^{-i2\pi f\tau_j} \qquad (17.5)$$

where τ_j is the wavefront travel time along the jth path, and $A_j(f)$ the relative amplitude that depends on the path length, volume absorption, number of turning points, number and angle of surface/bottom reflections, and surface/bottom reflection coefficients. Since both $\vec{x}_w$ and $S(f)$ are unknown in (17.4), it is clear that the reconstruction of $S(f)$ from the measurements $R_p(f)$ requires that the whale be located first. The uniqueness of the multipath interference, $H(f; \vec{x}_w, \vec{x}_p)$, to the location of the source is exploited in the localization.

With the signal bearing relative to the towed array known from horizontal plane-wave beamforming, the localization method introduced by Parvulescu (1995) can be adopted to estimate range and depth. With an array of multiple hydrophones at known relative positions $\vec{x}_p$, the method first correlates the received signals with a set of modeled transfer functions, each associated with a possible whale location (x, z) on a search grid. Based on the Fourier transform relations (17.2) and (17.3), these time correlations can be calculated in the frequency domain as

$$\phi_p(\tau; x, z) = \int_{-\infty}^{+\infty} R_p^*(f) H(f; x, z, \vec{x}_p) e^{i2\pi f\tau}df \qquad (17.6)$$

where the superscript * denotes complex conjugate. After averaging over the number of hydrophones M, only the largest correlation value for each of the possible locations is extracted to form the so-called ambiguity surface:

$$a(x, z) = \max_{\tau} \left\{ \frac{1}{M} \sum_{p=1}^{M} \phi_p(\tau; x, z) \right\} \tag{17.7}$$

The best location estimate $(\hat{x}, \hat{z})$ is where the ambiguity surface $a(x, z)$ attains its maximum.

With the whale location estimate $(\hat{x}, \hat{z})$, the source signal can then be reconstructed using least-squares estimation. The reconstruction involves the use of a measure of the misfit (cost function) Ψ, defined as the sum of the squares of the differences between the measured spectral densities by each of the hydrophones and the predicted (i.e., modeled) ones:

$$\Psi = \sum_{p=1}^{M} \left[R_p(f) - H(f; \hat{x}, \hat{z}, \vec{x}_p) S(f) \right]^2 \tag{17.8}$$

The value of $S(f)$ that minimizes the misfit between data and model is thus the best estimate of the source spectral density function in a least-squares sense.

As an illustration, ambiguity surfaces associated with three A calls over a nine-minute period are displayed in Fig. 17.4. These calls were recorded on a horizontal array during an experiment to characterize blue whale vocalizations in the Monterey Bay. During these nine minutes, the blue whale appeared to be vocalizing in relatively shallow water and was moving away from the towed array. These A-call ambiguity surfaces show "footprints" on the order of 200 m horizontal by 30 m vertical. It is important to note that the size of the footprint (or localization resolution) is strongly dependent on the bearing of the source relative to the orientation of the towed array. The largest footprint was attained when signals were received broadside (90 degrees) to the array, where the signals arrived at each of the hydrophones close to the same time. The smallest footprint (higher localization resolution) occurred at end-fire, when the arrival times of the signals were spread out in time over the hydrophone array. The ambiguity surfaces shown in Fig. 17.4 are for the cases where the signals were coming in close to broadside. This direction of arrival constituted a less-than-optimum geometry for target localization for which insufficient "independent" spatial information on the target location was distributed across the array. The insufficient spatial information, however, was supplemented adequately by the richer temporal information containing multipath arrivals in the combined space–time (or space–frequency) processing to resolve the source position unambiguously for those cases.

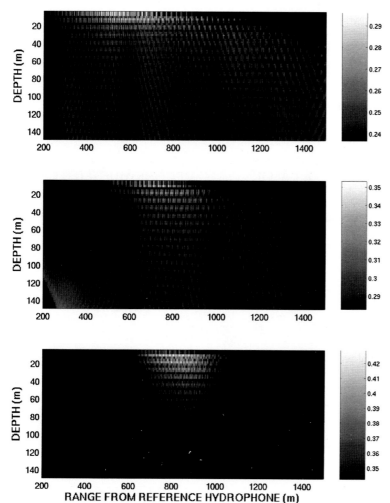

Fig. 17.4. Ambiguity surfaces calculated for three received blue whale "A" calls. These surfaces show the maximum correlation values between the measured and modeled sound fields. The range and depth showing the highest correlation (white) correspond to the best estimate of the whale location. During this 9-minute period, the whale appears to be moving away from the towed array. The gray scale on the right provides the correlation amplitude values.

RANGE FROM REFERENCE HYDROPHONE (m)

Based on a large set of deverberated A and B calls (five each are shown in Fig. 17.5) obtained in the Monterey Bay, some important signal parameters including source level, signal duration, vocalization depth, and call-to-call correlation were estimated. The mean and standard deviation of these estimates are summarized in Table 17.1. The source level and duration of both the A and B calls (made by the same whale or by different whales), and the shallow but variable depths of vocalization are very robust. Looks not withstanding, the B-call waveforms are actually highly correlated whereas the A-call waveforms are not. Strong similarity between the structure-rich A calls, however, are found in the magnitude of the waveforms. Clearly, these correlation results suggest that autodetection matched filters can be built based on signal models resembling

Table 17.1 *Summary of signal characteristics and call-to-call correlation estimates*

	Source level (dB re 1 μPa)	Vocalization depth (meters)	Duration (seconds)	Call-to-call correlation: waveform	Call-to-call correlation: magnitude of waveform
A call, 5th harmonic	158 ± 5	38 ± 36	14 ± 2	0.3 ± 0.1	0.9 ± 0.1
B call, 3rd harmonic	157 ± 6	38 ± 36	10 ± 2	0.7 ± 0.1	0.9 ± 0.1

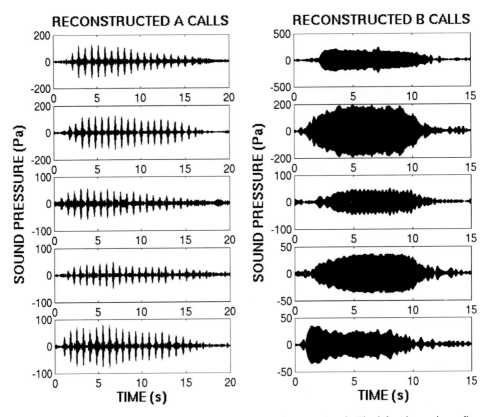

Fig. 17.5. Examples of reconstructed source signals. The left column shows five deverberated A calls in the 85–95 Hz band, whereas the right column shows five deverberated B calls in the 48–53 Hz band.

the mean of the deverberated B-call waveforms and the mean of the magnitude of the deverberated A-call waveforms, respectively. Although the magnitude of the B-call waveform is also robust, the time structure of the B-call magnitude lacks the complexity that can be uniquely discerned from the background noise.

17.5 Blue whale census results

Initialized, respectively, by the A and B call source signal models, the matched filter detector was applied to the available data from the NPS OAO. Figure 17.6 and Table 17.2 summarize the call distribution and totals from the matched filter detector. A variable threshold selection process was used to yield a constant false alarm rate of 0.3%, which yielded detection rates from 50–90%. Higher false alarm rates were observed when large ships, having considerable energy in the same frequency band as the whale calls, passed by the array. For this study, we picked a suitably low false alarm rate (and accepted that we were missing some call detections) to ensure a conservative measure of the whale vocalizations (not over-counting) to be useful for census monitoring.

In general, there are twice as many B calls as A calls for a given period. This is consistent with the Northeastern-Pacific blue whale call patterns A–B–B that have been previously documented by others. There

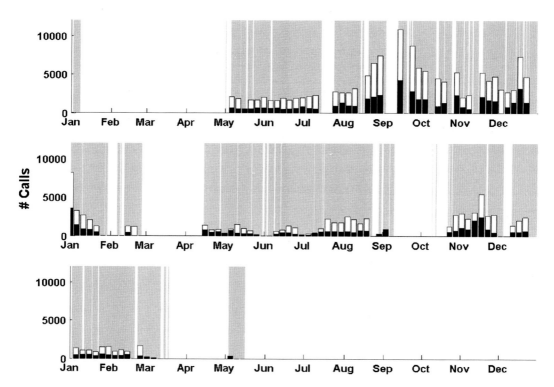

Fig. 17.6. Call abundance of blue whale A (black) and B (white) calls, heard from the NPS Observatory, during 1998 (top), 1999 (middle), and 2000 (bottom). Areas of available data are shown in gray; data outages are blank areas of white.

Table 17.2 *Blue whale call abundance totals for*
1998–2000. The total data availability (percentage)
for each year is also listed in column 1

	A call totals	B call totals
1998 (57%)	59036	108221
1999 (67%)	34404	57200
2000 (22%)	7260	11164

were 51% more A calls, and 45% more B calls detected during the summer (June–August) of 1998 (15 811 A calls, 29 338 B calls) than 1999 (7699 A calls, 16 134 B calls), despite the fact that there are several large gaps in the 1998 data set. This abundance during 1998 is consistent with independent acoustical surveys of Watkins *et al.* (2000), and biological surveys of Benson *et al.* (2002) that were done during the same period. Plankton levels along the eastern Pacific Ocean were drastically reduced, due to the warm waters of the 1998 El Niño event. The levels along the central California coast were less reduced than other areas, so still provided the most abundant supply of food for the blue whales. This resulted in a higher than normal seasonal whale population, even though the overall food supply was reduced.

17.6 The future

The number of calls is not the same as the number of whales, with the latter being a more important biological parameter. Acoustical censusing of the number of blue whales would require the localization and tracking of calls after detection. Furthermore, localization and tracking would also be required for the estimation of migration routes. To perform localization and tracking from a distance, multiple, real-time arrays installed at complementary geographic locations would be required. With these in place, the horizontal location of a vocalizing blue whale can be rapidly estimated based on simplistic triangulation schemes applied to beamformed bearing information and arrival times of the auto-detected signal. In order to apply acoustical techniques to monitor other types of whales, dedicated at-sea experiments would also be required in the future to characterize the deverberated signals of the other whales of interest.

Part IV
Studies of ocean dynamics

Chapter 18
Ocean acoustic tomography

ROBERT C. SPINDEL
University of Washington

Summary

In this chapter we introduce the fundamentals of tomographic imaging and its application to remote sensing of the ocean. We include a brief section on Radon transforms and the Fourier-Slice theorem as generalized background, but the chapter is complete without it, and that section can be skipped if the mathematics are too difficult. From its origins as a new method to study the ocean mesoscale, to its present application in measuring ocean-wide temperatures, this chapter describes how acoustic travel time tomography is practiced. Tomography offers several exciting advantages over conventional ocean measurement techniques. The interior of large ocean regions can be monitored efficiently using a relatively small number of instruments, because the number of measurements using N instruments grows as N^2 rather than linearly with N as is the case with conventional point measurements. The tomographic measurement is inherently spatially integrating, so it automatically yields average sound speeds (temperatures) and currents over long distances, thus smoothing over small scale features and internal wave "noise" and revealing underlying long-term and large-scale trends. Finally, because it uses sound rather than ships, and the speed of sound in water is 3000 knots, tomography can provide rapid (synoptic) and repeated measures of sound speed, temperature, and current.

Contents

18.1 Introduction

Ocean acoustic tomography is a new technique using underwater sound to measure the ocean's sound speeds and currents. It is analogous to medical computer aided tomography (CAT) scans. An object, the ocean in our case, the body in the medical case, is probed remotely from a series of locations around its periphery, and its interior structure is deduced by inverting the measured data. In medicine, X-ray and radio wave attenuation are the most common observables. In oceanography, the usual observable has been the travel time of acoustic signals, although other parameters such as Doppler shift, amplitude, and phase also have been considered.

Observing the large-scale ocean is difficult and challenging. Distances are large, access is limited, time scales are long, and ships are slow and expensive. Undersampling and aliasing have been the norm. To compound the problem, until recently most observations have been point measurements in space and time. Satellites have helped overcome these shortcomings since they offer essentially global coverage with rapid sampling in time, but they see only the ocean surface, and do not provide measurements of its interior.

Ocean acoustic tomography was proposed by Walter Munk and Carl Wunsch in 1979 as a tool to complement the new science of satellite oceanography (Munk and Wunsch, 1979). They focused on monitoring the ocean mesoscale, those ocean fluctuations with spatial scales of O (100 km) and temporal scales of O (100 days). (A useful analogy to the ocean mesoscale is atmospheric weather, although global weather scales, 1000 km spatially and 3–4 days temporally are very different.) During the 1960s and early 1970s our understanding of ocean dynamics evolved rapidly from believing the ocean was dominated by large, slow, stately flows such as the Gulf Stream – mostly a consequence of undersampling – to suspecting that perhaps as much as 90% of the ocean's kinetic energy is in the mesoscale. This remarkable transformation in thinking was the result of a few experiments whose scale and magnitude taxed the resources of the world's oceanography community. Other than these few observations, little was known about the mesoscale, and measuring it globally, or even on ocean basin scales, presented a formidable challenge. Munk and Wunsch's idea offered a potential solution to the problem.

As currently practiced, ocean acoustic tomography consists of measuring the travel time of acoustic pulses transmitted between sources and receivers, or ideally between transceivers (combined source and receiver), and interpreting them in terms of the interior sound speed and current fields. Because sound speed, c, is dependent to first order on temperature, T, through $\delta T(^\circ C) \approx \delta c(m/s)/4.6$ (see (2.31)), measurements of travel time change can also be interpreted in terms of temperature change. Figure 18.1 shows a notional ocean acoustic tomography system sampling the ocean in the horizontal and vertical dimensions. Horizontal spatial resolution is controlled by the number and distribution of instruments. Vertical resolution is controlled by the placement of instruments in the vertical plane as well as the background sound speed profile. In the deep ocean, as illustrated in the figure, multipath propagation resulting from the background vertical sound speed profile provides depth resolution with a single instrument placed near the axis of the deep sound channel.[1]

The attractive features of ocean acoustic tomography are first that the number of measurements using N instruments at locations on the border of a region grows as N^2, rather than linearly with N as it does for conventional point measurements. Second, as a consequence of the deep ocean sound speed channel, sound travels between a source and receiver along many different paths, or rays, each sampling different depths, so a single set of instruments can yield information on the depth as well as horizontal structure. Also, since the measurement is inherently spatially integrating – the travel time along a ray is a result of the average sound speed along that ray – the acoustical measurement automatically provides a spatial average that would be difficult to achieve by collecting and then averaging a large number of point measurements. Small-scale processes are averaged out, thus improving the resolution of large-scale features. Finally, since the measurement is made with sound whose

---

[1] See Chapter 2, specially Section 2.3, for a discussion about the deep ocean sound channel and multipath acoustic propagation. Basically, sound speed decreases with increasing depth as temperature decreases, until eventually the competing effect of increasing pressure causes the sound speed to increase. The result is a vertical sound speed profile with a minimum, called the axis of the deep sound channel, near a depth of 1500 m in tropical oceans, and shallower as one proceeds towards either pole. This profile produces a channel, or waveguide, such that propagating sound is refracted towards the axis, and it causes sound to travel along many distinct paths (eigenrays, or Fermat paths) between a source and receiver. Thus, a pulse emitted by a source is split into many pulses each traveling to the receiver along its unique path, and each arriving at a slightly different time.

Fig. 18.1. The upper panel is a notional ocean tomography system with acoustic transceivers around the periphery. In the deep ocean the background sound speed profile results in multipath propagation such that a pulse transmitted from one transceiver to another travels along many distinct paths, each sampling the ocean differently in the vertical, and each arriving at a different time (although all within a few seconds of the nominal 200-second gross travel time for a 300 km range) and at a different angle. The predicted arrival pattern is based on the average background sound speed profile shown. The differences between the measured and predicted patterns are a result of the departure of the present ocean from the historical average, and constitute the basic tomographic data. (The positive and negative numbers from −7 to +19 represent whether the ray left the source at a negative or positive angle with repect to the horizontal, and the number of times it cycled before reaching the receiver.)

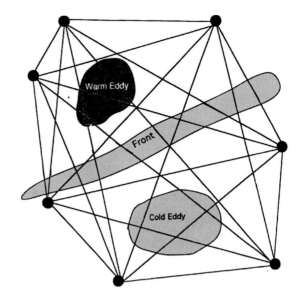

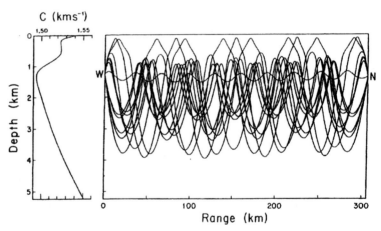

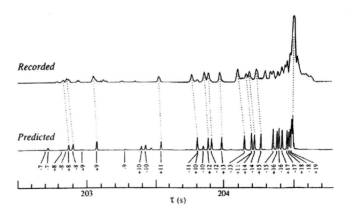

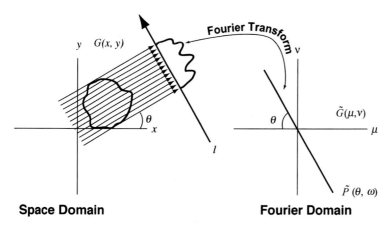

Space Domain   **Fourier Domain**

Fig. 18.2. The projection-slice theorem. The projection through the object at angle θ in physical space maps into a line at angle θ in Fourier space. If the object is probed at all angles over 180 degrees, thereby filling in Fourier space, the two-dimensional image of the object can be reconstructed with a two-dimensional inverse Fourier transform.

speed through the ocean is 3000 knots, rather than ships, tomography can provide rapidly repeated measurements and synoptic maps of sound speed, temperature, and current.

18.2 Theory

18.2.1 Radon transform

The underlying basis for tomography is the Radon transform which provides a mathematical method for reconstructing the interior of an object from geometric projections measured at many angles with respect to the object. It is possible to understand the principles of ocean acoustic tomography without using the Radon transform or the Fourier-Slice theorem which follows, and the reader so inclined can skip to Section 18.3 without loss of continuity. This section is provided for the student seeking to comprehend the fundamentals of the process.

Consider the two-dimensional unknown object $G(x, y)$ in Fig. 18.2, probed by energy along many paths at angle θ, producing the one-dimensional projection $P(\theta, r)$ along the line l defined by its distance from the origin, r, and angle θ,

$$P(\theta, r) = \int_l G(x, y) \, dl \qquad (18.1)$$

All points on the line satisfy,

$$r = x \sin \theta + y \cos \theta \qquad (18.2)$$

so we can write

$$P(\theta, r) = \int_{-\infty}^{\infty} \int_{-\infty}^{\infty} G(x, y) \delta(r - x \sin \theta - y \cos \theta) \, dx \, dy \qquad (18.3)$$

where $\delta(x)$ is the Dirac delta function defined as,

$$\delta(x) = 0 \qquad x \neq 0$$
$$\delta(x) = \infty \qquad x = 0$$

and

$$\int_{-\infty}^{\infty} \delta(x) = 1$$

Written this way, the projection function, $P(\theta, r)$, has values only along the line l. The collection of $P(\theta, r)$ over all angles θ is called the Radon transform of the image $G(x, y)$, and through the Fourier-Slice theorem the set of $P(\theta, r)$ can be used to reconstruct an image of the two-dimensional object.

18.2.2 Fourier-Slice theorem

The Fourier-Slice theorem states that the one-dimensional Fourier transform of the projection on line l is equal to the two-dimensional Fourier transform of the image evaluated on the line. We can see this by taking the Fourier transform of the projection,

$$\tilde{P}(\theta, \omega) = \int_{-\infty}^{\infty} P(\theta, r) e^{-i\omega r} \, dr \qquad (18.4)$$

And then substituting (18.3) into (18.4) to give,

$$\tilde{P}(\theta, \omega) = \int_{-\infty}^{\infty} \int_{-\infty}^{\infty} \int_{-\infty}^{\infty} G(x, y) \delta(r - x \sin\theta - y \cos\theta) e^{-j\omega r} \, dx \, dy \, dr \qquad (18.5)$$

Now, by invoking the "sifting" property of the delta function, i.e., the integral is zero except when $r = x \sin\theta + y \cos\theta$, (18.5) can be simplified to,

$$\tilde{P}(\theta, \omega) = \int_{-\infty}^{\infty} \int_{-\infty}^{\infty} G(x, y) e^{-i\omega(x \sin\theta + y \cos\theta)} \, dx \, dy \qquad (18.6)$$

Recalling the definition of the two-dimensional Fourier transform,

$$\tilde{G}(\mu, \nu) = \int_{-\infty}^{\infty} \int_{-\infty}^{\infty} G(x, y) e^{-i\omega(\mu x + \nu y)} \, dx \, dy \qquad (18.7)$$

it can be seen that $\tilde{P}(\theta, \omega)$ is just $\tilde{G}(\mu, \nu)$ evaluated at $\mu = \omega \sin\theta$ and $\nu = \omega \cos\theta$ which is the line that the projection $\tilde{P}(\theta, \omega)$ was taken on. Hence, if $\tilde{P}(\theta, \omega)$ is produced by probing an object at all possible angles θ and taking a one-dimensional Fourier transform of each projection, in principle the image $G(x, y)$ can be obtained by an inverse two-dimensional transform of $\tilde{P}(\theta, \omega)$. This procedure is the basis for tomographic imaging, and is the concept underlying ocean acoustic tomography although, as we shall see, the latter is not implemented in exactly this way.

18.3 Tomography

In medical X-ray tomography (CAT scans) the body (or test object in industrial non-destructive testing applications) is probed by a line of X-ray sources and a corresponding line of receivers which measure the radiation attenuation along a series of parallel paths through the body. The sources and receivers (or body) are then rotated a few degrees, another series of measurements is taken, and the process is repeated though 180 degrees of rotation. The data are then inverted as described above to yield an X-ray absorption image of a plane section of the body. Images of parallel sections obtained this way can be combined to produce a three-dimensional picture.

In the case of ocean tomography the procedure is basically the same except that the data consist of measured travel times of acoustic pulses between fixed points on the periphery of an ocean basin rather than X-ray absorption. The travel times, functions of the interior current and sound speed fields, are then inverted to obtain sound speed and current. Although similar, ocean tomography differs from the medical case in several important ways. First, we use the travel time data to measure departures from an assumed background ocean reference state, whereas the medical procedure assumes no prior information. Second, the amount of data collected is much less. The number of tomography instruments used in a given experiment is small, and the resulting number of paths is usually less than about 10. For a CAT scan 10^5 or more paths are typical. Third, the space and time scales are vastly different, and finally, because of natural multipath propagation in the deep ocean, a single set of measurements can provide both horizontal and vertical sampling simultaneously, so that multiple images at different depths (planes) need not be taken.

Tomography is fundamentally a two-step process. The first is data collection: the object is probed at many angles. The second is computational: the data are processed (inverted) to yield an image. A consequence of the second step is that tomography is computationally intensive. Also, since in practice it is impossible to probe the object at an infinite number of angles, there is an inherent limit to achievable image resolution. Nevertheless, the images that can be formed often surpass those obtained by conventional means, and of course there is the immense advantage of being able to image the interior of objects without actually having to insert instruments into them (or sail across them!).

18.4 Acoustic travel times

We consider the basic tomographic datum, the acoustic travel time between fixed sources and receivers in the ocean. We take the ocean sound speed field to be the sum of a background reference field $c_o(\bar{x})$ and a perturbation field $\delta c(\bar{x}, t) \ll c_o^2(\bar{x})$, where the overbar signifies a vector quantity. Then,

$$c(\bar{x}, t) = c_o(\bar{x}) + \delta c(\bar{x}, t) \tag{18.8}$$

The reference field can be thought of as an average background, which in practice is obtained from historical data or from more contemporary ocean surveys. Ocean acoustic tomography seeks to determine the perturbation field (which can then simply be added to the reference field to obtain the complete field). The acoustic travel time along a nearly horizontal ray path i in the presence of a current $\bar{u}(\bar{x}, t)$ is,[2]

$$\tau_i = \int_i \frac{ds}{c(\bar{x}, t) + \bar{u}(\bar{x}, t)} \tag{18.9}$$

Defining a reference travel time for the reference field as,

$$\tau_{oi} = \int_{oi} \frac{ds}{c_o(\bar{x})} \tag{18.10}$$

the perturbation in travel time over the reference travel time for transmission in the forward direction is,

$$\delta\tau_i^+ = \tau_i^+ - \tau_{oi} = -\int_{oi} \frac{\delta c(\bar{x}, t) + \bar{u}(\bar{x}, t)}{c_o^2(\bar{x})} ds \tag{18.11}$$

and for transmission in the opposite direction is,

$$\delta\tau_i^- = \tau_i^- - \tau_{oi} = -\int_{oi} \frac{\delta c(\bar{x}, t) - \bar{u}(\bar{x}, t)}{c_o^2(\bar{x})} ds \tag{18.12}$$

We compute the sum of (18.11) and (18.12),

$$\delta\tau_{ic} = (\delta\tau_i^+ + \delta\tau_i^-) = -2\int_{oi} \frac{\delta c(\bar{x}, t)}{c_o^2(\bar{x})} ds \tag{18.13}$$

---

[2] We express the travel times as integrals along the ray paths, because the paths are not straight lines. If the reader is uncomfortable with integral calculus, it is a simple matter to approximate the ray paths as straight lines, all traveling about the same distance R between source and receiver. Then (18.9) becomes $\tau_i = R/[c(\bar{x}, t) + \bar{u}(\bar{x}, t)]$, and (18.10) becomes, $\tau_{oi} = R/c_o(\bar{x})$, etc., and we can continue the development with algebraic equations to yield convenient approximations to (18.13) and (18.14) for the travel time perturbation due to sound speed changes alone, $\delta\tau_{ic} \approx -2R\delta c/c_o^2$, and those due to currents alone, $\delta\tau_{iu} \approx -2R\delta u/c_o^2$. The one-way travel time perturbations are of course half these quantities.

and the difference,

$$\delta\tau_{iu} = (\delta\tau_i^+ - \delta\tau_i^-) = -2\int_{oi}\frac{\bar{u}(\bar{x}, t)}{c_o^2(\bar{x})}ds \qquad (18.14)$$

and observe that these quantities are linearly related to the sound speed perturbation and current, respectively (hence the subscripts c and u). The sum and difference travel times constitute the basic tomographic data which are inverted to determine the fields δc and u. Obtaining sums and differences requires reciprocal transmissions, i.e., transmissions of pulses in opposite directions simultaneously, or nearly so. If the ocean current field is weak, i.e., $\bar{u}(\bar{x}, t) \ll \delta c(\bar{x}, t)$, which is often the practical case, then (18.11) and (18.12) become

$$\delta\tau_i = -2\int_{oi}\frac{\delta c(\bar{x}, t)}{c_o^2(\bar{x})}ds \qquad (18.15)$$

and one way transmissions can be used to obtain an estimate of the sound speed field.

To illustrate the one-way travel time measurement precision required in typical tomography experiments, we use the approximations in footnote 2 and a nominal sound speed of 1500 m/s to estimate order of magnitude quantities. Then a change of 9.6 m/s (which could be caused, for example, by a temperature change of 2 °C, consistent with a strong mesoscale eddy) would result in a travel time perturbation of 425 ms over a 100 km range which is easily measured. Smaller temperature variations of about 0.5 °C, representative of changes in the upper ocean from summer to winter season, produce travel time perturbations of about 100 ms over 100 km range, and 1 s over a 1000 km range, again, both easily measured. On the other hand, the travel time change due to the much smaller expected ocean temperature trends associated with climate change, a millidegree or less per year, is only a few milliseconds over 1000 km, and presents a challenging measurement problem. The effects of currents are also small and more difficult to measure. An average current of 1 cm/s along a 100 km range would produce a one-way travel time perturbation of about 0.5 ms and thus a difference between forward and reverse travel times of only about 1 ms; for a 1000 km range the difference is about 10 ms.

18.5 Forward problem

In order to use ocean acoustic tomography one has to first solve the so-called forward problem which, in essence, is the solution to the wave equation. Given the sound speed and current fields, the forward problem consists of predicting the arrival times of the received multipath signals, and identifying each with the path it took. The inverse problem, the object of ocean tomography, seeks to calculate the sound speed and current fields given the arrival times of an observed received signal structure. One of the outstanding successes of modern ocean acoustics has been the solution of the forward problem to the degree of accuracy required to implement a tomographic system.

Figure 18.1 shows a typical reference sound speed profile and predicted ray paths, and the corresponding predicted and measured multipath arrival pattern. The departures of the measured from the predicted travel times, highlighted by the dotted lines connecting corresponding arrivals, are the basic tomographic data, the δT_{oi}. Note that the travel time departures from the reference ocean are small, of order 10 ms, so for tomography to succeed the measurement of arrival time must be accurate to within a few ms. This implies that short pulses and high signal to noise (S/N) ratios are needed. Typical tomography signals have pulse widths less than 20–30 ms, and are received with $S/N > 15$ dB.

A key feature of deep ocean propagation illustrated in Fig. 18.1 is the clustering of late arrivals, which correspond to near axial paths, all traveling about the same distance and at the same average sound speed. They are very closely spaced and difficult to resolve given the limited bandwidths (< 100 Hz) obtainable at the low acoustic frequencies necessary for long-range ocean propagation. (Typical ocean tomography signals are in the 75 Hz–400 Hz band.) As a result, resolved early arrivals are the data of choice.

18.6 Inverse problem

The forward problem consisted of predicting the background sound speed field and reference ray paths and arrival patterns. The inverse problem uses the measured travel time data, δT, to estimate the interior fields, δc and u. The problem is underdetermined since only a finite number of observations are available to estimate continuous fields. To get around this difficulty, we represent the fields with models that are based on our *a priori* knowledge of the dynamics and kinematics of the ocean. We choose models that are parameterized by a small number of coefficients so the problem reduces to finding a limited set of coefficients for the model fields. A typical model construction is a linear set of "basis functions," $F_n(\bar{x})$, describing the horizontal and vertical sound speed structure,

$$\delta c(\bar{x}) = \sum_{A_n} F_n(\bar{x}) \tag{18.16}$$

where the coefficients A_n are what we seek to calculate from the data. For simple illustrative purposes Munk and Wunsch used rectangular boxes in all three dimensions as basis functions in their 1979 paper. However, the usual practice is to assume a simple horizontal dependence such as a truncated Fourier series or even just a series of boxes, because the ocean generally varies slowly in range, and a vertical dependence based on knowledge of the dominant vertical modes of ocean variability. In the first demonstration of ocean tomography in 1981, Cornuelle used empirical orthogonal functions, EOFs, and these remain the current choice (Behringer *et al.*, 1982). Other basis functions, such as Rossby modes, have also been used (Howe *et al.*, 1983).

Substituting (18.16) in (18.15) yields,[3]

$$\delta \tau_i = -\int_{oi} \frac{\delta c(\bar{x}, t)}{c_o^2(\bar{x})} ds = -\sum_{n=1}^{N} A_n \int_{oi} \frac{F_n}{c_o^2(x)} ds \qquad (18.17)$$

The coefficients A_n are obtained by inverting (18.17) using standard linear inverse techniques.

Figure 18.3, from Cornuelle *et al.* (1989), is a simple computer example that provides an intuitive feeling for the ocean tomographic process. A two-dimensional ocean consisting of a single horizontal slice is assumed, and rays are propagated through it along straight lines. The top panel is the "true" ocean. The rows beneath it each represent a single tomography experiment. The left hand panel of each row shows the acoustic paths for the experiment, and the right hand panel is the result of inverting the travel time data obtained along those paths. So, for example, the panels in (*a*) represent an experiment in which only horizontal paths ($\theta = 0°$) are available; only coarse horizontal information can be recovered. For (*b*), only vertical paths ($\theta = 90°$) are available. In (*c*) both horizontal and vertical paths are used and the recovered ocean has both horizontal and vertical structure. In (*d*) additional paths at $\theta = 45°$ and $\theta = 135°$ are used, and most of the features of the true ocean are recovered. As the ocean is probed from more and more directions, the reconstructed image continues to improve.

A reconstruction based on an actual tomography is shown in Fig. 18.4.

18.7 Tomography systems

An ocean acoustic tomography system requires acoustic sources and receivers capable of transmitting pulses narrow enough to resolve ray path arrivals and strong enough to be received above the background ambient

---

[3] Again, alternatively, assuming straight line ray paths for simplicity yields,

$$\delta \tau_i = \frac{-2R\delta c}{c_o^2} = -2R \sum A_n F_n / c_o^2$$

Fig. 18.3. In this computer experiment, the upper panel is meant to represent the true state of a 1000 x 1000 km section of the ocean. The panels beneath it show the progressive improvement in tomographic reconstruction as additional projections are acquired. The left panel in each row shows the acoustic paths used, and the right panel shows the resulting reconstructed image. Thus in (*a*) only horizontal paths are available, and the reconstruction contains only horizontal information. In (*b*) only vertical paths are used. In (*c*) both are used, and the map is much improved. Adding additional paths in (*d*) produces a map that closely resembles the true ocean.

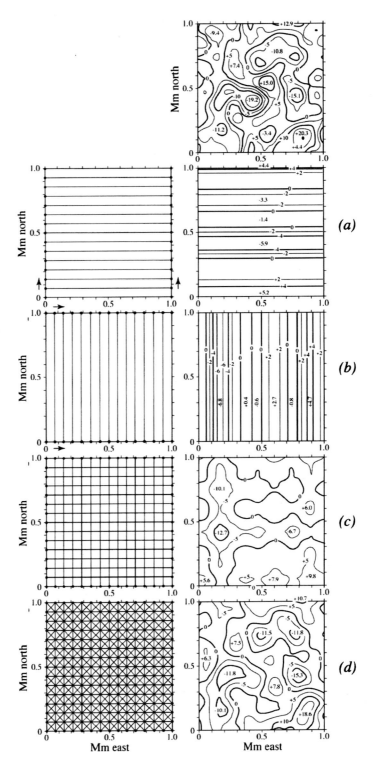

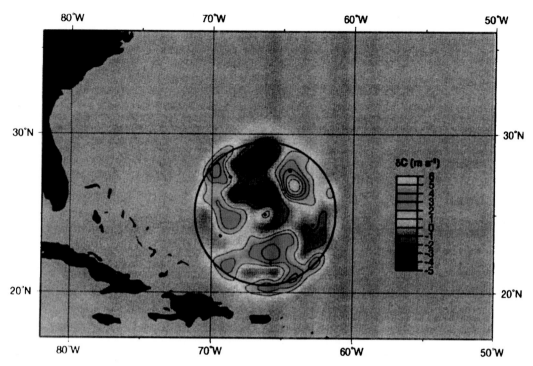

Fig. 18.4. This map of sound speed perturbations at 700 m depth was obtained during the 1991 Acoustic Mid-Ocean Dynamics Experiment. Five transceivers were moored for a year in a pentagonal array East of Florida, and a sixth was placed at the center, yielding a total of 15 paths. Twice during the year a ship circumnavigated the array stopping every 25 km to receive signals from the six moored instruments. During the period of ship circumnavigation, about 750 paths were available, thus providing much increased horizontal resolution over the rest of the year. The map shown was created from data collected during one of the periods when the additional ship data was available (AMODE-MST Group, 1994). (See colour plate section.)

noise. Timekeeping must be accurate to within a few ms to allow measurement of the small travel time perturbations caused by typical ocean sound speed fluctuations and currents. For short-range systems, say 10 km or so, high acoustic frequencies in the 10s of kHz range can be used, and short duration, high intensity pulses are relatively easy to generate. For long-range systems low frequencies must be used, and useful bandwidth is limited to about 100 Hz implying a minimum pulse duration of $\tau = 10$ ms. Because even the loudest acoustic sources do not provide enough energy in a 10 ms pulse to be received above the ambient noise at long ranges, tomography systems have relied upon coded signals which spread the transmitted energy in time. Phase modulated maximal length sequences (pseudorandom sequences) have been most common. Upon reception the signals are processed by replica cross correlation to achieve gains up to 30 dB. Additional gain can be had by incoherent averaging a number of repeatedly transmitted sequences. A representative system has a center frequency of 250 Hz and uses a 1023 digit phase modulated sequence, repeated ten times. Each digit, or pulse, consists of three cycles of the 250 Hz carrier. Thus the pulse length is $\tau = 12$ ms in duration, the replica cross correlation processing gain is $10 \log(1023) = 31$ dB, and the incoherent gain from averaging ten sequences is $10 \log(10) = 10$ dB giving a total gain of 41 dB. For a typical source level of 195 dB re 1 μPa, and an ambient noise level of about 90 dB over the nominal 83 Hz band $(1/\tau)$, this amount of processing gain brings the received signal to noise ratio for 1000 km range to about 20 dB.[4]

Multipaths whose arrival times are separated by less than roughly $\tau/\sqrt{2}$ cannot be resolved in time alone. However, additional resolution can be had by distinguishing them on the basis of arrival angle as well as time. Thus, tomography receivers often employ short vertical arrays (four elements with $\lambda/2$ spacing has proven to be practical) that allow discrimination between arrivals with positive and negative angles with respect to the horizontal.

Accurate timekeeping to within about 1 ms/yr is provided by atomic (rubidium) clocks in each instrument. Further, each instrument is tracked by a local acoustic navigation system with 1–2 m (1 ms) accuracy for the purpose of distinguishing travel time changes due to sound speed and current changes from those due to motion of the instruments which would change the path length. Even though the instruments are generally moored near the sound channel axis at a depth of about 1000 m, and the mooring lines are made very stiff (1000 lb or more tension) considerable motion is observed (see Fig. 18.5 and Spindel (1985); Worcester *et al.*, (1985); Spindel and Worcester (1986)). Figure 18.6 shows two typical tomography transceivers. The unit in the upper panel uses free-flooded resonant tube acoustic sources and operates at 400 Hz center frequency; the device in the lower panel uses a 250 Hz center frequency hydroacoustic source.

---

[4] The possible effects of low frequency sound at this level on marine mammals has been the subject of much recent research, especially during the development of acoustic thermometry described in Section 18.8.2. Extensive research involving aerial and visual observations in the vicinity of operating thermometry sources, experiments in which animals were tagged to record the sound field to which they were exposed and their responses to it, and experiments where tomography sounds were deliberately played near animals, found no overt or obvious changes in abundance, distribution, or behavior. Intensive statistical analyses of the data have revealed subtle changes in the spatial distribution and behaviors of some animals close to the sources, but these subtle effects are not thought to be biologically significant.

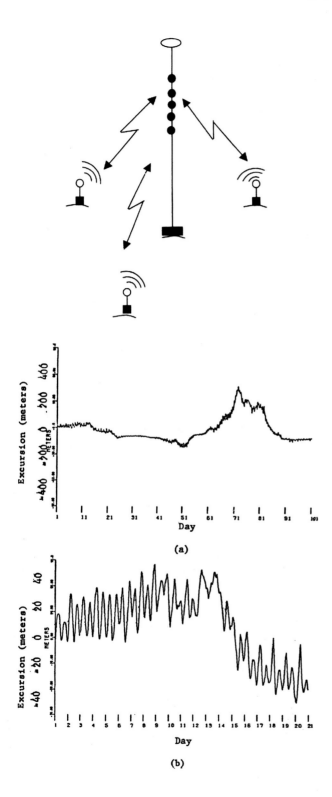

(a)

(b)

Fig. 18.5. The position of tomography instruments is tracked to allow discrimination between acoustic travel time changes due to changes in the ocean, and those due to changes in the acoustic path length resulting from motion of the moored instruments. The transceivers interrogate a local transponder net and record round trip travel times. These are converted to distances, thereby triangulating the instrument's position. The middle panel shows the motion of a typical mooring over a 3-month period. The lower panel is an expansion of the first twenty days showing peak excursions of about 30 m as a result of tidal current induced motion. 30 m represents a 20 ms travel time and, if not accounted for, would completely mask ocean sound speed and current effects.

Fig. 18.6. Two tomography transceivers. The upper panel uses two free-flooding resonant tube sources about 2 m long (Webb Research Corp., Falmouth, MA) which have no depth limitations. Its center frequency is 400 Hz. Batteries and electronics are housed in the central tube. The lower unit uses a 250 Hz center frequency hydroacoustic source (Hydroacoustics, Inc., Rochester, NY) and must be pressure compensated. High pressure gas bottles for this purpose can be seen in the upper part of the frame. Batteries and electronics are housed in the cylinders on either side. Two circular aluminum discs (one on each side) about 1 m in diameter are the radiating elements. They are driven by a hydraulically operated piston that pushes the centers of the discs outwards.

18.8 Applications

18.8.1 Ocean tomography

There have been numerous applications of ocean acoustic tomography. Examples of long-range applications are its use in the Greenland Sea to observe deep convection, in the Mediterranean Sea to study convection and heat content, in the North Pacific to measure barotropic and internal tidal currents and large-scale vorticity, in the western equatorial Pacific to study El Niño, in the Barents Sea to study polar fronts, and in the Monterey Canyon (CA) to study surface waves. Short-range systems operating at high frequencies have been used in the Strait of Gibraltar to monitor flow in and out of the Mediterranean (this application used acoustic phase in addition to travel time), and in coastal Japanese waters to monitor currents.

18.8.2 Acoustic thermometry

Ocean acoustic thermometry, an outgrowth of tomography studies conducted during the 1980s, uses long-range acoustic transmissions to study climate change. It takes advantage of the integrating feature of acoustic tomography to average over mesoscale features, thus greatly reducing measurement noise. An initial test in 1991 used very long-range transmissions (up to 20 000 km) from near Heard Island in the southern Indian Ocean to receiving stations scattered world-wide. It showed that signals could be received at these distances, but subsequent thermometry experiments have been at shorter ranges. (The same success in solving the forward problem at several thousand km range is not enjoyed at these trans-global ranges for a number of reasons. Perhaps the most important is that fidelity and spatial sampling of historic sound speed profile data over these large ranges is not sufficient to support accurate acoustic propagation predictions.) In perhaps the best demonstration of acoustic thermometry, 20 Hz signals transmitted across the Arctic Ocean in 1994 and 1999 experiments revealed a clear warming trend (Fig. 18.7).

The Acoustic Thermometry of Ocean Climate (ATOC) network in the north Pacific has been acquiring data since 1995, first with a source off the coast of California (from 1995 to 1998), and later with a source off the north coast of Kauai (from 1997 to present (November, 2004)). The signals have been received on US Navy SOSUS and other arrays distributed throughout the Pacific, as far south as Kiritamati Island and New Zealand. The ATOC signals are centered at 75 Hz and have a bandwidth of 30 Hz. These measurements have shown a variety of trends over the north Pacific basin, revealing the complexity of ocean climate change even over basin scales, and emphasizing the need for continuing

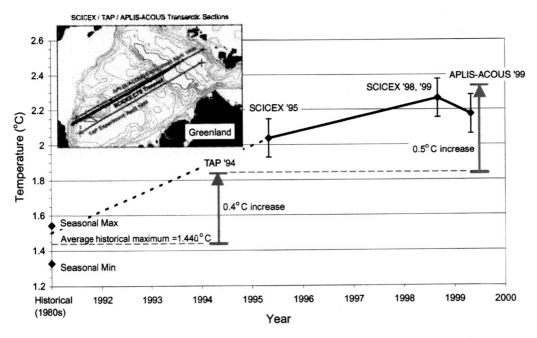

Fig. 18.7. Two trans-Arctic thermometry experiments, TAP '94 and APLIS-ACOUS '99 show a warming trend. SCICEX refers to data acquired with actual temperature sensors during submarine transects in 1995, 1998, and 1999 (courtesy P. Mikhalevsky).

measurements. Figure 18.8 shows the ATOC network, and compares ATOC temperature measurements to model predictions and those inferred from satellite altimeter data.

Extensive references to the above and other applications are available in Munk *et al.* (1995) and Dushaw *et al.* (2001). Information about the Japanese system is in Yamoaka *et al.* (2002). The Heard Island experiment is documented in Munk *et al.* (1994).

18.9 The future

Ocean acoustic tomography can be applied to a wide variety of ocean observations. Already it has been used to measure tides, heat content, temperature change, and transport, and to contribute to basin scale monitoring of ocean climate. Among its attractive features is its unique ability to obtain averages over long ranges easily, efficiently, and quickly. It would seem to be a natural component in the present and future implementation of global ocean observing systems. It complements the good horizontal resolution of satellite altimetry with depth resolution and good time resolution. However, in common with many other acoustical oceanography

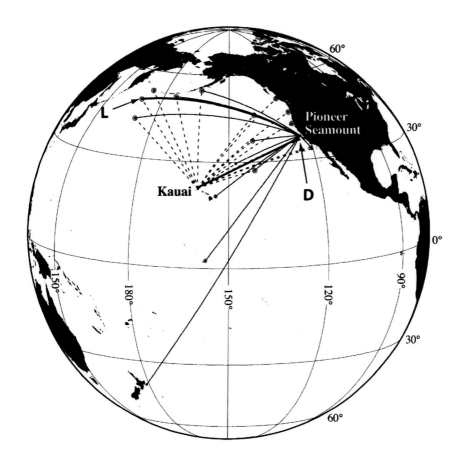

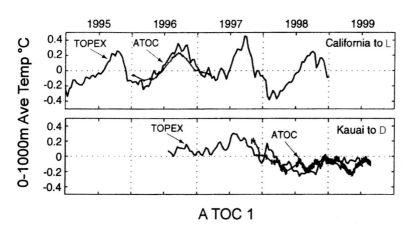

A TOC 1

Fig. 18.8. The Acoustic Thermometry of Ocean Climate (ATOC) network with acoustical sources off California and Kauai, and receivers scattered throughout the Pacific. The lower panels compare ATOC temperature measurements to temperature inferred from the TOPEX/POSEIDON satellite altimeter. Variations in sea surface height measured by the altimeter are converted to temperature by assuming that the height variations are a result of thermal expansion in the upper 1000 m of the ocean. The ATOC measurement is more direct.

tools, it is not a conventional oceanographic measurement, and is un-familiar to many practicing oceanographers. Today's ocean models are not (yet!) parameterized with the quantities obtained directly by tomo-graphic systems, so that data assimilation is not as simple as it could be. As a result, there are important education and translation hurdles that must be cleared before tomography becomes a standard tool in the oceanographer's kit. But given tomography's unique capabilities, and given the transparency of the ocean to sound, it is hard not to imagine a future where it will be employed widely and routinely.

Chapter 19
Acoustic time reversal in the ocean

DAVID R. DOWLING
University of Michigan

HEECHUN SONG
University of California

Summary

Time travel is a fascinating concept for anyone wishing to investigate past events or seeking to correct past mistakes. Imagine the efforts of a springboard diver that would otherwise be perfect except for a small technical mistake that generates a big splash. How wonderful it would be to reverse time in order to identify and correct the mistake! Although identification of the mistake is possible with a movie or video recording, true time reversal of the dive itself, with the splash and spray returning to the point of entry and the diver popping out of the water is impossible. In fact, time reversal of complex processes – like springboard diving – is not possible when organized mechanical energy is converted into disorganized thermal energy. Fortunately, such energy conversion may be minimal for acoustic waves traveling through the ocean. Thus, ocean sound waves often can be time reversed and such time reversed waves possess special and intriguing properties.

The everyday experience of two-way conversation establishes that airborne sound can travel in both directions between a source and receiver. Sound in the ocean has the same capacity for two-way travel. Acoustic time reversal is special type of two-way acoustic exchange in which the sound that travels in the forward direction, from the source to the receiver, is time reversed so it travels backwards, from the receiver to the source. Acoustic time reversal has attracted attention in ocean acoustics because it is a robust means for focusing sound in unknown underwater environments where scattering, diffraction, and multipath propagation cause other sound focusing technologies to fail. This chapter describes the basic elements of acoustic time reversal and illustrates them with simple calculations and experimental results from the ocean.

Contents

19.1 Introduction

Acoustic time reversal can be accomplished with a sound source and a special array of transducers that combine the functions of microphone (receiver) and loudspeaker (transmitter). This special transducer array is commonly called a time-reversing array (TRA) or a time-reversal mirror

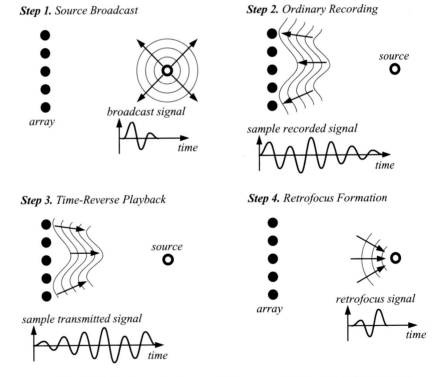

Step 1. Source Broadcast

array

broadcast signal

time

Step 2. Ordinary Recording

source

sample recorded signal

time

Step 3. Time-Reverse Playback

source

sample transmitted signal

time

Step 4. Retrofocus Formation

array

retrofocus signal

time

Fig. 19.1. The four steps in acoustic time reversal using a transducer array. A sample signal is shown for each step. The third step produces the time-reversed acoustic waves; it explicitly involves transmitting the recorded sound backwards.

(TRM). The process of acoustic time reversal can be described by four simple steps involving a sound source and the TRA (see Fig. 19.1). First, sound is generated by a source and travels through the environment to the transducer array. Second, the TRA records the sound in the usual manner. Third, the sound recorded by each transducer in the array is retransmitted from that transducer with the direction of time inverted; first-recorded sounds become last-transmitted sounds. And in the final step, these array-transmitted time-reversed sound waves travel backwards retracing their paths to converge at the location where they were generated. The location where these waves return and focus is called the *retrofocus*.

The reason that time-reversed sound waves travel backwards and focus at the location where they were generated is a direct consequence of the lossless linear wave equation for the acoustic pressure $p(\vec{x}, t)$

$$\nabla^2 p = \frac{1}{c^2} \frac{\partial^2 p}{\partial t^2} \tag{19.1}$$

The second-order time derivative in (19.1) ensures that if $p(\vec{x}, t)$ is a solution then $p(\vec{x}, -t)$ is too. Thus, if $p(\vec{x}, t)$ represents sound waves expanding *away* from a small sound source, then $p(\vec{x}, -t)$ represents sound waves converging *toward* the same source. Acoustic absorption losses could be included in (19.1) through the addition of a first-order time-derivative term. The addition of such a term would break the invariance of (19.1) to changes in the sign of t so that $p(\vec{x}, t)$ and $p(\vec{x}, -t)$ could not both be solutions. However, when absorption losses are present but mild – as is commonly the case in the ocean – their detrimental effects on the time-reversal process are mild as well.

As shown in Fig. 19.1, a TRA distributed in space records the outward propagating field $p(\vec{x}, t)$ and then attempts to construct the converging field $p(\vec{x}, -t)$ when it transmits. In an ideal situation, the array of transducers would completely surround the source. In practice, this isn't possible and time reversal is usually performed with a limited aperture array and the array's construction of $p(\vec{x}, -t)$ is not ideal. However, the fidelity of a limited-aperture array's construction of $p(\vec{x}, -t)$ does improve when the array's size and its transducer density are increased. Interestingly, the waveguide nature of ocean acoustic propagation effectively increases the aperture of a TRA and this has a beneficial impact on the array's construction of $p(\vec{x}, -t)$; construction fidelity increases as the number of propagation paths between the source and the array increases.

This array-size dependence in the construction of $p(\vec{x}, -t)$ is exactly analogous to the optical case where an imaging system with a large light-collection lens or mirror can generate brighter images and achieve finer resolution than an equivalent imaging system with a smaller light-collection lens or mirror. In fact, the basic principles of acoustic time-reversing arrays are similar to those of optical phase-conjugate mirrors

(Zel'dovichi *et al.*, 1985) which exhibit retroreflectance; they reflect light back toward its source.

The first acoustic time-reversal experiments in the ocean were conducted in the early 1960s (Parvulescu and Clay, 1965; Parvulescu, 1995). In Long Island Sound, signals were recorded with a single transducer 20 nautical miles from a sound source in water that was approximately one nautical mile deep. These signals were time-reversed, and retransmitted through the ocean to show the temporal stability of the multipath propagation. This point-to-point time-reversal process corresponds to temporal matched filtering with only environmental spatial focusing. The first acoustic time-reversal experiments utilizing a transducer array were conducted in a laboratory at ultrasonic frequencies (Fink *et al.*, 1989; Fink, 1999). The formal basis of acoustic time reversal for underwater applications (Jackson and Dowling, 1991) developed parallel to the ultrasound work. The true potential of acoustic time reversal in the ocean was demonstrated by landmark experiments in the Mediterranean Sea that showed stable retrofocusing for source-array ranges of many kilometers in coastal waters (Kuperman *et al.*, 1998). These experimental studies will be described in the third section of this chapter. A more thorough treatment of ocean acoustic time reversal is also available (Kuperman and Jackson, 2002).

The goals of this chapter are to illustrate acoustic time reversal of harmonic signals in simple situations and in ocean environments, to describe the limitations of acoustic time reversal, and to discuss some of the ways acoustic time reversal is being applied to the ocean. The automatic focusing properties of acoustic time reversal are well suited to some active sonar applications, underwater communication systems, and acoustic transponders.

19.2 Time reversal in a uniform unbounded medium

Acoustic time reversal can be used to focus or compress sound in both time and space. Consider how the spatial focusing works in an unbounded uniform medium (commonly called free space) with speed of sound c where a harmonic point source creates spherical waves. If r is the radial distance from the origin of coordinates, then the acoustic pressure field, $p(r, t)$, from a point source at the origin having radian frequency ω is given by (see Section 1.8.2)

$$p(r, t) = \hat{p}(r)\exp\{i\omega t\} = \frac{A\exp\{i(\omega t - kr + \phi)\}}{r} \tag{19.2}$$

where $\hat{p}(r)$ is the complex pressure amplitude, A is a real amplitude that sets the strength of the point source, ϕ is the phase angle of the source, and

$k = \omega/c$. The acoustic waves represented by (19.2) propagate *away* from the origin. When time is reversed in (19.2), that is, t is replaced by $-t$,

$$p(r, -t) = \hat{p}(r)\exp\{-i\omega t\} = \frac{A\exp\{i(-\omega t - kr + \phi)\}}{r} \qquad (19.3)$$

the temporal (ωt) and spatial (kr) phase terms no longer have opposite signs so the waves converge *toward* the origin. Consequently, a device that can produce a time-reversed wave front is able to retrodirect sound back to the location where that wave front was generated.

Before moving on to describe how such a device, a time-reversing array, can be made, it is worth noting that time-reversed waves can also be constructed by simultaneously changing the signs of kr and ϕ in (19.2) while leaving the sign of t unchanged. The combination $-kr + \phi$ is the phase of the complex pressure $\hat{p}(r)$ in (19.2). Therefore, time reversal and complex conjugation of $\hat{p}(r)$ (i.e., changing $-kr + \phi$ into $+kr - \phi$ in (19.2)) have the same meaning for pure harmonic signals so that time reversal is also called phase conjugation in the frequency domain. For example, it readily follows from (19.2) and (19.3) that:

$$\mathrm{Re}\{p(r, -t)\} = \mathrm{Re}\{\hat{p}(r)\exp\{-i\omega t\}\} = \frac{A}{r}\cos\{\omega t + kr - \phi\}$$
$$= \mathrm{Re}\{\hat{p}^*(r)\exp\{i\omega t\}\} \qquad (19.4)$$

This use of complex conjugation (designated by superscript *) to construct time-reversed harmonic signals is nearly universal and is adopted here as well. Results for broadband signals can be constructed by a Fourier superposition of single frequency results.

Although other formulations are possible, the most common mathematical description of a time-reversing array is given by a sum over array elements with each element transmitting an acoustic pressure proportional to the complex conjugate of the acoustic pressure it heard during the recording interval. This array transmission creates a time-reversed portion of the original wave front which then propagates back toward the source. When there are many array elements spread out in space, the fraction of the original wave front that is time-reversed increases and the effectiveness of retrofocusing improves.

These facts can be illustrated by setting down a few definitions, choosing a propagation model, and evaluating the results. Start with the situation depicted in Fig. 19.2, the source is located at $\vec{r}_s$ and array elements are located at $\vec{r}_n$ where $n = 1, 2, \ldots, N$. The output, p_{TRA}, of a time-reversing array may then be defined by:

$$p_{TRA}(\vec{r}, t) = \hat{p}_{TRA}(\vec{r})\exp\{i\omega t\} \quad \text{and} \quad \hat{p}_{TRA}(\vec{r}) = \sum_{n=1}^{N} B_n \hat{p}_n(\vec{r})\hat{p}_s^*(\vec{r}_n)$$
$$(19.5a,b)$$

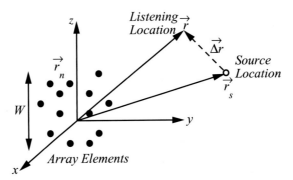

Fig. 19.2. Schematic of a time-reversing array having elements at $\vec{r}_n$ that responds to a harmonic point source at $\vec{r}_s$. The output of the array is listened to at a field point $\vec{r} = \vec{r}_s + \Delta \vec{r}$. The length scale W sets the aperture of the array.

where B_n is the real electronic amplification factor for the nth array element, $\hat{p}_n(\vec{r})$ is the complex pressure produced by the nth array element at location $\vec{r}$, and $\hat{p}_s(\vec{r}_n)$ is the complex pressure produced by the original source at the location of the nth array element. The B_n weighting factors carry the necessary units (Pa^{-1}) to ensure that p_{TRA} is a pressure and they allow the contribution of each array element to be weighted differently, if desired. The above mathematical form (19.5) for defining the time-reversing array can be also used when the acoustic environment is more complicated than an unbounded uniform medium.

Even at this general level, the form given by (19.5) can be used to deduce some of the intrinsic features of acoustic time reversal. For example, when the listening location $\vec{r}$ coincides with the original source location $\vec{r}_s$ then

$$\hat{p}_{TRA}(\vec{r}_s) = \sum_{n=1}^{N} B_n \hat{p}_n(\vec{r}_s)\hat{p}_s^*(\vec{r}_n) \propto \sum_{n=1}^{N} B_n \,|\hat{p}_n(\vec{r}_s)|^2 \qquad (19.6a,b)$$

where the principle of spatial reciprocity – here stated as $\hat{p}_s^*(\vec{r}_n) \propto \hat{p}_n(\vec{r}_s)$ – has been used to generate the proportionality (19.6b). Spatial reciprocity, the invariance of sound signals under the exchange of source and receiver locations, facilitates successful acoustic time reversal. Here (19.6b) shows that the pressure field from each array element is in-phase and real at the original source location. Therefore, the sum over array elements achieves its largest possible value because all of its terms are positive real numbers. When the listening point $\vec{r}$ does not coincide with $\vec{r}_s$ the terms in the sum in (19.6b) take on complex values and may partially or completely cancel with each other when summed together. Thus, (19.6b) already indicates that the TRA-produced pressure-field amplitude is likely to peak at $\vec{r}_s$, and that to achieve this peak amplitude it is not necessary to specify or know the locations of the array elements.

This spatial focusing can be illustrated when the original source and the array elements are monopole transducers that create spherical waves. This choice for the transducer characteristics is made only for the sake of simplicity; monopole transducers are not required for successful retrofocusing. So, when (19.2) is extended to sources away from the origin and combined with (19.5b), $\hat{p}_{TRA}$ becomes:

$$\hat{p}_{TRA}(\vec{r}) = A_s \sum_{n=1}^{N} A_n B_n \frac{\exp\{-ik|\vec{r} - \vec{r}_n| + ik|\vec{r}_n - \vec{r}_s|\}}{|\vec{r} - \vec{r}_n| \, |\vec{r}_n - \vec{r}_s|} \qquad (19.7)$$

where A_s and A_n (units Pa · m) set the broadcast strengths of the source and the array elements, respectively. To proceed with an evaluation of (19.7), a specific geometry must be chosen. For convenience, the coordinate axes shown in Fig. 19.2 can be rotated so that the source lies on the x axis, $\vec{r}_s = x_s\mathbf{i}$. In addition, assume that the TRA is linear and vertical, $\vec{r}_n = z_n\mathbf{k}$, a common array configuration for oceanic sound studies – and denote the vector distance from the source location by $\Delta\vec{r} = \Delta x\mathbf{i} + z\mathbf{k}$ with $\Delta x = x - x_s$. Furthermore, assume A_n and B_n are uniform across the array.

When all distances are scaled by the acoustic wavelength λ, (19.7) can be evaluated without specifying ω. Figure 19.3 shows such results for a 201-element time-reversing array centered at the origin of coordinates with transducers placed a half wavelength apart to avoid spatial

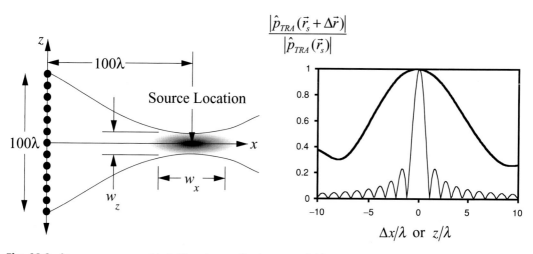

Fig. 19.3. Source-array geometry (left) and normalized pressure field amplitudes (right) for a vertical linear time-reversing array. The 201 array elements are $\lambda/2$ spaced along the z axis and the source is located on the x axis 100λ from the array. The two curves shown are the TRA field amplitude along the x axis ($\Delta\vec{r} = \Delta x\,\mathbf{i}$, dark line) and in the vertical direction ($\Delta\vec{r} = z\,\mathbf{k}$, thin line).

aliasing, yielding $W = 100\lambda$, when the sound source is located at $x_s = 100\lambda$. The two curves are for $\Delta\vec{r}$ aligned with x axis and the z axis.

Although the extent of the high amplitude retrofocus region is different in each direction, $\hat{p}_{TRA}$ clearly achieves its maximum value at or very near the location of the sound source. In this case, the minimum size and shape of the high-amplitude retrofocus region are set by the free-space diffraction limits for a linear vertical array. These limiting focal region sizes w_x and w_z are proportional to z_s^2/kW^2 and z_s/kW, respectively, when the source-array geometry satisfies the Fresnel approximation (see Section 1.5.2). These results are typical for a free-space environment where there is one straight acoustic path between any two points. The effect of multiple propagation paths on TRA retrofocusing is covered in the next section.

19.3 Acoustic time reversal in multipath sound channels

This section presents two examples of acoustic time reversal in multipath environments that illustrate how the characteristics of the array *and* the environment influence retrofocus properties. The first example involves the simplest possible environment that embodies multipath propagation, a uniform half space bounded by a flat reflecting surface. The second example is a presentation of results from actual time-reversal experiments conducted in shallow ocean waters. In both cases, time-reversal exploits the extra propagation paths to improve focusing.

To see how this is accomplished, consider what happens when a time-reversing array responds to a harmonic point source placed near a reflecting surface. For simplicity, the array is again taken to be linear and vertical, but the coordinate system is flipped over and chosen so that the reflecting surface, having complex reflection coefficient R, coincides with the x–y plane (see Fig. 19.4). The array is centered at depth z_a with elements at z_n, while the source is located at $\vec{r}_s = x_s\mathbf{i} + z_s\mathbf{k}$; for this geometry, only the x and z coordinates are needed to describe the array-produced focusing. Thus, the influence of the surface reflected path can be determined by evaluating (19.5b) in the environment shown in Fig. 19.4 for different values of R.

Results of such evaluations are shown in Fig. 19.5 when $x_s = 100\lambda$, $z_a = z_s = 15\lambda$, and there are 21 array elements placed one wavelength apart with A_n and B_n uniform across the array. This figure illustrates how acoustic time reversal uses multipath propagation to improve focusing. Two important phenomena are shown. First, the amplitude of the field produced by the array at the source location ($\Delta z = 0$ in Fig. 19.5) increases when the reflection coefficient increases. In fact

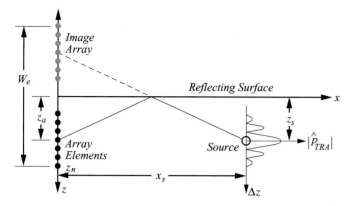

Fig. 19.4. Source–array geometry for a time-reversing array that responds to a sound source in a uniform environment bounded by a reflecting surface lying in the x–y plane. The amplitude of the field from the array on a vertical cut through the source location is illustrated.

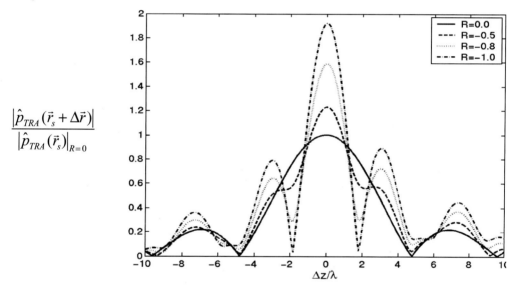

$$\frac{\left|\hat{p}_{TRA}\left(\vec{r}_s + \Delta\vec{r}\right)\right|}{\left|\hat{p}_{TRA}\left(\vec{r}_s\right)\right|_{R=0}}$$

Fig. 19.5. Normalized pressure field amplitudes along a vertical cut ($\Delta\vec{r} = \Delta z\,\mathbf{k}$) through the source location for a 21-element vertical linear time-reversing array centered 15λ below a reflecting surface with reflection coefficient R. The source is located 100λ from the array and is also 15λ below the reflecting surface. The field amplitude at the source location ($\Delta z = 0$) increases and the transverse size of the retrofocus region decreases as the strength of the surface reflection increases.

for this environment, it can be shown that $\hat{p}_{TRA}(\vec{r}_s)$ is almost proportional to $1 + |R|^2$, subject to some limitations imposed by source–array–surface geometry and diffraction. Therefore, as $|R|$ approaches unity in the present situation, $\hat{p}_{TRA}(\vec{r}_s)$ becomes nearly twice what it would be in an unbounded uniform environment where $|R| = 0$. This near doubling of retrofocus-amplitude occurs here because the back-propagating sound waves travel on two distinct paths having nearly equal transmission efficiency.

The second important phenomenon shown in Fig. 19.5 is the narrowing of the high-amplitude retrofocus region as $|R|$ increases. This occurs because the sound traveling on the reflected path increases the effective aperture, W_e, of the array. For example, when $|R| = 1$, the method of images suggests that the field at the source location in the bounded environment can be determined by super-imposing the field produced by the actual array plus that from a fictitious image array positioned above the $x–y$ plane (see Fig. 19.4). Therefore, as $|R|$ increases from zero to unity, the effective aperture of the array increases from 20λ to 50λ and the transverse focal size, w_z, decreases by a factor of $50/20 = 2.5$.

Both of these important phenomena persist when acoustic time reversal is conducted in ocean waveguides. The beneficial effects of multiple propagation paths are compounding so that TRA-produced retrofocus sizes may be much smaller than the free-space diffraction limit. Here, insights can again be gleaned from the method of images. Although it is only precisely applicable to range-independent constant sound-speed sound channels, the method of images well describes most of the shallow-water time-reversal results presented in the remainder of this section. In fact, when the ocean surface and bottom are treated as flat reflectors with R at the surface equal to -1 and R at the bottom determined by its geoacoustic properties, expressions for TRA retrofocus sizes can be derived from the method of images (Kim et al., 2001). Analysis of experimental data indicates that the measured TRA focal sizes approach the in-waveguide diffraction limit of the array.

The best way to illustrate these focusing improvements is by examining the experimentally measured focusing characteristics of a TRA placed in an ocean waveguide. From 1996 onward, a group of researchers led by Dr. William Kuperman, Scripps Institution of Oceanography, and Dr. Tuncay Akal, SACLANTCEN Undersea Research Center, conducted a series of ocean acoustic experiments in which a time-reversing array was deployed in coastal waters and retrofocus characteristics were measured for signals having center frequencies of 450 Hz and 3500 Hz. These experiments highlight the basic physics of acoustic time reversal in the ocean and allude to possible applications (see Section 19.5). In particular, the ocean experiments showed that time reversal is robust enough to produce diffraction-limited focusing out to a 30 km range. In addition, the TRA focus region was found to be stable at 450 Hz

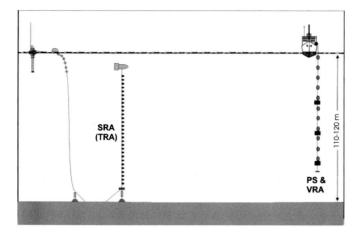

Fig. 19.6. Experimental set-up of the time-reversal experiments in the ocean.

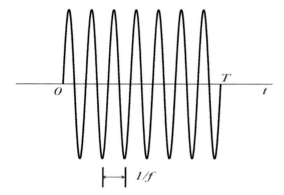

Fig. 19.7. Illustration of the signal broadcast by the probe source.

for repeated array transmissions covering time periods up to a week or more.

Figure 19.6 shows the components of these time-reversal experiments. Two arrays were needed; one for time-reversing measured sounds (the TRA; it was called the source-receive array or SRA in these experiments), and one for measuring the time-reversed field (a vertical receive array or VRA). Here, a sound source (called a probe source, PS, in the experiments) starts the process by emitting an acoustic signal that travels through the environment to the TRA. This source was located at three different depths indicated by the rectangles on the VRA.

The signal waveform broadcast from the source was a time-gated pure-tone pulse with a center frequency of 450 Hz or 3500 Hz as shown in Fig. 19.7. For the 450 Hz and 3500 Hz experiments, the pulse durations

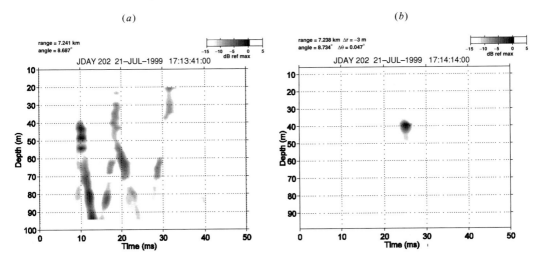

Fig. 19.8. Experimental results for the 2 ms, 3500 Hz signal at a range of about 8 km. (*a*) Probe pulse signal amplitude received on the TRA for a PS depth of 40 m. (*b*) The data received on the VRA from the time-reversed transmission of the time-dispersed signals shown in (*a*).

were 50 ms, and either 2 ms or 10 ms, respectively. The signal received by the TRA is illustrated in Fig. 19.8(*a*) for a 2 ms probe pulse at 3500 Hz transmitted from a source at a depth of 40 m located 8 km from the TRA. The gray-scale plots indicate signal envelope amplitude, the horizontal axis is time, and the vertical axis is depth for both parts of Fig. 19.8. At a given depth, 60 meters for example, the 2 ms original signal has been dispersed over approximately 20 ms and a half dozen or so propagation paths are apparent. When this time-dispersed signal at each hydrophone depth is time-reversed and retransmitted, the signal's multipath structure is undone, and a spatial and temporal retrofocus is formed at the original sound source position.

As shown in Fig. 19.8(*b*), this retrofocus is well defined and formed at the depth of the source. The size of the focal region depends on the signal frequency, the effective aperture of the TRA, and the waveguide characteristics. Here, the measured vertical retrofocus size is roughly six to seven times smaller than that for an equivalent TRA in an unbounded environment. In addition, the time-dispersed signal, recorded at the TRA, has been recompressed to the 2 ms duration of the original signal.

Such spatial and temporal signal compression persists at both shorter and longer ranges. Figure 19.9 shows the focal sizes at various ranges for two different frequencies. The spatial focus in depth broadens with range due to mode stripping caused by acoustic attenuation, but the vertical resolution increases with frequency. In addition, Figs. 19.9(*b*) and 19.9(*d*)

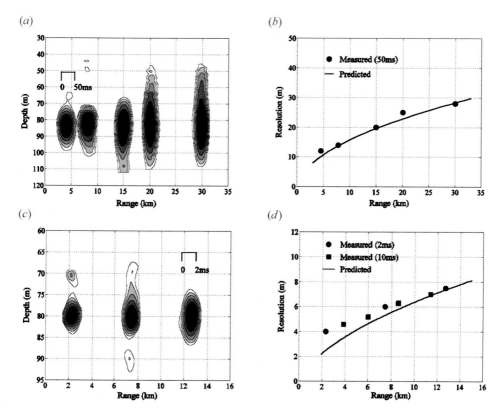

Fig. 19.9. Measured vertical focal sizes at different ranges from the TRA: (*a*)–(*b*) for 450 Hz and (*c*)–(*d*) for 3500 Hz. The contour interval is 2 dB, and (*b*) and (*d*) show that the method-of-images predictions agree with the measured sizes.

show that the method-of-images predictions are in good agreement with the measured focal sizes (Kim *et al.*, 2001).

The experiments also confirmed that the TRA-produced field focuses at the correct range. This was easily accomplished by moving the PS away from the VRA while fixing the VRA–TRA geometry. Figure 19.10(*a*) shows an example of these results when the PS is 700 m further away from the VRA and broadcasts the 450 Hz signal from a depth of 82 m at a nominal range of 15 km. The vertical extent of non-trivial signal amplitude clearly indicates that the TRA-produced field is out-of-focus at this VRA location when compared to the compact field amplitude distribution at the probe source location which is similar to Fig. 19.10(*b*) in this case. Additional special details of Fig. 19.10(*b*) are discussed in Section 19.5.

The experiments also investigated how long a retrofocus would persist if the same received signal was repeatedly retransmitted by the TRA as the ocean environment evolved. A basic requirement for

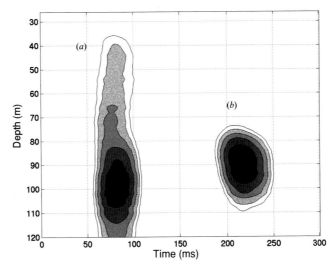

Fig. 19.10. TRA-produced field amplitudes for a 50 ms, 450 Hz PS at a depth of 82 m. (*a*) Out-of-focus results on the VRA when the VRA is 700 m outbound of the source location. (*b*) Same as (*a*) except a +30 Hz frequency shift has been applied to the field received at the TRA prior to retransmission.

successful TRA retrofocusing is that the acoustic environment should not change significantly between the source broadcast and the retro-focus formation. Consequently, TRA retrofocusing should degrade as this overall source-broadcast to focus-formation duration increases in time-dependent ocean environments. Although this phenomenon was observed, the time scales for degradation were surprisingly long at 450 Hz as shown in Fig. 19.11 where retrofocus field amplitudes are shown for overall time durations of 16 minutes, one day, and one week. These results imply that there is substantial temporal consistency (or reciprocity) in the multipath structure of this ocean sound channel that can be exploited by acoustic time reversal. However, stable focusing only lasted on the order of a few tens of minutes at 3500 Hz, implying that the effect of environmental change is not trivial for higher frequency sound propagation.

19.4 Limitations of acoustic time reversal

In underwater environments, the focusing properties of TRAs are limited by insufficient array aperture and/or number of elements, acoustic absorption, noise, reverberation from bottom and surface roughness, time-varying ocean characteristics, and source or array motion. In general, decreasing the array aperture or increasing the extent or severity of any of the other phenomena degrades the effectiveness of acoustic time reversal.

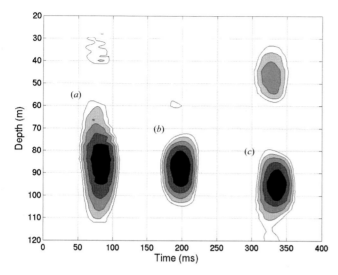

Fig. 19.11. Retrofocus amplitudes for a 50 ms, 450 Hz signal pulse emitted from a source 15.2 km from the TRA at a depth of 81 m for increasing time delay between source broadcast and retrofocus formation. The time delays are (*a*) 16 min, (*b*) 1 day, and (*c*) 1 week.

Reducing array aperture or element number prevents a TRA from fully sampling the information in the source-to-array propagating wave fronts. Thus, small aperture or low element-number TRAs, in any environment, will only time reverse a fraction of the acoustic wave fronts that are reversed by larger aperture or higher element-number TRAs under identical circumstances. Consequently, small TRAs with few elements do not perform as well as large arrays with many elements. However, a small aperture TRA with few elements may be more cost effective in some potential applications than a large TRA with many elements.

The influence of reduced TRA aperture and element number was measured in coastal waters in 1997 using a TRA composed of 23 elements spanning 77 m of a 125 m water column. Various subsets of elements were used for retransmitting the time-reversed signal (Kuperman *et al.*, 1998). With a single element TRA, the results matched the classic 1960s results; temporal signal compression was achieved but vertical spatial focusing was not. Results from less severe aperture and element number reductions are shown in Fig. 19.12. By using all of the elements (Fig. 19.12(*a*)), the TRA-generated field focuses well with side lobe levels above and below the focal region suppressed by more than 15 dB. Comparison of Figs. 19.12(*a*) and (*c*) shows the influence of decreasing the array aperture by a factor of approximately four. The temporal signal compression is still good but the vertical spatial focusing is degraded.

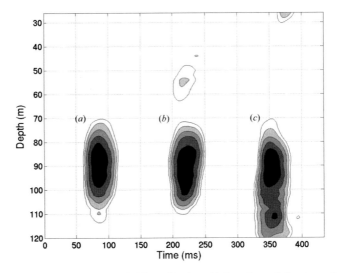

Fig. 19.12. Retrofocus field amplitudes with less than a full water column TRA at 450 Hz. (*a*) Full aperture (77 m) using all 23 elements, (*b*) same aperture as (*a*) with every other element used, and (*c*) 6-element array having one quarter the aperture of the array in (*a*).

Figure 19.12(*b*) shows what happens when the TRA is undersampled so that its aperture is the same as (*a*) but its element number is approximately halved and its element spacing is doubled. Significant spatial focusing was still achieved although the side lobe levels rose. These results suggest that the time reversal process can remain effective as a focusing procedure at 450 Hz when the aperture or number of elements is reduced in a downward refracting ocean environment as long as the array is optimally placed.

In ocean sound channels, limitations of acoustic time reversal also arise from the array orientation. For example, the focusing abilities of vertical and horizontal TRAs are not equal. However, disparities may be moderated or even eliminated by random scattering within the environment.

The effects of acoustic absorption on TRA retrofocusing are much like reducing the TRA's effective aperture; increased absorption during source-to-array propagation prevents wave front information from reaching the TRA, and prevents the time-reversed field created by the TRA from arriving undiminished at the source location. Figures 19.5 and 19.9 show this effect. Changing $|R|$ from unity to zero in Fig. 19.5 can be interpreted as changing the reflected-wave absorption loss from negligible to complete. Thus, the curves in Fig. 19.5 show that retrofocus amplitude decreases and retrofocus size increases as surface absorption increases. Similarly, absorption in an ocean sound channel leads to mode stripping and is a cumulative effect with increasing range. Therefore, the retrofocus degradation shown in Fig. 19.9 can be largely attributed to the increase in acoustic absorption at longer ranges (Kim *et al.*, 2001).

The level of environmental noise impacts acoustic time reversal because noise has the potential to masquerade as signal during the recording step at the array, so that it steals broadcast power from the intended signal during the array transmission step. Furthermore, any noise unintentionally transmitted by the array adds to the environmental noise and further corrupts the intended signal when the retrofocus is formed.

Time dependence in the acoustic environment (surface and internal waves, tides, currents, turbulence, etc.) or in the source–array geometry (i.e., source or array motion) also degrades a TRA's ability to focus sound. When the environment or geometry differs between the forward source-to-array propagation and the backward array-to-source propagation, unintended wave front distortion will arise because of scattering, diffraction, and other propagation differences induced by mismatch in the environment and the geometry, and because of Doppler shifts in the forward or backward propagating signals. A sample of potential environmental effects is displayed in Fig. 19.11 which shows that the retrofocus deteriorates and side lobes grow as environmental changes become more pronounced due to the passage of time. However, ocean dynamics are frequently slow or mild enough so as not to preclude effective use of acoustic time reversal over time periods of minutes to hours. Similarly, motion of the source, the array, or the medium in an ocean sound channel may cause the retrofocus location to shift and the retrofocus characteristics to degrade, depending on the geometry of the source, array, and direction(s) of motion.

19.5 Adaptations of acoustic time reversal

This section presents a few short descriptions of how acoustic time reversal is being adapted and implemented for use in practical applications. First of all, merely forming a retrofocus may not be sufficient for some active sonar tasks where a region of interest must be scanned. Acoustic time reversal can be adapted to this task in a waveguide through the waveguide invariant theory (Grachev, 1993) that relates the acoustic intensity in range and frequency. This capability is illustrated in Fig. 19.10(*b*) where the retrofocus is shifted 700 m in range to the VRA by a 30 Hz frequency shift at the TRA prior to retransmission. In pursuit of eventual applications in active sonar and underwater acoustic communication, the waveguide invariant ideas have been blended with non-linear adaptive beamforming techniques either to place a null or to broaden a null in the TRA-produced field. In addition, adaptive methods for robust time-reversal focusing in a fluctuating ocean environment have been developed using the waveguide invariant theory.

Because time reversal is an environmentally self-adaptive process, it can be applicable to localization and communication in complicated ocean environments. The fact that a TRA both spatially and temporally refocuses energy with the aid of a probe signal suggests that ocean self-equalization can mitigate the inter-symbol interference (ISI) caused by multipath dispersion. This self-equalization process in underwater communications has been demonstrated experimentally in both active and passive implementations of acoustic time reversal.

Typically, active sonar systems operate in reverberation-limited environments where target echoes can be masked by reverberation from bottom roughness. Here, time reversal can be used to enhance the echo-to-reverberation ratio as demonstrated in a recent experiment. On the other hand, reverberation returns from a rough sediment interface can be utilized to probe the waveguide rather than being considered as a source of clutter in sonar performance. In particular, the reverberation return can be focused back to the rough surface interface without a probe source.

For multiple scatterers in free space, the time-reversal process can be iterated in order to focus an active sonar system on the most reflective scatterer. When an ocean sound channel contains several scatterers, iterative time reversal focuses on the scatterer that provides the strongest return to the array. This scatterer is selected based on a combination of its scattering strength and the sound propagation characteristics between the scatterer and the array.

Another application of time reversal is an acoustic barrier concept that would detect the presence of an intruder who crosses a trip line or trip plane. The classic difficulty in constructing such an acoustic barrier is that the scattered field from the target must be extracted from the much more intense, and usually fluctuating, direct arriving beam – the acoustic equivalent of "looking into the sun" to find an airborne object. As shown in Fig. 19.8(b), a TRA's acoustic field away from the retrofocus is typically 15–20 dB less than the field at the retrofocus. Any disturbance of the sound propagation between the TRA and the retrofocus will partially fill-in quiescent regions. Thus, detection of sound in a quiescent region is the diagnostic of the trip-wire barrier. Such a barrier concept was demonstrated successfully in a time-reversal experiment at 3500 Hz.

19.6 The future

In the future, acoustic time reversal and its many varied implementations are likely to form the foundation for environmentally-adaptive sonar systems. In fact, the technique and its limitations are now understood well enough that applications of time reversal in other fields – beyond optics, bio-medical ultrasound, and ocean acoustics – are likely to emerge. The concept of time reversal can be readily applied to electromagnetic, seismic, and structural waves. Time reversal can also be applied to many mechanical systems, and even to water waves. Thus the impact of time reversal concepts may be felt in the areas of telecommunications, radar, remote sensing, surveillance, transduction, and non-destructive evaluation, to name a few of the many exciting possibilities.

Chapter 20

Studies of turbulent processes using Doppler and acoustical scintillation techniques

DANIELA DI IORIO
University of Georgia

ANN E. GARGETT
Old Dominion University, Virginia

Summary

Turbulent water motions, occurring over a wide range of space and time scales, affect practically every aspect of oceanography. Turbulence acts to transport momentum and energy from the scales of the largest turbulent eddies (typically a few to 10s of meters), down to smaller scales where turbulent energy can be dissipated into heat. At the same time, turbulence acts to homogenize fluid properties such as temperature or salinity. Recently developed high-frequency acoustic techniques are now able to provide measurements of temporally or spatially averaged turbulence quantities, continuously in time. This chapter describes two acoustical approaches (backscatter and forward-scatter) that allow estimation of important turbulence quantities – such as dissipation rates of turbulent kinetic energy (TKE) and scalar variance, vertical momentum stresses, and characteristics of the large-scale turbulent eddies – as well as determination of the mean flows that generate them. In the backscatter technique, acoustic Doppler current measurements made along four slanted beams allow direct measurement of turbulent stresses, hence the rate of production of TKE. A fifth vertical beam, providing estimates of vertical velocities and their corresponding length scales, allows determination of the individual components of TKE, as well as its dissipation rate. In the forward-scatter approach, amplitude and phase fluctuations at a fixed depth are inverted to give path-averaged estimates of the mean and turbulent velocity and temperature levels, as well as their dissipation rates, by assuming statistical models for the turbulent random medium. This new ability to make long-term continuous measurements of crucial turbulence quantities promises major advances, not only in the study of ocean turbulence itself, but also in many other areas of research that depend heavily upon the presence and impact of turbulent processes in the sea.

Contents

20.1 Introduction

Turbulent flows transfer mass, momentum, salt, heat, and other water properties from large scales where turbulence is generated to small scales where water properties are dissipated by molecular processes. Turbulence plays a key factor in resupplying nutrients to the surface layer so that plankton can grow and in planktonic predator/prey interactions. Turbulence affects gas flux across the air–sea interface, and transmission of heat in and out of the ocean reservoir, thus governing climate. Turbulence in bottom boundary layers affects sediment movement, deposition, and resuspension, and the flux of nutrients out of the benthic boundary layer. In the coastal ocean many substances (man made or natural) are discharged from land, become diluted by turbulent mixing processes and spread to the coastal sea by advective and dispersive processes. Recent advances in ocean acoustics are providing us with new means to quantify and monitor turbulent motions in the sea.

Turbulent fluid flow is random, strongly non-linear, and inherently three-dimensional. Existing theory (Tennekes and Lumley, 1972) suggests that energy and scalar variance are supplied to the largest eddies by instability of a mean flow and subsequently cascade to small scales through sequential instabilities. Energy and scalar variance are dissipated at the smallest available scales, determined by molecular characteristics of the fluid itself. The large-eddy scales, which contain most of the turbulent kinetic and potential energies, may be substantially anisotropic, i.e., their characteristics vary with the spatial angle at which they are observed. Anisotropy may result either from the properties of the mean flow, or through inhibition of vertical velocities by the stable stratification that is normal in the ocean interior. However, it is generally assumed that such anisotropy disappears by a sufficiently small scale, L_0. Within a range of scales between L_0 and dissipation scales, known as the inertial subrange, turbulent eddies may be considered isotropic. Values of L_0

typically range from a few centimeters in strong diurnal pycnoclines (a region of enhanced vertical gradient in density), through tens of centimeters to a few meters in the main pycnocline, and up to tens of meters in very energetic coastal and deep ocean flows.

In addition to its fundamental dissipative and diffusive characteristics, turbulence produces vertical fluxes of momentum and various scalars (such as temperature and chemical species), fluxes that are of major importance to problems as diverse as the stability of the thermohaline circulation of the ocean and the maintenance of marine ecosystems. In sheared flow, a simplified form of the steady-state equation for turbulent kinetic energy per unit mass (E) is

$$-\overline{v'w'} \frac{\partial V}{\partial z} - \overline{u'w'} \frac{\partial U}{\partial z} = \frac{g}{\rho_0} \overline{\rho' w'} + \varepsilon \qquad (20.1)$$

(Tennekes and Lumley, 1972), where z is the vertical direction, $U = [U(z), V(z), W(z)]$ is the mean flow, $u' = [u', v', w']$ is the turbulent velocity vector, ρ_0 is a reference density, ρ' is the turbulent density fluctuation, g is the gravitational acceleration, and an overbar represents a time average. The terms on the right-hand side represent losses of E: the first term is the vertical buoyancy flux, associated with work against gravity in a stably stratified fluid, and the second term ε represents loss of energy to heat by molecular viscosity (in a similar equation describing the steady-state balance of temperature variance, ε_T represents the dissipation rate of $(\frac{1}{2} \overline{T'^2})$ by molecular diffusion). The terms on the left-hand side represent production of E by interaction of the mean flow shear with Reynolds stresses, $-\overline{u'w'}$ and $-\overline{v'w'}$, which represent the vertical flux of horizontal momentum.

Observational turbulence studies attempt to quantify some of the terms in the turbulent kinetic energy (20.1), since this equation describes the dynamics of turbulent oceanic boundary layers and free shear flows. This chapter will summarize the increasingly important role that high frequency acoustical methods are playing in the measurement of these small- and large-scale processes in the ocean.

20.2 Measuring turbulence with an acoustic Doppler current profiler

Since the mid-1980s, oceanographers have increasingly used the acoustic Doppler current profiler (ADCP) frequency shift to measure vertical profiles of horizontal currents. Sections 2.6.1 and 2.6.2 have already discussed the general operating principles of Doppler sonars, in which the Doppler frequency shift associated with scattering from acoustic targets is used to determine the speed of the water containing them. A standard commercial ADCP, like that shown in Figs. 3.15 and 3.16 (see also

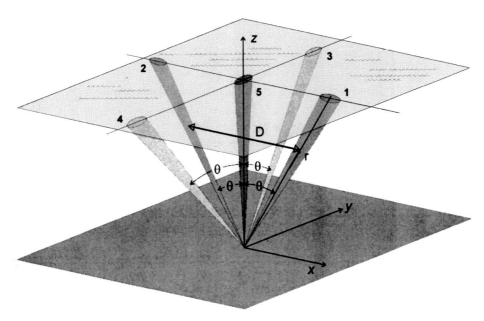

Fig. 20.1. A standard 4-beam ADCP is made up of two pairs $\{1,2\}$ and $\{3,4\}$ of sonar beams, shown here aligned along the x axis and y axis, respectively. Each beam in a pair makes an angle of θ with the vertical. Addition of a vertical fifth beam $\{5\}$ enables improved measurement of turbulence as well as mean flow quantities.

Fig. 20.1), has four slanted beams arranged so that the acoustic beam from each individual transducer makes the same angle (θ) with the local vertical (z), and pairs of slant beams $\{1,2\}$ and $\{3,4\}$ lie in the x–z and y–z planes, respectively. A short burst of transmitted sound is scattered back to the transducer with a Doppler shift determined by the component of water velocity that is parallel to the axis of the acoustic beam: we call this a "beam velocity." Since sound from further away takes longer to return, measuring the Doppler shift at the receiver as a function of time is the same as measuring beam velocity as a function of range r along the beam. How are the four beam velocities $B_q(r)$, $q = [1\ldots, 4]$ then used to calculate the three components of water velocity?

20.2.1 Determination of mean velocities

For now, suppose that the water velocity is constant everywhere. Then at any range, the slant beam velocities (defined as positive towards the transducer) are related to the constant fluid velocity $\boldsymbol{U} = [U, V, W]$ by

$$B_1 = -U \sin\theta - W \cos\theta \qquad (20.2)$$

$$B_2 = U \sin\theta - W \cos\theta \qquad (20.3)$$

$$B_3 = -V \sin\theta - W \cos\theta \qquad (20.4)$$

$$B_4 = V \sin\theta - W \cos\theta \qquad (20.5)$$

and we can solve (20.2)–(20.5) for the velocity components:

$$U = \frac{\overline{B}_2 - \overline{B}_1}{2\sin\theta} \qquad V = \frac{\overline{B}_4 - \overline{B}_3}{2\sin\theta} \qquad W = -\sum_{i=1}^{4}\frac{\overline{B}_i}{4\cos\theta} \qquad (20.6)$$

Now consider a more general case where the water velocity $U(z)$ varies only in the z direction. At range r, corresponding to height $z = r\cos\theta$ above bottom, the four slant beams lie on a circle of diameter $D(z) = 2\,r\sin\theta$ (the beam "spread," shown in Fig. 20.1). The flow field is uniform everywhere in the horizontal, so over the beam spread, $u_q(z) = U(z)$ for all q. Using the beam velocities at range r in (20.6) then provides an estimate of $U(z)$ at $z = r\cos\theta$. The collection of such estimates made over the full acoustic range provides the standard ADCP determination of velocity component profiles as functions of z, under the implicit assumption that the velocity field is horizontally uniform, at least over the beam spread.

20.2.2 Determination of turbulence quantities

In the more general case typical of the ocean, the instantaneous water velocity $u = U + u'$ consists of two parts, a time-mean field $U(z)$ that is a function only of z and a three-dimensional fluctuating field u' with zero time mean $\overline{u'} = 0$ and (turbulent) kinetic energy per unit mass $E = 1/2(\overline{u'^2 + v'^2 + w'^2})$. In this case, the beam velocities $B_q = \overline{B}_q + B_{qf}$ are also made up of two parts: a time-mean $\overline{B}_q(z)$ and a turbulent part B_{qf} with zero mean $\overline{B}_{qf} = 0$ and variance $\overline{B^2_{qf}}$. Because three-dimensional turbulent motions can have spatial scales smaller than the spread between slant beams (particularly as this spread increases with distance from the transducers), the turbulent part of the velocity field is *not* uniform over the horizontal beam spread, so cannot be determined from expressions like those in (20.6).

However as early as 1983, Lhermitte proposed that certain turbulent quantities, specifically the turbulent stresses $-\overline{u'w'}$ and $-\overline{v'w'}$, *could* be estimated from radial current measurements made with a standard slant-beam ADCP, using the much weaker assumption that second-order *statistics* of the turbulent field are the same over the beam spread, i.e., that

$$\overline{u'w'} = \overline{u'_1 w'_1} = \overline{u'_2 w'_2} = \ldots$$
$$\overline{(u')^2} = \overline{(u'_1)^2} = \overline{(u'_2)^2} = \ldots, \text{ etc.}$$

Subtracting time averages of (20.2)–(20.5) from the original equations, then squaring, forms equations for variances of the fluctuating part of the beam velocities. Using beam pair $\{1,2\}$ as an example yields

$$\overline{B^2_{1f}} = \overline{(B_1 - \overline{B}_1)^2} = \overline{(u')^2}\sin^2\theta + \overline{(w')^2}\cos^2\theta + \overline{u'w'}\sin 2\theta, \quad (20.7)$$
$$\overline{B^2_{2f}} = \overline{(B_2 - \overline{B}_2)^2} = \overline{(u')^2}\sin^2\theta + \overline{(w')^2}\cos^2\theta - \overline{u'w'}\sin 2\theta, \quad (20.8)$$

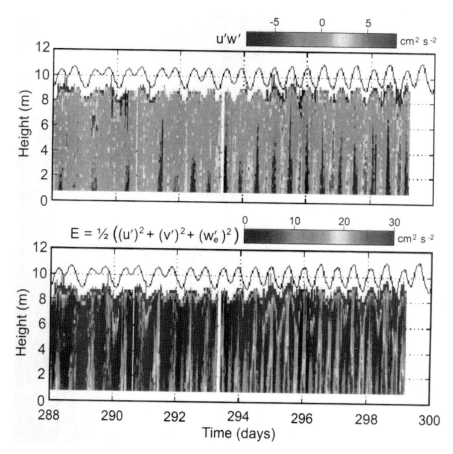

Fig. 20.2. Turbulent shear stress (upper panel) and turbulent kinetic energy (lower panel) determined over a period of several days by a bottom-mounted ADCP in a shallow tidal channel off San Francisco Bay (data courtesy of M. Stacey and S. Monismith). (See colour plate section.)

when Lhermitte's assumption is used. Thus the turbulent momentum stress $-\overline{u'w'}$ can be estimated as

$$-\overline{u'w'} = \frac{\overline{B_{2f}^2} - \overline{B_{1f}^2}}{2 \sin 2\theta},\tag{20.9}$$

and $-\overline{v'w'}$ can be determined from a similar expression using beam pair $\{3,4\}$.

Lhermitte's (1983) suggestion has been subsequently used in a variety of shallow coastal ocean regimes. As an example of such measurements, the upper panel of Fig. 20.2 illustrates the along-channel stress component measured with an upward-looking ADCP mounted on the bottom of a narrow, flat-bottomed estuarine channel at the north end of San Francisco Bay (Stacey et al., 1999). The upper black curve shows a predominantly semi-diurnal period of variation in surface water height, associated with the tidal flows that generate turbulence in this environment (in this protected location, there were no significant surface waves). Tidal currents increase with the increasing spring tidal amplitudes seen toward the end of the record. Both the degree of bottom-intensification of the stresses and the variation in their magnitude and sign with changes in tidal current speed and direction are as expected for flow over a flat solid boundary.

Other important turbulence parameters are the turbulent kinetic energy per unit mass, $E = {}^1\!/_2\overline{(u'^2 + v'^2 + w'^2)}$ and its rate of dissipation, $\varepsilon = dE/dt$ (with units of W/kg $= m^2\ s^{-3}$). With only four variance equations ((20.7) and (20.8) plus similar equations for $\overline{B_{3f}^2}$ and $\overline{B_{4f}^2}$) but five unknowns (the two stress components plus the three velocity variances), E can be determined from measurements with a standard ADCP only if assumptions can be made about the relative magnitudes of the turbulent velocity components. Reliable relationships among these three components exist in the well-studied case of boundary layer flow over a flat rough bottom, permitting the estimates of E that are shown in the lower panel of Fig. 20.2. However such relationships are not available for more complex (hence more realistic) situations involving bottom irregularities, water column stratification, etc. Thus for the general study of turbulence in the ocean, it is essential to add a fifth beam to the standard four-beam ADCP. Addition of a fifth beam, oriented vertically so that $\overline{B_{5f}^2} = \overline{w'^2}$ as shown in Fig. 20.1, allows all three components of the turbulent kinetic energy to be determined directly and individually.

Such a five-beam ADCP (hereafter called a VADCP) also enables estimation of ε via a "large-eddy" algorithm (Gargett, 1999),

$$\varepsilon = C_\varepsilon \frac{w^3}{\ell_v}, \tag{20.10}$$

where C_ε is a dimensionless constant, and both $w =$ root mean square vertical velocity and $\ell_v =$ vertical length scale associated with the energy-containing eddies of a turbulent field are determined from the vertical beam measurement of vertical velocity. This algorithm is based on a widely used scale relationship originally suggested by Taylor (1935) for isotropic turbulence characterized by single length and velocity scales. However the presence of stable stratification and/or horizontal boundaries acts to "squash" the large eddies in the vertical, so the energy-containing scales of turbulence may actually be substantially anisotropic, in the sense $\ell_v < \ell_h =$ the horizontal large-eddy length scale. In such situations, use of the vertical length scale ℓ_v in (20.10) is justified by arguing that ℓ_v is the largest scale that can possibly be isotropic, hence serve as the low wavenumber end of an inertial subrange, i.e., of the irreversible cascade of energy to dissipation scales that describes the effect of turbulence on the energy of a fluid flow.

Using a subset of data taken with a VADCP in coastal tidal fronts, a value for C_ε was determined by comparing the large-eddy estimate (20.10) with estimates of ε calculated from a totally different technique, simultaneous vertical profiles with an instrument carrying microscale shear probes (Osborn, 1974). The resulting algorithm was verified by comparing with other microscale profiler data taken on the same cruise, as well as independently during a subsequent cruise. Since the oceanic dissipation rate can vary over several orders of magnitude, it is customary to use its logarithm for display purposes; thus the color-coded background in the panels of Fig. 20.3 illustrates the two-dimensional field of $\log \varepsilon$ determined by this acoustic technique as a ship-mounted VADCP traversed a tidal front in the coastal waters of British Columbia. The superimposed profiles of $\log \varepsilon$ from microscale measurements exhibit magnitudes and vertical distributions that are consistent with local statistics of the field determined by the large-eddy technique. The VADCP technique delineates the complex spatial structure of dissipation in this turbulent front, as well as its temporal evolution as tides increase from neap towards spring, revealing a wealth of detail that cannot be achieved with the intermittent profiling technique.

20.3 Acoustical scintillation measurements of flow and turbulence

Doppler methods use backscattered sound. An alternate method, using forward scattered sound with the transmitter and receiver separately located (see Fig. 20.4(a)), is referred to as acoustical scintillation. Over short distances, using high frequencies and high transmission rates, amplitude and phase fluctuations measured over spaced receivers can be used to infer properties of both the mean and turbulent flow fields, averaged along the acoustic path over the range separating transmitter and receivers.

"High frequency" refers to signals ranging from 50 kHz to several hundred kHz with propagation range (L) generally a few 100 m to a few km. Acoustic amplitude fluctuations are sensitive to spatial structure in turbulence. For a given range, the acoustic frequency is chosen so that scales within the inertial subrange are sampled: spatial scales in various experiments have ranged from 1 m to 7 m. Typical sampling rates of 5 to 16 Hz ensure resolution of the small spatial scales of the inertial subrange under the Taylor hypothesis, which relates frequency and wavelength (f, ℓ) by $f = {}^U\!/_\ell$, under the assumption that turbulent eddies are essentially frozen as they are advected by the mean flow U.

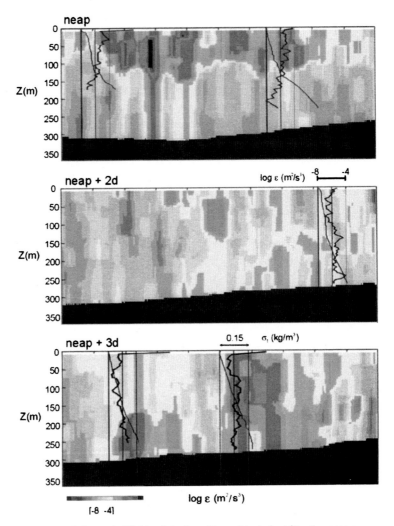

Fig. 20.3. Color coded fields of the logarithm of turbulent kinetic energy dissipation rate ε, determined from the vertical velocity field measured by the vertical beam of a VADCP crossing a tidal front in the coastal waters of British Columbia, as tides increase from neap (top) towards spring. Superimposed profiles of ε (heavy lines) are used to calibrate the acoustic technique, while those of density (σ_t, light lines) show the effect of turbulence in eroding the stratification of the water column. (See colour plate section.)

The application of acoustical scintillation to ocean measurement has its origin in the interpretation of electromagnetic or acoustic signal modulations from atmospheric refractive index irregularities convected across the path between transmitter and receiver (Tatarski, 1971). In the ocean, refractive index irregularities associated with variability in temperature and salinity distort the spherically spreading acoustical wave

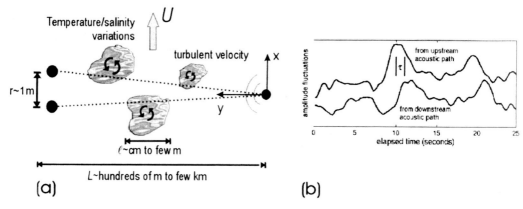

Fig. 20.4. (*a*) Acoustic propagation from a single source with reception from two horizontally spaced receivers, resulting in two diverging acoustical paths. Effective refractive index fluctuations arise from temperature/salinity variations and turbulent velocity and are advected by the mean flow *U*. (*b*) Amplitude fluctuations at the two receivers show a time lag τ between the upstream acoustic path and downstream acoustic path, a lag that corresponds to the time it takes for turbulence structures to cross the acoustic paths.

fronts as they travel away from the transmitter, resulting in amplitude and phase (time of arrival) fluctuations at the receivers. These fluctuations are referred to as scintillations and are analogous to the "twinkling" of stars as a result of the turbulent atmosphere. For acoustical propagation in strong coastal flows, small-scale fluctuations in the flow velocity can also produce an important scattering effect.

Fine structure variability in the medium transported by the mean current will create a space–time scintillation pattern over horizontally spaced receivers. Detection of the motion of this pattern is the essential concept exploited in acoustical scintillation. Thus a crucial requirement of the scintillation method for mean flow measurements is that there be sufficient correlation between signals detected at two horizontally spaced receivers a distance *r* apart. In the simplest configuration (Fig. 20.4(*a*)) a signal is transmitted from a single source and received at two points aligned with the mean flow (perpendicular to the acoustic path). Effective refractive index perturbations from temperature, salinity, and velocity fluctuations passing through a source/receiver path create amplitude and phase fluctuations at the receiver. For a given mean flow *U*, the scintillation pattern will first be detected at the upstream acoustic path, and then a small time τ later, at the downstream path, as shown in Fig. 20.4(*b*). The path averaged mean current speed is then estimated as $U = r/(2\tau)$, where $r/2$ is the path averaged separation between the two acoustic paths. Although this scintillation drift method requires the smallest number of components, the calculation is complicated by the beam divergence,

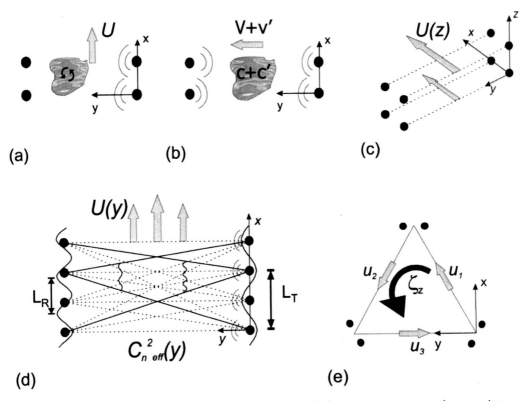

Fig. 20.5. (*a*) Parallel acoustic paths between two sources and two receivers. (*b*) Reciprocal transmission, where all transducers transmit and receive. (*c*) Two-dimensional arrays for shear and turbulent anisotropy measurements. (*d*) Linear arrays for spatial filtering of flow and turbulence parameters along the path. (*e*) Triangular array with paired transducers at each corner for horizontal velocity, turbulence and vorticity measurements.

which weights the sensitivity to flow as a function of range toward the receiver array. The parallel path configuration shown in Fig. 20.5(*a*) is simpler to interpret because it produces a symmetric path dependent weighting function for amplitude fluctuations (Tatarski, 1971).

In reciprocal transmission mode, signals are transmitted simultaneously in opposite directions (Fig. 20.5(*b*)), giving travel times (measured from phase) T_+ and T_-. Because of the current velocity component v that is aligned *along* the acoustic propagation path L, acoustic travel times will be shortened if propagation is in the direction of the current ($T_+ = L/(c + v)$) and will be lengthened if propagation is opposite to the direction of the current ($T_- = L/(c - v)$), where c is the sound speed. The cross channel flow speed and sound speed measurements are then based on travel time differences and sums respectively. Accurate phase (i.e., travel time) measurements result in a 1 mm/s sensitivity

in the measurements of cross channel flow and sound speed for each
acoustic transmission. Averaging over a few minutes (many transmissions) gives path averaged time-mean values. By combining the previous
scintillation drift method with this reciprocal transmission method, the
two-dimensional path averaged mean velocity field (U,V) is recovered.

The additional information available from reciprocal transmission
also allows the determination of various statistical characteristics of the
turbulent fields v' and c'. Sound speed fluctuations are related to temperature and salinity fluctuations in a complicated way but by assuming
a mean temperature–salinity relation, a transfer function is obtained that
converts sound speed fluctuations to temperature fluctuations. For a moving random medium with scale sizes, hence wavelengths, ℓ within the
inertial subrange, the one-dimensional spectrum of the cross channel
velocity fluctuations (v') is,

$$F_v(k) = 0.327\varepsilon^{2/3}k^{-5/3} \qquad (20.11)$$

and for temperature fluctuations (T') it is,

$$F_T(k) = 0.474\varepsilon_T\varepsilon^{-1/3}k^{-5/3} \qquad (20.12)$$

where $k = {}^{2\pi}/_\ell$ is defined as the along channel wave number (Tennekes
and Lumley, 1972). The spectra, corrected for path averaging, are obtained from time series of the path averaged measurements of v' and c'
(hence T'). Frequency spectra are transformed to wave number space
assuming the Taylor hypothesis $(f = {}^U/_\ell)$. From the measured spectra
indirect estimates of the dissipation rates ε and ε_T are obtained.

Comparison of various acoustically derived fields with independent
moored measurements of mean and dissipation scale quantities is shown
in Fig. 20.6. Figure 20.6(a) compares mean along-channel flow U, derived from scintillation drift, and the mean cross-channel flow V and
sound speed c, derived from reciprocal transmission, with values obtained from an ADCP and a current meter sampling at mid-path. Because of path averaging, the acoustic time series are smoother than either
of the point measurements, an advantage if water transport rather than
the detailed flow structure is the measurement of interest. Figure 20.6(b)
presents a comparison of acoustically derived dissipation rates with point
estimates made by a tethered autonomous microstructure instrument
(TAMI)(Lueck et al., 1997). Again, an advantage of the acoustic method
is that much of the temporal intermittency in the dissipation rates, visible in the point measurements from TAMI, is removed as a result of the
acoustic spatial and temporal averaging.

Although reciprocal propagation is the preferred method for turbulence measurements, one-way propagation can provide some information on turbulence in special environments. As measured by one-way

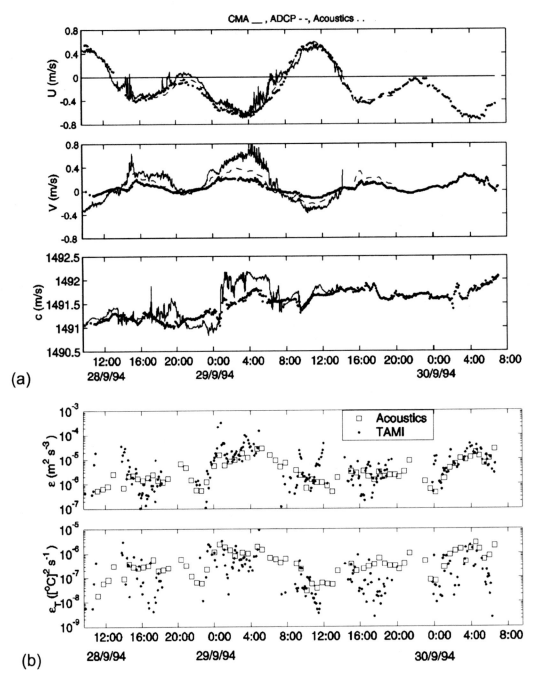

Fig. 20.6. (a) Mean current velocity (U, V) and sound speed (c) from acoustic scintillation drift and reciprocal transmission, together with *in situ* current meter and ADCP results. (b) Dissipation rate of turbulent kinetic energy and temperature variance (ε, ε_T) shown together with results from Tethered Autonomous Microstructure Instrument (TAMI), (data courtesy of R. Lueck).

propagation in isotropic turbulence, the effective refractive index fluctuations associated with both temperature/salinity fluctuations (sound-
speed) and turbulent velocity fluctuations is,

$$C^2_{n_{\text{eff}}} = \frac{F^2 C^2_T + \frac{11}{6} C^2_v}{c^2_0} \qquad (20.13)$$

where $C^2_{n_{\text{eff}}}$ is obtained from the acoustic log–amplitude variance, F
is the transfer function converting temperature fluctuations to sound
speed fluctuations, c^2_0 is the mean sound speed, $C^2_v = 1.97\varepsilon^{2/3}$ and
$C^2_T = 3.82\varepsilon_T\varepsilon^{-1/3}$. This theoretical relationship, developed by Ostachev
(1994), has been verified using simultaneous forward and reciprocal
transmissions (Di Iorio and Farmer, 1998). From one-way propagation
alone, it is impossible to determine which physical parameter contributes
to the total scattered signal and in what proportion, without independent
measurements of the temperature/salinity fluctuations or the turbulent
velocity structure. However, in special highly energetic and well mixed
environments, the scalar contribution can be negligible compared to that
associated with turbulent velocity. In these situations $C^2_{n_{\text{eff}}} = \frac{11}{6} C^2_v / c^2_0$,
and a path- and time-averaged measurement of ε can be obtained using
one-way propagation (Di Iorio and Farmer, 1996).

In these special situations where the current velocity fluctuations
dominate the acoustic scattering, one-way propagation with two-
dimensional arrays (Fig. 20.5(c)) can provide additional flow information. Vertical shear (dU/dz) of mean velocity can be estimated as a difference between the mean flows determined by the scintillation drift method
at two depths. Turbulent anisotropy can also be detected, through measurement of horizontal and vertical arrival angles (Di Iorio and Farmer,
1996). Horizontal arrival angle is proportional to the phase difference between horizontally separated receivers, thus related to $\partial v / \partial x$, the along
channel gradient of cross channel velocity. Vertical arrival angle is proportional to the phase difference between vertically separated receivers,
thus related to $\partial v / \partial z$, the vertical gradient of cross channel velocity. The
joint probability distribution of horizontal and vertical arrival angles, averaged over all receiver combinations, will be symmetric if horizontal
and vertical cross channel velocity gradients are statistically identical,
i.e., if the turbulence is isotropic. Asymmetry in the probability distribution reveals preferred directivity, hence anisotropy.

These concepts are illustrated by the observations in Fig. 20.7.
Figure 20.7(a) shows acoustic determinations of ε, U, and dU/dz; the
latter two fields compare well with estimates from a pair of current meters
moored in mid-channel at the depths of the two acoustic paths. The only
significant shear was observed during the strong flood tide that developed
between 04:00 h and 12:00 h. The dissipation rate rises and falls with the

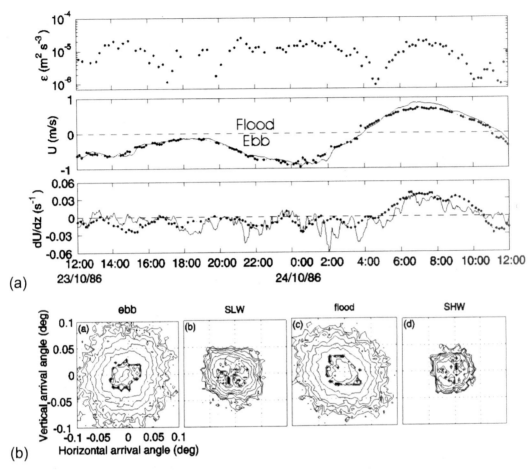

(a)

(b)

Fig. 20.7. (*a*) Estimates from one-way propagation of turbulent kinetic energy dissipation rate ε, the mean along channel current U, and the vertical shear dU/dz. Independent current measurements are from two vertically spaced current meters. (*b*) Angle of arrival distributions during ebb, slack low water (SLW), flood, and slack high water (SHW) taken between 00:00 h and 12:00 h.

tidal cycle, and has been shown to be consistent with a log–layer model of the bottom boundary layer. Figure 20.7(*b*) shows arrival angle distributions obtained during times of strong ebb, slack low water (SLW), flood, and slack high water (SHW) (left to right). Each contour represents the number of times the acoustic signal had a specific combination of horizontal and vertical arrival angles. The directivity pattern observed during the ebb is circular, suggesting isotropic acoustic scattering, hence isotropic turbulence (no preferred direction). However during the time of strong flood, also the time of strong shear, the directivity pattern becomes elliptical, implying anisotropic conditions (i.e., $\partial v/\partial x \neq \partial v/\partial z$) over scale sizes of a few meters. The enlarged directivity pattern under both flood and ebb conditions shows larger acoustic scatter, hence larger turbulence levels, than during slack water conditions.

The scintillation measurements described thus far were made with rigidly mounted transducers on shore-cabled tripods firmly fixed to the ocean bottom, allowing the precise phase (travel time) measurements required for reciprocal transmission. Autonomous moored systems, required for more general applications, will severely degrade the measurement of travel time, as a result of path length changes associated with mooring motion. However, acoustic amplitude fluctuations are insensitive to small path length changes, hence internally logging, battery operated acoustic scintillation systems can be used to measure path averaged flow and some turbulence characteristics during autonomous deployments. One such system (with vertically separated transmitters and receivers) was used to measure vertical flow and turbulent refractive index properties in the outflow from a hydrothermal vent on the Juan de Fuca Ridge. This same system was also used in the Strait of Istanbul (Di Iorio et al., 2004) (Bosporus) to measure the turbulent bottom boundary layer resulting from Mediterranean flow into the Black Sea (Di Iorio and Yuce, 1999). A vane attached to the instrument package, with swivels above and below, was necessary to keep the horizontal transmitter and receiver arrays aligned in the direction of flow.

An important further development in acoustical scintillation is the application of spatial aperture filtering, which uses linear transmitting and receiving arrays (Fig. 20.5(d)), to measure turbulence and flow parameters as a function of position along the path (Crawford et al., 1990). By using sinusoidal amplitude weighting functions on transmitters and receivers, the received signal can be made sensitive to a single path position y_s and a single spatial wavelength L_s of the intervening turbulent refractive index field. By changing the relative spatial wavelength of the transmitter and receiver filters, the system can be tuned to different path positions and can in principal yield cross channel profiles of $C^2_{n_{eff}}(y)$ and $U(y)$.

For a point transmitter, a sinusoidal spatially filtered receiver is only sensitive to one scale size of refractive variability and this scale size is $L_R = L_s(L/y)$. Similarly for a point receiver, a spatially filtered transmitter is only sensitive to $L_T = L_s(L/(L - y))$. Even though contributions are from the entire path, all but one refractive component at each path position has been filtered out. By combining spatially filtered transmitters and receivers, the system will be sensitive only to the single refractive index wavelength $L_s = L_T L_R/(L_T + L_R)$ at the single path position by $y_s = L_T L/(L_T + L_R)$, which satisfies the equations for L_T and L_R. In practice, the spatially filtered signal is generated by

$S_f = \sum_{i=1}^{4} \sum_{j=1}^{4} \alpha_i \alpha_j S_{ij}(t)$, where S_{ij} is the acoustic amplitude or phase for the jth receiver listening to the ith transmitter. The filter weights α for the transmitter, i, and receiver, j, vary between ± 1 to form the sinusoidal wavelengths for transmitter L_T and receiver L_R. The level of the effective refractive index fluctuations at location y_s is obtained from the variance of the spatially filtered signal ($C_{n_{eff}}^2(y_s) \propto \overline{S_f^2}$) and the velocity $U(y_s) = f_s L_s$ is obtained from the frequency f_s of the peak of the power spectral density of the spatially filtered signal.

An alternative triangular configuration with paired transducers, shown in Fig. 20.5(e), allows acoustical determination of the mean and turbulent quantities described previously, as well as an added determination of vertical vorticity (Menemenlis and Farmer, 1992). Transmissions from each transducer pair are made consecutively, while the remaining two pairs act as receivers. Once all signals from a paired transmission have been received, the next pair of transducers transmit. For a triangle with 200 m sides, the entire cycle around the triangle takes just under 1 s, resulting in reciprocal propagating paths along each side. From travel times along each path, path averaged values of current velocity (u_i in Fig. 20.5(e)) and sound speed are obtained along the three sides of the triangle. These three velocities can be used to obtain the mean two-dimensional current velocity (U, V), while estimates of the turbulent dissipations ε and ε_T can be made from the reciprocal transmissions along any side of the triangle (Menemenlis and Farmer, 1995). The unique contribution of this kind of configuration is the ability to measure relative vorticity at the scale of the array. The vertical component of vorticity ζ_z, a measure of fluid rotation in the horizontal plane, can be determined using Stoke's theorem for integration around the triangle,

$$\oint u \cdot dl = \iint (\nabla \times u)_z dx dy = \zeta_z A \qquad (20.14)$$

where A is the area of the triangle. The line integral on the left hand side of this equation can be evaluated from the sum of the three measured velocities, scaled by the respective path lengths. Changes in vorticity may arise from frictional effects within a boundary layer flow and from changes in mixed layer depth as internal waves pass.

20.4 The future

Much work remains to be done in extending both of the techniques discussed here to more general oceanic conditions. In the Doppler techniques, the turbulent stress algorithm must be verified in regimes other than the generally flat-bottomed and weakly stratified environments in which it has proven successful to date. Similarly, the large-eddy technique for ε has been verified only for strongly turbulent flows. In the scintillation techniques, anisotropic models should be used to address the sensitivity of various measurements to the underlying assumption of isotropic turbulence which has been successful in turbulent bottom boundary layer flows. For both techniques, possible contamination of measured turbulent parameters by platform motion must be considered in more general deployments. For measurements at shallow depths, effects of surface waves must be addressed.

Assuming that these challenges can be met, both applications promise major advances in our understanding of ocean turbulence. Many oceanic turbulent phenomena, particularly in coastal environments, span a range of space and time scales that severely challenge the capability of traditional measurement techniques. Our understanding of turbulent processes and their relation to mean flows and external forcing is presently limited by the relatively short time series achievable with microscale profiling techniques, and/or the point character of moored sensors. In contrast, measurements of flow and turbulence parameters obtained from acoustical methods can be made over extended time periods and provide either vertical (Doppler) or horizontal (scintillation) profiles. In addition, path averaged scintillation techniques average out the intermittent nature of turbulent dissipation. Deployed at cabled ocean "observatories," acoustical systems could furnish the long time series necessary to assess the causes and frequency of strong turbulent mixing events in the ocean. Mounted in specially designed ultra-stable towed bodies, Doppler systems could provide spatial swathe-maps of turbulent parameters from a moving ship, allowing us to locate spatially isolated regions of strong mixing. Extended deployments of acoustical systems designed for turbulence measurements thus promise to provide the quantitative information crucial to understanding oceanic turbulent processes and incorporating the effects of turbulence in predictive numerical models of the ocean and its embedded ecosystems.

Chapter 21
Very high frequency coastal acoustics

T. G. LEIGHTON
University of Southampton, UK

G. J. HEALD
DSTL, UK

Summary

Higher frequencies provide us with the short wavelengths required to re-solve the small features which provide much of the fascination of coastal waters, be they objects in or on the seabed (sand ripples, weed, shells, wreckage, military ordinance), sub-bottom features (geology, pipelines), or oceanographically significant entities in the water column (fauna, flora, suspended solids, bubbles). High frequency signals also provide us with the temporal resolution required to follow the rapid fluctuations which characterize many of these important oceanographic features. For example a breaking wave in the surf zone, along with its associated tur-bulence and currents, can dramatically alter the shallow water oceanog-raphy in under a second.

This chapter will discuss a scenario which is doubly unsolved, first because the environment will be coastal waters (which represent perhaps the least cataloged regions of the ocean) and, second, because the sensors will involve Very High Frequency (VHF) acoustics. For the purposes of this work, it is convenient to define VHF as 200 kHz and higher. Not only is this lower limit (200 kHz) greater than the frequencies used by the ma-jority of underwater acoustic sensors, but it is also close to the boundary between the "known" and "unknown" in a number of applications.

Contents

21.1 Introduction

It is ironic that the region of sea most close to us geographically should be one of the least explored acoustically. The main reasons for this are two-fold. First, in the Cold War period the emphasis was on deep water acoustics, primarily for the detection of nuclear submarines. This has changed since the fall of the Berlin Wall, when the likely arena for operation has been shallow water coastal areas. Second, the science of acoustics is very difficult to apply in near-shore (or "littoral") regions, where propagation is complicated not only by multiple reflections from the air/sea interface and the bottom, but also by the myriad acoustically active fluctuating inhomogeneities which are present in the water column in coastal waters.

In such waters there is a tendency to use higher frequencies. Consider, for example, that a 3 m separation between the top and bottom of the water column corresponds to a propagation time of around 2 ms. This not only means that 1 s of acoustic testing may incorporate hundreds of reflections. It also illustrates one of the reasons why coastal acoustics tend to exploit higher frequencies: whilst that 2 ms interval corresponds to only 1/5 of the cycle at 100 Hz, it corresponds to 200 cycles at 100 kHz. There are therefore many more cycles available for transmitting information. High frequencies also allow pulse lengths to be shortened, offering a reduction in the complicated multipath reverberation. Those multi-paths which remain are more likely to be clearly resolved, because the same number of cycles at high frequency covers a much smaller spatial range in the water than it does at low frequencies. Once resolved, multipaths and target interactions can be used to gain information about the environment.

Such apparent advantages may, however, be complicated or compromised by acoustic absorption, which increases with frequency (see Section 2.4.2). Whilst higher absorption might reduce the number of multipaths in the received signal, it will also reduce detection ranges. Then again, high absorption might be seen not just as a problem which limits signal range, but absorption itself might be the measure by which we monitor the ocean environment. Examples are given in Sections 21.4–21.6, where for example the highest frequencies can be so strongly absorbed in sediment that reflections from the seabed can primarily contain information from the seabed/water interface, a fact which can be exploited in their interpretation.

Higher frequencies also are further removed from some of the sources of ambient noise in coastal regions (see Section 5.5). One has only to swim through the surf to appreciate the level of noise from wave-breaking and saltation in the surf zone, as well as non-acoustic sources of audible or hydrophone "noise" such as turbulence; and from there to swim out over a reef and hear biological activity, and the noise from such human activities as industry and shipping. With some exceptions, many of these would not produce very high frequencies at source, and of course absorption during propagation will further reduce their high frequency contribution to the signal at a remote detector.

At the start of this chapter, 200 kHz was deemed to be close to the boundary between the "known" and "unknown" in a number of applications. Taking two examples which will be explored in the chapter, the models for acoustical scattering from sediment are only validated at lower frequencies (Section 21.5); and measurements of bubbles in coastal waters have been restricted to those having resonance frequencies lower than about 200 kHz (Section 21.6). Because the VHF frequency range is relatively unexplored in the ocean, this chapter is written from an acoustical point of view, rather than an oceanographic one, in which information can be derived from well-characterized acoustical tools.

The use of high frequencies brings with it a set of problems not encountered with the lower frequencies which are more often used in oceanography. Because the signal is absorbed to a far greater extent, transducers are required which can generate high amplitude fields. As a result, consideration must be given to the potential for non-linearities in both source and propagation (see Chapter 4). The complexity of the environment, with its small scale features and rapid fluctuations, has stimulated exploration of signal features (bandwidth, coherence) which have placed further requirements on the transducer design, signal processing, and understanding of the acoustics. These issues are explored in the rest of chapter.

21.2 Acoustics

Restating the development in Section 4.1, the characterization of the propagation of an acoustic wave in a fluid requires three fundamental inputs enshrined in equations reflecting: first, the motion of mass in a fluid; second, the fluid dynamic properties relating such motions to the pressure gradient which causes them; and third, an equation which shows that pressure gradient to be part of a longitudinal wave. Stated in one-dimensional form, these three equations are, respectively: the equation of continuity

$$\frac{\partial \rho}{\partial t} + \frac{\partial (\rho v)}{\partial x} = 0 \qquad (21.1)$$

where v is the particle velocity, and ρ the fluid density; Euler's equation for an inviscid fluid

$$v\frac{\partial v}{\partial x} + \frac{\partial v}{\partial t} = -\frac{1}{\rho}\frac{\partial p}{\partial x} \qquad (21.2)$$

where p is the sum of all steady and unsteady (assumed here to be purely acoustic) pressures; and the wavespeed equation

$$c^2 = \frac{\partial p}{\partial \rho} \qquad (21.3a)$$

Combination of these three equations allows formulation of the propagation of acoustic longitudinal waves in a fluid, linearization allowing the generation of what is generally termed the (linearized) plane wave equation. In the small amplitude regime where this is applicable, the sound speed is represented by the phase speed in the linear limit c_0, and the approximation made that $\rho^{-1} \approx \rho_0^{-1}$ where ρ_0 is the equilibrium density. In this linear limit (21.3a) gives

$$c_0^2\frac{\partial \rho}{\partial t} = \frac{\partial p}{\partial t} \qquad c_0^2\frac{\partial \rho}{\partial x} = \frac{\partial p}{\partial x} \qquad (21.3b)$$

and substitution of (21.3b) into (21.1) converts the term $(\partial \rho/\partial t)/\rho$ into $(\partial p/\partial t)/(\rho_0 c_0^2)$ to give:

$$\frac{\partial v}{\partial x} + \frac{v}{\rho_0}\frac{\partial \rho}{\partial x} = -\frac{1}{\rho_0 c_0^2}\frac{\partial p}{\partial t} \qquad (21.4a)$$

Similarly, if $\rho^{-1} \approx \rho_0^{-1}$ then Euler's equation (21.2) becomes:

$$v\frac{\partial v}{\partial x} + \frac{\partial v}{\partial t} = -\frac{1}{\rho_0}\frac{\partial p}{\partial x} \qquad (21.4b)$$

Linear acoustics is based on the assumption that two of the terms in the above series of equations are negligible, specifically that $|(v/\rho_0)(\partial \rho/\partial x)| \ll |\partial v/\partial x|$ when calculating $(\partial p/\partial t)/(\rho_0 c_0^2)$ using (21.4a); and that $|v(\partial v/\partial x)| \ll |\partial v/\partial t|$ when calculating $(\partial p/\partial x)/\rho_0$ using (21.4b). Differentiation of (21.4a) with respect to t, and of (21.4b) with respect to x, gives the one-dimensional linearized plane wave equation for pressure if the equivalence $(\partial^2 v/\partial x\partial t) = (\partial^2 v/\partial t\partial x)$ is made:

$$\frac{\partial^2 p}{\partial t^2} = c_0^2\frac{\partial^2 p}{\partial x^2} \qquad (21.5)$$

Equation (21.5) describes a linear pressure wave propagating at speed c_0, and has solutions $p = g(t \pm x/c_0)$. Similarly differentiation of (21.4a) with respect to x, and of (21.4b) with respect to t, with the same two terms deemed to be negligible, gives the one-dimensional linearized plane wave equation for particle velocity, with solutions $v = f(t \pm x/c_0)/(\rho_0 c_0)$, since

$$\frac{\partial v}{\partial t} = \frac{f'}{\rho_0 c_0} \qquad \frac{\partial^2 v}{\partial t^2} = \frac{f''}{\rho_0 c_0} \qquad \frac{\partial v}{\partial x} = \pm \frac{f'}{\rho_0 c_0^2}$$

$$\frac{\partial^2 v}{\partial x^2} = \frac{f''}{\rho_0 c_0^3} \qquad \text{satisfy} \qquad \frac{\partial^2 v}{\partial t^2} = c_0^2 \frac{\partial^2 v}{\partial x^2} \qquad (21.6)$$

where $f' = \partial f/\partial(t \pm x/c_0)$. If such linear solutions are valid, then the conditions which make them valid (i.e., which ensure $|v(\partial v/\partial x)| \ll |\partial v/\partial t|$ and $|(v/\rho_0)(\partial \rho/\partial x)| \ll |\partial v/\partial x|$) are readily found. Taking the first inequality, $|v(\partial v/\partial x)| \ll |\partial v/\partial t|$, substitution from (21.6) gives

$$v \frac{\partial v}{\partial x} \Big/ \frac{\partial v}{\partial t} = \pm \frac{v}{c_0} \qquad (21.7a)$$

Consider the second inequality, $|(v/\rho_0)(\partial \rho/\partial x)| \ll |\partial v/\partial x|$ in the same linear limit, where (21.3b) is valid. Since neglect of the non-linear term in (21.4b) implies $\partial v/\partial t = (-\partial p/\partial x)/\rho_0$, then substitution from (21.6) reduces the second ratio to

$$\frac{v}{\rho_0} \frac{\partial \rho}{\partial x} \Big/ \frac{\partial v}{\partial x} = -\frac{v}{c_0^2} \frac{\partial v}{\partial t} \Big/ \frac{\partial v}{\partial x} = \pm \frac{v}{c_0} \qquad (21.7b)$$

Hence the linear limit is approached as the acoustic Mach number becomes small and the two non-linear terms become negligible (see Section 1.6.2). As stated in the introduction, when VHF (≥ 200 kHz) acoustics are used in coastal waters, both the increased absorption seen at higher frequencies, and the extra losses due to absorption and scattering by inhomogeneities, mean that to obtain a good signal-to-noise ratio, the tendency is to use high amplitude signals at the source.

The consequence of finite acoustic Mach numbers is that (21.5) no longer holds, and a waveform which is initially sinusoidal will not remain so because of two non-linear effects which act co-operatively. Both can be readily understood through the realization that, if dissipation is small, then p and v would be in phase (Leighton, 1994). First, there is a *convection* effect. In simple terms, if v/c is not negligible, then parts of the wave tend to propagate as $c + v$. The particle velocity varies throughout the wave, and so the greater the local acoustic pressure, the greater the velocity of migration of that section of the wave. If then p and v are in phase, regions of compression (where the v and c are in the same direction) would tend to migrate faster than the regions of rarefaction (where v is opposite to c). The pressure peak travels with the greatest speed, the trough with the least (see Fig. 4.2).

Second, there is an effect which arises because, when a fluid is compressed, its bulk modulus and stiffness increase. This results in an increase in sound speed, and this effect too will cause the pressure peaks to travel at greater speed than the troughs, and tend to try to catch up and encroach upon them. See also Section 4.1.

A continuous wave that is initially sinusoidal will therefore distort in the way shown in Figure 21.1. The plots in *(i)* illustrate the time history of the wave, measured at increasing distances (*a–f*) from the source. The corresponding spectra are shown in *(ii)*. Close to source, the waveform is initially nearly sinusoidal (*a(i)*) and single frequency (*a(ii)*). As it propagates through the medium, each compressive region gains upon the preceding rarefactive half-cycle, the peak positive acoustic pressure appearing earlier and earlier in the time history compared to the peak rarefaction. An accumulated steepness of the waveform between the two develops (*b(i)*) and harmonics appear in the spectrum (*b(ii)*). After propagating a certain distance (the *discontinuity length*) the waveform includes a discontinuity: a shock wave develops (*c(i)*), with an associated continuum in the spectrum visible between the harmonic peaks (*c(ii)*). The contribution of higher harmonics to the waveform is clearly visible (*c(i)*). Note that the amplitude after this time decreases, because any further compressional advance leads to dissipation and results in a reduction in the amplitude of the shock. This is because the waveform distortion has been equivalent to transferring some of the energy of the initial wave to higher frequencies, which are more strongly absorbed (*d, e*). The energy transfer is not sufficient to maintain the shock, and the wave approaches a low-amplitude sinusoidal form (termed "old age") (*f*).

Hence during non-linear propagation energy is pumped from lower to higher frequencies, where it is preferentially absorbed. This means that the net attenuation over distance will be greater if non-linear propagation occurs, than if conditions were linear. Furthermore, a narrow band detector tuned to the frequency of the emitted pulse might fail to detect energy in the returned signal which is outside of its bandwidth (and hence "invisible" to it). Both of these will act to reduce the signal-to-noise of the received signal, a problem which may not be alleviated by simply increasing the amplitude of the emitted pulse (since this enhances the non-linear effects, a phenomenon known as "acoustic saturation" see Leighton (1998) and Chapter 4).

Propagation such as that described above is just one of the possible sources of non-linearity: others include the transducer itself, and entities within the water column (see Section 21.6). The degree to which the effects occur for a given acoustic signal in the coastal environment of course depends not only on the non-linearity (and hence the rate at which energy is transferred from lower to higher frequencies) but also on the absorption. Some coastal features, such as bubbles, increase both, and it may be that the absorption is so great that non-linearity is not present

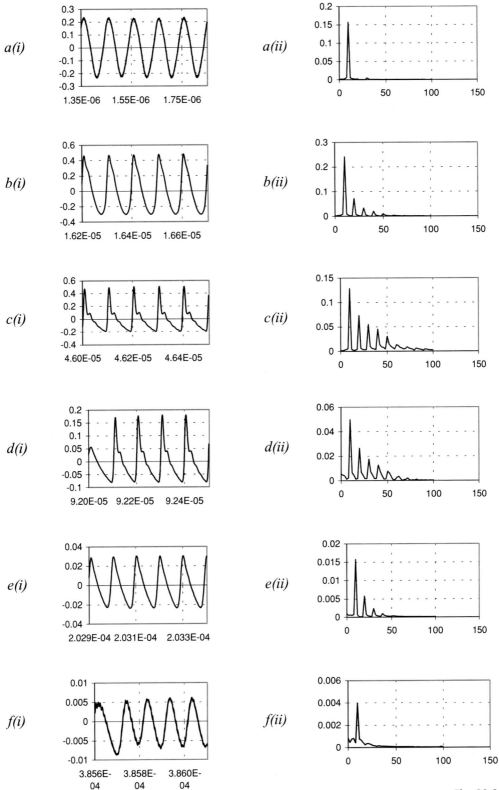

Fig. 21.1

in the received signal. However its possibility during propagation must be appreciated. This is not only because non-linear propagation might cause enhanced absorption, or may make some energy in the received signal "invisible" if it is outside the bandwidth of the detector, but also because it might be exploited. The ability of the propagating medium to generate multiple frequencies, as illustrated in Fig. 21.1, could not only be used to diagnose the properties of that medium (e.g., bubble water). It could also be used to generate a signal containing harmonics across a wide frequency range for simultaneous testing of, say, the scatter from the seabed or a target across a wide range of frequencies (the propagation after scattering, being of lower amplitude, would tend to be linear and so preserve the frequency characteristics of the target within the usual linear constraints of absorption, etc.).

A simple model for the transferring of some energy from lower to higher frequencies (which does not include the critical absorption component) can be found in a simple power series expansion relating, for example, the response $¥(t)$ of a fluid element to pressure,

$$¥(t) = s_0 + s_1 \cdot p(t) + s_2 \cdot p^2(t) + s_3 \cdot p^3(t) + s_4 \cdot p^4(t) + \ldots \quad (21.8)$$

where s_0 etc. are coefficients in the expansion. This is adequate to demonstrate the generation of harmonics through a non-linear process (including propagation). If the non-linear fluid element (in this case the liquid) is subjected to single-frequency insonification $p \propto \cos \omega t$, then the second and third harmonics are generated by the quadratic s_2 and cubic s_3 terms, and so on:

$$2 \cos^2 \omega t = 1 + \cos 2\omega t$$
$$4 \cos^3 \omega t = \cos 3\omega t + 3 \cos 3\omega t \quad (21.9)$$

Fig. 21.1. The signal from a 0.25″ diameter Panametrics plane transducer, driven by a 10-cycle 10 MHz tone burst. Part (i) (left column) shows the oscilloscope voltage (V) against the timebase (s). The overall system sensitivity (hydrophone plus amplifier, minus loading correction) at 10 MHz was 378.65 nV/Pa, so the peak positive pressures are around 1.2 MPa. Part (ii) (right column) shows the amplitude spectral level against frequency (MHz). Measurements were made with a 2×9 micron bilaminar membrane hydrophone (Marconi, with a high gain 100 MHz bandwidth preamplifier). This had an active element of diameter 0.5 mm, separated from the transducer face by (a) 2 mm, (b) 24 mm, (c) 68 mm (the position of the last axial maximum, i.e., the ratio of the square of the faceplate radius to the acoustic wavelength), (d) 136 mm, (e) 300 mm, (f) 570 mm, determined from the time relative to the transducer firing. Data was recorded by a Tektronix TDS 784D DPO oscilloscope (50 ns/div, 5000 point waveforms). The experimental system was aligned according to IEC 61102 prior to measurement: the water temperature was 18.7 °C. (Measurements taken specifically for this chapter by M. Hodnett and B. Zeqiri, National Physical Laboratory, UK.)

Note also that a non-linear element in the sea which is insonified by two coherent frequencies $p(t) = p_1 \cos \omega_1 t + p_2 \cos \omega_2 t$, can generate combination-frequency signals at $\omega_1 \pm \omega_2$, $\omega_1 \pm 2\omega_2$, $\omega_1 \pm 3\omega_2$ etc. since the quadratic term alone gives $2p_2 p_1 \cos \omega_1 t . \cos \omega_2 t = p_2 p_1 \{\cos(\omega_1 + \omega_2)t + \cos(\omega_2 - \omega_1)t\}$.

Whilst this scheme allows an easy appreciation of how harmonics come to be generated during non-linear propagation of high-amplitude acoustic beams through water, and for the operation of parametric sonar devices (Ostrovsky *et al.*, 2003), it is important to recognize that the above scheme is general. As such, any non-linear system can generate such signals, and this can include turbulence and bubbles (Section 21.6), and transducers, the topic of the next section.

21.3 Transducers and signals

The possibility of non-linear propagation is one of the reasons why VHF technology has moved towards the use of wideband transducers. Given that high output powers are required to counteract the acoustic losses at VHF, the use of transducers with high-quality factors would suggest itself. With these, the available energy would be concentrated in a narrow bandwidth, so that high amplitude signals could be projected at specific frequencies. From the preceding section, this has the clear disadvantage that some energy in the returned signal may not be detected if narrowband receivers, or assumptions of linearity, are used.

At low frequencies it is possible to form arrays using individual transducers at an appropriate spacing. Designers generally aim to have the individual element spacing set to less that half the wavelength of the highest frequency so that they can be electronically steered without the introduction of diffraction grating lobes. Half wavelength spacing is only a requirement if steering from broadside to 90° is required. If only smaller steer angles are required, then clearly the separation criterion can be relaxed. At very high frequencies it becomes difficult to produce arrays using individual transducers as the elements, so novel approaches are required.

One of the recent developments in the production of high frequency arrays has been the advent of the 1–3 composite transducers. Rather than using individual transducers as the elements, these arrays are formed from a single block of ceramic. There are two methods used in the production of these arrays, specifically by dicing the ceramic or injection moulding. In dicing, the ceramic is first cut to rods of length equal to a quarter of the wavelength of the desired center frequency. It is then diced, leaving a series of rods that are all cut to the same length, and hence are responsive to a characteristic wavelength. As these arrays are generally

Fig. 21.2. Photograph of a ~20 cm length of a 1–3 composite array showing the ceramic/epoxy grid (fine texture between diamonds) and the shaped elements (the raised cosine function giving them a diamond-like shape in the image), which provide mechanical shading and reduce the vertical sidelobes.

used for wideband signals, the choice of bonding material between each of the rods can be used to adjust the stiffness (and hence the damping coefficient). It is now commonplace to see high frequency arrays with a quality factor of 2 or less. The material used is generally an epoxy resin which has the advantage that once it has set, it keeps the individual rods at a fixed separation.

Having formed a matrix of rods, individual array elements may then be formed by screen printing conductive paint on the front and back of the array. In the simplest case the shape of the element can be made square or rectangular with only a small insulation gap between adjacent elements. In shallow water applications it is often desirable to reduce the scattering from the ocean boundaries. The shape of the element may therefore be tailored to reduce the vertical sidelobes and hence provide vertical shading (Stansfield, 1990). The example given in Fig. 21.2 shows a receiver array where a cosine shading has been applied to the element shape. Owing to the small size of high frequency arrays, it is possible to gather acoustic data from the individual elements. This allows the raw data from each element to be stored for later processing (beamsteering and beamforming) if required.

There are a number of ways that the array can be used to produce and receive broadband signals. In theory it should be possible to produce an impulse in the water, and by taking the Fourier transform to display the spectral content of the received signal. There are a number of reasons why this is not practical. Imperfections in the array mean that achieving perfect matching is not possible, and as a result the overall bandwidth of the array will have a finite limit. Using an impulse would mean that large amounts of power would be required, in a short space of time, and the commensurate emitter amplitudes would generally be above the cavitation limit of the transducers, resulting in low efficiency.

An additional factor of very high frequency arrays is the range to the near-field/far-field transition. If we use l^2/λ as the criterion (where l is the array length and λ is the acoustic wavelength) then a 1 m array would have a near-field region out to 133 m at 200 kHz, and to 667 m at 1 MHz. Bearing in mind the higher absorption that occurs as the frequency increases, it is possible that all the useful ranges at very high frequencies would be in the near-field. This leaves the sonar designer with two options: either correct for the near-field curvature using Fresnel correction/focusing; or use a shorter array which will result in wider beamwidths and loss of angular resolution.

Prior to the introduction of wideband signals into underwater acoustics, systems were generally limited to pulse transmissions that were modulated with a single frequency (CW). The resolution of these sonars was limited by the minimum pulse length used. A more practical solution to achieve a wideband system is to spread the frequency information over time to produce a time/frequency chirp. High spatial resolution can be achieved by passing the received wideband signal through a replica correlator (Lynn, 1985) so that pulse compression is achieved. The function of a simple finite correlator function may be written as:

$$r_{xy}(t_d) = \int_{-\infty}^{\infty} f_1(t) \cdot f_2(t + t_d) \cdot dt \qquad (21.10)$$

where $f_1(t)$ is the amplitude function of the source signal, $f_2(t)$ is the received signal and t_d represents the time shift between the transmit and receive functions.

At high frequencies the coherence of the received signal is likely to become degraded owing to a number of factors. The small wavelength means that the minor disturbances due to the water column or the ocean boundaries can have a significant impact on the signal. A small roughness on the seabed or the ocean surface will mean that the signal can become incoherent. The shape of the wavefront can be affected by spatial or temporal variability, turbulence, moving waves, scintillation, or sediment in transport, and this can limit the ability to beamform over the full array. At lower frequencies attempts have been made to overcome wavefront distortion by using time reversal or phase conjugation techniques (Edelmann et al., 2002). These methods rely on a point source so that the time of flight to each element along an array can be stored. Using reciprocity, the array can then be used to transmit a signal with the same time (or phase) delays introduced. By definition this means that there will be a focal point where the original source was located, provided that the environment has remained stationary. This type of technique originated in focusing for medical transducers, and so it has been applied in laboratory and biomedical environments at the VHF frequencies of interest in this chapter. The authors are not aware of any coastal water experiments in time reversal or phase conjugation that have been reported using VHF acoustics (Kustov *et al.* (1985) exploited signals of 60 and 100 kHz in a laboratory pool). Because the environment must be stable over the timescale between the point source transmission and the reception back at that point, it is possible that coastal waters may not be stable enough to show any improvement when VHF time reversal or phase conjugation techniques are attempted. The environmental factors presented in the remainder of this chapter cover some of the issues that will impact on the coherence, distortion, and attenuation of VHF acoustic signals. This is a relatively new area of ocean acoustics so there is still much research that is required.

A wideband facility opens up other opportunities, however. Acoustical oceanography differs from active sonar in that the sources of signal loss which hindered target detection in sonar might be the very oceanographic features which, in acoustical oceanography, one is trying to observe. Sections 21.5 and 21.6 detail how scattering from inhomogeneities in the water column can be used to diagnose the nature of those inhomogeneities: in target detection by active sonar, such oceanographic scattering would simply be viewed as a signal loss which one would try to avoid; in acoustical oceanography that "loss" is the signal itself. Similarly, whilst the temptation is to view the introduction of non-linearity as a problem, it can in fact be exploited: Section 21.6 shows how the non-linearity itself can be used to diagnose the bubble population. Before that however, we will consider the bottom of the water column, and work upwards, beginning with the measurement of bottom properties.

21.4 The seabed

The increase in attenuation with increasing frequency is a characteristic, not just of sea water, but also of the seabed. When an acoustic signal is scattered from the sea floor, some of the energy returns from the seabed/water column interface (the "top" of the seabed) and some is reflected from regions beneath the sea floor. Therefore, the use of VHF acoustics opens up the possibility of emphasizing scattering from the top of the seabed, since the energy which propagated the greater depths will be that much reduced. In addition, with a reduction in wavelength comes the ability to resolve fine detail on the sea floor, such as sand ripples, shells, and other fauna and flora (Williams *et al.*, 2001).

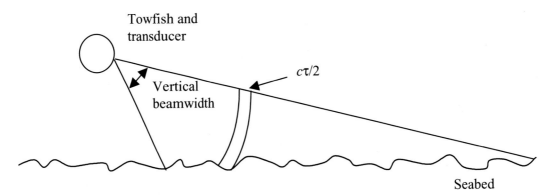

Fig. 21.3. Schematic of sidescan sonar geometry (single side), where c is the sound speed and τ is the pulse duration.

High frequency seabed acoustic systems have been employed for a number of decades through the application of sidescan sonar (Fig. 21.3). These sonars are generally housed in a towed body and are "flown" relatively close to the seabed, and the transducer axis is perpendicular to the forward motion of the host platform. The acoustic beam is incident on the seabed at a low grazing angle. This means that the undulations due to the sediment roughness have high intensities on the facing slopes, but the backslopes of the ripples are in an acoustic shadow. As the sonar moves through the water, pings emitted at regular intervals are used to build up a 2D image of the seafloor features. Figure 21.4 shows an output from a 325 kHz sidescan sonar from a region where patches of different sediment can be seen in the different bedform texture of various parts of the image. Where the parameters of the sonar and geometry are well known, it is possible to use image texture processing algorithms to discriminate the regions of different sediment type. With ground truth information, it is then possible to provide aided classification of the seabed (Fig. 21.5).

Commercial sidescan sonars are available covering the frequency range from about 100 kHz, for wide swath widths, up to about 600 kHz. At these higher frequencies the range is significantly reduced, but fine resolution can be obtained in images of the sea floor, and of objects on or in it. With the advent of unmanned underwater vehicles, it is anticipated that the upper frequency limit may move even higher (although issues regarding coherence and platform stability may prove to be a limiting factor).

Treating the seabed backscatter as an image tends to ignore the physical processes of the actual sea floor scattering that are taking place. It also does not take account of the image sensitivity to other changes in the

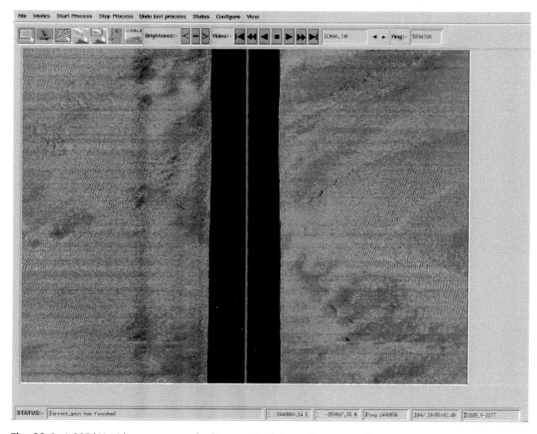

Fig. 21.4. A 325 kHz sidescan sonar seabed image. The left panel corresponds to 100 m (swath width) of seabed to the left of the towfish, and the right panel corresponds to the 100 m swath width to the right. (Image provided using the Classiphi™ system courtesy of QinetiQ Ltd.)

environment, or in the system being used to gather the data. It is clear that very few publications have included the physics of sea floor scattering in the frequency range above 200 kHz, and current high frequency sediment scattering models have only been validated up to about 150 kHz.

The concept of seabed "roughness" (as exploited in the classification scheme of Fig. 21.5) is no simple matter. Physically, there is of course the morphology or "surface relief," which can take an anisotropic 2D form over the seabed, but for which a simple example is a uniform set of 1D seabed ripples. In addition to this, there is the roughness imparted by the individual grains, the largest of which is gravel (of which there are four categories based on the range of particle diameters, $2a$: boulders $2a > 256$ mm; cobbles 256 mm $> 2a > 64$ mm; pebbles 64 mm $> 2a > 4$ mm; granules 4 mm $> 2a > 2$ mm). Smaller grains

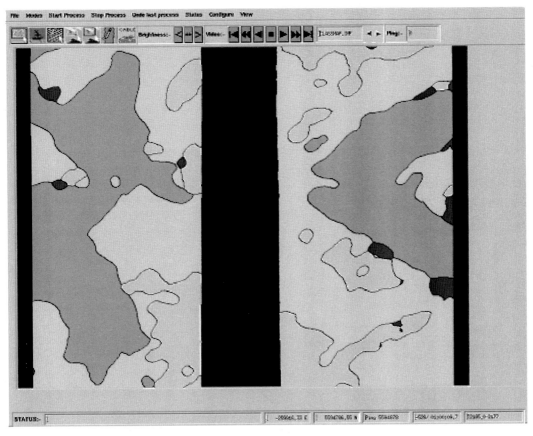

Fig. 21.5. Sidescan sonar sediment classification using image processing algorithms on the data of Fig. 21.4. The four different grayscales correspond to a classification of the sea bed, using texture image processing of sidescan sonar, from darkest gray to lightest, as follows: gravel, sand, muddy sand, mud. The left and right panels correspond to the geometries described in the caption for Fig. 21.4. (Image provided using the Classiphi^{TM} system courtesy of QinetiQ Ltd.)

are termed sand (2 mm $> 2a >$ 0.0625 mm), silt (0.0625 mm $> 2a >$ 0.0039 mm), and clay (0.0039 mm $> 2a$). However the "effective roughness" of a surface does not depend only on the physical size of the surface relief or grains, but also on the wavelength of the radiation used to resolve them. The high spatial resolution imparted by VHF acoustics, compared to that available at lower frequencies, means that fine roughness features become resolvable. Consider when the Rayleigh criterion is used to determine the roughness of the surface, with respect to the incident acoustic wave number (k). It is based on the Rayleigh parameter, $R_a = kh \sin \theta_g$ where h is the rms height of bed ripples with respect to a mean plane, and θ_g is the source angle with respect to grazing incidence. The Rayleigh criterion states that if $R_a < \pi/4$ then the surface is regarded as smooth, whereas if $R_a > \pi/4$ then the surface is regarded as rough. It is clear that the wavelength of interest at VHF frequencies would start to be comparable with the roughness dimension of the sediment (from shell-size at 200 kHz to sand-grain size at 1 MHz). For a plane sediment surface insonified with VHF acoustics, surfaces made of gravel would start to look very rough. It is, however, interesting to note that sediments with finer grains (sand, mud, silt, and clay) would still fall below the Rayleigh criterion. In reality the sediment always contain some surface relief, so that the contribution to roughness from this would also need to be taken into account. In principle these surfaces exceed the Rayleigh criterion, and the surface should be treated as a rough surface. This is as would be expected for decreasing wavelength.

Close inspection of the sidescan sonar image provided in Fig. 21.4 reveals that the texture (highlights and shadows) is directly related to the undulations on the sediment, so that the front face of the ripple gives a high backscatter which gradually reduces as the local grazing angle decreases, with a shadow region behind the ripple. Whilst it is true that the surface would look rough if the insonification were over a number of ripples, hence giving incoherent backscatter, in the case where a short pulse length is used (which is most often the case at VHF) the surface spectrum must be truncated to take account of this effect. Brothers *et al.* (1999) have applied this truncation scheme to the composite roughness model and made good comparisons with experimental data. Their scheme was applied to both short pulse length and pulse compressed wideband signal. As the insonified area decreases it must, by definition, remove some of the lower frequency components from the surface spectrum. This must be taken into account in modeling or when conducting experiments, if representative scattering levels are to be achieved. If this argument is taken to its extreme, then the insonified area becomes a single point on the surface, which is effectively smooth, and hence the scattered levels would be a function of the angle of the facet and the reflection coefficient.

In summary, VHF scattering has the potential to offer acoustical oceanography new tools for exploration of the sea floor. As a first example, given that VHF beams might only penetrate a few centimeters into the seabed, comparison of VHF and lower frequency scatter might allow discrimination between information on the water/bed interface and that which scatters from deeper in the seabed. This point will be expanded upon in the following section with discussion of a commercial device. Second, consider that at lower frequencies sediments have been modeled as a frame structure (Stoll, 1974) based on the Biot theory. If this theory holds at high frequencies, then it is possible that resonances would be observed at frequencies much lower than would be expected for the individual grain resonances. The ability to observe any such resonance will depend on the amount of damping which, if high, would suppress any effect. The question of whether models will be able accurately to predict sound speed in sediment is raised by the suggestion that the apparent velocity of sound in the sediment might decrease as the frequency rises above 350 kHz (an effect which also appears to happen in artificial seabeds comprising glass beads). Because each sediment grain (or glass bead) is not necessarily connected to its neighbors, the energy which passes though the sediment alone has to take a complex route. This increases the path length, thus giving the impression of reduced velocity at lower velocity. Finally, there is seldom a clear distinction between pure seabed and the water column. Boreholes, gas bubbles in sediment, and gas bladders on weeds, shells, and other fauna and flora all have the potential to scatter VHF signals significantly, perhaps even enough to swamp the background level that would normally be seen from the sediment alone. The next section, however, details more subtle phenomena which produce a finite transition zone between seabed and water column.

21.5 Flow-related processes in the bed/water column transition zone

The transition region between two media always has the potential to be complicated (see, for example, Section 21.6). The acoustical implications of the boundary between a well-defined seabed and the water column were discussed in Section 21.4, from a viewpoint that, even with such "clutter" as shells and weed on the seabed or bubbles beneath it, the two regions could to a degree be well-defined because those items tend to present discrete scatterers to VHF signals. This section discusses the more subtle ectopia which characterize the transition between the seabed and the water column.

The most obvious of these features which subtly blend the seabed into the water column, is the suspended particulate matter. Much of this first rises from some bed (sea, river, or estuary) as the result of turbulence, tidal action, waves, etc. and then tends to settle back towards it under gravity (Fig. 21.6). In still water there would therefore be a concentration gradient of such matter, it becoming less concentrated as one moves away from the seabed, a tendency which is reduced by turbulence, etc. The importance of such "sediment in transport" has been recognized for thousands of years: its deposition during the annual flooding of the Nile determined the fertility of the land, a dominant feature in the religion, culture, and politics of Egypt from 4500 BC until the last century.

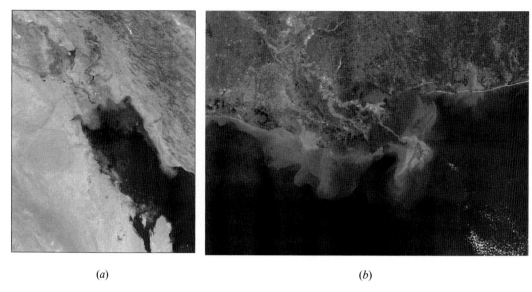

(a)   (b)

Fig. 21.6. True-color satellite images (from the Moderate Resolution Imaging Spectroradiometer MODIS carried by NASA) of sediment carried by rivers into the sea. (*a*) As a result of flooding by the confluence of the Tigris and Euphrates Rivers (at center), the sediment-laden waters of the Persian Gulf (November 1, 2001) appear light brown where they enter the northern end of the Persian Gulf and then gradually dissipate into turquoise swirls as they drift southward. (Image courtesy Jacques Descloitres, MODIS Land Rapid Response Team at NASA GSFC.) (*b*) The Mississippi River carries roughly 550 million metric tonnes of sediment into the Gulf of Mexico each year. Here (March 5, 2001 at 10:55 am local time) the murky brown water of the Mississippi mixes with the dark blue water of the Gulf two days after a rainstorm. The river brings enough sediment from its 3 250 000 square km (1 250 000 square miles) basin to extend the coast of Louisiana 91 m (300 ft) each year. (Image courtesy Liam Gumley, Space Science and Engineering Center, University of Wisconsin-Madison and the MODIS science team.) Source: NMASA, reproduced with the permission of the Lunar and Planetary Institute. (See colour plate section.)

Today the presence of suspended sediment reduces the ability of active sonar systems to detect mines and torpedoes (the sonar equivalent of the effect of fog on car headlamps). Conversely, the acoustic scattering and absorption it produces can be used to monitor this environmentally important feature.

Because typical particle diameters $2a$ are in the range 0.1–100 μm, such suspensions scatter most strongly at VHF. Up to $ka \sim 1$ (which for $c = 1500$ m s^{-1} occurs at frequency f of 2.4 MHz for $a = 100$ μm; and at higher frequencies for the smaller particles), Rayleigh scattering occurs: the scattered power increases as both particle size and frequency increase (eventually taking an oscillatory form in the geometric regime

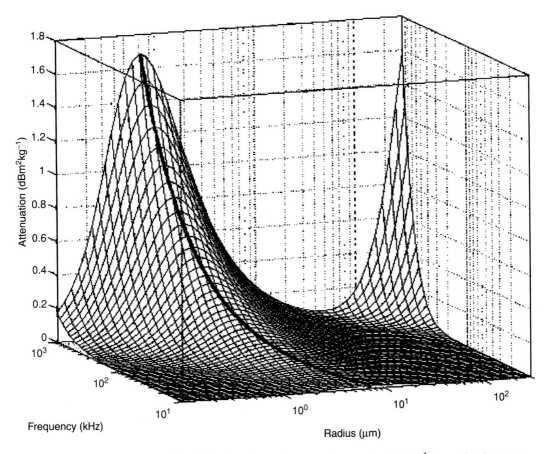

Fig. 21.7. Plot of acoustic attenuation constant dB m^{-1} (normalized to mass concentration kg m^{-3}) for suspensions of quartz spheres in sea water. The bold line tracks the local maximum of the absorption contribution. (Image courtesy S. D. Richards.)

where $a/\lambda \gtrsim 1$). The surface plotted in Fig. 21.7 is still undergoing this Rayleigh-regime growth as it passes through the rear right corner (i.e., large-radius, high-frequency limits) of the figure. Within this Rayleigh regime, at a fixed frequency the cross-sectional "target" area πa^2 "seen" by the ultrasonic beam increases quadratically with radius.

However, Fig. 21.7 plots, not just the scatter, but the attenuation. There is a second contributing factor to this, namely the acoustic absorption. This occurs because the density of the solid is different from that of the water. As a result, when an acoustic wave propagates through the water (the particle velocity reversing every half-cycle), the motion of the latter is not in phase with that of the particles. There is net flow around the particle, and viscous losses occur in a viscous shear boundary layer which extends around each particle to a "skin depth" of about

$\sqrt{\eta/\pi\rho f}$ (i.e., O[1 micron] in water at 500 kHz), for frequency f in a liquid of density ρ and shear viscosity η. The contribution of these losses to attenuation exhibits a local maximum for fixed frequency (see the bold line in Fig. 21.7). This is because very small particles move almost in phase with the fluid, whilst very large particles hardly move at all. Hence both activities are almost loss-less. Maximum absorption losses occur when the particle radius is a comparable size with the skin depth for shear waves, resulting in a maximum phase difference between the motion of the water and that of the particle.

To use either scatter, absorption, or both to measure suspended sediments at sea requires development, testing, and calibration of those instruments in the laboratory. Because of this, measurement of backscatter from suspended sediment has been far simpler than use of absorption, because the backscattered signal is so much easier to detect in controlled laboratory systems than the absorption loss. Whilst the latter can be significant in the ocean, where propagation lengths of km or more are used (Richards and Leighton, 2001), the finite size of laboratories makes it difficult to generate controlled, sizeable absorption losses. This difficulty increases as the acoustic frequency decreases, because the absorption is greater at higher frequencies. Hence whilst Urick (1948) compared absorption loss due to kaolin and fine sand at 1–15 MHz with theory for spherical particles, it is only in recent years that the dilute suspensions typical of the ocean have been compared with predictions of an appropriate theory (of, for example, non-spherical particles) at O[100 kHz] (Richards *et al.*, 2003). This has allowed feasibility studies of how absorption might be used to monitor mass flux by large rivers (Richards and Leighton, 2001). Crucially the technique would allow the measurement to be made right across the length of the river, averaging over 100 m or more, to monitor the entire mass flux.

In contrast, whilst the stronger laboratory signal allowed systems based on particulate backscatter to be developed more easily, they have been restricted to local measurements, with small (ml) sample volumes O[1 m] from the sensor. This is because, as the magnitude of the relevant effect (i.e., backscatter or attenuation) illustrated by Fig. 21.7 suggests, scatter systems should typically exploit higher frequencies (typically O[1 MHz]) than do absorption systems (O[100 kHz]). Clear sea water has an absorption coefficient of ∼0.5 dB/m at 1 MHz (depending on temperature, salinity, pH etc.), so 100 m of river mouth would degrade the two-pass backscattered signal by up to 100 dB. In contrast, the ∼3 dB degradation seen in a one-pass bistatic absorption device (sea water absorbing at ∼0.03 dB/m at 100 kHz) would provide the base-line absorption against which the enhanced absorption due to particles would need to be observed (Leighton, 1998).

One might ask why the backscatter technique does not exploit the O[100 kHz] frequencies and thereby extend the range over which it might measure. The problem is that, as the frequency decreases in this range, the scattering strength becomes small because $ka \ll 1$ (Fig. 21.7). If however the process of estimating particle sizes and concentrations from backscatter (inversion) is restricted to the MHz range, an additional benefit accrues. Inversion requires calculation of the acoustic attenuation of the beam as it passes through the particulate suspension, and the accuracy with which the contribution of particulate absorption needs to be described need not be too exacting. Figure 21.7 shows that scatter, rather than absorption, dominates attenuation at 1 MHz for populations of large particles (e.g., 50–100 microns radius), such as sand. The effects on the predicted attenuation resulting from a simplified expression for the particulate absorption are small (Holdaway et al., 1999). If however micron-sized particles dominate the populations (as in fine clay and silt), then the contribution of absorption becomes more important, and calculation of their contribution to attenuation might require such complexities as consideration of particle shape (Richards et al., 2003).

To add perspective however, whilst absorption techniques alone offer the prospect of measuring over long paths without using multiple sensors, they have yet to be used in the field to measure sediment in transport. In contrast, whilst scatter techniques are not as widespread as optical ones, they are commercially available, and are becoming increasingly accepted as valuable both in their own right, and as providing a useful synergy with optical and conventional techniques (which include bottle/pumped sampling; laser diffraction; optical backscatter; spectral reflectance; and the optical transmittance and absorption of water, usually as measured at several wavelengths using the "transmissometer") (Holdaway et al., 1999; Thorne and Haynes, 2002). Whilst the backscatter technique lacks the ability to integrate spatially over large volumes without multiple hardware, the compensation is that the suspended sediment can be measured with high spatial resolution. The sampling volumes are formed by dividing the returned signal into time bins. The time limits of each bin demarcate the start and end ranges respectively of the volume sample. For the AquaScat™ (a commercial acoustic backscatter sensor) these volume elements have lengths of 10 mm (with 5 mm and 2.5 mm options) at ranges of typically 1 m. The cross-sectional area of the volume element is controlled by the beamwidth (the AquaScat™-3dB half beamwidth being 1.4°–0.9° in the range 3–5 MHz, assuming a 10 mm diameter disk transducer).

However, the small wavelength also means the bed profile can be measured and even imaged, in the manner described in Section 21.4. The AquaScat™ also has a "long range" (O[10 m]) mode, with volume elements 20 mm–160 mm long, capable of measuring morphology (e.g., seabed ripples, Section 21.4). In addition to small sample volumes and the ability to measure morphology (with images of up to 10–30 m^2; Thorne and Haynes, 2002), the O[MHz] frequencies employed in backscatter offer a third advantages over the O[100 kHz] range. Because they are more strongly attenuated by the sediment than by range, the higher frequency backscatter in principle contains more information about the surface (uncluttered by returns from depth) than the lower frequencies (see Section 21.4). Hence most of the information in the returned signal comes for the water/bed interface, and not from the depths of the seabed, and the "blindness" of the high frequencies to the depths of the seabed simplifies their interpretation when used as an interface sensor.

Not only can backscatter measure suspended sediments and bed profile simultaneously, VHF systems in the bed/water transition zone can also measure flow. Absorption systems are relatively insensitive to flow and turbulence, which is beneficial in that it gives the assurance that flow-induced losses are not being interpreted as indicating the presence of spurious particles (Richards et al., 2003). However, this also means that they would not make good sensors for flow. Backscatter systems are however sensitive to flow and turbulence, and can be used to monitor it, providing there is none of the above ambiguity mentioned above. The Coherent Doppler Velocity Profiler (CDVP) exploits the effect of Doppler shift on the phase coherence between consecutive backscattered pulses to determine the component of flow velocity in the direction of the beam axis. Alternatively, the Cross-Correlation Velocity Profiler (CCVP) examines the temporal coherence in the suspension as it is advected, for example, through two vertical ultrasound beams, transmitted by a pair of closely spaced, horizontally aligned transducers. If the backscattered signal is divided into time bins, each bin represents data from a further range than the bin previously. Cross-correlating the returns from each emitter allows identification of the time interval it takes for the suspension to advect from the first beam to the second, allowing the velocity profile to be determined (Thorne and Haynes, 2002).

The changeover between Sections 21.4 and 21.5 introduced a variety of features which can make up the transition zone between the bed and the water. Whether they are associated with biology, trapped in sediment, or present in the form of free gas, a relatively small number of bubbles can contribute so much to acoustic scatter and absorption, that sensors of suspended sediment based on either system can be compromised (Richards and Leighton, 2001). However the strength of bubble scatter can be exploited, for example by using the bubbles themselves as tracers for flow (Thorpe, 1982). This is because, notwithstanding the ability of bubbles associated with the bed/water interface to scatter strongly, in most coastal waters the greatest acoustical effects are generated by those bubbles which are entrained at the top of the water column and move through it. The VHF effects of these bubbles will be the topic of the next section.

21.6 Bubbles

Both the acoustical and near-shore aspects of VHF bubble effects in coastal waters are fascinating and, to a large part, undiscovered. Bubbles are formed through a variety of mechanisms in the oceans, ranging from methane seeps to plunging breakers. In the coastal waters of this chapter, mechanisms which are less important in deeper water may become more important. These include the production of bubbles by biology and the flux of gas in sediment which is exposed at low tide. Most impressive, however, and most potent acoustically, are the effects of coastal breakers (Fig. 21.8), which can generate huge void fractions (the proportion of a sample of bubbly water which, by volume, is free gas) see Chapter 6.

Fig. 21.8. A 10 ft wave breaks over a combination-frequency bubble detector, which can be seen as a small black circle strapped between the cross-bars on the scaffolding. The plunging breakers on the water's edge were consistently far more active than any of the waves further from shore (Leighton *et al.*, 1996).

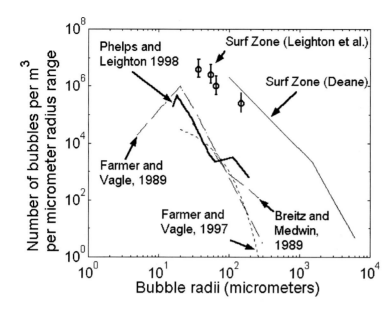

Fig. 21.9. The number of bubbles per cubic meter per micrometer increment of radius ($n(R_0)$ – see text for details), shown for a range of oceanic bubble populations, both in the surf zone (Leighton *et al.* 1996 *c*, Deane 1997 *d*) and out of it (Farmer and Vagle 1989 *a* and 1997 *b*, Breitz and Medwin 1989 *b*; Phelps and Leighton 1998 *c*). Labels *a–d* relate to experimental techniques described in Fig. 21.10, as explained in the text (from Leighton *et al.*, 2001).

Bubbles can enter into the VHF aspect of coastal acoustics in two ways. The first is through the pulsation resonance frequency f_0 of small bubbles. As discussed in Chapter 6, f_0 varies inversely with bubble size, so long as surface tension effects are neglected. This is valid down to a bubble radius of $R_0 \sim 7$ microns, where the inverse relationship fails because, for air bubbles in "dirty" ocean water (Thorpe, 1982) having an assumed surface tension $\sigma = 3.6 \times 10^{-2}$ N m^{-1}, surface tension contributes to the bubble gas pressure a component $p_\sigma = 2\sigma/R_0$ equal to 0.1 atm (Leighton, 1994). For air bubbles of 7 microns radius under 1 atm, f_0 is about 400 kHz. Hence the transition from the HF (high frequency) to the VHF regime is very interesting in bubble acoustics. VHF values of f_0 represent the region where surface tension on the bubble, which we do not know well, becomes important. This in turn affects the validity of our inversions of acoustical data to estimate bubble numbers. Moreover, having obtained these bubble counts, our estimates of, for example, atmosphere/ocean mass flux (which for atmospheric carbon alone may exceed 10^9 tonnes per year globally) are compromised by uncertainties in surface tension. Furthermore, the VHF range for f_0 also represents the lower limit where bubble population estimates have been made in the ocean, as will now be shown.

Do such small bubbles, resonant at frequencies greater than 200 kHz, exist in the ocean? Figure 21.9 plots a histogram of the bubble population, as measured in the oceans by various researchers (a small sample of the data sets available). The abscissa shows the bubble radius divided into bin sizes of 1 micron width, and $n(R_0)$ plots the number of bubbles per cubic meter of sea water contained within that bin. One extraordinary feature illustrated by Fig. 21.9 echoes the sentiment at the start of this chapter, that the waters closest to us are in some respects perhaps the least explored. The two "surf zone" data sets in Fig. 21.9 represent only a handful taken close to shore, in contrast to a multitude of data sets taken further from land (some of which are shown in Fig. 21.9). Note also the scarcity of small-bubble data: the smallest bubble radius shown in Fig. 21.9 corresponds to a bubble resonance frequency of 200 kHz; and the smallest taken to the authors' knowledge had a 9 micron radius, corresponding to a bubble resonance of 360 kHz (see Leighton *et al.*, 2001 and pp. 235–241 of that volume). Hence the lower limits of the VHF range represent the beginning of the "undiscovered" region of oceanic bubble size distribution, where data are scarce and, as discussed above, surface tension effects can dominate.

The second way in which bubbles can enter into the VHF aspect of coastal acoustics is through the combination-frequency techniques which are used by some researchers to obtain $n(R_0)$. Figure 21.10 illustrates the

Fig. 21.10. The four main families of active acoustical techniques for obtaining the bubble size distribution, $n(R_0)$. Transmitters and their waveforms are shown in dark gray; receivers and received waveforms are shown in light gray. (*a*) Propagation techniques; (*b*) Resonator method; Also shown are bubble-mediated generation of (*c*) modulation frequencies and (*d*) combination frequencies. Note that *n* and *m* correspond to non-zero integers. (From Leighton, http://www.isvr.soton.ac.uk/fdag/UAUA/INDEX.HTM)

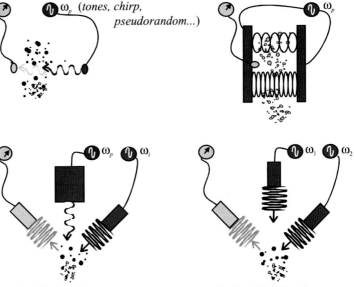

(a) **Propagation**
Monitors scatter or changes in signal amplitude (for absorption), travel time (for group velocity), phase and frequency (Doppler, 2nd harmonic)

ω_p (tones, chirp, pseudorandom...)

(b) **Resonator**
Monitors changes in frequency and damping of resonator modes

ω_p

ω_p ω_i

ω_1 ω_2

(c) **Modulation frequency**
Interprets signals at $\omega_i +/- \omega_p$ and $\omega_i +/- (n\omega_p/m)$ as being generated by bubbles resonant close to ω_p

(d) **Combination frequency**
Interprets signals at $\omega_1 - \omega_2$ as corresponding to the bubble resonance frequency

four broad categories of acoustical techniques for measuring $n(R_0)$, the dark gray corresponding to transmitters and their waveforms, and the light gray corresponding to receivers and received waveforms. Methods a–c all transmit a signal which is figuratively indicated by ω_p, the "pump" signal, which in general terms is meant to drive bubbles into pulsation and elicit the strongest response from those resonant at ω_p. Implementation of this technique, however, needs to take into account the fact that other bubbles (e.g., having radii much larger than the radius which is resonant with ω_p) might also contribute to some measured effects, such as attenuation at ω_p (Commander and McDonald, 1991): such big bubbles may not pulsate to large amplitude when driven at ω_p, but they can nevertheless scatter strongly because they represent physically large targets of gas in water. To avoid such ambiguities, the tendency is to move to effects which such large "inert" bubbles cannot produce, which often means exploiting the non-linearities which bubbles can generate when they pulsate at high amplitudes. Since high amplitude pulsation usually occurs in bubbles having radii which are close to resonance at ω_p, the non-linear effect can be taken as diagnosing the presence of resonant bubbles. One example is the non-linear generation of signals $2\omega_p$ by bubbles insonified at ω_p (Ostrovsky $et\ al.$, 2003). The generation of effects at $2\omega_p$ can readily be appreciated if the non-linear fluid element in (21.8) is the bubble. If, however, it represents, say, turbulence, then one can understand how even this signal may be unreliable if used to count bubbles under a turbulent breaking wave, if the sensor interprets all the $2\omega_p$ as being generated by near-resonant bubbles. Similar comments apply to the techniques shown in Figs. 21.10 c and d, where non-linear effects such as turbulence, as well as bubbles, may generate signals at $\omega_1 \pm \omega_2$, $\omega_i \pm \omega_p$, $\omega_i \pm 2\omega_p$, etc. All these are VHF issues, since if (as shown earlier) there are enough small bubbles in the ocean for ω_p to be in the VHF range, certain harmonics of it will be.

The techniques in Figs. 21.10 c and d are, however, VHF issues for another reason. As with a, b, in the modulation frequency method and (c) a pump signal is emitted in order to drive the bubble to pulsate. However a VHF "imaging" signal (at ω_i) is then scattered off the pulsating bubble: on the timescale of the ω_i signal, the wall of a bubble which is pulsating at ω_p is moving relatively slowly, and so the amplitude of the scattered ω_i is modulated on the slower timescale of ω_p (as the bubble expands and contracts; Fig. 21.11 a). This is interpreted in the spectra as the appearance of energy at $\omega_i \pm \omega_p$ in addition to the original ω_i signal (an alternative and equivalent description for the generation of these signals can be seen in (21.9)). If the pulsation of the bubble has a second harmonic component, signals at $\omega_i \pm 2\omega_p$ will also be detected, and so on (Ostrovsky $et\ al.$, 2003). Noting that in principle such signals

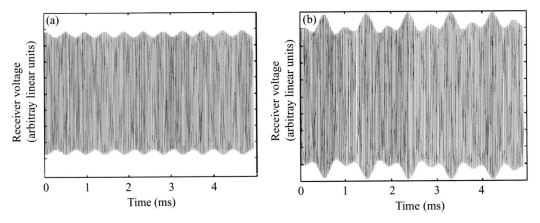

Fig. 21.11. Imaging signal modulated at (*a*) ω_p and (*b*) $\omega_p/2$. The period of the imaging signal is so short on the timescale of this plot that its sinusoid plots closely enough to appear as a solid black entity (Leighton, 1994).

could also be produced by any other source of non-linearity which could be described by power-series of (21.8), Leighton *et al.* (1996) exploited a signal which could not be generated through (21.8), specifically the generation of signals at $\omega_i \pm \omega_p/2$. These can also be visualized as scattering of the imaging signal by a target whose strength varies on a much slower timescale, as described above, only here the slow timescale corresponds to the subharmonic of the pump frequency (Fig. 21.11*b*). The shape mode on a bubble wall which requires the lowest threshold driving pressure to excite is the Faraday wave (Fig. 21.12), which oscillates at the subharmonic and was shown to be the source of the emission at $\omega_i \pm \omega_p/2$ (Phelps and Leighton, 1996, 1997).

Unlike the other techniques, the combination frequency method (Fig. 21.10*d*) does not emit a pump signal to drive the bubble into pulsation. Instead, two VHF signals (ω_1, ω_2) are transmitted into the water, and signals generated at $\omega_1 - \omega_2$ are taken to reflect the presence of bubbles resonant at that difference frequency (Didenkulov *et al.*, 2001; Ostrovsky *et al.*, 2003).

In the caption to Fig. 21.9, each of the bubble size data sets is labeled with a letter, *a–d*, to indicate which method (of Figs. 21.10*a–d*) was used to make the measurement.

In summary therefore, VHF enters bubble acoustics through the high natural frequencies of the large number of small bubbles prevalent in coastal waters, and consequently through the high value taken by ω_p and related signals ($2\omega_p$, $\omega_i \pm \omega_p$, $\omega_i \pm 2\omega_p$, etc.) used in determining the bubble numbers. In addition, two types of active acoustical techniques (Figs. 21.10*c* and *d*) transmit VHF signals (ω_1, ω_2, ω_i) into the water,

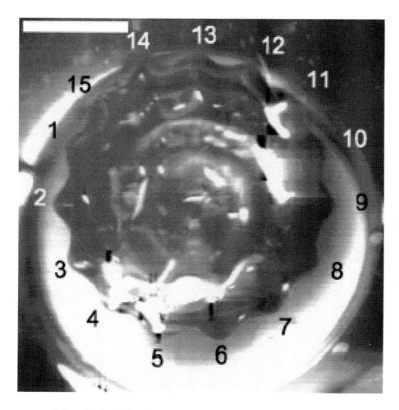

Fig. 21.12. A Faraday wave is stimulated on the wall of a pulsating bubble, driven at 1.29 kHz. The white line at the top of the figure is a 2 mm length scale bar. Just as the pulsation can be described as a perturbation on the bubble wall corresponding to the zero-order spherical harmonic, so this Faraday wave corresponds to its perturbation by a spherical harmonic of order 15 (as can be seen by counting the peaks and troughs on the bubble's wall). (Photograph: P. R. Birkin, T. G. Leighton and Y. E. Watson.)

not to drive the bubbles into pulsation, but rather to enable them to be detected through processes which may be described as non-linear.

21.7 Conclusions

Space limitations have meant that only a few VHF acoustical issues in coastal waters have been covered in this chapter, and those not to great depth. Additional issues are mentioned below.

VHF acoustic signals have sufficiently small wavelengths to allow the acoustic scatter from many features to be understood by analogy to optics. Both the top and bottom of the water column can be reflective. Figure 21.13 illustrates how, as a result of this, an observer will receive a sound field as if the reverberation were generated by images (with appropriate time/phase delays). The case when the observer is itself the source of sound is particularly fascinating. A sound source in idealized wedge-shaped coastal waters can "perceive" image sources; one can, for example, imagine how a single cetacean might see this as a ring of "twins." Whilst of course an intelligent creature might recognize such reverberation for what it is, other sound sources may not. For a bubble, such reverberation will alter the natural frequency and damping expected

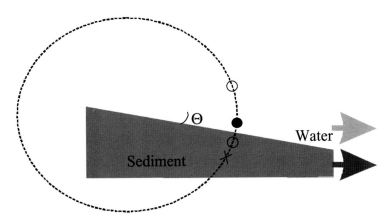

Fig. 21.13. If a coastal zone can be approximated by a wedge shape of ocean, with a bottom which reflects acoustic pressure waves with no phase change, and an air/water interface which reflects them with π phase change, then the net sound field built up in the water by an object (•) emitting sound will be that which would be produced were the object in free-field, and sound were in addition emitted by image sources either in phase (○) with the original source, or in antiphase (×). The sediment and atmosphere boundaries of the water column being flat acoustic mirrors in this model, in the 2D plane passing vertically through the source these images will be distributed around the circle shown by the dashed line. The first few image sources are shown (○,×). For certain wedge angles Θ (such as the 15° used for Fig. 21.14) the sources map on to discrete sites on the dashed circle.

in free-field conditions (see Chapter 6), as illustrated in Fig. 21.14 for the ideal wedge of Fig. 21.13 (Leighton *et al.*, 2002).

Of course most coastal regions do not resemble the flat-sided wedge of Fig. 21.13, but these complications make the coastal region even more fascinating for sensors of VHF acoustics. The optical equivalent would be stranger than a carnival "hall of mirrors," its floor covered by a fluctuating "dry-ice fog" (the optical equivalent of suspensed sediment particles), its wedge-shape complicated by ripples on the mirrored floor. Its ceiling would be an undulating, highly reflecting mirror, in some places focusing the sound in moving "hot spots" within the water column and floor, and in other places producing areas of dark, absorbing bubble clouds covered with a bright speckle of resonant bubble scatterers. Imagine those clouds being explosively generated by a breaking wave, then spreading over time.

Since most VHF acoustics intend to deploy discrete beams in a monostatic or bistatic scenario, the optical equivalent might involve one or more people with flashlights in this otherwise dark "hall of mirrors." As the frequency increases to the upper VHF, the absorption reduces the amplitude transmitted to, and scattered back from, a distance away. In

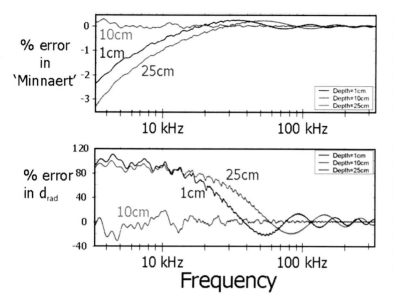

Fig. 21.14. The percentage error between the actual natural frequency of a bubble and the radiation damping constant, respectively, if the free-field formulations of Minnaert and Devin are used in the 15° wedge shown in Fig. 21.13. The result is shown for three bubble depths, for a bubble situation 1 m from the water's edge (where the water depth is 26 cm). (Figure by T. G. Leighton and P. R. White.) (See colour plate section.)

the monostatic optical equivalent it would be as if the flashlight emitted a dim red light in the carnival hall of mirrors; when, to compensate for this, the brightness of the flashlight is increased, blue new colors might be generated (the optical production of second-harmonic production described in Section 21.2, though of course the frequencies of blue light in the optical analog are not twice that of red).

21.8 The future

The move into research at VHF frequencies, over 200 kHz, offers the excitement of a transition into the "unknown." With respect to bubbles, sediments, and a range of other oceanographic phenomena, whilst we cannot say that all is well-understood at frequencies below 200 kHz, we are sure that the VHF regime offers the excitement of discovery and the potential for new acoustical tools. This is especially so in coastal waters which, whilst being the most explored by our vessels, are also the least understood acoustically.

Part V
Studies of the ocean bottom

Chapter 22

Acoustical imaging of deep ocean hydrothermal flows

DAVID R. PALMER
NOAA-Miami

PETER A. RONA
Rutgers University

Summary

Some years ago we proposed the use of high-frequency sonars to acoustically image and to study hydrothermal plumes (Palmer, Rona, and Mottl, 1986). The original idea is illustrated in Fig. 22.1. The backscattered intensity measured by a sonar system is related to the concentration of the particles suspended in the plume. Using Doppler techniques, the radial component of the velocity of the particles, representing the component of the flow velocity of the discharging hydrothermal fluid along the line from the sonar to the plume, can be estimated. In this chapter we summarize the results that have been obtained from the program that grew from our original idea and from an expansion of the program in the 1990s to include development of acoustical techniques for locating and mapping regions of low-temperature diffuse flow. After describing hydrothermal flows, we discuss the justification for using a sonar to study their properties. We then present the physical arguments that demonstrated the feasibility of the idea. We then discuss the results obtained from several sea trials. With one exception these sea trials involved small, three-person, deep-diving submersibles. The most recent used the Jason/Medea Remotely Operated Vehicle (ROV) System. This system is perhaps best known for having been used to investigate the ill-fated passenger ship HMS Titanic. We provide examples of the three-dimensional digital images that were constructed from the gridded data and how those images are helping to understand the role of hydrothermal flows as agents of the dispersal of heat and chemicals in the ocean. Finally, we present a view of the future.

Contents

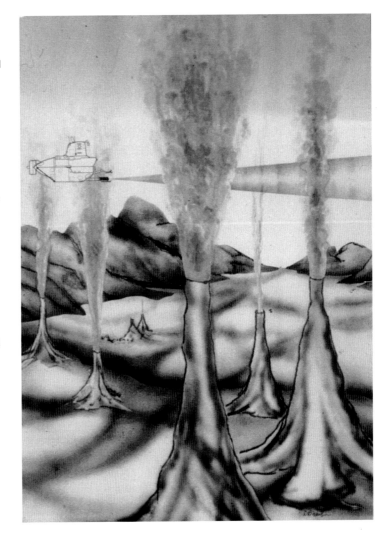

Fig. 22.1. Original conception of how a black smoker plume could be imaged with a sonar mounted on a submersible (Palmer, 1995). Most plume imaging data have been collected while the platform was stationary on the ocean bottom. We continue to explore the original concept of imaging plumes with a hovering or moving platform. During VIP 2000 we conducted a hovering experiment designed to investigate the extent to which a plume cast an acoustic shadow and we attempted to image a complete plume by pinging while the ROV slowly rose. Such experiments are difficult because it is not possible to average data from multiple pings to reduce statistical variability.

22.1 Hydrothermal flows

The outer shell of the Earth, the lithosphere, consists of ten major and numerous minor rigid tectonic plates that essentially float on the plastic asthenosphere. They move relative to one another at rates of centimeters per year driven by convection of the mantle. The globally distributed undersea volcanic mountain chains that comprise the ocean ridge system correspond to locations where two adjacent tectonic plates are separating. Sea floor spreading centers are regions between the separating plates where hot magma rises from the mantle to form new ocean crust. Cracks in the basaltic rock at the plate boundaries allow cold, dense sea water to penetrate kilometers deep into the lithosphere where it is heated by magma, expands, and buoyantly rises while leaching metals from the rock. The hydrothermal fluid discharges at chimney-like vents on the sea floor at flow rates as high as several m/s (comparable to the flow rate from a fire hose) and having temperatures as high as 400 °C. Within a few centimeters above the vent the acidic, metal-rich fluid mixes with the cold, alkaline sea water resulting in the precipitation of fine metallic sulfides that are then transported in the flow. These precipitates give the hot springs the appearance of "black smoker" plumes. Black smoker plumes turbulently mix with surrounding sea water as they rise until a height of neutral buoyancy is reached at typically 200 m where they then spread horizontally, similar to the behavior of smoke rising from a factory smokestack.

The traditional methods for studying the characteristics of hydrothermal plumes are visual observations, video and photographic records, and the placement of measurement instruments directly in the plume, that is, point sampling. In the absolute darkness of the deep-ocean sea floor, visual observations and video and photographic imagery are only possible within the immediate vicinity (i.e., within a few meters) of the plume chimneys where illumination can be provided by lights on the instrumentation platform. Point sampling is limited by lack of knowledge of the plume boundaries needed for the placement of the instrumentation and the asynchronous nature of the measurements at individual positions in the plume.

In proposing in 1986 the use of a sonar system to study black smoker plumes, we believed that an acoustical system could provide images of plume boundaries as a function of height above the ocean floor far beyond the limits of visual, photographic, or video images and could provide data from which estimates could be made of particle concentrations and fluid velocities within the plume. To get an idea of the advantage of acoustics, think of trying to predict weather by watching clouds at night with a spotlight. Weather radar is far superior. Acoustical imaging has the advantage over traditional point sampling in that it is not limited by the number of locations that can be sampled in the time available or by the problems associated with positioning sampling equipment within the plume. If physical or chemical sampling is still needed then the acoustical data could be used beneficially to guide the placement of sampling equipment. There are several important reasons for having the improved characterization of a plume that acoustical techniques can offer. Knowledge of the ascent and horizontal spreading of a plume can help provide an understanding of how a plume interacts with prevailing tides and currents, couples to mid-depth ocean circulation, and disperses heat, chemicals, and biological material into the surrounding ocean. Not all hydrothermal systems result in the concentrated, high-velocity, high-temperature flow of black smokers. If the hydrothermal fluid mixes with ambient sea water before it discharges from the sea floor, minerals may be partially precipitated beneath the sea floor, and lower temperature fluids ($\sim$200 °C) may discharge from chimneys as white smoker plumes. Diffuse flow is another type of low temperature hydrothermal discharge from patchy, irregular areas in vent fields. Most of the vent marine animals live in areas of diffuse flow and the cumulative thermal flux of the diffuse flow may exceed that from black smokers (Rona and Trivett, 1992). The temperature of the diffuse flow typically ranges up to tens of degrees C. While diffuse flow is clear because it lacks precipitates, it shimmers when illuminated due to refraction effects related to temperature fluctuations. Diffuse flow may result from the mixing of high-temperature hydrothermal fluid with sea water within the fractured basaltic rock and precipitation of all metallic minerals below the ocean bottom.

The fluxes from regions of diffuse flow are significant and need to be measured to have a complete picture of the transfer of heat and chemicals associated with hydrothermal systems. Our initial program to use sonar systems to image black smokers was expanded in the 1990s to include acoustical techniques for locating and mapping regions of diffuse flow. These techniques are based, essentially, on the shimmering of the water in these regions.

22.2 Demonstrating the feasibility of a plume imaging sonar

A question we immediately faced was what type of platform could be used for the sonar. Only two types of sonar platforms had been used in the past to image naturally occurring plumes – surface ships and instrumented sleds. The most important platform requirement is that the sonar be operated near the sea floor close to the plume. To provide images of value, the sonar must have a footprint (i.e., the cross-sectional area insonified by the sonar) at the location of a plume small compared to the plume's dimensions. Also, high-resolution sonars operate at frequencies above tens of kilohertz and have operation ranges of the order of hundreds of meters. For deep-water, black smoker plumes this requirement of proximity to the plume rules out surface ships. The instrumentation sleds that existed in the mid 1980s did not have the navigational or stability capabilities to serve as platforms. The only remaining possibility was to use a deep-diving submersible. Since a submersible had never been used as a platform for a plume imaging sonar, a host of engineering problems had to be addressed. How would the sonar be mounted on the submersible so as to provide an unimpeded view? How would the electrical cables be fed into the small, pressurized sphere that provides working space for the three-person crew? What recording and display equipment could be placed in the sphere that did not take up much space, would not use much power, and would not give off even infinitesimal amounts of toxic gaseous emissions.

More importantly, a deep-diving submersible and its support ship are scarce and valuable resources in great demand. They would not be made available for an experiment unless we could provide convincing arguments that there was a high probability of success. At first glance it might seem that black smoker plumes could not be imaged. Measurements had been made indicating the suspended particles in a plume had equivalent spherical radii less than 35 microns. Hence, for sonars with carrier frequencies less than 500 kHz, ka is less than 0.07 indicating scattering takes place in the Rayleigh frequency region.

For Rayleigh scattering, the differential backscattering cross-section for a fluid sphere is given by (6.51) with $\theta = 0°$,

$$\Delta\sigma_{bs}(f) = (ka)^4[(e-1)/3e + (g-1)/(2g+1)]a^2 \qquad (22.1)$$

Even though the precipitates in black smoker plumes have very irregular shapes, not at all similar to spheres, detailed analysis indicates that, in the Rayleigh region, they can be approximated by spheres when determining the feasibility of a plume imaging sonar (Palmer, 1996). Since the factor in square brackets is of order unity, the cross-section is smaller than the geometrical cross-section by a factor of $(0.07)^4 = 2.4 \times 10^{-5}$ or less. Clearly the plume particles are so small they are not very good scatterers of sound having a frequency of 500 kHz or below.

What is important, however, is not the cross-section for scattering from a single particle but rather the volume scattering coefficient (7.9)

$$s_v = <n_b\Delta\sigma_{bs}> \qquad (22.2)$$

where n_b is the number of scatterers per unit volume and the angular brackets indicate an average over a number of pings. If we take the average equivalent spherical radius of the particles to be 15μm, their average volume is $4\pi(15\mu m)^3/3 = 1.06 \times 10^{-6}$ cm^3. The particles that comprise black smoker plumes are primarily iron and copper sulfide minerals having densities that average about 4.8 g/cm^3. The average mass of the particles is then $(4.8\,g/cm^3) \times (1.06 \times 10^{-6}\,cm^3) = 5.09 \times 10^{-6}$ g. Estimates of the mass concentration in plumes are as much as 150 g/m^3. Therefore, $n_b = (150\,g/m^3)/(5.09 \times 10^{-6}g) = 2.9 \times 10^7$ particles/m^3. This large value for n_b in a sense cancels the small value for $(ka)^4$ and suggests it might be possible to image plumes acoustically.

Detailed analysis taking into account all the characteristics of high-frequency sonars and of the environment, as well as the specific properties of the plume particles, confirmed this simple calculation (Palmer, Rona, and Mottl, 1986). The analysis indicated a black smoker plume could be detected at ranges of hundreds of meters using a conventional sonar.

A search was made for observational data that would confirm the theoretical feasibility study. While no submersible had ever been equipped with a sonar designed for imaging plumes, they are equipped with obstacle avoidance sonars. We arranged with a colleague (M. J. Mottl) to use such a sonar on a dive by the Deep Submergence Research Vehicle (DSRV) Alvin in 1984 at a vent site at the East Pacific Rise off Baha California. These data provided the needed confirmation. The vent site, at a depth of 2500 m, called the Feather Duster vent field in recognition of the high concentration of feather-like worms living there, was being surveyed by DSRV Alvin. An examination of photographs taken manually using a camera mounted in front of the sonar display screen indicated the presence of plumes rising at least 16 m from the ocean floor while the submersible was at a range about 35 meters from the vent complex. The theoretical analysis together with the data from DSRV Alvin demonstrated that a plume imaging sonar was feasible. The stage was set to build and test a prototype sonar.

22.3 Testing of a prototype sonar

As National Oceanic and Atmospheric Administration (NOAA) scientists at that time, we arranged for the development of a prototype plume-imaging sonar in collaboration with the Naval Research Laboratory by modifying an existing commercial sonar belonging to the U.S. Navy Submarine Development Group One. It had a carrier frequency of 333 kHz, a pulse duration of 0.1 ms, and a conical beam with a beam width of 1.7°. The sonar provided azimuthal coverage by motion of the sonar head using a stepping motor in steps of 0.9°. A second axis allowed the sonar head to be moved in elevation in steps of 1.7°, again, using a stepping motor. The analog pressure amplitude was digitized at a sampling rate of at least 10 kHz providing range resolution of at least 0.15 m.

The sonar was mounted on the forward part of the sail of the U.S. Navy Deep Submergence Vehicle (DSV) Turtle. While imaging plumes, Turtle sat on the sea floor at horizontal distances from the plumes varying from a few meters to about 70 meters. The dive site chosen was the Southwest Vent Field at a depth of 2635 m at latitude 21°N on the East Pacific Rise. This is the site where black smokers were first discovered. It was chosen because the topography there is flat. Sonar returns from topography would not be present to interfere with the returns from the black smokers.

Our attention focused on two adjacent black smoker plumes emanating from chimneys 3.5 m apart. One chimney was about 5 m high and had two orifices on the top each about 2 cm in diameter from which hydrothermal fluid discharged. The second chimney was about 7 m high and had five orifices each about 2 cm in diameter discharging from the top and side of the chimney. Discharge rates and temperatures of the fluids were estimated to be 1 m/s and 350°C, respectively. Using this information, we estimated that the plumes would buoyantly rise to about 200 m where they would become neutrally buoyant and spread laterally. This estimate was obtained using an equation that relates the total weight deficiency produced by a volume of the expanding hydrothermal solution per unit time, i.e., the buoyancy flux, to the density stratification in the ocean.

Figure 22.2 is a three-dimensional reconstruction of an acoustical image of the lower 40 m of the buoyant plumes discharging from the two chimneys (Rona et al., 1991). The sonar was located about 7 m from the nearest, smaller chimney (right). The image clearly defines the boundaries of the plumes, shows their merging into a single plume at about 8–12 meters above the chimneys, and bending of the plumes in the direction of the prevailing oceanic current.

This sea trial demonstrated the value of acoustical remote sensing for defining the boundaries of buoyant hydrothermal plumes, for characterization of the plume structure on spatial scales of meters to tens of meters and time scales of seconds to hours, and for detection of plumes at ranges of tens of meters (Rona and Palmer, 1993).

22.4 Plume modeling and reconstruction

The next step was to develop the capabilities to advance from spectacular images to quantitative measurements of plume parameters from the images. Working in the Laboratory for Visualization and Modeling at Rutgers University, we developed computer programs to reconstruct

Fig. 22.2. Buoyant black
smoker hydrothermal plumes
discharging from adjacent
chimneys reconstructed in 3D
from volume backscattering
data at latitude 21° N on the
East Pacific Rise (Rona *et al.*,
1991). The rectilinear grid on
the sea floor is 5 m by 5 m by
5 m (*x, y, z*). (See colour plate
section.

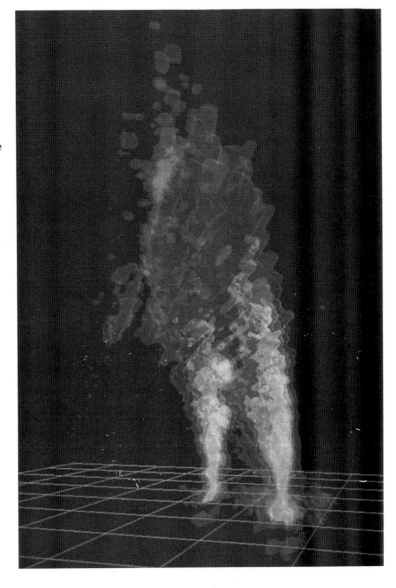

Fig. 22.2. Buoyant black smoker hydrothermal plumes discharging from adjacent chimneys reconstructed in 3D from volume backscattering data at latitude 21° N on the East Pacific Rise (Rona *et al.*, 1991). The rectilinear grid on the sea floor is 5 m by 5 m by 5 m (*x, y, z*). (See colour plate section.

and view the buoyant plumes in three dimensions and in cross-section from volume backscatter data, and to measure plume parameters from the reconstructions (Rona *et al.*, 2002a), which enable description of plume behavior (Bemis *et al.*, 2002) and comparison with buoyant plume model predictions. The buoyant plumes are reconstructed as isosurfaces of various percentages of maximum-recorded backscatter intensity, with decreasing intensity outward from the centerline and upward along the centerline of a plume. Table 22.1 contains some of the scalar parameters

Table 22.1 *Plume centerline position and radius*
(Fig. 22.2, larger plume)

Height (m)	E–W (m)	N–S (m)	[a] Backscatter Intensity	[b] Radius (m)
0 (vent)	0.00	0.00	1.000	1.200
5	−0.87	−0.44	0.979	1.700
10	−2.12	0.36	0.313	2.100
15	−3.52	0.77	0.058	2.600
20	−4.86	0.35	0.018	–
25	−3.48	−0.93	0.008	–

[a] Backscatter intensity normalized.
[b] Radius at $1/e$ of centerline intensity.

measured at one threshold (0.25 percent of maximum backscatter intensity near the vent) for the larger buoyant plume shown in Fig. 22.2. These parameters include time averages of the centerline position, expansion with height above vent, volume, summed intensity, center of intensity, and surface area. We also measure bending of the plume centerline by prevailing ocean currents. We apply Doppler methods to estimate particle flow velocity in plume cross-sections. These estimates are converted to estimates of the mean vertical rise velocity, thus enabling the calculation of the volume flux (Jackson *et al.*, 2003).

22.5 Acoustical imaging of diffuse flow

In the mid 1990s we expanded the plume imaging program to include acoustical remote sensing of diffuse flow. This required using a different set of physical principles than those used to image black smokers. The transparent fluid in the diffuse flow lacks particles that scatter sound and would indicate the presence of the flow. The flow originates from cracks and fissures that exist in irregular, patchy areas that range in size from a few square meters to a few hundred square meters, rather than from small orifices. While a black smoker plume may extend hundreds of meters above the ocean floor, the diffuse flow generally is no longer observable above a few meters because of dilution with ambient sea water, lateral advection by prevailing ocean currents, and entrainment in black smoker plumes. In addition, the purposes for obtaining images of diffuse flow are different. Images of black smokers are valuable because they provide information about plume shape, particle properties, plume dynamics, and the relationship of the plume to its ocean environment. The purpose for imaging the diffuse flow is to provide maps indicating where it exists for the purpose of studying the distribution of vent animals and guiding the placement of instruments for making the point measurements of temperature and flow rate needed to calculate fluxes. The method chosen to acoustically map diffuse flow was adapted by Rona *et al.* (1997) from one developed by Jackson and Dworski (1992) for measuring ocean temperature variations near the sea floor. It uses the topography in the vicinity of the vent field as a stationary, strong backscatterer. The sonar is mounted on a platform sitting on the ocean floor or hovering just above it. Pings in a narrow conical beam are transmitted forward and downward. Some of the sound scattered by the rocky sea floor travels back to the sonar. If the sea water along the transmission path does not change with time, the signals received from two consecutive pings will be similar, that is, they will be *correlated*. If, however, diffuse flow exists near the sea floor along the path traveled by the sound, the same temperature fluctuations that result in the shimmering of the water will cause fluctuations in the speed of sound. These fluctuations will result in changes in the received signal that differ from ping to ping. Thus the signals from two consecutive pings will be partially *uncorrelated*. By measuring the degree of correlation of the returns from consecutive pings for different ranges and elevation and azimuthal settings, one can map regions of diffuse flow.

We tested the technique in 1996 at Monolith Vent Field, a well-studied vent site located on the Juan de Fuca Ridge about 250 km southwest of Vancouver Island, British Columbia, Canada, at a depth of about 2250 m. The same sonar used in the pioneering experiment in 1990 to image black smokers was mounted on the U.S. Navy DSV Sea Cliff. Results are shown in Fig. 22.3. Three separate patches of uncorrelated signal returns stand out from a background of returns that exhibit ping-to-ping correlation. A video survey indicated these patches coincided with areas of diffuse flow thus confirming the mapping technique. We named the technique Acoustic Scintillation Thermography (AST) after the physical process on which it is based. Subsequently, others have applied our AST technique to map diffuse flow in large areas (km^2) of the sea floor in coordination with *in situ* measurements of temperature and flow rate to estimate the heat flux associated with the diffuse flow.

22.6 Vent Imaging Pacific (VIP) 2000

The Vent Imaging Pacific (VIP) 2000 cruise took place in July, 2000, at the Main Endeavour Field at a depth of about 2200 m on the Endeavour spreading segment of the Juan de Fuca Ridge. The vent site is within the Canadian Exclusive Economic Zone about 250 km southwest of

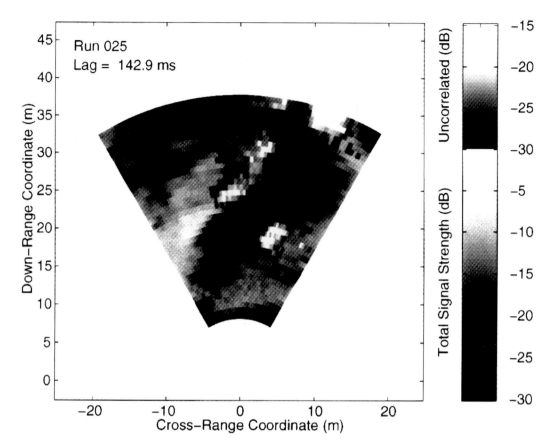

Fig. 22.3. Mapping of diffuse flow on the Juan de Fuca Ridge using acoustical correlation techniques (Rona *et al.*, 1997). The brightly colored patches are areas of decorrelation corresponding to diffuse flow. (See colour plate section.)

Vancouver Island and is being designated a Marine Protected Area by the Canadian Government. It is also referred to as the National Science Foundation Ridge Program Endeavour Observatory. The operational area was only about 350 m long and 150 m wide, a typical size for a sea floor hydrothermal field. The vent complexes had all been named by previous expeditions.

For the first time we used a remotely operated vehicle as a platform for the plume imaging sonar rather than a human-occupied, deep-diving submersible. By any measure the Jason/Medea ROV, operated by the Deep Submergence Laboratory of the Woods Hole Oceanographic Institution, is a remarkable system. Medea is a towed survey vehicle about the size of an office desk that is connected to a surface ship by an 8000 m long, 1.7 cm diameter, armored electro-optical cable. The ship in this case was the Research Vessel (R/V) Thomas G. Thompson operated by the University of Washington. ROV Jason is a precision imaging and sampling platform connected to MEDEA by a 35 m long fiber optic tether. It was equipped with five cameras of various types and two sonars including our plume imaging one. It has a precise navigational system and can move along any axis with the aid of seven thrusters. While ROV JASON has all the positioning capability of a manned submersible, it has far greater endurance. During VIP 2000 a new dive record was set for ROV JASON of 112 hours, equivalent to about 20 dives using a manned submersible.

VIP 2000 represented a significant advance in the capability of the sonar used to image plumes and diffuse flow. The sonar used in the 1990 and 1996 experiments obtained backscattered data for different values of the elevation and azimuthal angles by physically moving the sonar head. The new 200 kHz sonar used in VIP 2000 had a multiple beam capability that allowed it to obtain data simultaneously for all values of the azimuthal angle. A typical black smoker plume imaged with the old sonar in about 30 min was imaged with the new sonar in 2 min.

The improved capability of the sonar together with the endurance of ROV JASON provided an opportunity to conduct a number of different types of experiments during VIP 2000. The first long-term acoustical imaging of an underwater plume was achieved when ROV JASON sat on the ocean floor and continuously imaged a plume complex for 24 hours or 2 tidal cycles (Rona et al., 2002b). Flow rates were estimated using the Doppler shift of the backscattered signal (Jackson et al., 2003). Various methods (stop and go, hovering, and continuous forward motion) were tested to determine how best to survey for diffuse flow. At the end of the cruise a great sense of accomplishment was felt by all the participants because of the large amount of high-quality data that was collected. We now use the VIP acronym to connote Vent Imaging and Processing to describe our whole project.

22.7 The future

In a few short years acoustical imaging of deep-ocean hydrothermal plumes and diffuse flow has evolved from a speculative proposal that many people did not think possible to a routine scientific activity. Based on an analysis of all the scientific work that had been done to acousti- cally image naturally occurring plumes in the ocean, we speculated on the future for this type of research (Palmer, 1995). We felt commercial sonars, with modifications, would continue to be used to image under- water plumes and that a sonar designed specifically for imaging plumes would not be built in the foreseeable future. Today a plume imaging sonar is being built by our collaborators at the Applied Physics Laboratory of the University of Washington. It was clear in 1995 that there was a need to determine the flow characteristics of plumes using Doppler techniques, but a Doppler plume imaging sonar did not exist. Today isosurfaces of the vertical velocity of plumes can be obtained almost as readily as can isosurfaces of intensity (Jackson *et al.*, 2003). Plans are now being made to solve the "inverse problem," that is, to use acoustical imaging to estimate particle concentration in a hydrothermal plume. In 1995 we did not see any solution to this problem and felt that to spend time trying to solve it represented misplaced priorities. We anticipated in 1995 that the use of bottom-mounted sonar platforms could provide data that would open up a new research field – long-term characterization of plume dynamics. The 24-hour time series collected during VIP 2000 should help in the design of a bottom-mounted platform. In fact much of our research program is aimed in this direction. Long term characterization should be valuable in detecting the linkages between hydrothermal flow (plumes and diffuse flow) and external forcing factors such as currents, tides, earthquakes, and magnetic activity. Progress has been more rapid than we anticipated. We expect the coming years will be even more exciting than the past ones.

Chapter 23
Remotely imaging underwater mountain ranges in minutes

NICHOLAS C. MAKRIS
Massachusetts Institute of Technology

Summary

Is there a way to rapidly image underwater mountain ranges over hundreds of kilometers in mere minutes? Yes, that's what the current chapter is about, but there are pros and cons to the method as we will see. One of the big pros is that it's not tedious and time-consuming like the traditional method for measuring bathymetry, where a research vessel must laboriously pass almost directly over every patch of ocean to be imaged by a downward or near downward looking echo-sounder. That's about as exciting as watching somebody mow a lawn. Why not send the sound waves out horizontally and cover large areas at once? People have been known to actually do this when they're in the mountains to range a distant canyon just by clapping their hands, knowing the sound speed of air and clocking the echo return's travel time. Well, let's see how well this idea works in looking for underwater mountain ranges.

Contents

23.1 Introduction

In the summer of 1993, a group of scientists including myself set out on two research vessels to test the horizontal echo sounding idea on the underwater mountains of the Mid-Atlantic Ridge. Both ships met their

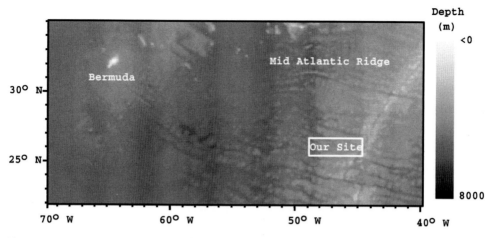

Fig. 23.1. Deep ocean bathymetry including part of the Mid-Atlantic Ridge in the vicinity of our experiment.

scientific parties at the Azores, a group of beautiful lush volcanic isles owned by Portugal in the mid-Atlantic that were once prominent stops in the age of sail between North America and Europe and where time seems to have stood still since then. In fact, many a Spanish Galleon laden with gold and other treasures from the New World met a watery grave at the hands of British pirates in these waters. Perhaps some of you future underwater explorers will uncover these treasures. The Azores also attract bio-acousticians from around the world because they are a prime habitat for many kinds of marine mammals.

After leaving the Azores, our research vessels cruised a couple of days southwest to the western flank of the Mid-Atlantic Ridge in the box shown in Fig. 23.1, at about the latitude of southern Florida. Geophysicists were really interested in this area because they wanted to test hypotheses about how the Earth's crust spreads at mid-ocean ridges. They needed higher resolution bathymetry than was available to do this. We also needed higher resolution bathymetry as "ground truth" to see how well our long-range echo-sounding approach could resolve the mountains. This turned out to be a win–win situation for everybody.

Now the question is, how much higher in resolution should the bathymetry be? Well, presently the majority of the world's oceans are only sampled at roughly 5 km horizontal resolution. That's pretty bad when you realize that we have optically imaged the surfaces of many distant moons and planets in our solar system, or at least large portions of these surfaces, to better than 1 km scale resolution and sometimes even the meter scale (Weissman *et al.*, 1999). But light and electromagnetic waves do not penetrate very deeply into the ocean, that's why we see only darkness when we peer down into them. With only 5 km resolution, you can miss entire mountain ranges, and in fact, we will see that this is exactly what happens. Incidentally, this 5 km resolution has been obtained independently by years of standard downward looking echo-sounding and years of satellite altimetry measurements. The satellite measurements extract bathymetry from its gravitational effect on local sea surface elevation. While this surface displacement is much smaller than the height of passing surface waves and tides, the surface wave and tidal changes can be averaged out over time to obtain the static eleva-tions. This method has an inherent 5 km resolution limit corresponding to the size of an unavoidable gravitational "footprint," and requires the satellite to pass over the area to be imaged many times to average out the time-dependent sea-surface changes.

Well, on a purely intuitive level, one would like the bathymetry to be sampled densely enough that its essential shape or morphology can be distinguished. This type of sampling should make it possible to see the basic structure of all the peaks, ridges, escarpments, and valleys of a mountain range, but maybe not all the cliff faces and gullies. If you think about this more quantitatively in terms of Fourier transforms and the Nyquist theorem, it would mean that you are properly sampling the dominant spatial frequencies of the bathymetry, but are aliasing the higher frequencies.

The agreed upon sampling resolution was 200 m over an area of roughly 400 km by 200 km. This was also fine enough for the acousticians to see if the long-range echo sounding technique was indeed able to image the existing features accurately. It was also fine enough to enable the geophysicists to figure out how the crust was evolving, but not so fine that it would take "forever" to do the job. Actually, it took about 1 month of tedious lawn mower tracks in the summer of 1992. From this you can see why it would take many vessels many decades to sample all the oceans of the world at 200 m resolution.

The 200 m resolution was still not fine enough to answer an important question. What physical mechanisms cause the sound waves to return back to the receiver? Clearly, it is some kind of scattering from the sea floor, but what kind of scattering?

In fact, two opposing views had begun to polarize the scores of acousticians involved. This was in fact a large program involving a good portion of the basic research community in ocean acoustics. One view held that the most prominent returns could be used to image prominent geomorphology. The idea here was that returns striking sea floor features that are large compared to the wavelength would be more intense as they struck closer to normal incidence. In fact, the way we perceive shape from shading in the natural world hinges on this in our sense of vision. The other held that smaller scale roughness on the order of the wavelength would cause all returns to interfere so randomly that little correspondence to geomorphology would be found. More intense or prominent returns in this case would result from random interference and not from prominent bathymetric features. This type of argument goes on all the time in remote sensing circles. We'll soon see how it was resolved in this case.

Now this whole debate can lead to some unsettling questions like, what are humans seeing when they look at each other? What are the fundamental scattering mechanisms that cause a person's face to look the way it does when we see it with scattered light? Strangely enough, most people use their sense of sight to lead very happy lives without ever caring about this issue! (So do most researchers in the field of Computer Vision.) Why? And why are we so concerned about knowing the fundamental scattering mechanisms of sound from the sea floor and not of light from our own faces? Well, maybe there is such a strong geometric relationship between the optical image and facial morphology that we don't have to care. Maybe, given images where the geometric relationship is less obvious, we would be forced to care. Maybe we already know what flesh is made of so we don't worry about deducing it from the scattering properties of optical images.

On the other hand, maybe we sometimes do know something about the facial scattering process and make deductions about material properties from it. For example, it is often pretty obvious when somebody is sweating, and you don't have to see beads of sweat dripping down a face to notice it. Sweat becomes obvious when the usual matte appearance of flesh that diffusely scatters light due to roughness on the optical wavelength scale develops a shiny specularly reflecting liquid coat. It's easy to test this remote deduction by reaching over and touching the sweaty brow.

It's not so easy to obtain such "ground truth" about sea floor properties when the sea floor is many kilometers below! We care about sea floor scattering mechanisms because the actual geometry and material properties of the sea floor can only be unambiguously inferred from remote measurements if we have an accurate physical and statistical model that relates the measured data to the parameters to be estimated from it. We care about "ground truth" because it validates the physical and statistical models we need to infer sea floor properties from remote sensing data.

So, in order to establish "ground truth" for the remote sensing, two sites were selected to do "fine scale surveys" where bathymetry was sampled at the wavelength scale and finer. Here the wavelength is that of the horizontally looking long-range echo sounding system which operates in the frequency range of 200–300 Hz to minimize attenuation from absorption. The wavelength scale is then a few meters.

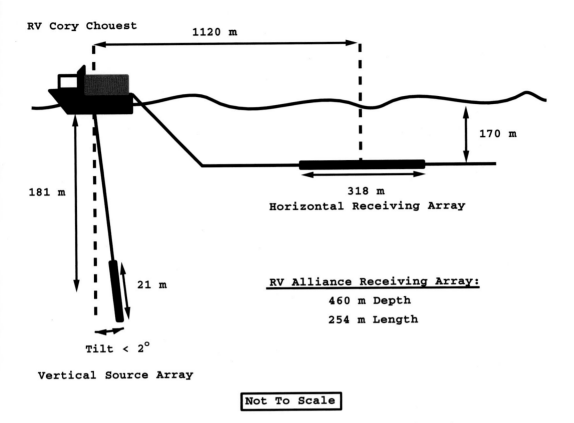

RV Cory Chouest

1120 m

170 m

181 m

318 m
Horizontal Receiving Array

21 m

RV Alliance Receiving Array:
460 m Depth
254 m Length

Tilt < 2°

Vertical Source Array

Not To Scale

Fig. 23.2. The research vessel we used with towed acoustic arrays.

Each research vessel was equipped with both a source and receiver array as shown in Fig. 23.2. Note the large apertures of the receiving arrays to obtain high azimuthal resolution images of sea floor returns by beamforming. The US research vessel Cory Chouest served as the home base for the experiment. It had a huge center well from which the primary source array used in the experiment was deployed. The source array steered the sound into a roughly 10 degree beam directed horizontally. This was to maximize the long-range sensing. Note in Fig. 23.2 the significant depths at which the source and receiver arrays were deployed. These depths were chosen to try to optimize the ability to sense at long ranges.

Consider the typical sound speed profiles measured during the experiment as shown in Fig. 23.3. By Snell's Law, rays propagating with a specific elevation angle from the source will propagate with the same elevation angle at depths that have the same sound speed as the source. The source is at roughly 180 m. The depths where the sound speed is the same as at the source are known as *conjugate depths*. The only conjugate

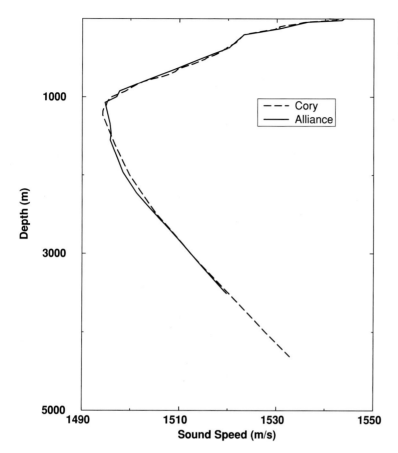

Fig. 23.3. Typical sound speed profile measured during the experiment.

depth for this case is at roughly 3800 m. This means that rays propagating horizontally from the source will also propagate horizontally at 3800 m.

In fact, when we compute the field propagated out from the source array using a range-dependent model called the parabolic equation (PE), shown in Fig. 23.4, with bathymetry determined from our 200 m resolution bathymetric survey, that's exactly what we see. These figures show the time harmonic field at a single frequency, 255 Hz, as a function of range and depth. A main beam is seen to propagate horizontally at very close range from the source and then bend downward due to refraction where it either turns at the conjugate depth, roughly 33 km in horizontal range from the source, and continues to propagate into convergence zones at the source depth every 66 km, or it is cut off by prominent sea floor bathymetry somewhere in between. From now on, for $n = 0, 1, 2, 3, \ldots$ we will refer to horizontal ranges from the source of $n \times (66 \text{ km})$ as convergence zone ranges, CZ ranges, for short and $(33 \text{ km}) + (n \times 66 \text{ km})$ ranges as $n + \frac{1}{2}$ CZ ranges. When the sea floor

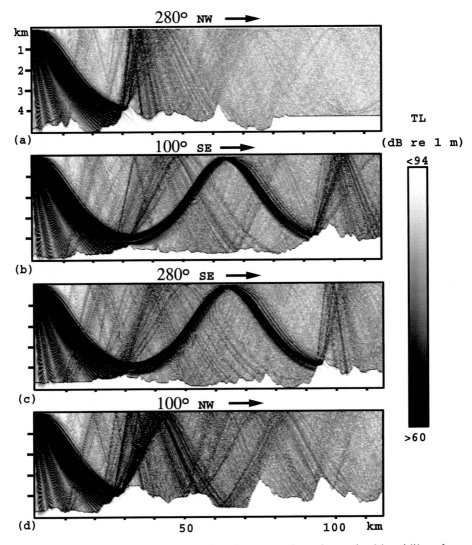

Fig. 23.4. Transmission loss along range-dependent paths, (a) and (b) are from the center of the Western Star of Fig. 23.6 while (c) and (d) are from the center of the Eastern Star.

is deeper than the conjugate depth at a particular location, it is said to have *excess depth* and will not intersect the path of main beam propagation. This is an important concept because the main beam is closest to the bottom at ½ CZ ranges where it is at conjugate depth and is most susceptible to bottom interaction. These deep turning points of the ray are known as the *deep vertex* locations.

See if you can identify the sidelobes of the source array and distinguish these from Lloyd's Mirror interference patterns from the reflection of the source at the free surface in Fig. 23.4. Also, see if you can explain why the ray paths sometimes look like straight lines and sometimes are curved. Note, the best way to insonify the sea floor at ranges between deep vertex locations is through paths that must first interact with the sea floor. (It's a funny thing that we have recently published a paper on using naturally occurring sound waves to remotely sense the interior of Jupiter's icy moon Europa which likely harbors a roughly 100 km deep ocean under about 20 km of ice (Lee *et al.*, 2003) and so is a key spot for possible extra-terrestrial life. We assumed the sound speed structure in the Europan ocean is constant as a function of depth for lack of a better model. This leads to boring straight ray paths in the ocean. (If we had predicted a sound speed structure similar to Fig. 23.3 that leads to the crazy convergence zone propagation seen on Earth (e.g. Fig. 2.9) we probably would have been laughed at for being too presumptuous!)

These ideas of excess depth and convergence zone propagation came in handy in designing the experiment. Given the constraint that our sources could not be deployed deeper than about 200 m, we wanted to find an area that had enough excess depth that sound waves could propagate out over long distances without being cut-off by the bottom, but still interact with the bottom enough at $n + \frac{1}{2}$ CZ ranges to image it. While searching through the newly acquired 200 m bathymetry back in 1992, I came across what has become known as the B′–C′ corridor shown in Fig. 23.5. It is part of a geologic segment valley that has newer more irregular upturned "inside corner" crust to the north and older highly lineated "outside corner" crust to the south. The corridor stood out because it had excess depth for exactly 2 CZs and was capped at either end by a prominent "outside corner" seamount to the west, now know as B′, and a prominent "inside corner" seamount to the east, now known as C′. One could place a source at $\frac{1}{2}$ CZ from B′ that is also $1\frac{1}{2}$ CZs from C′, and another source $\frac{1}{2}$ CZ from C′ that is also $1\frac{1}{2}$ CZs from B′ as shown in Fig. 23.5 and Fig. 23.4. This way scattering from these two types of features, which span the primary class of seamounts in mid-ocean ridges, could be efficiently studied at two important ranges in a kind of "two for one deal." (This kind of salesmanship actually proved to be pretty effective in getting the plan approved. Marketing is important even in science.) It is interesting to see how poorly the standard bathymetry for the world's oceans, an 11 km resolution as shown in Fig. 23.5(b), represents the area. Major mountain ranges simply don't appear or are so blurred they appear as continuous blobs often spanning 50 to 100 kilometers or more.

The concentric circles show the places where sound waves would have their deepest turning depths at $\frac{1}{2}$ CZ and $1\frac{1}{2}$ CZs from the two source locations. It is possible to see if sound can make it beyond

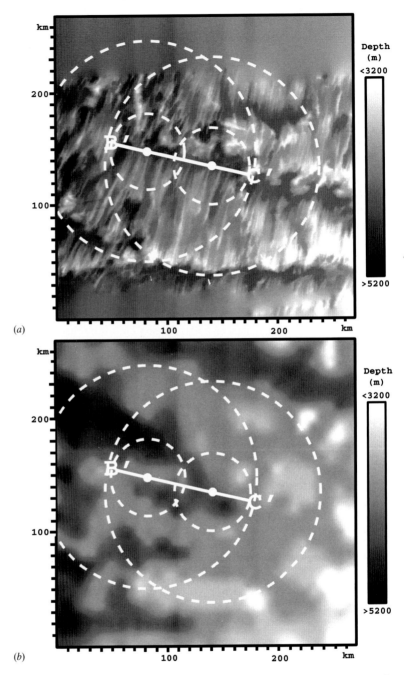

Fig. 23.5. Illustration of the "two-for-one deal." Excess depth bathymetry and ¹/₂ and 1¹/₂ CZs circles indicating deep ray vertex about the western and eastern focal points of the experiment design in the B′–C′ corridor. (*a*) (above) Our special 200-m resolution bathymetry; (*b*) (below) same thing with the more typical 11-km resolution bathymetry.

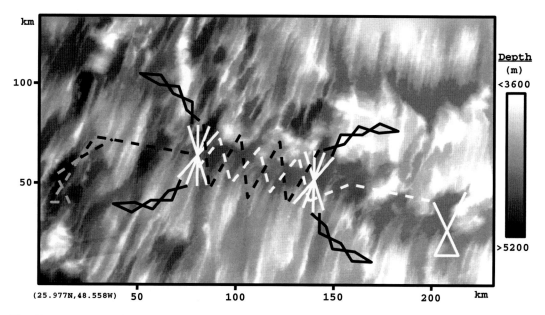

Fig. 23.6. Experimental tracks with eastern and western stars at focal points of Fig. 23.5. Cory tracks are in white, Alliance tracks are in black.

½ CZ from the source at a particular azimuth by seeing if the ½ CZ circle has excess depth at the corresponding location. The same goes for 1½ CZs. After some inspection, the Fig. 23.5 shows that sound will likely travel out beyond ½ and 1½ CZs at many azimuths. It also shows which bathymetric features could possibly be imaged at these ½ CZ and 1½ CZs ranges from each source location.

The bistatic ship tracks for the experiment are shown in Fig. 23.6. The white star patterns, at the two focal points of the corridor, were traced by the Cory Chouest. The redundant angles of the stars were needed to break the inherent right–left ambiguity in azimuth of the beamformed output of a horizontal line array. Similarly, the Alliance ran intersecting zig–zag patterns to break right–left ambiguity while obtaining bistatic scattering information from B′ and C′. In this chapter due to space limitations we will focus on only the monostatic results of the Cory Chouest.

A wide area remote acoustic image of the sea floor taken from the center of the eastern star with the Cory Chouest's source and receiver array is shown in Fig. 23.7. To make this image, a single sinusoidal pulse of ½ second duration was transmitted by the source array at 268 Hz. Travel time along the ray path was used to chart the range of the returns radially outward from the source. It took only about 150 seconds to collect the data used in this image. That's roughly the time it took the

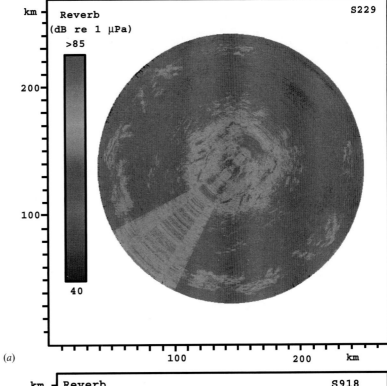

(a)

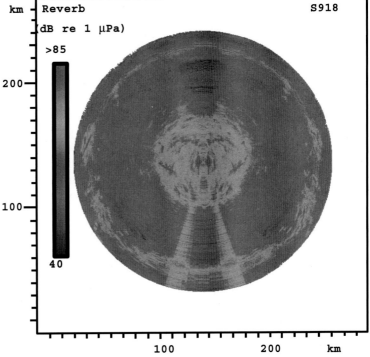

Fig. 23.7. (*a*) Wide area acoustical image taken with single transmission from Cory at center of Eastern Star. (*b*) Same as (*a*) but with a different array orientation to break ambiguity about receiver array axis. (*c*) Prominent reverberation in white overlain on color directional derivative of bathymetry in the direction of monostatic source-receiver towed from Cory. Red indicates steep slopes facing source-receiver, blue indicates steep slopes facing away from source-receiver. (*d*) Zoom in of white reverberation overlain on color directional derivative at B' seamount over box shown in (*c*) but for a different transmission. (See colour plate section.)

(*b*)

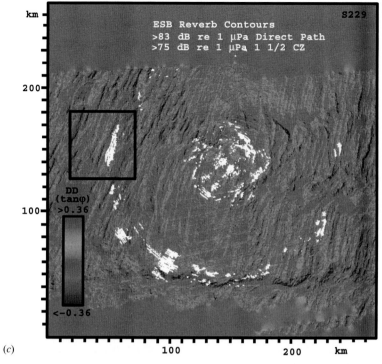

Fig. 23.7. (*Continued*)

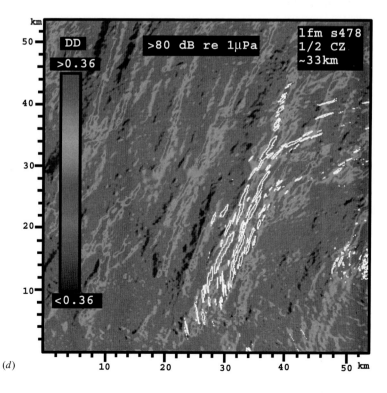

acoustic signal sent underwater to travel out to 110 km horizontal range from the source and back to the receiver array.

Beamforming on the horizontal receiving array was used to chart the azimuth of the returns. The image is symmetric about a southwesterly running axis that cuts the image in half. This is due to the right–left ambiguity inherent in a horizontal line array imaging system towed by a moving ship. The higher levels in the forward sector are from the noise of the moving ship. The array's azimuthal resolution is best when looking normal to the array from its center, known as the *broadside* direction. It is worst looking in directions along the array to the front and back, known as the *endfire* directions. Azimuthal resolution gets worse as you scan from broadside to endfire because the projected aperture of the array decreases. In fact, the angular resolution away from endfire is roughly the wavelength over the projected aperture of the array due to diffraction limitations. For the given array length and frequency, this corresponds to a roughly 1 degree azimuthal resolution at broadside, or roughly 1 km cross-range resolution at $1/2$ CZ from the ship and 3 km cross-range resolution at $1\frac{1}{2}$ CZs from the ship. The radial resolution of the images equals half the spatial extent of the transmitted sinusoidal pulse, or about 375 m.

Note in Fig. 23.7(*a*), that the high level returns tend to appear as discrete lineated features. They correspond to prominent sea floor features such as extended escarpments on the underwater seamounts and appear mostly within $1/2$ CZ and at $1\frac{1}{2}$ CZs because those are the ranges where waterborne paths to and from the sea floor exist. In fact, the location and morphology of the returns can be predicted using Fig. 23.5.

Fig. 23.7(*b*) shows another monostatic view of the same sea floor taken by the Cory Chouest along a different track at the center of the eastern star. The imaging system sees it differently because of the receiver array's changing orientation. Right–left ambiguity can be eliminated by comparing Figs. 23.7(*a*) and (*b*). The location of the true return will not change but the virtual or image return will move as the array orientation changes and always be symmetric to the true return about the array's axis. A method to remove the right–left ambiguity by optimally combining images taken from different array orientations has been developed and applied to the data (Makris, 1993; Makris *et al.*, 1995).

After breaking the right–left ambiguity, prominent returns at $1\frac{1}{2}$ CZs and within $1/2$ CZ are overlain on the radial *directional* derivative of the bathymetry, which is the dot product of gradient of the bathymetry with the radial directional vector from the monostatic source-receiver location of the ship at the eastern star center to a point in the horizontal plane. The 200 m sampled bathymetry was used to obtain the gradient. The radial directional derivative can help us understand how sound interacts with the sea floor because it shows the sea floor slopes that sound will encounter as it travels radially outward from the source. Steep slopes facing the source-receiver are seen as red while steep slopes facing away from the source are seen as blue. The blue slopes tend to be shadowed. Red slopes register extremely well with prominent returns within ½ CZ and at azimuths where there is excess depth at ½ CZ and seamounts at 1½ CZs that rise above the conjugate depth.

The remarkably high correlation between the long-range echo sounding and bathymetric slopes facing the monostatic sonar can be seen in better detail in Fig. 23.7(d), which shows a zoomed version of the box around B' in Fig. 23.7(c). Contours of prominent returns in white are again overlain on color directional derivatives of the 200 m sampled bathymetry. This time the observation is from the Western Star at ½ CZ from B' using a broadband waveform of 55 Hz bandwidth with range resolution of roughly 15 m. Range averaging was performed to reduce the variance of the image leading to an effective range resolution of roughly 50 m. The cross-range resolution of the system is along the axis

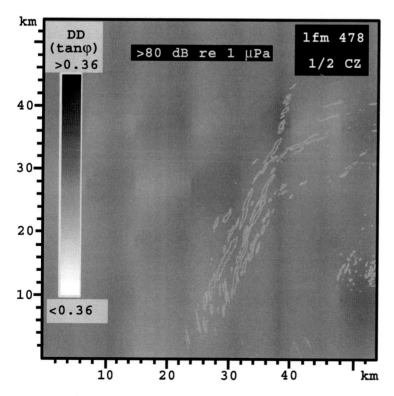

Fig. 23.8. Same as 23.7 (d) but with 11 km resolution bathymetry used to compute color directional derivative rather than 200 km bathymetry. The long-range reverberation finds this underwater mountain, the size of New York City, that does not appear in the standard bathymetry database.

of the scarps in the center of B' where correlation is best, but tends to cross the scarp axis to the north and south.

Fig. 23.8 shows a complete lack of correlation between the same long-range returns and the directional derivative of the standard 5 km sampled bathymetry of the region. This shows that long-range echo sounding provides a synoptic view of the detailed morphology of underwater mountain ranges that does not exist in standard bathymetry of the world's oceans. As we mentioned earlier, one of the pros of this technique is that it is very rapid. But as we see from the images there are some cons. Some of the primary cons are that (1) ambiguity in the line array measurements must be broken by varying the array orientation, (2) the technique always works well within $\frac{1}{2}$ CZ, but may or may not work at $n+\frac{1}{2}$ CZ ranges depending on bottom limitation at $n-\frac{1}{2}$ CZ for a given azimuth, and (3) sea floor imaging between $n+\frac{1}{2}$ CZ ranges is not so good since it requires bottom-interacting paths that distort the returns.

But what about the fundamental scattering process? Why did the returns look like the essential sea floor morphology and not like random interference patterns resulting from wavelength scale roughness on the sea floor? Well, the random effect due to wavelength scale roughness is there and gives the images a *speckly* appearance. But the acoustic waves scatter from many of these random roughness elements within a single resolution footprint of the sonar. Each of these random components of the acoustic field adds independently at the receiver so that the field returning from the sonar's footprint becomes a zero-mean Gaussian random variable according to the central limit theorem of probability theory. Such random variables have a 5.6 dB standard deviation (Makris, 1996; Makris *et al.*, 1995). The prominent deterministic features of the bathymetry in the acoustic images of Fig. 23.7(*a*), (*b*) stand above the background levels by many tens of decibels. The deterministic variations of the image of tens of decibels are then much larger than the random variations typically of 5.6 dB. That's why we see deterministic features in the images.

Is the 200 m sampling of bathymetry sufficient to pin down the fundamental scattering process? Have a look at Fig. 23.9 that compares the upper scarp of the B' ridge for 200 m sampled bathymetry and 5 m bathymetry obtained with the same deep towed system used to find and explore the Titanic. Overlain on the scarp is the resolution footprint of the Cory Chouest's long-range sonar from $\frac{1}{2}$ CZ away for the broadband signal of Fig. 23.7(*d*), which contributes to a single pixel of a long-range image. Many steep canyons on the scarp have slopes ranging from 60 to 90 degrees as can be seen only in the 5 m sampled bathymetry.

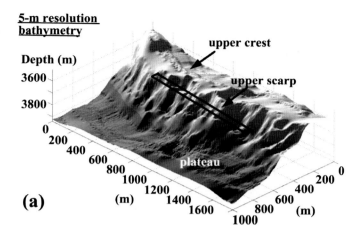

5-m resolution bathymetry

Depth (m)

Fig. 23.9. (*a*) Portion of upper scarp on B′ from 5 m resolution bathymetry data, (*b*) same as (*a*) with 200 m bathymetry data.

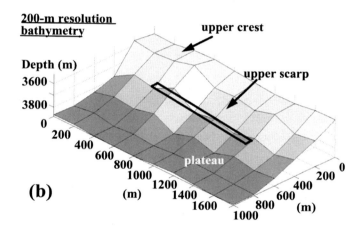

200-m resolution bathymetry

Depth (m)

The canyons have a typical scale of roughly 200 m and so are lost to the 200 m sampled bathymetry, which can only resolve slopes of roughly 20 degrees.

Since the amplitude of scattered returns from even a locally planar surface patch are a strong function of geometry such as the area of the patch, as well as the in- and out-going angles of waves to and from the patch, the 200 m bathymetry is *extremely under sampled* when it comes to studying the fundamental scattering processes. Also, the wide cross-range extent of the long-range imaging system makes it difficult to study this fundamental process because it blurs over many surfaces in a single pixel of a remote image. It also blurs over variations in material composition that are not determined by bathymetry but are important to the scattering process.

After investigations of bistatic scattering from B′ using the 5 m sampled bathymetry, and similar analysis for C′, we found that the escarpments on underwater seamounts can be adequately modeled as *Lambertian* surfaces with a surface *albedo* of roughly $\pi/10^{1.7}$, where *albedo* is the ratio of incident to scattered energy for a surface. This is much higher than the $\pi/10^{2.7}$ albedo typical of sedimented sea floor implying that the scarps are composed of exposed rock outcrops, which is supported by deep submergence data we collected at the sea floor both acoustically and optically. We still do not fully understand the detailed processes that govern the scattering of sound from underwater seamounts, just as we still do not completely understand the detailed processes that govern light from human faces, but what we do know is good enough to allow us to use sound to remotely image sea floor geomorphology.

23.2 The future

We have shown that there is a way to rapidly image underwater mountain ranges over hundreds of kilometers in minutes. The method is very fast and can rapidly find and image undiscovered mountain ranges. The primary problem with the method is that it does not image the sea floor well at certain regularly spaced intervals known as convergence zones from the source-receiver ship, due to refractive propagation in the ocean that shadows the sea floor from waterborne arrivals at these ranges. Many of the results of this work, including the bistatic imaging of the mountains, have been published in some of the references.

We have recently conducted long-range acoustic imaging experiments in continental shelf waters that typically have depths of roughly 100 m. We have some very interesting and surprising new results from this work that show many amazing environmental features all around our vessels out to tens of kilometers.

Chapter 24

Acoustical remote sensing of the sea bed using propeller noise from a light aircraft

MICHAEL J. BUCKINGHAM
University of California, San Diego

Summary

The main source of sound from a light aircraft is the propeller, which produces a fundamental tone, typically around 80 Hz, with a dozen or so harmonics at multiples of the fundamental. When the aircraft flies over the ocean, some of the propeller sound penetrates the air–sea interface, passes through the water column and enters the seabed. Thus, the aircraft acts as a high-speed (50 to 75 m/s), low-frequency ($\approx$ 80 to 800 Hz) sound source, which has potential application in ocean-acoustics experiments. In this chapter, the basic physics of aircraft sound in the ocean is discussed, including the Doppler effect, associated with the high-speed of the source, which introduces a significant shift in the frequencies of the propeller harmonics as the aircraft flies over the receiver station. An inversion technique, currently under development, for obtaining the speed of sound in the sediment from the Doppler-shifted propeller harmonics is described and suggestions for future applications of aircraft sound in ocean-acoustics experiments are offered.

Contents

24.1 Introduction

It may seem that an aircraft would not have much of a role to play in underwater acoustics, where most experiments and applications involve submerged sources and receivers. Usually, this underwater instrumentation is deployed from a surface ship or, in near-coastal waters, fixed installations such as towers and piers. In certain circumstances, however, a conventional deployment platform is not practicable, for instance, in remote ice-covered seas, which are often hazardous if not impenetrable to a surface ship. Aircraft offer a solution to this problem and indeed have been used successfully in polar regions to drop acoustic sources and receivers (sonobuoys) into leads of open water between ice floes (Diachok and Winokur, 1974; Buckingham, 1991). After deployment, the data from the sonobuoy sensors are relayed back to the aircraft over a radio link, where they are recorded and analyzed.

Apart from serving as an airborne instrument-launch platform above hostile seas, an aircraft has potential application as a source of sound in ocean-acoustics experiments. Fixed-wing aircraft come in a variety of types, ranging from propeller-driven light aeroplanes powered by a single piston-engine to multi-engine turbo-props and jets, all of which generate sound as they fly. As anyone living near an airport will attest, the sound of an aircraft in flight is clearly audible on the ground.

It is less well known that a fixed-wing aircraft can also be heard beneath the sea surface. This was confirmed by Urick (1972), who used hydrophones (underwater microphones) at various depths in the sea to record the passage of a US Navy P-3C Orion, a four-engine turbo-prop aircraft, flying at 200 knots ($\approx$ 100 m/s) at altitudes between 80 and 330 m. On either side of the sensor station, the sound of the P-3 was detected underwater for just a few seconds, which translates into a horizontal detection range of about 500 m.

Similar observations by Richardson *et al.* (1995) have been reported for several types of twin-engine, propeller-driven aircraft. Certain jet aircraft, notably military types and some older commercial jets, are extremely noisy in the atmosphere, although little has been reported on their underwater sound levels. However, during airborne deployments of sonobuoys in the Greenland Sea from a British BAC 1–11 research aircraft (Buckingham, 1991), the highpitched whine from the twin jet engines was regularly detected on the underwater sensors as the aircraft circled over the drop site. Helicopters tend to be particularly noisy in the atmosphere, often producing a low-frequency beating sound from the rotors, and, like fixed-wing aircraft, are known to be audible underwater (Richardson *et al.*, 1995; Medwin *et al.*, 1973). (For a commentary on the transmission of sound through a rough air–sea interface, including data on sub-surface sound levels from a Sikorsky SH-3D helicopter, see the discussion in Section 9.6 of this book.)

Aircraft-generated sound in the sea is the topic of this chapter. The principal aim of acoustical oceanography is to take received sound signals, which contain information about the source and the environment, and extract that information by performing an appropriate inversion. In this context, we focus on the underwater sound from a fixed-wing, propeller-driven light aircraft powered by a single piston-engine. As we shall discuss, the sound from the aircraft is detectable not only in the atmosphere and the water column but also on a hydrophone buried about 1 m deep in a (fine-sand) seabed.

At present, the ocean-acoustics research community is interested in characterizing the geoacoustic properties of marine sediments, which typically consist of varying proportions of clay, silt, and sand, often with an admixture of shell fragments. The US Office of Naval Research (ONR) is the sponsor of two sediment acoustics experiments (Simmen *et al.*, 2001; Thorsos and Richardson, 2002) in the northern Gulf of Mexico, one of which was conducted in 1999 (SAX99) whilst the second was planned for 2004 (SAX04). An important aim of SAX is to relate the wave properties (e.g., sound speed and attenuation) to the mechanical properties (e.g., porosity, density, and grain size) of a sediment. In a granular material such as a marine sand, the speed and attenuation of sound both depend on the frequency of the acoustic waves. The sediment sound-wave measurements in SAX are intended to cover as wide a frequency range as practicable; but few techniques are capable of returning reliable information at low frequencies, below 1 kHz. Yet this low-frequency part of the spectrum is critically important to our understanding of the physical mechanisms governing wave propagation in saturated granular materials such as marine sediments.

Fortuitously, the sound from a light aircraft is concentrated at low frequencies, between 50 Hz and 1 kHz. As we shall see, aircraft sound is detectable not only in the water column but also on acoustic sensors buried in the seabed, making it appealing as a measurement tool for obtaining the low-frequency wave properties of a sediment. Using a simple procedure, the acoustic data from a buried sensor may be inverted to obtain the speed of sound in the sediment, which is arguably the wave property of the seabed that is of greatest interest.

24.2 Propeller sound

As a means of aircraft propulsion, the propeller has a long history, dating back to the beginnings of powered flight early in the twentieth century. Propellers consist of a number of blades, usually two or three in the case of light aircraft. A propeller blade is a rotating aerofoil that creates a positive pressure difference between the rear and forward surfaces, which results in thrust.

In addition to thrust, a propeller blade also generates sound (Magliozzi *et al.*, 1995), the sources of which are categorized as either steady or unsteady. Steady sources would be perceived as constant by an observer rotating with the blade and are the most important in the context of this discussion. An example of a steady source is so-called thickness noise, arising from the displacement (forwards and backwards relative to the plane of the propeller) of air by the finite volume of the rotating blade. To a stationary observer, steady sources are periodic with a fundamental frequency that is given by the rotation rate R of the blade. For a propeller with N blades, the fundamental frequency is the rotation rate times the number of blades, NR. Usually in light aircraft, there is a direct drive from the engine to the propeller, in which case the rotation rate R is the same as the rotation rate of the engine, which is traditionally expressed in revolutions per minute (rpm). In cycles per second, the frequency f_1 of the fundamental tone emitted by the propeller is therefore

$$f_1 = \frac{NR}{60} \text{ Hz} \tag{24.1}$$

Although the steady sources are periodic, they are not pure sine waves, which means that, in addition to the fundamental, higher-frequency tones are also emitted. Any periodic signal can be represented as a Fourier sum of sinusoidal terms, each representing a tone, more usually called a harmonic. The frequencies of the harmonics are multiples of the fundamental frequency, a relationship which is expressed by writing the frequency of the nth harmonic as

$$f_n = nf_1 = \frac{nNR}{60} \text{ Hz} \tag{24.2}$$

Thus, the first harmonic and the fundamental are one and the same.

The harmonic structure of the sound from a two-blade propeller is illustrated in the panel on the right of Fig. 24.1, which shows a time–frequency plot, or spectrogram, of the signal from a microphone near the port wing tip of a stationary light aircraft with the engine running at 2000 rpm. According to (24.2), the harmonics should occur at multiples of approximately 67 Hz, which does indeed match the sequence of prominent tones (bright horizontal lines) in the spectrogram. Between the principal harmonics, and especially noticeable at lower frequencies, below 200 Hz, are additional tones (sub-harmonics) from the engine, which are

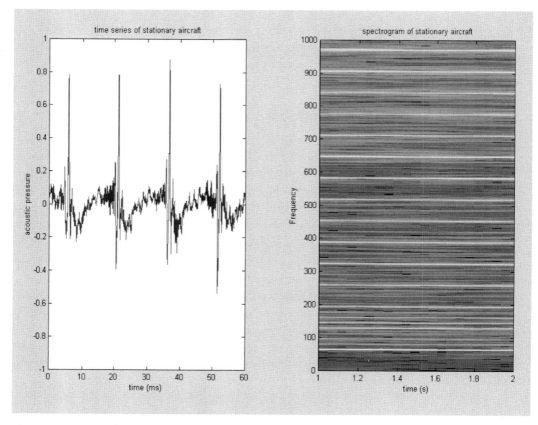

Fig. 24.1. Acoustic data from a stationary Tobago TB10 taken with a microphone at the port wing tip, about 20° behind the plane of the two-blade propeller, which was turning at 2000 rpm. Pressure time-series (left), exhibiting four uneven cylinder-firing impulses, and spectrogram (right), showing the harmonics from the propeller at frequency intervals of approximately 67 Hz.

a result of uneven firing of the cylinders. This lack of uniformity is visible in the pressure time-series data in the panel on the left of Fig. 24.1, where the four detonation impulses (one from each cylinder) exhibit slightly differing amplitudes. As the sub-harmonics are unimportant in the present context, they are excluded from the following discussion.

When the aircraft is stationary on the ground with the engine running, propeller harmonics are clearly detected with a local microphone up to a frequency well above 1 kHz. This was the situation when the data in Fig. 24.1 were recorded. With the aircraft in flight, the propagation path to the receiver is much greater, one effect of which is that the higher-frequency harmonics suffer heavy attenuation. Typically, when the aircraft is airborne, the detectable harmonics on a ground-based microphone extend no higher than about 800 Hz. Still, from a practical point of view, with sediment-acoustics applications in mind, this means that the aircraft propeller is a source of sound with highly desirable characteristics: it has a comb of ten or so discrete tones in the low-frequency band between say 50 and 800 Hz, which is a spectral region of particular interest for sediment acoustics experiments.

24.3 Doppler frequency shifts

Most sound sources used in underwater-acoustics experiments are essentially stationary, at least compared with a light aircraft, which moves very rapidly, typically in the region from 100 to 150 knots (50 to 75 m/s). As a result of this motion, the harmonics from the propeller will be subject to the Doppler effect (Pierce, 1981), a common-place example of which is the change in pitch of an ambulance siren as the vehicle speeds past a stationary observer. (For a general account of sources and receivers in motion, including the Doppler effect, see the discussion by Medwin in Section 2.6.1 of this book.)

Ahead of an aeroplane in flight, the forward motion "compresses" the acoustic wavefronts from the propeller, which raises the frequency of each harmonic, whereas to the rear the wavefronts are "dilated" and the frequencies of the harmonics are reduced. These effects are illustrated in Fig. 24.2, which shows successive wave crests radiating from the moving source. The wave crests are separated in time by the wave period, Δt, and each is centered on the position of the source at the earlier instant when the wavefront was transmitted. Note that the wave crests are depicted as circular because, in all directions, they propagate away from the point of origin with the same speed, namely the speed of sound, which is governed by the density and compressibility of the medium.

To an observer on the ground, the frequency shifts associated with the compression and dilation of the wave crests will be heard as a down-sweep in pitch as the aircraft flies through the zenith. The *change* in frequency from approach to departure depends on the speed of sound in the transmission medium (air for an observer on the ground) and, as described below, it is this *difference frequency* which will be exploited in measurements of the speed of sound in marine sediments.

The Doppler-shifted frequency of a given harmonic may be derived either by considering the phase of the sound field radiated by the propeller or by making a so-called Galilean transformation from the moving frame of reference of the aircraft to the stationary frame of the observer (Pierce, 1981). Whichever approach is taken, the following expression is found for the Doppler-shifted frequency of the nth harmonic:

$$f_{nD}(\theta) = \frac{f_n}{1 - \frac{V}{c_a}\cos\theta} \qquad (24.3)$$

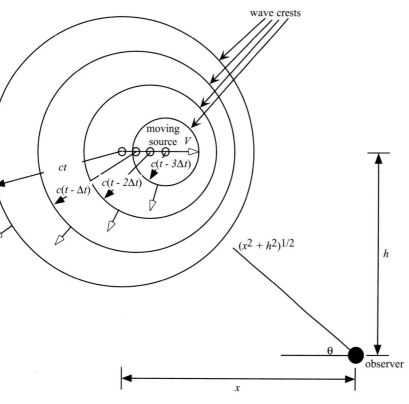

Fig. 24.2. Ground-based observer listening to an aircraft approaching with horizontal speed V at altitude h. The circles represent wave crests radiated from the moving source at intervals of one wave period, Δt, and traveling through the medium with speed c. At time $t = 0$, the horizontal range to the source is x and the angle of elevation is $\theta = \cos^{-1}(x/\sqrt{x^2 + h^2})$. (Adapted from Pierce, 1981.)

where V is the horizontal speed of the aircraft, c_a is the speed of sound in air, and θ is the angle of elevation, that is the angle between the horizontal and the line of sight to the aircraft. Notice that the Doppler shift depends on the component of velocity, $V \cos \theta$, along the line of sight but is independent of the normal component of velocity, $V \sin \theta$. With the aircraft approaching the observer, the angle of elevation lies in the interval $0 < \theta < \pi/2$, in which case $\cos \theta$ is positive and, from (24.3), the frequency is upshifted; on departure, $\pi/2 < \theta < \pi$, $\cos \theta$ is negative and the frequency is downshifted. There is no Doppler shift on the harmonics when the component of velocity along the line of sight is zero, which occurs only when the aircraft is directly overhead (i.e., $\theta = \pi/2$ and $\cos \theta = 0$).

When the aircraft is a long way out, the line of sight is essentially horizontal with an angle of elevation on approach of $\theta \approx 0 \,(\cos \theta \approx 1)$

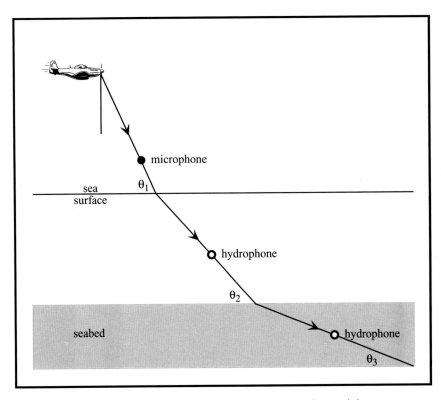

Fig. 24.3. Refraction of sound rays across the sea surface and the water–sediment interface.

or on departure $\theta \approx \pi$, $(\cos \theta \approx -1)$. For these two extreme situations, the Doppler difference frequency is maximal and from (24.3) is given by

$$\Delta f_{nD} = f_{nD}(0) - f_{nD}(\pi) \approx \frac{2V f_n}{c_a} \tag{24.4}$$

where we have taken the speed of the aircraft V to be much smaller than c_a, the speed of sound in air. According to (24.4), the difference frequency scales linearly with the speed of the aircraft, the unshifted frequency of the harmonic, and the reciprocal of the speed of sound in the medium. Conversely, we could re-write (24.4) as an expression for the sound speed c_a:

$$c_a \approx \frac{2V f_n}{\Delta f_{nD}} \tag{24.5}$$

On the right, V, f_n, and Δf_{nD} may be assumed known, since all three are fairly easily measured, and hence (24.5) provides a simple basis for recovering the speed of sound, c_a, in the medium.

24.4 Refraction

Of course, in our ocean-acoustics application of propeller sound, the medium in which we are interested is not the atmosphere but the sediment beneath the water column. A sound ray from the aircraft must penetrate the sea surface and the sea floor to reach a sensor buried in the sediment. At both these interfaces the ray will be refracted, or bent, as shown in Fig. 24.3, because of the difference in sound speed on either side of the boundaries. A familiar example of (optical) refraction is the apparent bending of a stick when one end is placed in water. What effect does refraction have on the Doppler difference frequency of a propeller harmonic in the water column or the sediment?

This question may be answered by returning to (24.3), the basic expression for the Doppler shifted frequency. The first point to note is that the angle of elevation governs the Doppler frequency of a ray; the steeper the ray (i.e., the greater the value of θ), the lower the frequency. Second, as a ray with elevation angle θ crosses from air into water, its frequency remains unchanged, from which it follows that the term $(\cos\theta)/c_a$ appearing in the denominator of (24.3) must be the same on either side of the boundary. A similar argument applies to the interface between the water column and the sediment, allowing us to write

$$\frac{\cos\theta}{c_a} = \frac{\cos\theta_w}{c_w} = \frac{\cos\theta_s}{c_s} \tag{24.6}$$

which is well-known as Snell's Law for refraction at boundaries between media of differing sound speed.

Now we see that the Doppler difference frequency of the nth harmonic, as observed in air, water, or sediment, may be generally expressed as

$$\Delta f_{nDi} \approx \frac{2Vf_n}{c_i} \tag{24.7}$$

where the subscript i denotes a, w, or s, according to whether the sensor is in the atmosphere, the water column, or buried in the seabed. Since $c_w > c_a$ and, for sand sediments, $c_s > c_w$, the Doppler difference frequency will be greatest when observed on a sensor in the air and least on a sensor buried in the sediment. Conversely, the sound speed inverted from the difference frequency,

$$c_i \approx \frac{2Vf_n}{\Delta f_{nDi}} \tag{24.8}$$

will be least in the air and greatest in the sediment.

24.5 Aeroplane experiments

Equation (24.8) is our essential tool for measuring the speed of sound in the atmosphere, sea water, and sediment using sound from the propeller of a light aircraft. For the Doppler difference-frequency technique to be successful, it is necessary for the aircraft to be detectable by acoustic sensors in the water column and, more importantly, buried in the sediment. Before mid-2002, we did not know whether a single-engine light aircraft would be detectable on a hydrophone in the water column, let alone by a sensor in the seabed. Although underwater detection of sound from certain types of multi-engine, fixed-wing aircraft had previously been reported, all were relatively large and noisy: a Lockheed P-3C Orion (Urick, 1972) (four-engine turbo-prop), a de Havilland Twin Otter (Richardson *et al.*, 1995) (twin turbo-prop), a Grumman Turbo Goose (Richardson *et al.*, 1995) (twin turbo-prop), and a Britten Norman Islander (Richardson *et al.*, 1995) (twin piston-engine).

As essentially nothing was known about the coupling of the sound of a single-engine, light aircraft from air-to-water-to-sediment, a series of exploratory flying experiments (Buckingham *et al.*, 2002a,b) was conducted in the summer of 2002 over the Pacific Ocean, off the coast of southern California about 2 km north of La Jolla. At this near-shore location, the water is about 15 m deep. A four-seat light aircraft, a Tobago TB10 with two-blade propeller and four-cylinder, 180 hp Lycoming engine, was flown parallel to the coast at altitudes between 33 and 330 m at a nominally constant air speed of 106 knots (53 m/s). The propeller speed was adjustable and held fixed at 2500 rpm. Accordingly, from (24.2), the frequency of the fundamental tone from the propeller was 83.3 Hz, with harmonics at 166.7 Hz, 250 Hz, 333.3 Hz, ...

A sensor station (Fig. 24.4), with a microphone 1 m above the sea surface, a vertical string of hydrophones in the water column, and an additional hydrophone buried 75 cm deep in the fine-sand sediment, was located 1 km or so off-shore. Using a sea water jet, divers buried the sediment phone about one week before the experiments began, allowing time for the sand to settle, providing good coupling to the sensor.

During a single flight, the aircraft flew many tracks back and forth over the sensor station on approximate north–south headings. Flight parameters, including ground speed, lateral position, and altitude, were recorded using a GPS satellite navigation system. The duration of a typical flight was 1.8 hours from take-off to touch-down.

Supporting environmental data were collected during the experiments, to help later in interpreting the acoustical measurements. Most importantly, a SeaBird temperature profiler was regularly deployed in the water column, returning the temperature of the sea-water as a function of

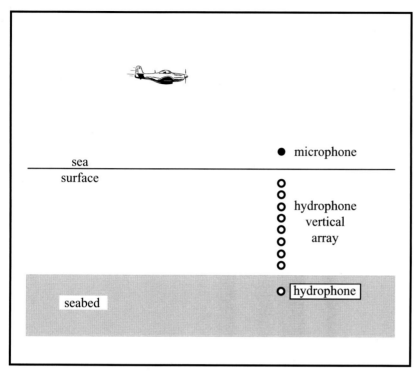

Fig. 24.4. Configuration of the flying experiments conducted north of La Jolla, California in July 2002.

depth. From the temperature profile, the speed of sound as a function of depth was computed from an empirical formula due to Mackenzie (1981) expressing sound speed in sea water as a function of three variables, temperature, pressure (which scales with depth), and salinity. Since the salinity was not measured as part of our experimental procedure, it was assumed to be constant at 35%, which is reasonable for the experiment site, where no fresh water inflow is present. (Elsewhere, for instance, near an estuary or in polar regions close to melting ice floes, the salinity may differ significantly from 35%, which should not be ignored when computing sound speed.)

24.6 Detection of aircraft sound

As the Tobago flew over the sensor station, the sound from the propeller was detected not only in the air but also below the sea surface in the water column and in the sediment. Figure 24.5 shows calibrated spectrograms from sensors in the three media from an overflight at an altitude of 66 m on July 2, 2002. Also included in the figure is the sound speed profile in

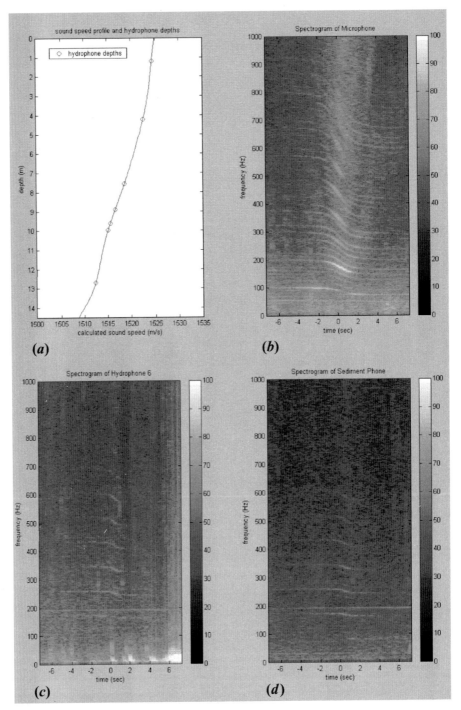

Fig. 24.5. (*a*) Average sound speed profile from SeaBird, with small circles depicting depths of hydrophones in water column; (*b*)–(*d*) Calibrated spectrograms from flight of Tobago at altitude of 66 m on July 2, 2002 showing Doppler shifted propeller harmonics (all color bars: dB re 1 μPa^2/Hz): (*b*) microphone 1 m above sea surface; (*c*) hydrophone at depth of 10 m in water column; and (*d*) hydrophone buried 75 cm deep in sediment. (See colour plate section.)

the channel (averaged over several deployments of the SeaBird profiler), superimposed upon which are the positions of the hydrophones in the water column on that particular day.

The propeller harmonics appear in all three spectrograms as bright yellow lines, which are visible for approximately 6 seconds on either side of the closest point of approach (CPA) to the sensor station. The intensity of the harmonics may be read from the color bars alongside the spectrograms, which are scaled in dB relative to a standard reference level in underwater acoustics, 1 μPa^2/Hz. Notice that the harmonics in the water column and the sediment are an order of magnitude ($\approx$ 10 dB) less intense than those in the atmosphere.

A prominent feature of all the harmonics is a sweep to lower frequency in a brief time interval centered on $t = 0$ as the aeroplane flies over the sensors. The Doppler difference-frequency between approach and departure is easily seen to be greatest on the microphone data, as expected since the speed of sound in air ($\approx$ 340 m/s) is less than that in sea water ($\approx$ 1500 m/s), and fine-sand sediment ($\approx$ 1650 m/s) by a factor of at least 4.4. Harmonics are visible up to frequencies close to 800 Hz in the microphone data but only to 600 Hz or thereabouts in the water column and the sediment. If anything, the harmonics are a little more stable in the sediment than in the water column because the latter is a somewhat dynamic environment, due to local currents and turbulence.

The least stable harmonics are those from the microphone in air, showing *intensities* which vary noticeably with time and frequency (i.e., harmonic number). This variability in intensity is largely due to the fact that the microphone is about 1 m above the sea surface, which amounts to a significant fraction of a wavelength at frequencies within the band of the propeller harmonics. As discussed by Medwin in Section 1.5.5 in this book, this situation gives rise to the Lloyd's Mirror effect in which the direct ray from the source to the receiver interferes with the ray reflected from the sea surface. Since the source-receiver geometry of the aircraft experiment is constantly changing, due to the source motion, the interference may be constructive or destructive, with each harmonic exhibiting its own, unique intensity pattern. This accounts for the observed intensity fluctuations in time and harmonic number.

Although the intensities of the airborne harmonics fluctuate, it is important to note that the Doppler-shifted *frequency* of a given harmonic is not significantly affected by the presence of two arrivals, the direct and reflected rays. The frequency is governed by the elevation angle of a ray arrival. For the geometry of the aircraft experiments (i.e., the altitude of the source much greater than that of the receiver), the elevation angles of the direct and reflected arrivals at the microphone are much the same and so too are their frequencies.

Beneath the waves, the frequencies of the harmonics in the water column and in the sediment are stable, which may seem surprising at first sight, given that the sea surface is a rough, moving boundary. When the data in Fig. 24.5 were recorded, a light breeze was blowing with a wind speed of approximately 5 knots (2.5 m/s), which created a Sea State 1 (small wavelets). In addition, a light swell was running. Although not very rough, the sea surface was far from being a smooth, horizontal plane.

However, sea-surface roughness has little effect on the harmonic structure in either the water column or the sediment for several reasons, including the fact that surface slopes are small, just a few degrees, and, compared with the speed of the aircraft, surface movement is very slow, introducing negligible Doppler shifts. At the worst, these surface-induced frequency shifts will lead to a very slight broadening of the harmonic lines but should leave the peak frequencies unaffected. The broadening, if present at all, occurs because the aircraft will have flown several tens of meters during the time interval used in computing the component spectra in spectrograms such as those of Fig. 24.5. Throughout this analysis interval, the acoustic arrival at a sensor in the water column or the sediment may, with equal probability, be randomly up-shifted and down-shifted in frequency by the surface motion. The broadening is so small, however, that the aircraft motion may be considered simply to average out the effects of surface roughness.

24.7 Sound speed estimates

Perhaps the simplest approach to estimating the local speed of sound from the spectrograms in Fig. 24.5 would be to take a ruler and read off the frequencies of a given harmonic when the aircraft was far out on the approach, directly over the sensor, and far out on departure. Given that the speed of the aircraft V is known from the GPS data (24.8) could then be used to evaluate the speed of sound in air, sea water, and sediment.

In fact, a somewhat different procedure was used in which an equation for the spectral shape of the harmonics was fitted to the spectrogram data. This equation contained the required sediment parameters as well as flight parameters such as aircraft speed and altitude, all of which were varied until the best match to the data was obtained. The comparison between the equation and the data was performed by a desk-top computer with a program written in MATLAB.

From the microphone data from twenty or so overflights of the Tobago on July 2, 2002, the average ground speed of the aircraft was estimated to be 54.5 m/s, compared with 54.8 m/s from the GPS. The speed of sound in air, as derived from the acoustic Doppler technique, is 342.3 m/s, which compares favorably with an independent estimate, based on the temperature conditions of the day, of 343.5 m/s. Such close agreement is encouraging, indicating that inversions of the Doppler data yield an accurate estimate of the speed of sound in the atmosphere.

For the same sequence of overflights, Doppler difference-frequency inversions were performed on the data from two of the sub-surface hydrophones, one at a depth of 10 m in the water column and the other buried 75 cm deep in the sediment. The difference-frequency technique returned a mean sound speed of 1529.5 m/s in the water column, which is about 1% higher than the value of 1512.4 m/s obtained from the SeaBird temperature profiler. In the sediment, the Doppler estimate of the speed of sound was 1649 m/s, which cannot be assessed directly against any independent measurement in the seabed at the experiment site because none exists at the low frequencies produced by the aircraft's propeller. However, at higher frequencies (3.5, 7, and 14 kHz), Hamilton (1972) used *in-situ* probes to measure the speed of sound in fine-sand sediments off San Diego, finding values clustering around 1685 m/s.

The slightly higher sediment sound speed measured by Hamilton, as compared with that returned by the Doppler technique, may be largely due to small differences in the geophysical parameters of nominally similar sediments. In particular, the porosity (defined as the volume of pore space between grains per unit total volume) may lie anywhere from 0.4 to 0.45 for fine-sand sediments, corresponding to a spread of sound speeds between 1640 m/s and 1720 m/s, respectively. It is also possible that sediments show weak dispersion, that is, a sound speed that varies slightly with frequency. A recent theory (Buckingham, 2000) of wave propagation in saturated granular media such as marine sediments predicts dispersion of about 2% per decade of frequency, which is consistent with the difference between the low-frequency measurements of sound speed obtained using propeller harmonics and Hamilton's higher-frequency results.

24.8 The future

Ocean-acoustic inversion techniques based on propeller sound from light aircraft, including those techniques aimed at returning the wave properties of marine sediments, are still in their infancy. Our flights during the summer of 2002 demonstrated that the acoustic signal from the propeller of a Tobago is detectable in the air, the water column, and the sediment and that the observed Doppler shift depends on the speed of sound in the medium in which the sensor is located.

Using an elementary argument based on the refraction of acoustic rays, an inversion scheme for obtaining the speed of sound in all three

Fig. 24.6. Schematic of different types of waves in the air–sea–sediment environment.

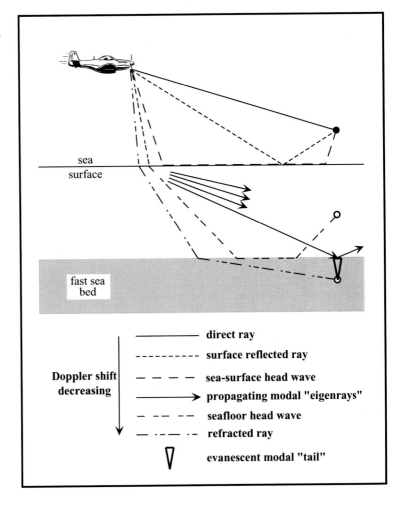

media has been introduced in this chapter. At the low frequencies of the propeller harmonics, however, the water column acts as a waveguide in which the acoustic field is not adequately described in terms of just a single refracted ray. A full solution of the wave equation is required, which will depend on the boundary conditions at the surface and bottom. This wave-theoretic solution contains several new types of wave (Fig. 24.6), including propagating normal modes, which may be thought of as acoustic resonances between the surface and bottom, evanescent modes, which leak energy from the water column into the bottom, and the lateral wave, sometimes called the head wave, which propagates along the bottom, continually re-radiating energy back into the water column.

Future inversions of aircraft sound to obtain wave speeds will be based on a full wave-theoretic solution for the acoustic field in the atmosphere, the sea water, and the sediment. The computation of the full wave field involves all the unknown parameters of the problem, including the wave speed in the sediment, which appears through the boundary conditions that apply at the sea water–sediment interface. To invert for these parameters, a technique adapted from Matched Field Processing (MFP) (Tolstoy, 1993) appears to hold some promise, although has yet to be investigated. As the aircraft approaches towards, overflies, and departs from the sensor station, it creates a "synthetic aperture." At a succession of positions within this aperture, a comparison between the data and the computed field is made for all reasonable combinations of parameter values until an optimal fit is obtained. The set of parameters giving the overall "best fit," including the speed of sound in the sediment, is taken to be representative of the environment. Although dispersion in the sediment is expected to be weak over the frequency band of the propeller harmonics, it may be detectable with the aid of this optimal fit technique.

As we have seen in this chapter, the Doppler shift observed on the propeller harmonics as the aircraft flies over the sensor station lends itself naturally to a determination of sediment sound speed. A wave property of comparable importance is the attenuation of sound in the sediment and, in particular, the frequency dependence exhibited by the attenuation. In principle, this could be recovered in two ways: by taking the ratio of the intensity of sound from the aircraft, as observed on a hydrophone buried in the sediment at two instants during the approach; or by taking the ratio of the sound intensities on two horizontally separated, buried hydrophones as the aircraft flies towards the sensor station. Future experiments will include two or more buried hydrophones, allowing the two techniques for obtaining the attenuation to be compared.

References

Abegg, F., Anderson, A., Buzi, L., Lyons, A. P., and Orsi, T. H. (1994). Free methane concentration and bubble characteristics in Eckernfoerde Bay, Germany. *Proceedings of the Gassy Mud Workshop*, 11–12 July, Kiel, Germany.

AMODE-MST Group: Birdsall, T. G., Boyd, B. D., Cornuelle, B. D., *et al.* (1994). Moving ship tomography in the northeast Atlantic Ocean. *EOS, Trans. Am. Geophys. Union*, **75**, 17, 21, 23.

Anderson, A. L. and Bryant, W. R. (1989). Acoustic properties of shallow seafloor gas, *Offshore Technology Conference Proceedings*, **5955**.

(1990). Gassy sediment occurrence and properties: Northern Gulf of Mexico. *Geo-Marine Letters*, **10**, 209–20.

Anderson, V. C. (1950). Sound scattering from a fluid sphere. *J. Acoust. Soc. Am.*, **22**, 426–31.

Andreeva, I. B. (1964). Scattering of sound by air bladders of fish in deep sound scattering ocean layers. *Sov. Phys. Acoust.*, **10**, 17–20.

Andrew, R. K., Farmer, D. M., and Kirlin, R. L. (2001). Broadband parametric imaging of breaking ocean waves. *J. Acoust. Soc. Am.*, **110**(1), 150–62.

Andrew, R. K., Howe, B. M., Mercer, J. A., and Dieciuch, M. A. (2002). Ocean ambient sound: Comparing the 1960s with the 1990s for a receiver off the California coast. *Acoust. Res. Lett.*, On-line, **3**, 65–70.

Arnone, R. W., Nero, R. W., Jech, J. M., and De Palma, I. (1990). Acoustic imaging of biological and physical processes within Gulf Stream meanders. *EOS*, **71**, 982 (July 17).

Asaki, T. J. and Marston, P. L. (1995). Free decay of shape oscillations of bubbles acoustically trapped in water and sea water. *J. Fluid Mech.*, **300**, 149–67.

Au, W.W. (1993). *The Sonar of Dolphins*. New York: Springer-Verlag.

(2000). *Hearing by Whales and Dolphins*, ed. W. W. Au, A. N. Popper, R. R. Fay. The Netherlands: Springer-Verlag.

Baggeroer, A. B., Kuperman, W. A., and Mikhalevsky, P. N. (1993). Matched field processing in ocean acoustics. In J. M. F. Moura and I. M. G. Lourtie (eds.), *Proceedings of the NATO Advanced Study Institute on Signal Processing for Ocean Exploration*. Dordrecht, Netherlands: Kluwer.

Baker, B. B. and Copson, E. T. (1950). *The Mathematical Theory of Huygens' Principle*. Oxford: Oxford University Press.

Ball, E. C. and Carlson, J. A. (1967). Acoustic backscatter from a random rough water surface. M.S. thesis, Naval Postgraduate School, Monterey, CA.

Baltzer, W. and Pickwell, G. (1970). Resonant acoustic scattering from gas-bladder fishes. In *Proceedings of an International Symposium on Biological Scattering in the Ocean*, B. Farquhar (ed.), Washington, DC: Maury Center for Ocean Science.

Barham, E. G. (1963). Siphonophores and the deep scattering layer. *Science*, **140**(3568), 826–8.

Barnhouse, P. D. C., Stoffel, M. J., and Zimdar, R. E. (1964). Instrumentation to determine the presence and acoustic effect of microbubbles near the sea surface. M.S. thesis, US Naval Postgraduate School, Monterey, CA.

Barraclough, W. E., Le Brasseur, R. J., and Kennedy, O. D. (1969). Shallow scattering layer in the Subarctic Pacific Ocean. Detection by high frequency echo sounder. *Science*, **166**, 611–13.

Bass, A.H. and Baker, R. (1990). Sexual dimorphisms in the vocal control system of a teleost fish: morphology of physiologically identified neurons. *J. Neurobiol.*, **21**, 1155–68.

Beckmann, P. and Spizzichino, A. (1963). *The Scattering of Electromagnetic Waves from Rough Surfaces*. New York: Macmillan.

Bemis, K. G., Rona, P. A., Jackson, D. R., Jones, C. D., Silver, D., and Mitsuzawa, K. (2002). A comparison of black smoker hydrothermal plume behavior at Monolith Vent and at Clam Acres vent field: dependence on source configuration, *Marine Geophys. Res.*, **23**, 81–96.

Berkson, J. M. and Clay, C. S. (1973). Microphysiography and possible iceberg grooves on the floor of Western Lake Superior. *Geol. Soc. Bull.*, **84**, 1315–28.

Berkson, J. M. and Mathews, J. E. (1983). Statistical properties of seafloor roughness. In *Acoustics of the Sea-Bed*, N. G. Psace (ed.), Bath, England; Bath University Press, pp. 215–23.

Berktay, H. O. and Muir, T. G. (1973). Arrays of parametric receiving arrays. *J. Acoust. Soc. Am.*, **53**, 1377–83.

Beyer, R. T. (1975). *Nonlinear Acoustics*, Naval Sea Systems Command, 1974; stock No. 0-596–215, Government Printing Office; Washington, DC.

Bies, D. A. J. (1955). Attenuation in magnesium sulfate solutions. *Chem. Phys.*, **23**, 428.

Biot, M. A. and Tolstoy, I. (1957). Formulation of wave propagation in infinite media by normal coordinates with an application to diffraction. *J. Acoust. Soc. Am.*, **29**, 381–91.

Blanchard, D. C. and Woodcock, A. H. (1957). Bubble formation and modification in the sea and its meteorological significance. *Tellus*, **9**, 145–58.

Blaxter, J. H. S. and Batty, R. (1990). Swimbladder behaviour and target strength, Rapp. P. -v. *Reun. Cons. Int. Explor. Mer.*, **189**, 233–44.

Bowles, F. A. (1997). Observations on attenuation and shear-wave velocity in fine-grained, marine sediments. *J. Acoust. Soc. Am.*, **101**, 3385–97.

Brandt, S. B. (1989). Application of knowledge-based systems to fisheries management and acoustic abundance measures, in J. Palmer (ed.), *BNOAA Workshop on Knowledge-Based Systems and Marine Sciences*, Fairfax, VA., VSG 89-03, May 21–9, pp. 93–117.

Breitz, N. D. and Medwin, H. (1989). Instrumentation for *in-situ* acoustical measurements of bubble spectra under breaking waves. *J. Acoust. Soc. Am.*, **86**, 739–43.

Bremhorst, J. H. (1978). Impulse wave diffraction by rigid wedges and plates. M.S. thesis, Naval Postgraduate School, Monterey, CA. (December). See also abstract, same title, Bremhorst, J. H. and Medwin, H. (1978). *J. Acoust. Soc. Am.*, **64**, S1, S64 (A).

Brothers, R. J., Thomson, P. A. G., and Heald, G. J. (1999). Coherent modelling of wideband backscatter from a rough anisotropic seabed at minehunting frequencies. In *Stochastic Volume and Surface Scattering (Recent Developments in Underwater Acoustics)*. B., Uscinski and P. F. Dobbins (eds.), Institute of Acoustics, *Proc. Inst. Acoust.*, **21**(9), 1–8.

Brown, D. A. (1994). Flextensional hydrophone. *J. Acoust. Soc. Am.*, **96**, 3208.

Browne, M. J. (1987). Underwater acoustic backscatter from a model of arctic ice open leads and pressure ridges. M.S. thesis, Naval Postgraduate School, Monterey, CA.

Buckingham, M. J. (1987). Theory of three-dimensional acoustic propagation in a wedgelike ocean with a penetrable bottom. *J. Acoust. Soc. Am.*, **82**, 198–210.

  (1991). On acoustic transmission in ocean-surface waveguides. *Phil. Trans. Roy. Soc. Lond., Ser. A*, **335**, 513–55.

Buckingham, M. J. and Potter, J. R. (1994). Acoustic daylight imaging: vision in the ocean. *GSA Today*, **4**, 99–102.

Buckingham, M. J. and Tolstoy, A. (1990). An analytical solution for benchmark problem 1: The 'ideal' wedge. *J. Acoust. Soc. Am.*, **87**, 1511–13. Wedge benchmark paper.

Buckingham, M. J., Berkhout, B. V., and Glegg, S. A. L. (1992). Imaging the ocean with ambient noise. *Nature*, **356**, 327–9.

Buckingham, M. J., Giddens, E. M., Simonet, F., and Hahn, T. R. (2002a). Propeller noise from a light aircraft for low-frequency measurements of the speed of sound in a marine sediment. *J. Comp. Acoust.*, **10**, 445–64.

Buckingham, M. J., Giddens, E. M., Pompa, J. B., Simonet, F., and Hahn, T. R. (2002b). Sound from a light aircraft for underwater acoustics experiments? *Acta Acustica united with Acustica*, **88**, 752–5.

Busnel, R.-G. and Fish, J. F. (1980). *Animal Sonar Systems*. New York: Plenum Press.

Butler, J.L. and Pearcy, W. G. (1972). Swimbladder morphology and specific gravity of myctophids off Oregon. *J. Fish. Res. B. Can.*, **29**, 1145–50.

Buxcey, S., McNeil, J. E., and Marks, Jr. R. H. (1965). Acoustic detection of microbubbles and particulate matter near the sea surface. M.S. thesis, Naval Postgraduate School, Monterey, CA.

Carey, W. M. and Bradley, M. P. (1985). Low-frequency ocean surface noise sources. *J. Acoust. Soc. Am.*, **78**(S1), S1–2 (A).

Carstensen, E. L. (1947). Self-reciprocity calibration of electroacoustic transducers. *J. Acoust. Soc. Am.*, **19**, 702 (L).

Cartmill, J. W. and Su, M.-Y. (1993). Bubble size distribution under saltwater and freshwater breaking waves. *Dynamics of Atmospheres and Oceans*, **20**, 25–31.

Cato, D. H. (1998). Simple methods of estimating source levels and locations of marine animal sounds. *J. Acoust. Soc. Am.*, **104**(3), 1667–78.

Cerveny, V. and Ravindra, R. (1971). *Theory of Seismic Head Waves*; Toronto; University of Toronto Press.

Chambers, J. B. and Berthelot, Y. H. (1994). Time domain experiments on the diffraction of sound by a step discontinuity. *J. Acoust. Soc. Am.*, **96**, 1887–92.

Chamuel, J. R. and Brooke, G. H. (1988a). Transient Scholte wave transmission along rough liquid-solid interfaces. *J. Acoust. Soc. Am.*, **83**, 1336–44.

  (1988b). Transient Rayleigh wave transmission along periodic and random grooved surfaces. *J. Acoust. Soc. Am.*, **84**, 1363–72.

  (1988c). Shallow-water acoustic studies using an air-suspended water waveguide model. *J. Acoust. Soc. Am.*, **84**, 1777–86.

Chapman, N. R. (1988). Source levels of shallow explosive charges. *J. Acoust. Soc. Am.*, **84**, 697–702.

  (1985). Measurement of the waveform parameters of shallow explosive charges. *J. Acoust. Soc. Am.*, **78**, 672–81.

Chapman, N. R. and Ebbeson, G. R. (1983). Acoustic shadowing by an isolated seamount. *J. Acoust. Soc. Am.*, **73**, 1979–84.

Chia, C. S., Makris, N. C., and Fialkowski, T. (2000). A comparison of bi-static scattering from two geologically distinct abyssal hills. *J. Acoust. Soc. Am.*, **108**, 2053–70.

Chiu, C.-S., Lynch, J., and Johannessen, O. M. (1987). Tomographic resolution of mesoscale eddies in the marginal ice zone: a preliminary study. *J. Geophys. Res.*, **92**, C7, 6886–902.

Chiu, C.-S., Miller, J. H., Denner, W. W., and Lynch, J. F. (1995). Forward modeling of the Barents Sea tomography: Vertical line array data and inversion highlights. In *Full-Field Inversion Methods in Ocean and Seismic Acoustics*, O. Diachok, A. Caiti, P. Gerstoft, and H. Schmidt (eds.), Dordrecht, Netherlands: Kluwer Academic Publishers, pp. 237–42.

Chiu, C.-S., Miller, C. W., Moore, T. C., and Collins, C. A. (2002). Detection and censusing of blue whale vocalizations along the central California coast using the decommissioned Point Sur SOSUS array, *USN J. Underwater Acoust.*, **52**.

  (1990). Exact solution for a density contrast shallow-water wedge using normal coordinates. *J. Acoust. Soc. Am.*, **87**, 2442–50.

Chu, D., Stanton, T. K., and Wiebe, P. H. (1992). Frequency dependence of backscattering from live individual zooplankton. *ICES J. Mar. Sci.*, **49**, 97–106.

Claque, D. A., Davis, A. S., and Dixon, J. E. (2003). Submarine Strombolian eruptions on the Garda mid-ocean ridge. *Geophysical Subcaveous Volcanism, Geophys. Monograph* **140**, Am. Geophys. Union.

Clark, C. W. (1980). A real-time, direction finding device for determining the bearing to the underwater sounds of Southern Right Whales, *Eubalaena australis*. *J. Acoust. Soc. Am.*, **68**, 508–11.

Clark, C. W. (1994). Blue deep voices: insights from the Navy Whales 93 program. *Whalewatcher, J. Am. Cetacean Soc.*, 6–11.

Clark, C. W. (1995). Application of US Navy underwater hydrophone arrays for scientific research on whales. *Rept. Int. Whal. Commission* **45**, 210–12.

Clark, C. W. and Fristrup, K. M. (1997). Whales '95: A combined visual and acoustic survey of blue and fin whales off Southern California, *Rept. Int. Whal. Commin.*, **47**, 583–600.

Clay, C. S. (1966a). Use of arrays for acoustic transmission in a noisy ocean. *Rev. Geophys.*, **4**, 475–507.

Clay, C. S. (1966b). Coherent reflection of sound from the sea bottom. *J. Geophys, Res.*, **71**, 2037–45.

Clay, C. S. (1987). Optimum time domain signal transmission and source location in a waveguide. *J. Acoust. Soc. Am.*, **81**, 660–4.

(1991). Low-resolution acoustic scattering models: fluid-filled cylinders and fish with swimbladders. *J. Acoust. Soc. Am.*, **89**, 2168–79.

(1992). Composite ray-mode approximations for backscattered sound from gas-filled cylinders and swimbladders. *J. Acoust. Soc. Am.*, **91**, 2173–80.

Clay, C. S. and Horne, J. K. (1994). Acoustic models of fish: the Atlantic cod (*Gadus morhua*). *J. Acoust. Soc. Am.*, **96**, 1661–8.

(1995). Analysis of rather high-frequency sound echoes from ensembles of fish. *J. Acoust. Soc. Am.*, **98**, 2881.

Clay, C. S. and Kinney, W. A. (1988). Numerical computations of time-domain diffractions from wedges and reflections from facets. *J. Acoust. Soc. Am.*, **83**, 2126–33.

Clay, C. S. and Medwin, H. (1964). High-frequency acoustical reverberation from a rough sea surface, *J. Acoust. Soc. Am.*, **36**, 2131–4.

(1970). Dependence of spatial and temporal correlation of forward scattered underwater sound on the surface statistics: part I – theory. *J. Acoust. Soc. Am.*, **47**, 1412–18. See also Medwin and Clay (1970), part II – experiment.

(1977). *Acoustical Oceanography*. New York: John Wiley.

Clay, C. S. and Sandness, G. A. (1971). Effect of beam width on acoustic signals scattered at a rough surface. *Adhesions Group for Aerospace R. D. North Atlantic Treaty Organization Conf. Proc.*, **21**(90), 1–8.

Clay, C. S., Wang, Y. Y. and Shang, E. C. (1985). Sound field fluctuations in a shallow water waveguide. *J. Acoust. Soc. Am.*, **77**, 424–8.

Clay, C. S., Medwin, H., and Wright, W. M. (1973). Specularly scattered sound and the probability density function of a rough surface. *J. Acoust. Soc. Am.*, **53**, 1677–82.

Clay, C. S., Chu, D., and Li, C. (1993). Specular reflection of transient pressures from finite width plane facet. *J. Acoust. Soc. Am.*, **94**, 2279–86.

Cole, R. H. (1948). *Underwater Explosions*. Princeton, NJ: Princeton University Press.

Colladon, J. D. and Sturm, J. K. F. (1827). The compression of liquids (in French), *Ann. Chim. Phys.* Series 2(36), part IV, Speed of sound in liquids, 236–57.

Commander, K. and McDonald, R. J. (1991). Finite-element solution of the inverse problem in bubble swarm acoustics. *J. Acoust. Soc. Am.*, **89**, 592–7.

Commander, K. and Moritz, E. (1989). Off-resonance contributions to acoustical bubble spectra. *J. Acoust. Soc. Am.*, **85**, 2665–9.

Cornuelle, B., Munk, W., and Worcester, P. (1989). Ocean acoustic tomography from ships, *J. Geophys. Res.*, **92**, 11 680–92.

Cox, C. S. and Munk, W. (1954). Statistics of the sea surface derived from sun glitter, *J. Marine Res.*, **13**, 198–227.

Crawford, G. B., Lataitis, R. J., and Clifford, S. F. (1990). Remote sensing of ocean flows by spatial filtering of acoustic scintillations: theory. *J. Acoust. Soc. Am.*, **88**(1), 442–54.

Crowther, P. (1988). Bubble noise creation mechanisms, *in Sea Surface Sound*. Dordrecht, Netherlands: Kluwer, pp. 131–50.

Cummings, W. C. and Holliday, D. V. (1987). Sounds and source levels from bowhead whales off Pt. Barrow, Alaska. *J. Acoust. Soc. Am.*, **82**, 814–21.

Daniel, A. C., Jr. (1989). Bubble production by breaking waves. M.S. thesis, Naval Postgraduate School, Monterey, CA.

D'Asaro, E. A., Farmer, D. M., Osse, J., and Dairiki, G. T. (1996). A Lagrangian float. *J. Atmos. Ocean Technol.*, **13**(6), 1230–46.

Davis, A. M. J. and Scharstein, R. W. (1997). The complete extension of the Biot–Tolstoy solution to the density contrast wedge. *J. Acoust. Soc. Am.*, **101**, 1821–35.

Deane, G. B. (1997). Sound generation and air entrainment by breaking waves in the surf zone. *J. Acoust. Soc. Am.*, **102**, 2671–89.

Del Grosso, V. A. (1974). New equation for the speed of sound in natural waters (with comparisons to other equations). *J. Acoust. Soc. Am.*, **56**, 1084–91.

de Moustier, C. (1986). Beyond bathymetry: mapping acoustic backscattering from the deep seafloor with Sea Beam. *J. Acoust. Soc. Am.*, **79**, 316–31.

Denny, P. L. and Johnson, K. R. (1986). Underwater acoustic scatter from a model of the Arctic ice canopy, M.S. thesis, Naval Postgraduate School, Monterey, CA.

DeSanto, J. A. (1977). Relations between solutions of the Helmholtz and parabolic equation for sound propagation. *J. Acoust. Soc. Am.*, **62**, 295–7.

Devin, Charles, Jr. (1959). Survey of thermal, radiation and viscous damping of pulsating air bubbles in water. *J. Acoust. Soc. Am.*, **31**, 1654–67.

Diachok, O. (1999). Effects of absorptivity due to fish on transmission loss in shallow water. *J. Acoust. Soc. Am.*, **105**, 2107–28.

(2001). Interpretation of the spectra of energy scattered by dispersed anchovies. *J. Acoust. Soc. Am.*, **110**, 2917–23.

Diachok, O. I. and Winokur, R. S. (1974). Spatial variability of underwater ambient noise at the Arctic ice-water boundary. *J. Acoust. Soc. Am.*, **55**, 750–3.

Diachok, O. I., Smith, P., and Wales, S. (2004). Classification of bioacoustic absorption lines at low frequencies: the Van Holliday connection. *J. Acoust. Soc. Am.*, **115**, 2521.

Didenkulov, I. N., Muyakshin, S. I., and Selivanovsky, D. A. (2001). Bubble counting in the subsurface ocean layer. In *Acoustical Oceanography*, T. G. Leighton, G. J. Heald, H. Griffiths, and G. Griffiths (eds.), Institute of Acoustics, *Proc. Inst. Acoust.*, **23**(2), 220–6.

Di Iorio, D. and Farmer, D. M. (1996). Two-dimensional angle of arrival fluctuations. *J. Acoust. Soc. Am.*, **100**(2), 814–24.

(1998). Separation of current and sound speed in the effective refractive index for a turbulent environment using reciprocal acoustic transmission. *J. Acoust. Soc. Am.*, **103**(1), 321–9.

Di Iorio, D. and Yuce, H. (1999). Observations of Mediterranean flow into the Black Sea. *J. Geophys. Res.*, **104**(C2), 3091–108.

Di Iorio, D., Lemon, D., and Chave, R. (2004). A self contained acoustic scintillation instrument for path-averaged measurements of flow and turbulence with application to hydrothermal vent and bottom boundary layer dynamics. *J. Atoms. Ocean. Tech.*, (in press).

Ding, L. and Farmer, D. M. (1992). A signal-processing scheme for passive acoustical mapping of breaking surface waves. *J. Atmos. Ocean. Technol.* **9**(4), 484–94.

(1994a). Observations of breaking surface wave statistics, *J. Phys. Oceanogr.*, **24**(6), 1368–87.

(1994b). On the dipole acoustic source level of breaking waves. *J. Acoust. Soc. Am.*, **96**(5), 3036–44.

D'Spain, G. L., Berger, L. P., Kuperman, W. A., and Hodgkiss, W. S. (1997). Summer night sounds by fish in shallow water. In *Shallow Water Acoustics*, R. Zhang and J. Zhou (eds.), Beijing: China Ocean Press, pp. 379–84.

Duncan, J. H. (1981). An investigation of breaking waves produced by a towed hydrofoil, *Proc. Roy. Soc. London Ser.* A, **377**, 331–48.

Dushaw, B. D., Bold, G., Chiu, C.-S., *et al.* (2001). Observing the ocean in the 2000's: A strategy for the role of acoustic tomography in ocean climate observation. In *Observing the Oceans in the 21st Century*, C. J. Koblinsky and N. R. Smith (eds.), (GODAE Project Office and Bureau of Meteorology, Melbourne), pp. 391–418.

Ebbeson, G. R. and Turner, R. G. (1983). Sound propagation over Dickins Seamount. *J. Acoust. Soc. Am.*, **73**, 143–52.

Eckart, C. (1948). Vortices and streams caused by sound waves. *Phys. Rev.*, **73**, 68–76.

(1953). The scattering of sound from the sea surface. *J. Acoust. Soc. Am.*, **25**, 566–70.

(1968). (ed.), *Principles and Applications of Underwater Sound*, Department of the Navy, NAVMAT P-9674: Government Printing Office; Washington, DC.

Edds, P. L. Characteristics of finback *Balaneoptera physalus* vocalizations in the St. Lawrence Estuary. *Bioacoustics*, **1**, 131–49.

Edelmann, G. F., Akal, T., Hodgkiss, W. S., Kim, S., Kuperman, W. A., and Song, H. C. (2002). An initial demonstration of underwater acoustic communication using time reversal, *IEEE J. Ocean. Eng.*, **27**, 602–9.

Ehrenberg, J. E. (1974). The dual-beam system: a technique for making in situ measurements of the target strengths of fish. *Oceans '74: Proc. IEEE Int. Conf. on Engineering in the Ocean Envr.*, IEEE; New York, pp. 152–5.

Elder, S. A. (1959). Cavitation microstreaming. *J. Acoust. Soc. Am.*, **31**, 54–64.

Eller, A. I. (1970). Damping constants of pulsating bubbles. *J. Acoust. Soc. Am.*, **47**, 1469–70.

Everest, F. A., Young, R. W., and Johnson, M. W. (1948). Acoustical characteristics of noise produced by snapping shrimp. *J. Acoust. Soc. Am.*, **20**, 137–42.

Ewing, M. and Worzel, J. L. (1948). Long-range sound transmission. In *Propagation of Sound in the Ocean*, Memoir 27: Geological Society of America; New York, Fig. 5, p. 19.

Ewing, W. M., Jardetzky, W. S., and Press, F. (1962). *Elastic Waves in Layered Media*. New York: McGraw-Hill.

Faran, J. J. (1951). Sound scattering by solid cylinders and spheres. *J. Acoust. Soc. Am.*, **23**, 405–18.

Farmer, D. M. and Ding, L. (1992). Coherent acoustical radiation from breaking waves. *J. Acoust. Soc. Am.*, **92**(1), 397–402.

Farmer, D. M. and Lemon D. (1984). The influence of bubbles on ambient noise in the ocean at high wind speeds. *J. Phys. Oceanogr.*, **14**, 1761–77.

Farmer, D. M. and Vagle, S. (1989). Waveguide propagation of ambient sound in the ocean-surface bubble layer. *J. Acoust. Soc. Am.*, **86**, 1897–908.

  (1997). Bubble measurements using a resonator system. In *Natural Physical Processes Associated with Sea Surface Sound*, T. G. Leighton (ed.), University of Southampton, UK, pp. 155–62.

Farmer, D. M., Vagle, S., and Booth, A. D. (1998). A free-flooding acoustical resonator for measurement of bubble size distributions. *J. Atmos. Ocean. Technol.*, **15**(5), 1132–46.

Farmer, D., Ding, L., Booth, D., and Lohmann, M. (2002). Wave kinematics at high sea states. *J. Atmos. Ocean. Technol.*, **19**, 225–39.

Farquhar, G. B. (1970). *Proceedings of an International Symposium of Biological Sound Scattering in the Ocean*. US Government Printing Office.

Fernandes, P. G., Brierley, A. S., and Simmonds, E. J. (2000). Fish do not avoid survey vessels. *Nature*, **404**, 35–6.

Feuillade, C. (1995). Scattering from collective modes of air bubbles in water and the physical mechanism of superresonances. *J. Acoust. Soc. Am.*, **98**, 1178–90.

Feuillade, C. and Nero, R. W. (1998). A viscous-elastic swimbladder model for describing enhanced-frequency resonance scattering from fish. *J. Acoust. Soc. Am.*, **103**(6), 3245–55.

Fink, M. (1999). Time-Reversed Acoustics. *Scientific American*, **281**(5), 91–7.

Fisher, F. H. and Levison, S. A. (1973). Dependence of the low frequency (1 kHz) relaxation in seawater on boron concentration. *J. Acoust. Soc. Am.*, **54**, 291.

Fisher, F. H. and Simmons, V. P. (1977). Sound absorption in sea water. *J. Acoust. Soc. Am.*, **62**, 558–64.

Flatte, S. M. and Stoughton, R. B. (1986). Theory of acoustic measurement of internal wave strength as a function of depth, horizontal position and time. *J. Geophys. Res.*, **91**(C6), 7709–20.

Flynn, H. G. (1964). Physics of acoustic cavitation in liquids, In W. P. Mason (ed.), *Physical Acoustics*, vol. 1, part B, New York: Academic Press. pp. 57–172.

Foote, K. G. (1980a). Importance of the swimbladder in acoustic scattering by fish: a comparison of gadoid and mackerel target strengths. *J. Acoust. Soc. Am.*, **67**, 2084–9.

  (1980b). Effect of fish behaviour on echo energy: the need for measurement of orientation distributions. *J. Cons. Int. Explor. Mer.*, **39**, 193–201.

  (1983). Linearity of fisheries acoustics, with addition theorems. *J. Acoust. Soc. Am.*, **73**, 1932–40.

Foote, K. G. (1985). Rather high-frequency sound scattering by swimbladdered fish. *J. Acoust. Soc. Am.*, **78**, 688–700.

Foote, K. G. and Francis, D. T. I. (2002). Comparing Kirchhoff-approximation and boundary-element models for computing gadoid target strengths. *J. Acoust. Soc. Am.*, **111**, 1644–54.

Foote, K. G. and Traynor, J. J. (1988). Comparison of walleye pollock target strength estimates determined from in situ measurements and calculations based on swimbladder form. *J. Acoust. Soc. Am.*, **83**, 9–17.

Forbes, S. T. and Nakken, O. (1972). (eds.), *Manual of Methods for Fisheries Resource Survey and Appraisal*, part 2. *The Use of Acoustic Instruments for Fish Detection and Abundance Estimation*. Food and Agricultural Organization of the United Nations; FAO Manuals in Fisheries Science No. 5; Rome.

Fortuin, L. and de Boer, J. G. (1971). Spatial and temporal correlation of the sea surface. *J. Acoust. Soc. Am.*, **49**, 1677.

Fox, F. E. and Herzfeld, K. F. (1950). On the forces producing the ultrasonic wind. *Phys. Rev.*, **78**, 156–7.

François, R. E. and Garrison, G. R. (1982a). Sound absorption based on ocean measurements. Part I: pure water and magnesium sulfate contributions. *J. Acoust. Soc. Am.*, **72**, 896–907.

(1982b). Sound absorption based on ocean measurements. Part II: boric acid contribution and equation for total absorption. *J. Acoust. Soc. Am.*, **72**, 1879–90.

Frankel, A. S., Clark, C. W., Herman, L. M., and Gabriele, C. M. (1995). Spatial distribution, habitat utilization, and social interactions of humpback whales, *Megaptera novaeangliae*, off Hawaii, determined using acoustics and visual techniques. *Can. J. Zool.*, **73**, 1134–46.

Freitag, L. E. and Tyack, P. L. (1993). Passive acoustic localization of the Atlantic bottlenose dolphin using whistles and echo location clicks. *J. Acoust. Soc. Am.*, **93**, 2197–205.

Frisk, G. V., Lynch, J. F., and Rajan, S. D. (1989). Determination of compressional wave speed profiles using modal inverse techniques in a range-dependent environment in Nantucket Sound. *J. Acoust. Soc. Am.*, **86**, 1928–39.

Gagnon, G. J. and Clark, C. W. (1993). The use of US Navy IUSS passive sonar to monitor the movement of blue whales. Tenth Biennial Conference on the Biology of Marine Mammals; Galveston, Texas, Abstract, p. 50.

Gargett, A. E. (1999). Velcro measurement of turbulence kinetic energy dissipation rate ε. *J. Atmos. Ocean. Technol.* **16**(12), 1973–93.

Garre, C., Li, M., and Farmer, D. M. (2000). The connection between bubble size spectra and energy dissipation rates in the upper ocean. *J. Phys. Oceanogr.*, **30**, 2163–71.

Gaunaurd, G. C. (1986). Sonar cross sections of bodies partially insonified by finite sound beams. *IEEE J. Ocean Eng.*, **OE-10**, 213–30.

Gazanhes, C. and Garnier, J. L. (1981). Experiments on single mode excitation in shallow water propagation. *J. Acoust. Soc. Am.*, **69**, 963–9.

Gee, J. H. (1968). Adjustment of buoyancy by longnose dace (*Rhinichtys cataractae*) in relation to velocity of water. *J. Fish. Res. B. Can.*, **25**, 1485–1496.

Gerstein, E., Gerstein, L., Forsythe, S., and Blue, J. (1999). The underwater audiogram of the West Indian manatee (*Trichechus manatus*). *J. Acoust. Soc. Am.*, **105**(6), 3575–3583.

(2001). It's all about SOUND Science: Manatees, masking and boats. *J. Acoust. Soc. Am.*, **110**, 2722.

Glotov, V. P., *et al.* (1962). Investigation of the scattering of sound by bubbles generated by an artificial wind in seawater and the statistical distribution of bubble sizes. *Sov. Phys.: Acoust.*, **7**, 341–5.

Gray, J. M. and Greeley, D. S. (1980). Source level model for propellor plade rate radiation for the world's merchant fleet. *J. Acoust. Soc. Am.*, **67**, 516–22.

Greenlaw, C. F. (1979). Acoustical estimation of zooplankton populations. Limnol. Oceanogr., **24**, 226–42.

Greenlaw, C. F. and Johnson, R. K. (1982). Physical and acoustical properties of zooplankton, *J. Acoust. Soc. Am.*, **72**, 1706–10.

Hagy, J. D., Jr. (1970). Transmission of sound through a randomly rough air–sea interface. M.S. thesis, Naval Postgraduate School, Monterey, CA.

Hall, M. V. (1989). A comprehensive model of wind-generated bubbles in the ocean and predictions of the effects on sound propagation at frequencies up to 40 kHz. *J. Acoust. Soc. Am.*, **86**, 1103–17.

Hamilton, E. L. (1963). Sediment sound velocity measurements made *in situ* from the bathyscaphe *Trieste*. *J. Geophys. Res.*, **68**, 5991–8.

(1971). The elastic properties of marine sediments. *J. Geophys. Res.*, **76**, 579–604.

(1972). Compressional wave attenuation in marine sediments. *Geophysics*, **37**, 620–46.

(1985). Sound velocity as a function of depth in marine sediments. *J. Acoust. Soc. Am.*, **78**, 1355–84.

Haslett, R. W. G. (1962). Determination of the acoustic scatter patterns and cross sections of fish models and ellipsoids. *Br. J. App. Phys.*, **13**, 611–20.

Helbig, R. A. (1970). The effects of ocean surface roughness on the transmission of sound from an airborne source. M.S. thesis, Naval Postgraduate School, Monterey, CA.

Hersey, J. B. and Backus, R. H. (1962). Sound scattering by marine organisms in M. N. Hill (ed.), *The sea.* vol. 1, New York: John Wiley. pp. 499–507.

Hickling, R. (1958). Frequency dependence of echoes from bodies of different shapes. *J. Acoust. Soc. Am.*, **30**, 137–9.

Hill, M. N. (1963). (ed.) *The Sea*, vol. 3. New York: Wiley-Interscience.

Holliday, D. V. (1972). Resonance structure in echoes from schooled pelagic fish. *J. Acoust. Soc. Am.*, **51**, 1322–33.

Holliday, D. V. and Pieper, R. E. (1980). Volume scattering strengths and zooplankton distributions at acoustic frequencies between 0.5 and 3 MHz. *J. Acoust. Soc. Am.*, **67**, 135–46.

Holliday, D. V., Pieper, R. E., and Klepple, G. S. (1989). Determination of zooplankton size and distribution with multifrequency acoustic technology. *J. Cons. Int. Explor. Mer.*, **46**, 52–61.

(1995). Bioacoustical oceanography at high frequencies. *ICES J. Mar. Sci.*, **52**, 279–96.

Horne, J. K. and Clay, C. S. (1998). Sonar systems and aquatic organisms: matching equipment and model parameters. *Can. J. Fish. Aquat. Sci.*, **55**, 1296–306.

Horne, J. K. and Jech, J. M. (1999). Multi-frequency estimates of fish abundance: constraints of rather high frequencies. *ICES J. Mar. Sci.*, **56**, 184–99.

Horne, J. K., Walline, P. D., and Jech, J. M. (2000). Comparing acoustic model predictions to *in situ* backscatter measurements of fish with dual-chambered swimbladders. *J. Fish Biol.*, **57**, 1105–21.

Horton, C. W. Sr. and Mellon, D. R. (1970). Importance of the Fresnel correction in scattering from a rough surface, II. Scattering coefficient, *J. Acoust. Soc. Am.*, **47**, 299–303.

Huffman, T. B. and Zveare, D. L. (1974). Sound speed dispersion, attenuation and inferred microbubbles in the upper ocean. M.S. thesis, Naval Postgraduate School, Monterey, CA.

Isakovich, M. A. (1952). The scattering of waves from a statistically rough surface. (in Russian). *Zhurn. Eks., Teor. Fiz.*, **23**, 305–14.

Jackson, D. R. and Dowling, D. R. (1991). Phase conjugation in underwater acoustics, *J. Acoust. Soc. Am.*, **89**, 171–81.

Jackson, D. R. and Dworski, J. G. (1992). An acoustic backscatter thermometer for remotely mapping seafloor water temperature. *J. Geophys. Res.*, **97**, 761–7.

Jackson, D. R., Winebrenner, D. P., and Ishimaru, A. (1986). Application of the composite roughness model to high-frequency backscattering. *J. Acoust. Soc. Am.*, **79**, 1410–22.

Jackson, D. R., Jones, C. D., Rona, P. A., and Bemis, K. G. (2003). A method for Doppler acoustic measurement of black smoker flow fields. *Geochem. Geophys. Geosyst.*, **4**.

Jacobson, P. T., Clay, C. S., and Magnuson, J. J. (1990). Size, distribution, and abundance of pelagic fish by deconvolution of single beam acoustic data. *J. Cons. Int. Explor. Mer.*, **189**, 404–11.

Jacobus, R. W. (1991). Underwater sound radiation from large raindrops. M.S. thesis, Naval Postgraduate School, Monterey, CA.

Jebsen, G. M. (1981). Acoustic diffraction by a finite barrier: theories and experiment. M.S. thesis, Naval Postgraduate School, Monterey, CA.

Jebsen, G. M. and Medwin, H. (1982). On the failure of the Kirchhoff assumption in backscatter. *J. Acoust. Soc. Am.*, **72**, 1607–11.

Jech, J. M. and Horne, J. K. (2002). Three-dimensional visualization of fish morphometry and acoustic backscatter. *Acoust. Res. Letters Online*, **3**, 35–40 (http://www.ops.aip.org/ARLO).

(2003). Three dimensional visualization of fish morphometery and acoustic backscatter. *Acoust. Res. Lett. Online.*

Johnson, B. D. and Cooke, R. C. (1979). Bubble populations and spectra in coastal water: a photographic approach. *J. Geophys. Res.*, **84**, 3761–6.

Jordan, E. A. (1981). Acoustic boundary wave generation and shadowing at a seamount. M.S. thesis, Naval Postgraduate School, Monterey, CA.

Kaczkowski, P. J. and Thorsos, E. I. (1994). Application of the operator expansion method to scattering from one-dimensional, moderality rough Dirichlet random surfaces. *J. Acoust. Soc. Am.*, **96**, 957–72.

Kasputis, S. and Hill, P. (1984). Measurement of mode interaction due to waveguide surface roughness. M.S. thesis, Naval Postgraduate School, Monterey, CA.

Keiffer, R. S., Novarini, J. C., and Norton, G. V. (1994). The impulse response of an aperture: numerical calculations within the framework of the wedge assemblage method. *J. Acoust. Soc. Am.*, **95**, 3–12.

Kibblewhite, A. C. (1989). Attenuation of sound in marine sediments: a review with emphasis on new low-frequency data. *J. Acoust. Soc. Am.*, **86**, 716–38.

Kibblewhite, A. C. and Wu, C. Y. (1991). The theoretical description of wave–wave interactions as a wave source in the ocean. *J. Acoust. Soc. Am.*, **89**, 2241–52.

Kim, S., Edelmann, G., Hodgkiss, W. H., Kuperman, W. A., Song, H. C., and Akal, T. (2001). Spatial resolution of time reversal arrays in a shallow water. *J. Acoust. Soc. Am.*, **110**, 820–9.

Kinney, W. A. and Clay, C. S. (1985). Insufficiency of surface spatial power spectrum for estimating scattering strength and coherence–numerical studies. *J. Acoust. Soc. Am.*, **78**, 1777–84.

Kinney, W. A., Clay, C. S. and Sandness, G. A. (1983). Scattering from a corrugated surface: comparison between experiment, Helmholtz–Kirchhoff theory, and the facet-ensemble method. *J. Acoust. Soc. Am.*, **73**, 183–94.

Kinsman, B. (1960). Surface waves at short fetches and low wind speeds: a field study. Chesapeake Bay *Inst. Tech. Rep.* **129**, Johns Hopkins.

Knowlton, Amy and Russell, Bruce (2002) co-chairs Ship Strike Subcommittee Report to the Northeast Implementation Team. A review of the issue of vessel speed and how it relates to vessel/whale collisions.

Knudsen, V. O., Alford, R. S., and Emliing, J. W. (1948). Underwater ambient noise. *J. Marine Res.*, **7**, 410–29.

Kuperman, W. A. (1975). Coherent component of specular reflection and transmission at a randomly rough, two-fluid interface. *J. Acoust. Soc. Am.*, **58**, 365–70.

Kuperman, W. A. and Jackson, D. (2002). Ocean acoustics, matched-field processing and phase conjugation. In *Imaging of Complex Media with Acoustic and Seismic Waves*, Fink *et al.* (ed.), Berlin: Springer-Verlag.

Kuperman, W. A., Hodgkiss, W. S., Song, H. C., Akal, T., Ferla, C., and Jackson, D. R. (1998). Phase-conjugation in the ocean: experimental demonstration of an acoustic time reversal mirror, *J. Acoust. Soc. Am.*, **103**, 25–40.

Kurgan, A. (1989). Underwater sound radiated by impacts and bubbles created by raindrops. M.S. thesis, Naval Postgraduate School, Monterey, CA.

LaFond, E. C. and Dill, J. (1957). Do invisible microbubbles exist in the sea? Navy Electronics Lab., San Diego, CA.

Laist, D. W., Knowlton, A. R., Mead, J. G., Collet, A. S., and Podesta, M. (2001). Collisions between ships and whales, *Marine Mammal Sci.* **17**(1), 35–75.

Lamarre, E. and Melville, W. K. (1991). Air entrainment and dissipation in breaking waves, *Nature*, **351**, 472–9.

(1994). Sound speed measurements near the ocean surface, *J. Acoust. Soc. Amer.*, **96**, 3605–16.

Lee, S., Zanolin, M., Thode, A., Pappalardo, R., and Makris, N. (2003). Probing Europa's interior with natural sound sources. *Icarus*, **165**, 144–67.

Leighton, T. G., Lingord, R. J., Walton, A. J., and Field, J. E. (1992). Bubble sizing by the nonlinear scattering of two acoustic frequencies. In *Natural Physical Sources of Underwater Sound*, B. R. Kerman (ed.), pp. 453–466. Dordrecht, The Netherlands: Kluwer.

(1994). *The Acoustic Bubble* (paperback edition). London, San Diego: Academic Press. pp. 4, 67, 120, 449.

(1998). Fundamentals of underwater acoustics and ultrasound. In *Noise and Vibration (Vol. 1)*, F. J. Fahy and J. G. Walker (eds.), E & F Spon (an imprint of Routledge, London), p. 395.

Leighton, T. G., Lingard, R. J., Walton, A. J., and Field, J. E. (1991). Acoustic bubble sizing by the combination of subharmonic emissions with an imaging frequency. *Ultrasonics*, **29**, 319–23.

Leighton, T. G., Phelps, A. D., and Ramble, D. G. (1996). Acoustic bubble sizing: from laboratory to the surf zone trials. *Acoustic Bulletin*, **21**, 5–12.

Leighton, T. G., Ramble, D. G., and Phelps, A. D., (1997). The detection of tethered and rising bubbles using multiple acoustic techniques. *J. Acoust. Soc. Am.* **101**, 2626–36.

Leighton, T. G., Meers, S. D., Simpson, M. D., *et al.* (2001). The Hurst Spit experiment: The characterization of bubbles in the surf zone using multiple acoustic techniques. In *Acoustical Oceanography*, T. G. Leighton, G. J. Heald, H. Griffiths and G. Griffiths (eds.), Institute of Acoustics. *Proc. of Institute of Acoustics* **23**(2), 227–34.

Leighton, T. G., White, P. R., Morfey, C. L., *et al.* (2002). The effect of reverberation on the damping of bubbles. *J. Acoust. Soc. Am.*, **112**, 1366–76.

Leonard, R. W. (1948). The attenuation of ultrasonic waves in water. *J. Acoust. Soc. Am.*, **20**, 224 (Abstract).

Leonard, R. W., Combs, P. C., and Skidmore, L. R. (1949). Attenuation of sound in sea water. *J. Acoust. Soc. Am.*, **21**, 63.

Leong, W. L. (1973). Use of acoustic scattering theory to interpret marine geophysical data. Ph.D. dissertation. University of Wisconsin, Madison.

Leroy, C. C. (1967). Sound propagation in the Mediterranean Sea, in *Underwater Acoustics, Vol. 2*, V. M. Albers (ed.), New York: Plenum Press. pp. 203–41.

Levin, F. K. (1962). The seismic properties of Lake Maracaibo. *Geophysics*, **27**, 35–47.

Lhermitte, R. (1983). Doppler sonar observation of tidal flow. *J. Geophys. Res.*, **88**, 725–42.

Li, S. and Clay, C. S. (1988). Sound transmission experiments from an impulsive source near rigid wedges. *J. Acoust. Soc. Am.*, **84**, 2135–43.

Li, S., Chu, D., and Clay, C. S. (1994). Time domain reflections and diffractions from facet-wedge constructions: acoustic experiments including double diffractions. *J. Acoust. Soc. Am.*, **96**, 3715–20.

Liebermann, L. N. (1949). Second viscosity of fluids. *Phys. Rev.*, **75**, 1415–22.

Livingston, E. and Diachok, O. (1989). Estimation of average under-ice reflection amplitudes and phases using matched-field processing. *J. Acoust. Soc. Am.*, **86**, 1909–19.

Lockwood, J. C. and Willette, J. G. (1973). High-speed method for computing the exact solution for the pressure variations in the near field of a baffled piston. *J. Acoust. Soc. Am.*, **53**, 735–41.

Loewen, M. R. and Melville, W. K. (1991). A model of the sound generated by breaking waves. *J. Acoust. Soc. Am.*, **90**, 2075–80.

Longuet-Higgins, M. S. (1962). The Statistical geometry of random surfaces. *Proc. Symp. Appl. Math.*, **13**, Am. Math. Soc.

  (1990). Bubble noise spectra. *J. Acoust. Soc. Am.*, **87**, 652–61.

  (1992). Non-linear damping of bubble oscillations by resonant interaction. *J. Acoust. Soc. Am.*, **91**, 1414–22.

Love, R. H. (1969). Maximum side-aspect target strength of an individual fish. *J. Acoust. Soc. Am.*, **46**, 746–60.

  (1971). Measurements of fish target strength: a review. *Fish. Bull.*, **69**, 703, 715.

  (1978). Resonant scattering by swimbladder bearing fish. *J. Acoust. Soc. Am.*, **64**, 571–80.

Løvik, A. (1980). Acoustic measurement of the gas bubble spectrum in water, in *Cavitation and Inhomogeneities in Underwater Acoustics*, W. Lauterborn (ed.). Berlin: Springer.

Løvik, A. and Hovem, J. (1979). An experimental investigation of swimbladder resonances in fishes. *J. Acoust. Soc. Am.*, **66**, 850–4.

Lucas, R. J. and Twersky, V. (1986). Inversion of data for near-grazing propagation over rough surfaces. *J. Acoust. Soc. Am.*, **80**, 1459–72.

Lyons, A. P., Duncan, M. E., Anderson, A. L., and Hawkins, J. A. (1996). Predictions of the acoustic scattering response of free methane bubbles in muddy sediments. *J. Acoust. Soc. Am.*, **99**, 163–72.

McCammon, D. F. and McDaniel, S. T. (1985). The influence of the physical properties of ice on reflectivity. *J. Acoust. Soc. Am.*, **77**, 499–507.

McCartney, B. S. and Bary, B. McK. (1965). Echosounding on probable gas bubbles from the bottom of Saanich Inlet, British Columbia. *Deep-Sea Res.*, **123**, 285–94.

McCartney, B. S. and Stubbs, A. R. (1970). Measurement of the target strength of fish in dorsal aspect, including swimbladder resonance. In *Proc. Int. Symp. Biol. Sound Scattering in Ocean*, ed. G. B. Farquhar, US Government Printing Office; Washington, DC, pp. 180–211.

McCauley, R. D. (2001). *Biological sea noise in northern Australia: patterns in fish calling*. Ph.D. thesis, James Cook University Queensland, 290 pp.

McCauley, R. D. and Cato, D. H. (2000). Patterns of fish calling in a nearshore environment in the Great Barrier Reef. *Phil. Trans. R. Soc. Lond. B.*, **355**, 1289–93.

McCauley, R. D., Jenner, C., Bannister, J. L., Cato, D. H., and Duncan, A. J. (2000). Blue whale calling in the Rottnest Trench, Western Australia, and low frequency noise. *Proc. Acoustics 2000*, Australian Acoustical Society, Perth, November 2000, pp. 245–50.

McDaniel, S. T. (1982). Mode coupling due to interaction with the sea bed. *J. Acoust. Soc. Am.*, **72**, 916–23.

(1987). Vertical spatial coherence of backscatter from bubbles. *IEEE J. Ocean. Eng.* **12**, 349–56.

McDonald, M. A. and Fox, C. G. (1999). Passive acoustic methods applied to fin whale population density estimation. *J. Acoust. Soc. Am.*, **105**, 2643–51.

Mackenzie, K. V. (1981). Nine term equations for sound speed in the ocean. *J. Acoust. Soc. Am.*, **70**, 807–12.

McNaught, D. C. (1968). Developments in acoustic plankton sampling. Proc. 12th Conf. Great Lakes Res., 76–84.

Acoustical zooplankton distributions. Proc. 11th Conf. Great Lakes Res., 61–8.

Makris, N. C. (1993). Imaging ocean-basin reverberation via inversion. *J. Acoust. Soc. Am.*, **94**, 983–93.

(1996). The effect of saturated transmission scintillation on ocean-acoustic intensity measurements. *J. Acoust. Soc. Am.*, **100**, 769–83.

Makris, N. C. and Berkson, J. M. (1994). Long-range backscatter from the Mid-Atlantic Ridge. *J. Acoust. Soc. Am.*, **95**, 1865–81.

Makris, N. C. and Ratilal, P. (2001). A unified model for reverberation and submerged object scattering in a stratified ocean waveguide. *J. Acoust. Soc. Am.*, **109**, 909–41.

Makris, N. C., Avelino, L., and Menis, R. (1995). Deterministic reverberation from ocean ridges. *J. Acoust. Soc. Am.*, **97**, 3547–74. (Also appears in full in a special volume commemorating ONR's 50th Anniversary.)

Makris, N. C., Chia, C. S., and Fialkowski, L. T. (1999). The bi-azimuthal scattering distribution of an abyssal hill. *J. Acoust. Soc. Am.*, **106**, 2491–512.

Marshall, N. B. (1970). Swimbladder development and the life of deep-sea fish. In G. Brooke Farquhar (ed.), *Proceeding of an International Symposium on Biological Sound Scattering in the Ocean*, Maury Center for Ocean Science, US Government Printing Office; Washington, DC, stock No. 0851-0053, pp. 69–78.

Martin, H. W. (1924). Decibel: the name for the transmission unit. *Bell Syst. Tech. J.*, **1**, 1–2. See *ibid*, **3**, 400–8.

Matsumoto, H., Dziak, R. P., and Fox, C. G. (1993). Estimation of seafloor microtopographic roughness through modeling of acoustic backscatter data recorded by multibeam sonar systems. *J. Acoust. Soc. Am.*, **94**, 2776–87.

Maynard, G. L., Sutton, G. H., Hussong, D. M., and Kroenke, L. W. (1974). The seismic wide angle reflection method in the study of ocean sediment velocity structure. In L. L. Hampton (ed.), *Physics of Sound in Marine Sediments*. New York: Plenum Press. pp. 89–118.

Mayo, N. H. (1969). Near-grazing, specular scattering of underwater sound from sea and swell. M.S. thesis, Naval Postgraduate School, Monterey, CA.

Medwin, H. (1954). Acoustic streaming experiment in gases. *J. Acoust. Soc. Am.*, **26**, 332–40.

Medwin, H. (1965). Design and use of an acoustic spectrometer for the detection of particulate matter and bubbles in the sea. In *Proc. 5th Int. Congr. Acoust.*, Liege, Belgium.

(1967). Specular scattering of underwater sound from a wind-driven surface. *J. Acoust. Soc. Am.*, **41**, 1485–95.

(1970). *In-site* acoustic measurements of bubble populations in coastal waters. *J. Geophys. Res.*, **75**, 599–611.

(1974). Acoustic fluctuations due to microbubbles in the near-surface ocean. *J. Acoust. Soc. Am.*, **56**, 1100–4

(1975). Speed of sound in water: a simple equation for realistic parameters. *J. Acoust. Soc. Am.*, **58**, 1318–19.

(1977a). *In situ* measurements of microbubbles at sea. *J. Geophys. Res.*, **82**, 971–6.

(1977b). Acoustical determinations of bubble-size spectra. *J. Acoust. Soc. Am.*, **62**, 1041–4.

(1977c). Counting bubbles acoustically: a review. *Ultrasonics.* **15**, 7–13.

(1981). Shadowing by finite noise barriers. *J. Acoust. Soc. Am.*, **69**, 1060–4.

Medwin, H. and Beaky, M. M. (1989). Bubble sources of the Knudsen sea noise spectrum. *J. Acoust. Soc. Am.*, **83**, 1124–30.

Medwin, H. and Breitz, N. D. (1989). Ambient and transient bubble spectral densities in quiescent seas and under spilling breakers, *J. Geophys. Res.*, **94**, 12751–9.

Medwin, H. and Clay, C. S. (1970). Dependence of spatial and temporal correlation of forward scattered underwater sound on the surface statistics: part II – experiment. *J. Acoust. Soc. Am.*, **47**, 1419–29. See also Clay and Medwin (1970), part I – theory.

Medwin, H. and Daniel, A. C., Jr. (1990). Acoustical measurements of bubble production by spilling breakers. *J. Acoust. Soc. Am.*, **88**, 408–12.

Medwin, H. and D'Spain, G. L. (1986). Near-grazing, low-frequency propagation over randomly rough, rigid surfaces. *J. Acoust. Soc. Am.*, **79**, 657–65.

Medwin, H. and Hagy, J. D., Jr. (1972). Helmholtz–Kirchhoff theory for sound transmission through a statistically rough plane interface between dissimilar fluids. *J. Acoust. Soc. Am.*, **51**, 1083.

Medwin, H. and Novarini, J. C. (1981). Backscattering strength and the range dependence of sound scattered from the ocean surface. *J. Acoust. Soc. Am.*, **69**, 108–11.

(1984). Modified sound refraction near a rough ocean bottom, *J. Acoust. Soc. Am.*, **76**, 1791–6.

Medwin, H. and Rudnick, I. (1953). Surface and volume sources of vorticity in acoustic fields. *J. Acoust. Soc. Am.*, **25**, 538–40.

Medwin, H., Clay, C. S., Berkson, J. M., and Jaggard, D. L. (1970). Traveling correlation function of the heights of windblown water waves. *J. Geophys. Res.*, **75**, 4519–24

Medwin, H., Helbig, R. A., and Hagy, J. D., Jr. (1973). Spectral characteristics of sound transmission through a rough sea surface. *J. Acoust. Soc. Am.*, **54**, 99–109.

Medwin, H., Fitzgerald, J., and Rautmann, G. (1975). Acoustic miniprobing for ocean microstructure and bubbles. *J. Geophys. Res.*, **80**, 405–13.

Medwin, H., Childs, E., and Jebsen, G. M. (1982). Impulse studies of double diffraction: a discrete Huygens interpretation. *J. Acoust. Soc. Am.*, **72**, 1005–13.

Medwin, H., Childs, E., Jordan, E. A., and Spaulding, R. J., Jr. (1984a). Sound
    scatter and shadowing at a seamount: hybrid physical solution in two and three
    dimensions. *J. Acoust. Soc. Am.*, **75**, 1478–90.

Medwin, H., D'Spain, G. L., Childs, E., and Hollis, S. J (1984b). Low-frequency
    grazing propagation over periodic, steep-sloped, rigid roughness elements.
    J. Acoust. Soc. Am., **76**, 1774–90.

Medwin, H., Reitzel, K. J., and Browne, M. J. (1987). Elements of Arctic surface
    scatter: Part III, the head wave. *J. Acoust. Soc. Am.*, **82**(SI), S31 (A).

Medwin, H., Browne, M. J., Johnson, K. R., and Denny, P. L. (1988). Low
    frequency backscatter from Arctic leads *J. Acoust. Soc. Am.*, **83**, 1794–803.

Medwin, H., Kurgan, A., and Nystuen, J. A (1990). Impact and bubble sound from
    raindrops at normal and oblique incidence. *J. Acoust. Soc. Am.*, **88**,
    413–18.

Medwin, H., Nystuen, J. A., Jacobus, P. W., Ostwald, L. H., and Synder, D. E. (1992).
    The anatomy of underwater rain noise. *J. Acoust. Soc. Am.*, **92**, 1613–23.

Mellen, R. H. and Browning, D. G. (1977). Variability of low-frequency sound
    absorption in the ocean: pH dependence. *J. Acoust. Soc. Am.*, **61**,
    704–6.

Mellen, R. H., Browning, D. G., and Simmons, V. P. (1983). Investigation of
    chemical sound absorption in sea water. Part IV. *J. Acoust. Soc. Am.*, **74**,
    987–93.

Mellinger, D. K. (2002). ISHMAEL: 1. 0 User's Guide. NOAA Technical
    Memorandum OAR PMEL-120, Pacific Marine Laboratory, Seattle.
    (http://centus.pmel.noaa.gov/cgi-bin/MobySoft.pl)

Menemenlis, D. and Farmer, D. M. (1992). Acoustical measurement of current and
    vorticity beneath ice, *J. Atmosph. Oceanic Tech.*, **9**(6), 827–49.
  (1995). Path-averaged measurements of turbulence beneath ice in the Arctic,
    J. Geophys. Res., **100**(C7), 13655–63.

Messino, D., Sette, D., and Wanderlingh, F. (1963). Statistical approach to
    ultrasonic cavitation. *J. Acoust. Soc. Am.*, **35**, 1575–83.

Mikeska, E. E. and McKinney, C. M. (1978). Range dependence of underwater
    echoes, from randomly rough surfaces. *J. Acoust. Soc. Am.*, **63**, 1375–80.

Miller, G. A., Sr. (1992). Underwater sound radiation from single large raindrops at
    terminal velocity: The effects of a sloped water surface at impact. M.S. thesis,
    Naval Postgraduate School, Monterey, CA.

Miller, J. H., Lynch, J. F., and Chiu, C.-S. (1989). Estimation of sea surface spectra
    using acoustic tomography. *J. Acoust. Soc. Am.*, **86**, 326–45.

Minnaert, M. (1933). On musical air bubbles and the sounds of running water, *Phil.
    Mag.*, (7)**16**, 235–48.

Mitson, R. B. (1995). Underwater noise of research vessels. *Cooperative Research
    Report 209*; International Council for the Exploration of the Sea, Copenhagen,
    Denmark.

Moffett, M. B., Westervelt, P. J., and Beyer, R. T. (1970). Large-amplitude pulse
    propagation – a transient effect. *J. Acoust. Soc. Am.*, **47**, 1473–4.

Monahan, E. and O'Muircheartaigh, I. G. (1986). Whitecaps and the passive remote
    sensing of the ocean surface. *Int. J. Remote Sensing*, **7**, 627–42.

Moore, S. E., Stafford, K. M., Dahlheim, M. E., *et al.* (1998). Seasonal variation in reception of fin whale calls at five geographic areas in the north Pacific. *Mar. Mamm. Sci.*, **14**(3), 617–27.

Muir, T. G. (1974). Non-linear acoustics and its role in the sedimentary geophysics of the sea. In *Physics of Sound in Marine Sediments*, L. L. Hampton (ed.), New York: Plenum Press. pp. 241–87.

Mulhearn, P. J. (1981). Distribution of microbubbles in coastal waters. *J. Geophys. Res.*, **86**, 6429–34.

Munk, W. (1986). Acoustic monitoring of ocean gyres. *J. Fluid Mech.*, **173**, 43–53.

Munk, W. and Wunsch, C. (1979). Ocean acoustic tomography: a scheme for large scale monitoring. *Deep-Sea Res.*, **26A**, 123–61.

Munk, W. H., Spindel, R. C., Baggeroer, A., and Bridsall, T. G. (1994). The Heard Island feasibility test. In *The Heard Island Papers, J. Acoust. Soc. Am.*, **94**(4), 2330–42.

Nero, R. W., Magnuson. J. J., Brandt S. B., Stanton T. K., and Jech, J. M. (1990). Finescale biological patchness of 70 kHz acoustic scattering at the edge of the Gulf Stream-EchoFront 85. *Deep-Sea Res.*, **37**, 999–1016.

Nichols, R. H. (1987). Infrasonic noise source: wind versus waves. *J. Acoust. Soc. Am.*, **82**, 1395–402 and **82**, 2150 (E).

Noad, M. J. (2002). The use of song by humpback whales (*Megaptera novaeangliae*) during migration off the east coast of Australia. Ph.D. Thesis, University of Sydney, Australia.

Noad, M. J. and Cato, D. H. (2001). A combined acoustic and visual survey of humpback whales off southeast Queensland. *Mem. Qd. Mus.*, **47**, 507–26.

Northrop, J. and Colborn, J. G. (1974). SOFAR channel axial sound speed and depth in the Atlantic Ocean. *J. Geophys. Res.*, **79**, 5633–41.

Novarini, J. C. and Medwin, H. (1978). Diffraction, reflection, and interference during near-grazing and near-normal ocean surface backscattering. *J. Acoust. Soc. Am.*, **64**, 260–8.

  (1985). Computer modeling of resonant sound scattering from a periodic assemblage of wedges: comparison with theories of diffraction gratings. *J. Acoust. Soc. Am.*, **77**, 1754–9.

Novarini, J. C., Keiffer, R. S., and Caruthers, J. W. (1992). Forward scattering from fetch-limited and swell-contaminated sea surfaces. *J. Acoust. Soc. Am.*, **92**, 2099–108.

Nyborg, W. (1965). Acoustic streaming. In W. P. Mason (ed.), *Physical Acoustics*, vol. 2, part B, New York: Academic Press, pp. 265–331.

Nystuen, J. A. (1998). Temporal sampling requirements for autonomous rain gauges. *J. Atmos. Ocean. Techno.*, **15**, 1254–61.

  (2001). Listening to raindrops from underwater: an acoustic disdrometer. *J. Atmos. Ocean. Techno.*, **18**, 1640–57.

Nystuen, J. A. and Farmer, D. (1989). Precipitation in the Canadian Atlantic Storm Program. Measurements of the acoustic signature. *Atmos. Ocean*, **27**, 237–57.

Nystuen, J. A. and McPhaden, M. J. (2001). The beginnings of operational marine weather observations using underwater ambient sound. In *Acoustical Oceanography, Proceedings of the Institute of Acoustics*, **23**, 135–41.

Nystuen, J. A. and Selsor, H. D. (1997). Weather classification using passive acoustic drifters. *J. Atmos. Ocean. Technol.*, **14** 656–66.

Nystuen, J. A., Ostwald, L. H., Jr., and Medwin, H. (1992). The hydroacoustics of raindrop impact. *J. Acoust. Soc. Am.*, **92**, 1017–21.

(1993a). An explanation of the sound generated by light rain in the presence of wind. In *Natural Physical Source of Underwater Sound*, B. R. Kerman (ed.) Dordrecht, Netherlands: Kluwer, pp. 659–69.

Nystuen, J. A., Mc Glothin, C. C., and Cook, M. S. (1993b). The underwater sound generated by heavy precipitation. *J. Acoust. Soc. Am.*, **93**, 3169–77.

(1996). Acoustical rainfall analysis: rainfall drop size distribution using the underwater sound field. *J. Atmos. Ocean. Technol.*, **13**, 74–84.

Nystuen, J. A., McPhaden, M. J., and Freitag, H. P. (2000). Surface measurements of precipitation from an ocean mooring: The acoustic log from the South China Sea. *J. Appl. Meteor.*, **39**, 2182–97.

O'Connell, C. P. (1955). The gas bladder and its relation to the inner ear in *sardinops caeruela* and *engraulis mordax*. *Fish. Bull.* **104**, *Fishery Bulletin of the Fish and Wildlife Service*, **56**, United States Government Printing Service.

O'Hern, T. J. *et al.* (1988). Comparison of holographic and Coulter Counter measurements of cavitation nuclei in the ocean. *J. Fluids Eng.*, **110**, 200–7.

Ona, E. (1990). Physiological factors causing natural variations in acoustic target strengths of fish. *J. Mar. Biol. Assn. UK*, **70**, 107–27.

Osborn, T. R. (1974). Vertical profiling of velocity microstructure. *J. Phys. Oceanogr.*, **4**, 109–15.

Ostachev, V. (1994). Sound propagation and scattering in media with random inhomogeneities of sound speed, density, and medium velocity. *Waves in Random Media*, **1**, 1–26.

Ostrovsky, L. A., Sutin, A. M., Soustova, I. A., Matveyev, A. L., and Potapov, A. I. (2003). Nonlinear scattering of acoustic waves by natural and artificially generated subsurface bubble layers in sea. *J. Acoust. Soc. Am.*, **113**, 741–9.

Pace, N. G., Al-Hamdani, Z. K. S., and Thorne, P. D. (1985). The range dependence of normal incidence acoustic backscatter from a rough surface. *J. Acoust. Soc. Am.*, **77**, 101–12.

Palmer, D. R. (1995). Acoustic imaging of naturally occurring plumes. In *Acoustic Imaging Vol. 21*, J. Jones (ed.). New York: Plenum.

(1996). Rayleigh scattering from nonspherical particles. *J. Acoust. Soc. Am.*, **99**, 1901–12.

Palmer, D. R., Rona, P. A., and Mottl, M. J. (1986). Acoustic imaging of high-temperature hydrothermal plumes at seafloor spreading centers. *J. Acoust. Soc. Am.*, **80**, 888–98.

Parsons, A. R., Bourke, R. H., Muench, R., *et al.* (1996). The Barents Sea polar front in summer. *J. Geophys. Res.*, **101**(C6) 14201–21.

Partridge, C. and Smith, E. R. (1995). Acoustic scattering from bodies: range of validity of the deformed cylinder method. *J. Acoust. Soc. Am.*, **97**(2), 784–95.

Parvulescu, A. (1995). Matched-signal (MESS) processing by the ocean. *J. Acoust. Soc. Am.*, **98**, 943–60.

Parvulescu, A. and Clay, C. S. (1965). Reproducibility of signal transmissions in the ocean. *Radio Elec. Eng.*, **29**, 223–8.

Payne, R. S. and McVay, S. (1971). Songs of humpback whales. *Science*, **173**, 585–97.

Pedersen, M. A. (1961). Acoustic intensity anomalies introduced by constant sound velocity gradients. *J. Acoust. Soc. Am.*, **33**, 465–74.

Pekeris, C. L. (1948). Theory of propagation of explosive sound in shallow water, in *Propagation of Sound in the Oceans*, Geological Society of America Memoir 27, New York.

Perkins, J. B. III. (1974). Amplitude modulation of acoustic signals by ocean waves and the effects on signal detection. M.S. thesis, Naval Postgraduate School, Monterey, CA.

Phelps, A. D. and Leighton, T. G. (1996). High resolution bubble sizing through detection of the subharmonic response with a two frequency excitation technique, *J. Acoust. Soc. Am.*, **99**, 1985–92.

(1997). The subharmonic oscillations and combination- frequency emissions from a resonant bubble: their properties and generation mechanisms, *Acta Acust.*, **83**, 59–66.

(1998). Oceanic bubble population measurements using a buoy-deployed combination frequency technique. *IEEE J. Ocean. Eng.*, 400–10.

Phillips, O. M. (1977). *The Dynamics of the Upper Ocean*, 2nd edn. New York: Cambridge University Press.

Pierce, A. D. (1981). *Acoustics: An Introduction to its Physical Principles and Applications*, New York: McGraw-Hill.

Pierson, W. J. and Moskowitz, L. (1964). A proposed spectral form for fully developed wind seas based on the similarity theory of S. A. Kitaigorodskii. *J. Geophys. Res.*, **69**, 5181–90.

Pinkel, R. and Smith, J. (1987). Open ocean surface wave measurement using doppler sonar. *J. Geophys. Res.*, **92**(C12), 12967–73.

(1992). Repeat sequence codes for improved performance of Doppler sounders. *J. Atmos. Oceanic Technol.*, **9**, 149–63.

Pinkel, R., Plueddemann, A., and Williams, R. (1987). Internal wave observations from *FLIP* in MILDEX. *J. Phys. Oceanogr.*, **17**, 1737–57.

Popper, A. N., Webb, J. F., and Fay, R. R. (2002). (eds). Special Issue on Fish Bioacoustics. *Bioacoustics*, **12**, 339 pp.

Porter, M. B. (1993). Acoustic models and sonar system. *IEEE J. Ocean. Eng.*, **18**(4), 425–37.

Prosperetti, A. (1988). Bubble-related ambient noise in the ocean. *J. Acoust. Soc. Am.*, **84**, 1024–54.

Prosperetti, A., Lu, N. Q., and Kim, H. S. (1993). Active and passive acoustic behavior of bubble clouds at the ocean's surface, *J. Acoust. Soc. Am.*, **93**, 3117–27.

Pumphrey, H. C., Crum, L. A., and Bjorno, L. (1989). Underwater sound produced by individual drop impacts. *J. Acoust. Soc. Am.*, **85**, 1518–26.

Rayleigh, Lord [J. W. Strutt] (1945). *The Theory of Sound* vols. 1 and 2 (edns. 1894 and 1896), New York: Dover.

Reeder, D. B., Jech, J. M. and Stanton, T. K. (2004). Broadband acoustic backscatter and high-resolution morphology of fish: measurement and modeling. *J. Acoust. Soc. Am.*, **116**, 729–46.

Rice, S. O. (1954). Mathematical analysis of random noise. In *Selected papers on Noise and stochastic processes*. N. Wax (ed.), New York: Dover, pp. 133–294.

Richards, E. J. and Mead, D. (1968). *Noise and Acoustic Fatigue in Aeronautics* London: John Wiley.

Richards, S. D. and Leighton, T. G. (2001). Acoustic sensor performance in coastal waters: solid suspensions and bubbles. In *Acoustical Oceanography.*, T. G. Leighton, G. J. Heald, H. Griffiths, and G. Griffiths, Institute of Acoustics. *Proc. Inst. Acoustics*, **23**(2), 399–406.

Richards, S. D., Leighton, T. G., and Brown, N. R. (2003). Visco-inertial absorption in dilute suspensions of irregular particles. *Proceedings of the Royal Society, Series A*, **459** (2037), 2153–67.

Richardson, W. J., Greene, C., Malme, C. I., and Thompson, D. H. (1995). *Marine Mammals and Noise*. San Diego: Academic Press.

Roderick, W. I. and Cron, B. F. (1970). Frequency spectra of forward-scattered sound from the ocean surface. *J. Acoust. Soc. Am.*, **48**, 759–66.

Rogers, P. H. (1977). Weak-shock solution of underwater explosive shock waves. *J. Acoust. Soc. Am.*, **62**, 1412–19.

Rona, P. A. and Palmer, D. R. (1993). Imaging plumes beneath the sea. *J. Acoust. Soc. Am.*, **93**, 569–70.

Rona, P. A. and Trivett, D. A. (1992). Discrete and diffuse heat transfer at ASHES vent field, Axial Volcano, Juan de Fuca Ridge. *Earth Planet. Sci. Lett.*, **109**, 57–71.

Rona, P. A., Palmer, D. R., Jones, C., Chayes, D. A., Czarnecki, M., Carey, E. A., and Guerrero, J. C. (1991). Acoustic imaging of hydrothermal plumes, East Pacific Rise, 21° N, 109° W. *Geophys. Res. Lett.*, **18**, 2233–6.

Rona, P. A., Jackson, D. R., Wen, T., Jones, C., Mitsuzawa, K., Bemis, K. G., and Dworski, J. G. (1997). Acoustic mapping of diffuse flow at a seafloor hydrothermal site: Monolith Vent, Juan de Fuca Ridge. *Geophys. Res. Lett.*, **24**, 2351–4.

Rona, P. A., Bemis, K. G., Silver, D., and Jones, C. D. (2002a). Acoustic imaging, visualization, and quantification of buoyant hydrothermal plumes in the ocean. *Marine Geophys. Res.*, **23**, 147–68.

Rona, P. A., Jackson, D. R., Bemis, K. G., Jones, C. D., Milsuzawa, K., Palmer, D. R., and Silver, D. (2002b). Acoustics advances study of sea floor hydrothermal flow. *EOS Trans. Am. Geophys. Union*, **83**(44), 497–502.

Ross, D. (1976). *Mechanics of Underwater Noise*. New York: Pergamon Press. (1987). *Mechanics of Underwater Noise*. Peninsula Publishing, Box 867, Los Altos, CA 94023.

Rudstam, L. G., Clay, C. S., and Magnuson, J. J. (1987). Density estimates and size of cisco, *coregonus artedii* using analysis of echo peak PDF from a single transducer sonar. *Can. J. Fish Aquat. Sci.*, **44**, 811–21.

Rusby, J. S. M. (1970). The onset of sound wave distortion and cavitation in seawater. *J. Sound Vib.*, **13**, 257–67.

Sarkar, K. and Prosperetti, A. (1994). Coherent and incoherent scattering by oceanic bubbles. *J. Acoust. Soc. Am.*, **96**, 332–41.

Schulkin, M. and Marsh, H. W. (1962). Sound absorption in seawater. *J. Acoust. Soc. Am.*, **35**, 864–5.

Scofield, C. (1992). Oscillating microbubbles created by water drops falling on fresh and salt water: Amplitude, damping and the effects of temperature and salinity. M.S. thesis, Naval Postgraduate School, Monterey, CA.

Shaw, P. T., Watts, D. R., and Rossby, H. T. (1978). On the estimation of oceanic wind speed and stress from ambient noise measurements. *Deep-Sea Res.*, **25**, 1225–33.

Shields, R. B., Jr. (1977). Signal enhancement of specularly scattered underwater sound. M.S. Thesis, Naval Postgraduate School, Monterey. CA.

Shooter, J. A. *et al.* (1974). Acoustic saturation of spherical waves in water. *J. Acoust. Soc. Am.*, **55**, 54–62.

Simmen, J. A., Stanic, S. J., and Goodman, R. R. (2001). Guest editorial, special issue on high-frequency acoustics, *IEEE J. Ocean. Eng.*, **26**, 1–3.

Simmons, V. P. (1975). Investigation of the 1 kHz sound absorption in sea water. Ph.D. dissertation, University of California, San Diego.

Sims, C. C. (1960). Bubble transducer for radiating high power low frequency sound in water. *J. Acoust. Soc. Am.*, **32**, 1305.

Skretting, A. and Leroy, C. C. (1971). Sound attenuation between 200 Hz and 10 kHz, *J. Acoust. Soc. Am.*, **49**, 276–82.

Smith, J. A. (2001). Phased-array Doppler sonar measurements of near-surface motion: Langmuir circulation, surface waves, and breaking. In *Acoustical Oceanography*, T. G. Leighton, H. Griffiths, and G. Griffiths (eds.), Southampton, UK. *Proc. Inst. Acoust.*, **23**(2), 163–70.

Snyder, D. E. (1990). Characteristics of sound radiation from large raindrops. M.S. thesis, Naval Postgraduate School, Monterey, CA.

Spaulding, R. P., Jr. (1979). Physical modeling of sound shadowing by seamounts. M.S. thesis, Naval Postgraduate School, Monterey. CA.

Spiesberger, J. L., Terray, E., and Prada, K. (1994). Successful ray modeling of acoustic multipaths over a 3000-km section in the Pacific. *J. Acoust. Soc. Am.*, **95**, 3654–7.

Spindel, R. C. (1985). Signal processing in ocean acoustic tomography, In *Adaptive Methods in Underwater Acoustics*, H. G. Urban (ed.), Dordrecht: Reidel, pp. 687–710.

Spindel, R. C. and Schultheiss, P. M. (1972). Acoustic surface reflection channel characterization through impulse response measurements. *J. Acoust. Soc. Am.*, **51**, 1812.

Spindel, R. C. and Worcester, P. F. (1986). Technology in ocean acoustic tomography, *J. Marine Tech. Soc.*, **20**(4), 68–72.

Stacey, M. T., Monismith, S. G., and Burau, J. R. (1999). Measurements of Reynolds stress profiles in unstratified tidal flow. *J. Geophys. Res.*, **104**(C5), 10 933–49.

Stafford, K. M., Nieukirk, S. L., and Fox, C. G. (2001). Geographic and seasonal variation of blue whale calls in the North Pacific. *J. Cetacean Res. Managem.*, **3**(1), 65–76.

Stansfield. D. (1990). *Electroacoustic Transducers*, Institute of Acoustics. Bath University Press.

Stanton, T. K. (1984). Sonar estimates of seafloor microroughness. *J. Acoust. Soc. Am.*, **75**, 809–18.

(1985). Density estimates of biological sound scatters using sonar echo peak PDFs. *J. Acoust. Soc. Am.*, **78**, 1868–73.

(1989). Sound scattering by cylinders of finite length. III. Deformed cylinders. *J. Acoust. Soc. Am.*, **86**(2), 691–705.

Stanton, T. K., Clay, C. S., and Chu, D. (1993). Ray representation of sound scattering by weakly scattering deformed fluid cylinder: simple physics and applications to zooplankton. *J. Acoust. Soc. Am.*, **94**, 3454–62.

Stanton, T. K., Wiebe, P. H., Chu, D., and Goodman, L. (1994). Acoustic characterization and discrimination of marine zooplankton and turbulence. *ICES. J. Mar. Sci.*, **31**, 469–79.

Stanzial, D., Prodia, N., and Schiffrer, G. (1996). Reactive acoustic intensity for general fields and energy polarization. *J. Acoust. Soc. Am.*, **99**, 1868–76.

Starritt, H. C., Duck, F. A., and Humphrey, V. F. (1989). An experimental investigation of streaming in pulsed diagnostic ultrasound beams, *Ultrasound in Med. Biol.*, **15**, 363–73.

Stenzel, H. (1938). On the disturbance of a sound field brought about by a rigid sphere, in German, *Elektr. Nachr. Tech.*, **15**, 71–8. Transl. G. R. Barnard and C. W. Horton, Sr. (1959) and republished as *Technical Report No. 159*. Defense Research Laboratory, University of Texas, Austin.

Stephens, R. W. B. (1970). (ed.) *Underwater Acoustics*, New York: Wiley. (Chapter 3, Scattering from the Sea Surface, by H. Medwin.)

Stokes, G. G. (1849). On the dynamical theory of diffraction. *Trans. Camb. Phil. Soc.*, **9**, 1. Reprinted in Stokes, (1883), *Mathematical and Physical Papers*, vol. 2, Cambridge, Cambridge University Press, pp. 243–328.

Stoll, R. D. (1974). Acoustic waves in saturated sediments. In L. L. Hampton (ed.), *Physics of Sound in Marine Sound Sediments*. New York: Plenum Press, pp. 19–40.

Stoll, R. D. (1989). *Sediment Acoustics (Lecture notes in Earth Science)*. New York: Springer-Verlag.

Strasberg, M. (1956). Gas bubbles as sources of sound in liquids. *J. Acoust. Soc. Am.*, **28**, 20.

Stroud, J. S. and Marston, P. L. (1994). Transient bubble oscillations associated with the underwater noise of rain detected optically and some properties of light scattered by bubbles. In *Bubble Dynamics and Interface Phenomena*, J. R. Blake *et al.* (eds.), Amsterdam: Kluwer Academic Publishing, pp. 161–9.

Strutt, J. W. [Lord Rayleigh] (1945). *The Theory of Sound*, vol. 2 (2nd edn. 1896) New York: Dover, pp. 89, 282.

Szczucka, J. (1989). Acoustic detection of gas bubbles in the sea. *Oceanologia*, **28**, 103–33.

Tavolga, W. N. (1964). (ed.) *Marine Bioacoustics.* New York: Pergamon Press.
    (1967). *Marine Bioacoustics.* Vol. 2. New York: Pergamon Press.

Taylor (1935). Statistical theory of turbulence. *Proceedings of the Royal Society of London*, Series A, **151**, 421.

Thorne, P. D. and Haynes, D. M. (2002). A review of acoustic measurement of small scale sediment processes. *Continental Shelf Research*, **22**, 1–30.

Thorne, P. D. and Pace, N. G. (1984). Acoustic studies of broadband scattering from a model rough surface. *J. Acoust. Soc. Am.*, **75**, 133–44.

Thorne, P. D., Pace, N. G., and Al-Hamdani, Z. K. S. (1988). Laboratory measurements of backscattering from marine sediments. *J. Acoust. Soc. Am.*, **84**, 303–9.

Thorp, W. H. (1965). Deep-ocean sound attenuation in the sub- and low-kilocycle-per sec region. *J. Acoust. Soc. Am.*, **38**, 648–54.

Thorpe, S. A. (1982). On the clouds of bubbles formed by breaking wind waves in deep water, and their role in air-sea gas transfer. *Phil. Trans. R. Soc. London*, Ser. A, **304**, 155–210.

Thorsos, E. I. (1988). The validity of the Kirchhoff approximation for rough surface scattering using a Gaussian roughness spectrum. *J. Acoust. Soc. Am.*, **83**, 78–92.
    (1990). Acoustic scattering from a Pierson–Moskowitz sea surface. *J. Acoust. Soc. Am.*, **88**, 335–49.

Thorsos, E. I. and Broschat, S. L. (1995). An investigation of the small slope approximation for scattering from rough surfaces. Part I: Theory. *J. Acoust. Soc. Am.*, **97**, 2082–93.

Thorsos, E. I. and Richardson, M. D. (2002). Guest editorial, special issue on high-frequency sediment acoustics. *IEEE J. Ocean. Eng.*, **27**, 341–5.

Tindle, C. T., Hobaek, H., and Muir, T. G. (1987a). Downslope propagation in a shallow water wedge. *J. Acoust. Soc. Am.*, **81**, 275–86.
    (1987b). Normal modes filtering for downslope propagation in a shallow water wedge. *J. Acoust. Soc. Am.*, **81**, 287–94.

Tjøtta, S. and Tjøtta, J. N. (1993). Acoustic streaming in ultrasonic beams. *Proceedings of the 13th International Symposium on Non-linear Acoustic.* London: World Scientific, pp. 601–7.

Tolstoy, A. (1993). *Matched Field Processing for Underwater Acoustics.* Singapore: World Scientific.

Tolstoy, I. (1989). Exact, explicit solutions for diffraction by hard sound barriers and seamounts. *J. Acoust. Soc. Am.*, **85**, 661.

Traynor, J. J. and Ehrenberg, J. E. (1990). Fish and standard-sphere, target-strength measurements obtained with a dual-beam and split-beam echo-sounding system. *Rapp. P.-V. Reun. Cons. Int. Explor. Mer.*, **189**, 325–35.

Trevorrow, M. V. (1996). Multifrequency acoustic investigation of juvenile and adult fish in Lake Biwa, Japan. *J. Acoust. Soc. Am.*, **100**, 3042–52.

Trevorrow, M. and Farmer, D. M. (1992). The use of Barker codes in Doppler sonar measurements. *J. Atmos. Oceanic Technol.*, **9**, 699–704.

Trorey, A. W. (1970). A simple theory for seismic diffractions. *Geophysics*, **35**, 762–84.

Tyce, R. C. (1986). Deep seafloor mapping systems – a review. *Mar. Tech. J.*, **20**, 4–16.

Updegraff, G. E. and Anderson, V. C. (1991). Bubble noise and wavelet spills recorded 1 m below the ocean surface. *J. Acoust. Soc. Am.*, **89**, 2264–79.

Urick, R. J. (1948). The absorption of sound in suspensions of irregular particles. *J. Acoust. Soc. Am.*, **20**, 283–9.

(1972). Noise signature of an aircraft in level flight over a hydrophone in the sea. *J. Acoust. Soc. Am.*, **52**, 993–9.

Vagle, S. and Farmer, D. M. (1992). The measurement of bubble size distributions by acoustical backscatter. *J. Atmos. Ocean. Technol.*, **9**(5), 630–44.

(1998). A comparison of four methods of bubble measurements. *IEEE Ocean. Eng.*, **23**(3), 211–22.

Vagle, S., Large, G. W., and Farmer, D. M. (1990). An evaluation on the WOTAN technique of inferring oceanic winds from underwater ambient sound. *J. Atmos. Ocean. Technol.*, **7**(4), 576–95.

Voronovitch, A. (1985). Small slope approximation in wave scattering by rough surfaces. *Soc. Phys. JETP*, **62**, 65–70.

Walker, R. A. (1963). Some intense, low-frequency, underwater sounds of wide geographic distribution, apparently of biological origin. *J. Acoust. Soc. Am.*, **35**(11), 1816–24.

Wang, P. C. C. and Medwin, H. (1975). Stochastic models of the scattering of sound by bubbles in the upper ocean. *Quart. J. Appl. Math*, 411–25 (January).

Wang, T.-C. and Shang, E.-C. (1981). *Underwater Acoustics*. China: Science Press.

Watkins, W. A. (1981). The activities and underwater sounds of fin whales. *Scientific Reports of the Whale Research Institute*, **33**, 83–117.

Watkins, W. A. and Schevill, W. E. (1972). Sound source location by arrival-times on a non-rigid three-dimensional hydrophone array. *Deep-Sea Res.*, **19**, 691–706.

Westervelt, P. J. (1963). Parametric acoustic array. *J. Acoust. Soc. Am.*, **35**, 533–7.

Weston, D. E. (1967). Sound propagation in the presence of bladder fish. In *Underwater Acoustics*, ed. V. M. Albers, pp. 55–88. New York, NY: Plenum Press.

Whitehead, P. J. P. and Blaxter, J. H. S. (1964). Swimbladder form in clupeoid fishes. *Zool. J. Limn. Soc.*, **97**, 299–372.

Wiebe, P. H., Greene, C. H., Stanton, T. K., and Burczynski, J. (1990). Sound scattering by live zooplankton and micronekton: empirical studies with a dual-beam acoustical system. *J. Acoust. Soc. Am.*, **88**, 2346–60.

Williams, K. L., Richardson, M. D., Briggs, K. B., and Jackson, D. R. (2001). Scattering of high-frequency acoustic energy from discrete scatterers on the seafloor: Glass spheres and shells. In *Acoustical Oceanography*, T. G. Leighton, G. J. Heald, H. Griffiths, and G. Griffiths (eds.), Institute of Acoustics. *Proc. Inst. Acous.*, **23**(2), 383–90.

Winn, H. E., Perkins, P. J., and Poulter, T. C. (1971). Sounds of the humpback whale. *Proceedings of the Seventh Annual Conference on Biological Sonar*, **7**, 39–42.

Wilson, O. B. and Leonard, R. W. (1954). Measurements of sound absorption in aqueous salt solutions by a resonator method. *J. Acoust. Soc. Am.*, **26**, 223–6.

Woodhead, P. (1966). The behaviour of fish in relation to light in the sea. *Oceanogr. Mar. Biol. Ann. Rev.*, **4**, 337–403.

Worcester, P. F., Peckham, D. A., Hardy, K. R., and Dormer, F. O. (1985). AVATAR: second generation transceiver electronics for ocean acoustic tomography. Inst. of Elec. and Electronic Engineers, New York. *Proc. OCEANS, 85*, 654–62.

Yamoaka, H., Kaneka, A., Park, J.-H., *et al.* (2002). Coastal acoustic tomography system and its field application. *IEEE J. Ocean Eng.*, **27**, 283–95.

Yang, C. T. (1993). Broadband source localization and signature estimation. *J. Acoust. Soc. Am.*, **93**, 1797–806.

Ye, Z. (1997). Acoustic dispersion and attenuation in many spherical scatterer systems and the Kramers–Kronig relations. *J. Acoust. Soc. Am.*, **101**(6), 3299–305.

Yeager, E., Fisher, F. H., Miceli, J., and Bressel, R. (1973). Origin of the low-frequency sound absorption in seawater. *J. Acoust. Soc. Am.*, **53**, 1705–7.

Zimdar, R. E., Bamhouse, P. D., and Stoffel, M. J. (1964). Instrumentation to determine the presence and acoustic effect of microbubbles near the sea surface. M.S. thesis, Naval Postgraduate School, Monterey, CA.

Bibliography

Abramowitz, A. and Stegun, I. A. (1964). *Handbook of Mathematical Functions.*
Washington, DC: US Government Printing Office.

Au, W. W. L. (1993). *The Sonar of Dolphins.* New York: Springer-Verlag.

Beckmann, P. and Spizzichino, A. (1963). *The Scattering of Electromagnetic Waves
from Rough Surfaces.* New York: Pergamon Press.

Bobber, R. J. (1970). *Underwater Electroacoustic Measurements.* Washington, DC:
Naval Research Lab.

Born, M. and Wolf, E. (1965). *Principles of Optics.* New York: Pergamon Press.

Brekhovskikh, L. M. and Godin, O. A. (1990). *Acoustics of Layered Media 1.*
Berlin: Springer-Verlag.

Buckingham, M. J. and Potter, J. R., eds. (1995). *Sea Surface Sound '94.* Singapore:
World Scientific.

Busnel, R.-G. and Fish, J. F. (1980). *Animal Sonar Systems.* New York: Plenum
Press.

Cerveny, V. and Ravindra, R. (1971). *Theory of Seismic Head Waves.* Toronto:
University of Toronto Press.

Chernov, L. A. (1960). *Wave Propagation in a Random Medium.* New York:
McGraw-Hill.

Clay, C. S. (1990). *Elementary Exploration Seismology.* Englewood Cilffs, NJ:
Prentice-Hall.

Clay, C. S. and Medwin, H. (1977). *Acoustical Oceanography.* New York: Wiley.

Cole, R. H. (1948). *Underwater Explosives.* Princeton, NJ: Princeton University
Press.

Eckart, C. (1968). (ed.) *Principles and Applications of Underwater Sound.*
Reissued (1968). NAVMAT, P9674, US Department of the Navy.

Ewing, M., Worzel, J. L., and Pekeris, C. L. (1948). *Propagation of Sound in the
Ocean, the Geological Society of America Memoir 27,* New York.

Farquhar, G. B. (1970). (ed.) *Proceedings of an International Symposium on
Biological Sound Scattering in the Ocean.* Washington, DC: US Government
Printing Office.

Flatte, S. M. (1979). (ed.) *Sound Transmission through a Fluctuating Ocean.*
Cambridge: Cambridge University Press.

Hampton, L. (1974). (ed.) *Physics of Sound in Marine Sediments.* New York:
Plenum Press.

Jensen, F. S., Kupermann, W. A., Porter, M. B., and Schmidt, H. (1994). *Computational Ocean Acoustics*. New Jersey.

Kerman, B. R. (1993). (ed.) *Natural Physical Sources of Underwater Sound; Sea Surface Sound 2; Proceedings of the 1990 Conference on Natural Mechanisms of Surface Generated Noise in the Ocean*, Lerici, Italy. Dordrecht, Netherlands: Kluwer.

Kuperman, W. A. and Jensen, F. B. (1980). (eds.) *Bottom-Interacting Ocean Acoustics*. New York: Plenum Press.

Landau, L. D. and Lifshitz, E. M. (1959). *Fluid Mechanics*, English trans. Reading, MA: Addison-Wesley.

Leighton, T. G. (1994). *The Acoustic Bubble*, New York: Academic Press.

Leighton, T. G., Heald, G. J., Griffiths, H., and Griffiths, G., (eds.) (2001). Acoustical Oceanography, *Proc. Inst. Acoust. UK*, **23**(2), 424 pp.

Kinsman, B. (1965). *Wind waves*. Englewood Cliffs, NJ: Prentice-Hall.

Lighthill, J. (1978, 1979). *Waves in Fluids*. Cambridge, UK: Cambridge University Press.

MacLennan, D. N. and Simmonds, E. J. (1992). *Fisheries Acoustics*. Van Nostrand-Reinhold.

Medwin, H. and Clay, C. S. (1998). *Fundamentals of Acoustical Oceanography*. San Diego: Academic Press.

Monahan, E. C. and MacNiocaill, G. (1986). (eds.). *Oceanic Whitecaps and Their Role in Air-Sea Exchange Processes*. Dordrecht, Netherlands: D. Reidel Publishing Co.

Morse, P. M. (1948). *Vibration and Sound*, 2nd edn. Woodbury, NY: Acoustical Society of America.

Munk, W., Worcester, P., and Wunsch, C. (1995). *Ocean Acoustic Tomography*. New York: Cambridge University Press.

National Academy of Sciences. (1963). *Ocean Wave Spectra: Proceedings of a Conference*. Englewood Cliffs, NJ: Prentice-Hall.

Naugolnykh, S. A. and Ostrovsky, L. A. (1997). *Non-Linear Wave Processes in Acoustics*. New York: Cambridge University Press.

Neubauer, W. G. (1986). *Acoustic Reflections from Surfaces and Shapes*. Washington, DC: Naval Research Laboratory.

Novikov, B. K., Rudenko, O. V., and Timoshenko, V. I. (1987). *Nonlinear Underwater Acoustics*. Woodbury, NY: Acoustical Society of America.

Olson, H. F. (1947). *Elements of Acoustical Engineering*, 2nd edn. New York: D. Van Nostrand.

Pace, N. G. (1983). (ed.). *Acoustics and the Seabed*. Bath, UK: Bath University Press.

Phillips, O. M. (1980). *The Dynamics of the Upper Ocean*, 2nd edn. Cambridge: Cambridge University Press.

Pierce, A. D. (1991). *Acoustics*. Woodbury, NY: Acoustical Society of America.

Potter, J. and Warn-Varnas, A., eds. (1991). *Ocean Variability and Acoustic Propagation*. Boston: Kluwer Academic Publishers.

Rayleigh, Lord [Strutt, J. W.] (1945). *The Theory of Sound* (2nd edn. 1896). New York: Dover Publications.

Rona, P. A. (1980). *NOAA Atlas 3. The Central North Atlantic Ocean Basin and Continental Margins: Geology, Geophysics, Geochemistry, and Resources Including the Trans-Altantic Geotraverese (TAG)*; U.S. Department of Commerce, National Oceanic and Atmospheric Administration, Environmental Research Laboratories, Miami.

Ross, D. (1987). *Mechanics of Underwater Noise*. Los Altos, CA: Peninsula Publishing.

Stoll, R. D. (1989). *Sediment Acoustics*. New York: Springer-Verlag.

Tolstoy, I. (1973). *Wave Propagation*. New York: McGraw-Hill.

Tolstoy, L. and Clay, C. S. (1966). *Ocean Acoustics*. New York: McGraw-Hill. (1987) Reprinted by the Acoustical Society of America, Woodbury, NY.

Urick, R. J. (1983). *Principles of Underwater Sound*, 3rd edn. Los Altos, CA: Peninsula Publishing.

Wang, T.-C. and Shang, E.-C. (1981). *Underwater Acoustics*. China: Science Press.

Wildt, R., ed. (1946). Acoustic properties of wakes, part 4 of *Physics of Sound in the Sea*, vol. 8; Summary Technical Report of Div. 6, National Defense Research Committee, Department of the Navy; Washington, D.C. Reissued by Naval Material Command, 1969.

Wilson, O. B. (1985). *An Introduction to the Theory and Design of Sonar Transducers*. Los Altos, CA: Peninsula Publishing.

Wood, A. B. (1955). *A Textbook of Sound*. New York: Macmillan.

Worzel, J. L. and Ewing, M. (1948). The propagation of sound in the ocean. In *The Geological Society of America Memoir 27*. New York: Geological Society of America.

Ziomek, L. J. (1985). *Underwater Acoustics: A Linear Systems Theory Approach*. New York: Academic Press.

(1995). *Fundamentals of Acoustic Field Theory and Space–Time Signal Processing*. Boca Raton, FL: CRC Press.

Symbols

Different ocean acoustic specialties such as wave propagation, scattering by objects and rough surfaces, transducers, non-linear propagation, oceanography, geophysics, and fishery acoustics use some of the same symbols in different contexts. Where possible, our expressions use symbols appropriate to the different technical areas; thus some of the same symbols and parameters appear in different contexts. We employ Italicized Times Roman, Greek, and Zapf Chancery fonts for symbols. When the meaning is clear, and in order to simplify the notation, the functional dependence may be omitted. When the units of the quantity are not obvious, we specify them. A&S refers to symbols and sections in Abramowitz and Stegun (1964).

a	radius of a sphere or cylinder
a	the ray parameter for refraction
a_{ec}	radius of an equivalent volume cylinder
a_{es}	radius of an equivalent volume sphere
a_n	displacement amplitude of the nth transducer element
A	dispersion constant for the boundary wave; dimensionless; insonified area
A_m	amplitude of the mth cylindrical mode at the source
A_s	amplitude factor proportional to source power in modal analysis
A_n	amplitude of the sum of hydrophone signals
A/D	analog to digital conversion
b	real component of the ratio of specific heats, γ, dimensionless; separation between adjacent elements of a line source
b_n	gradient of sound speed, (m/s)/m
B/A	parameter of non-linearity, dimensionless
BW	bandwidth, Hz
BSS	backscattering srength

B	pressure amplitude for impulse diffraction, pascal $s^{1/2}$
c	sound speed or velocity, subscripts identify the medium or type of wave; c_R, Rayleigh wave; c_s, shear wave; c_p, compressional wave
$C(k)$	covariance of a random signal; C_{uk}, unknown (sample) covariance matrix
$C(\tau)$	autocorrelation of time varying surface
$C(\xi)$	spatial correlation in the x direction of a rough surface
$C_{xx}(\tau)$	autocovariance of a signal
$C(x)$	Fresnel cosine integral; also see $S(x)$
d	a distance or separation; imaginary component of the complex ratio of specific heats, γ, dimensionless
dB	decibel, $20 \log_{10}$ of ratio of pressures, voltages, and so forth to appropriate reference values; $10 \log_{10}$ of ratio of intensities, or powers to appropriate reference values, subscript $_{10}$ is often omitted
dS	element of area
DSL	deep scattering layer (biological)
D	transducer and array directional responses; also with subscripts and functional dependence as in $D(\theta)$, D_t, and D_r directivity factors for transmitter and receiver, dimensionless; diffraction terms in the wedge solution; periodic separation of wedges
$\mathcal{D}$	product of transmit and receive directivities
DI_t, DI_r	directivity index for transmission, reception, dB, $10 \log_{10}(D_t)$, and $10 \log_{10}(D_r)$
DF	detection factor
DS	diffraction strength, dB
D_b	damping rate, s^{-1}
e	base of natural logarithms, 2.71828; error matrix for matrix inversion; ratio of bulk elasticities in two media; $e_n(t)$ amplitude of envelope of a ping
e_G	amplitude of the envelope within "gate open" interval
esr	equivalent spherical radius

E_m	message energy passing through an area in a given time, joules
E_n	bulk modulus of elasticity of medium n
$E_{xx}(f_j)$	energy spectral density at frequency f_j
f	sound frequency, Hz
f_m	frequency of the mth harmonic; subscripts give specific harmonic
f_c	carrier frequency, Hz; waveguide cut-off frequency
$f(t), f(x)$	general functions of time or position
f_r	relaxation frequency of a molecular process
f_b	simple bubble breathing frequency, Hz
f_R	corrected bubble breathing frequency, Hz
f_d	frequency of damped oscillation, s^{-1}; heterodyne-shifted frequency
f_H	heterodyning frequency
f	a generalized vector
F	ratio of near-surface bubble frequency to free-space frequency; frequency of an ocean surface gravity wave, s^{-1}
g	acceleration of gravity, m/s^2; ratio of densities in two media
g_r	volume flow per unit length, m^2/s
g_R	acoustical roughness for scatter from a randomly rough surface
g_t	acoustical roughness for transmission through a randomly rough surface
g_A	linear sonar receiver gain factor
$g(t)$	a function of time
g_{rcvr}	gain of receiver
g_{TVG}	time-varying gain function
$g_2 = \pm[(c_2/c_1)^2 - 1]^{1/2}$	a function for incident angles greater than critical
G	sonar receiver gain in dB; pressure transmission function in a waveguide; scattering angular function
$\mathcal{G}$	imaginary value for transmission; beam geometry term in rough-surface scatter
$G^* = G + G'$	dynamic shear modulus, $(newtons/m^2)/(m/s)$
h	ratio of sound speeds for two media; height; $h(t)$ filter response

$H_n^{(1)}$ and $H_n^{(2)}(x)$	cylindrical Hankel functions (*A&S*, Chap. 9)
$i = \sqrt{-1}$	imaginary number; also a summation index;
i or $i(t)$	instantaneous intensity: the energy passing through a perpendicular unit area in unit time, watts/m^2; a vector
$\mathbf{i}, \mathbf{j}, \mathbf{k}$	unit vectors along coordinates x, y, and z;
I	time average intensity, watts/m^2; current in transducer calibration, amperes; $I\,(\dots)$ spectral intensity in 1 Hz band (watts/m^2)/ Hz
I_n	an integral in the wedge diffraction solution
$\mathbf{j}$	unit vector along coordinate y
$j_n(x)$	spherical Bessel function of first kind and order n (*A&S* Chap. 10); see $y_n(x)$ for companion function
$J_n(x)$	cylindrical Bessel function of first kind and order n (*A&S* Chap. 9); see $Y_n(x)$ for companion function
J	transducer reciprocity factor, watts/Pa^2
k	wave number or propagation constant, m^{-1}, subscripts give the medium; an integer in the image construction of the wedge problem
k_0	a reference wave number
k_R	grazing propagation constant over a rough surface
k_S	grazing propagation constant over a smooth surface
k_B	grazing propagation constant for boundary wave over a rough surface
$\mathbf{k}$	unit vector along coordinate z
k_{LW}	empirical constant for fish weight, g/cm^3
K	ocean surface wave number, radians/meter; compressibility, reciprocal of bulk elasticity, kg s^{-2} m^{-1}
$\mathcal{K}$	ocean bottom spatial cycle frequency, cycles/ meter; thermal conductivity of a gas; horizontal component of wave number in waveguide
$K_{v/p}$	transducer voltage to pressure conversion factor, volts/µPa
$K_{p/v}$	transducer pressure to voltage conversion factor, at a defined position, µPa/volt
l	dipole element separation

L	length of a cylinder, or a rectangular transducer face, or a path; folding depth, distance to decrease to e^{-1} of reference value; spatial correlation length, where subscripts indicate direction
L_{ec}	equivalent scattering length of a cylinder, subscripts ebc for equivalent bent cylinder; subscript x for projection on x axis and so on
$\mathcal{L}(\theta, \phi, f)$	acoustic scattering length or amplitude (spectral) in θ, ϕ direction, meters; subscripts give particular case – for example, $blad$ for bladder, bod for body, and K for Kirchhoff method
$\mathcal{L}_{bs}(f)$	acoustic backscattering length
$\mathcal{L}_{gs}$	acoustic scattering length for geometrical scatter
$\|\mathcal{L}_{bs}(L/\lambda)\|/L$	relative, or reduced, backscattering length
$\ln$	natural logrithm to base e
$\log, \log_{10}, \log(.)$	logarithm to base 10
m	lumped mass of an oscillating acoustical system; source mass injection/volume; statistical moment of a surface roughness, subscript indicates order; mode number in a waveguide; order of diffraction by a grating
M	number of modes; mass per unit area per unit time
$\mathcal{M}$	acoustical Mach number
n	an integer; the number of sources; index of refraction
$n(a)\,da$	number of bubbles of radius a, in increment da, per unit volume, m^{-3}
$\boldsymbol{n}$	vector normal to an area, dS; matrix vector for density components of scatterers, m^{-3}
n_b	number of scattering bodies per unit volume, m^{-3}; noise output voltage of an array
$n_n(x)$ or $y_n(x)$	spherical Bessel function of order n (A&S Chap. 10)
n_{cdf}	critical density of fish, number per unit volume, m^{-3}
N	correction for spherical wave reflection; number of elements of a transducer
N_m	bandpass-filtered noise voltage of the mth mode
N_{it}	number of independent trials
$N_n(x)$ or $Y_n(x)$	cylindrical Bessel function of order n (A&S Chap. 9)

N_{pings}	number of pings
$\mathcal{N}_{\text{em}}$	number of echoes in the mth amplitude bin per ping
p	instantaneous acoustic pressures, pascals or micropascals; subscripts give name – *inc* for incident, s or *scat* for scattered; p_d, acoustic pressure radiated by a dipole; p_0, instantaneous acoustic pressure at the reference distance R_0; p_A, ambient pressure, pascals or micropascals; p_T, total pressure; p_{ns}, backscattered pressure with narrow-beam sonar; p_{ws}, backscattered pressure with wide-beam sonar; p_δ, impulse of pressure, pascals
P	amplitude of acoustic pressures; P_0 at the reference distance R_0; P_{ax}, axial acoustic pressure amplitude; P_2, pressure amplitude of second harmonic
P_{RL}	Langevin radiation pressure, pascals
$P(f)$	spectral acoustic pressure, pascals/Hz
$P(z)$	polynomial generating function of acoustic pressure
[*path*]	acoustic pressure and time history, dimensionless
[*paf*]	path amplitude factor, dimensionless
Pa	pascal, unit of pressure: newton/m^2
$\mathcal{P}[x_n] = w(x_n)\Delta x$	probability of observing x_n in $x_n \pm \Delta x/2$, subscripts give specific functions
$\mathcal{P}_E(e_n)$	joint probability function
$\mathcal{P}(\Delta V_G)$	probability of body being in gated volume
PDF	probability density function
PSD	power spectral density, watts/Hz
q_m	mode excitation, pascals
q_n	normal coordinate in the wedge solution
Q	quality of a resonant system, $Q = f_c/\Delta f$ or $Q = 1/\delta$
Q_t	directivity factor, subscript t for transmitter, r for receiver, dimensionless
r	radial distance in polar or cylindrical coordinates, r, z, and ϕ; r_{uk}, unknown range; r_{tr}, trial range in matched filter; r_0, range source to diffracting wedge; r, range from diffracting wedge to receiver

R	range in spherical coordinates, R, θ, and ϕ; R_s, range source to scatterer; R_0, reference range, usually 1 meter; R_c, "critical range" for far-field of transducer
R_m	lumped mechanical resistance of an oscillating system, newtons/(m/s)
$\mathcal{R}(R)$	radial function solution of the wave equation
$\mathcal{R}_{12}$	plane wave pressure reflection coefficient at the 1–2 interface
$\mathcal{R}_{ss}$	coherent reflection coefficient in specular scatter
$\mathcal{R}_a$	apparent reflection coefficient
s	lumped stiffness of an oscillating system, newton/m
$s(t)$	sum of pressure or voltage signals of all receivers in a waveguide
s	rms slope of a surface; s_w, rms slope in windward direction; s_c, rms slope in the cross wind direction
$s_v(f)$	volume scattering coefficient, m²/m³
$\mathbf{s}$	matrix vector of volume scattering coefficients
S_{bs}	backscattering cross-section per unit volume, m⁻¹
S_v	volume backscattering strength in dB
S	transducer source response at a designated position, μPa/ampere
$S(\omega)$	waveguide source function, Hz⁻¹
$S(x)$	Fresnel sine integral
S	salinity, parts per thousand; cross-sectional area – for example, of a raytube
$\mathcal{S}$	surface scattering coefficient
SL	source level, dB
SPL	sound pressure level, dB
t	time, subscripts give particular times
t_0	time between samples
$t_n = nt_0$	the digital time of a sample
t_p	actual ping duration
t_d	effective ping duration
t_{dir}	direct travel time from source to receiver
τ_r	relaxation time of a molecular process
[tips]	time integral of pressure squared; subscript GV for the gated volume

[*ties*]	time integral of an echo squared (voltage^2 s or pressure^2 s)
[*trans* (*r, f*)]	transmission fraction, dimensionless
T	period of a periodic wave, s
T	temperature, degrees centigrade
$T(t)$	temporal function for propagation of acoustic pressure, dimensionless
TL	transmission level, dB
TC	transmission change of level, dB
T_{12}	plane wave pressure transmission coefficient at the 1–2 interface
TS	target strength, dB re 1 m^2
TS_{re}	"reduced" target strength, various references
TVG	time-varying gain
u, v, w	rectangular components of particle velocity in x, y, and z directions, m/s
u_R	radial particle velocity at range R
$u(r, z)$	function of range and depth in waveguide propagation
u	group velocity of ocean surface wave
u_{gm}	group velocity of mth waveguide mode
u_{gmm}	group velocity minimum of mth waveguide mode
U_a	amplitude of radial particle velocity at $R = a$
U	void fraction, ratio of bubble volume to total volume
$U(Q)$	scattered (spectral) acoustic field at a point Q
U_s	incident acoustic field at the surface, in scattering calculations
U_E	Laplace generating function
U	a scalar solution of the wave equation
$U(r)$	radial dependence of field in cylindrical co-ordinates, dimensionless
U_s	propagation speed of a shock wave
v	voltage at a transducer; v_n, voltage peak for narrow-beam receiver; v_w, voltage peak for wide-beam transducer
v	particle velocity in the y direction
v_{rm}	phase velocity of mth mode propagating in the r direction
v	surface wave phase velocity
$v(r)$	the range function in the parabolic equation
V_0	transducer input voltage amplitude

V_{AB}	a voltage during transducer calibration
V	volume
$\dot{V}$	rate of volume flow, m^3/s
Var	the statistical variance
$\boldsymbol{w}$	mass of explosive, kg
w	particle velocity in the z direction, m/s
$w(x_n)$	probability density function, PDF; $w(x_n)\Delta x$ is the probability of observing a value of x_n between $x_n - \Delta x/2$ and $x_n + \Delta x/2$; w_{Rayl}, Rayleigh probability density function; w_E, PDF of an echo; w_F, PDF of fish scattering length
W_f	fish mass
W	width, of a rectangular face
W_T, W_E, W_F	Laplace generating functions
$\mathcal{W}$	characteristic function for scatter from a two-dimensional rough surface
x, y, z	rectangular coordinates in X, Y, Z system
x_{crit}	critical (minimum) distance for far-field approximation
$x(n)$	the nth digital input amplitude
$x(t)$	filter input source signal at time t; a signal
X	a distance
$X(x)$	a function of coordinate x
$X(f)$	input amplitude spectral density
$X_{ffi}(m)$	the mth spectral component in an FFT
y	a rectangular coordinate
$y(t)$	filter output in time domain
$y_M(j)$	convolution of signals
$Y(y)$	a function of coordinate y
$Y_n(x)$	spherical Bessel function of order n ($A\&S$ Chap. 10)
$Y(f)$	output amplitude spectral density
Y	non-shock wave energy in an explosion, joules
Y	ratio, resonance frequency divided by insonification frequency
z	a rectangular coordinate; z_{uk}, unknown depth; z_{tr}, trial depth for matched filter
$Z(z)$	a function of coordinate z; $Z_m(z)$ eigenfunction of z dependence of mode m in cylindrical coordinates

z_i	source depth vector matrix eigenfunction for waveguide
z_j	receiver depth vector matrix eigenfunction for waveguide
α_e	Naperian logarithmic attenuation rate for plane waves, nepers/m
α	attenuation rate for plane waves, dB/m
α_b	plane wave logarithmic attenuation rate due to bubbles, dB/m
α	a constant in the Pierson–Moskovitz wind wave spectrum, dimensionless
α	function of angles in rough surface scatter
α	ratio, gate open duration/signal duration
β	ratio of bubble interior pressure to ambient pressure
β	a parameter in nonlinear propagation
β	a constant in the Pierson–Moskovitz wind wave spectrum, dimensionless
β	a function of angles in rough surface scatter
β_+	a parameter in wedge diffraction
γ	a function of angles in rough surface scatter
γ	vertical component of wave number in cylindrical waveguide propagation
γ_m	eigenvalue of mth mode in z direction
γ	ratio of specific heats of a gas; $\gamma = 1.4$ for air, dimensionless
Γ	exponential power in the adiabatic relation
δ	an increment
δ	an empirical damping constant for dissipation in wedge diffraction
δ	total damping constant of a bubble, dimensionless
δ_r	bubble damping constant due to reradiation (scattering), dimensionless
δ_t	bubble damping constant due to thermal conductivity, dimensionless
δ_v	bubble damping constant due to shear viscosity, dimensionless
δ_R	total bubble damping constant at resonance, dimensionless
δ_{Rr}	bubble damping constant due to reradiation at resonance, dimensionless
δ_{Rt}	bubble damping constant due to thermal conductivity at resonance, dimensionless

δ_{Rv}	bubble damping constant due to shear viscosity at resonance, dimensionless
$\delta(t - t_n)$	Dirac delta function, everywhere 0 except infinite at $t = t_n$
$\delta(n - m)$	Kronecker delta function; value unity when $m = n$; otherwise zero
δ_f	finite, discrete delta function
δ_m	attenuation rate of mth mode
Δ	a finite increment, as in Δf for a narrow frequency band; ΔR for a range difference; Δz for virtual surface displacement; ΔE for energy passing through an incremental area ΔS; $\Delta \sigma_s$ for differential scattering cross-section; Δt for digital duration of a delta function; Δt, duration of impulse source; Δt_{ww}, duration of "water wave" in a waveguide
ΔT	time between impulses
Δf	effective frequency bandwidth between half-power or half-intensity points
$\Delta = kL \sin \chi$	a phase shift in scattering from cylinders
∇	gradient operator
$\nabla \cdot$	divergence operator
∇^2	Laplacian operator
ε	a small quantity, as in length of sagitta of the arc
ε	roughness parameter for boundary wave
ε	angular displacement from diffraction shadow boundary
ε_R	energy per unit area at range R
ζ	surface displacement
η	a phase shift
η	a function of time and distance in wedge diffraction
η	angle of incidence at a cylinder
θ	polar coordinate in spherical coordinates; angle with the axis of radiation; angle with the normal to the surface in plane wave reflections; angle between diffracting wedge face and receiver; angle between sound beam axis and wind wave system direction
θ_0	angle between diffracting wedge surface and source

θ_w	wedge angle; angle in fluid between sides of a diffracting wedge
θ_i	angle of incidence at a scatterer
θ_s	angle of scatter at a body
θ_c	critical angle for total reflection
$\mathcal{K}$	thermal conductivity of a gas, cal/(cm s °C)
$\mathcal{K}$	wave number of an ocean wave component, subscripts designate direction
λ	acoustic wavelength
λ_a	apparent acoustic wavelength in doppler shift
λ_s	spatial distance between samples
Λ	ocean wavelength
Λ	waveguide mode interference wavelength
μ	dynamic coefficient of shear viscosity, pascal s
μ_b	dynamic coefficient of bulk viscosity, pascal s
μ_m	waveguide parameter that includes unknown source power, depth, and range
v	a parameter in the wedge diffraction solution
v_m	a quantity proportional to the energy flux in mode m
ξ	a parameter in the wedge diffraction solution
Π	power passing through an area, watts
Π_{gs}	geometrically scattered power, watts
Π_M	source message power, watts
$\Pi_{xx}(f_m)$	spectral density (1 Hz band) of a source at frequency f_m, volt^2/Hz, or pascal^2/Hz
ρ	acoustic density, kg/m^3; subscript numbers identify the medium
ρ_A	ambient density of the medium, kg/m^3
ρ_T	total density of the medium, kg/m^3
σ_e	total extinction cross-section of a body, m^2
σ_s	total scattering cross-section of a body, m^2
σ_a	total absorption cross-section of a body, m^2
$\sigma_{bs}(f)$	obsolete form of backscattering (differential) cross-section (Clay and Medwin, 1977); see $\Delta\sigma_{bs}(f)$
$\sigma_c(f)$	concentrated (mean) component of the backscattering (differential) cross-section
$\sigma_d(f)$	distributed (variable) component of the backscattering (differential) cross-section

τ	a delay time; a decay time; surface tension; travel time excess beyond the least time in wedge diffraction
τ_0	travel time from source to wedge crest to receiver
τ_{12}	pressure transmission coefficient at 1–2 interface
ϕ	azimuthal spherical coordinate
Φ	phase angle; phase shift
$\Phi(\Omega)$	ocean wave spectral density, m²/Hz
χ	tilt angle of a cylinder
ψ	empirical phase shift or amplitude adjustment; spatial part of the solution to the wave equation in cylindrical coordinates
Ψ_D	integrated beam pattern
ω	angular frequency, radians/s; subscript number is harmonic number; subscript letter is descriptive
Ω	element of solid angle
Ω	angular frequency of a time-varying ocean surface displacement; Ω_m is the frequency of the maximum

Index

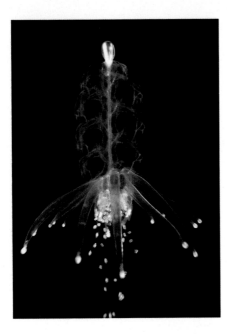

Fig. 7.7b. A siphonophore at sea. Note the ellipsoidal flotation gas bubble, of major axis approximately 6.5 mm. The total length of this particular specimen was about 8 cm; it was captured at 150 m depth. Siphonophores have been seen with lengths up to 40 meters and at depths greater than 3000 m. Siphonophore bubbles may be detected, and non-destructively measured at a distance, by their acoustical resonance. (Photo courtesy of Dr. S. Haddock, MBARI.)

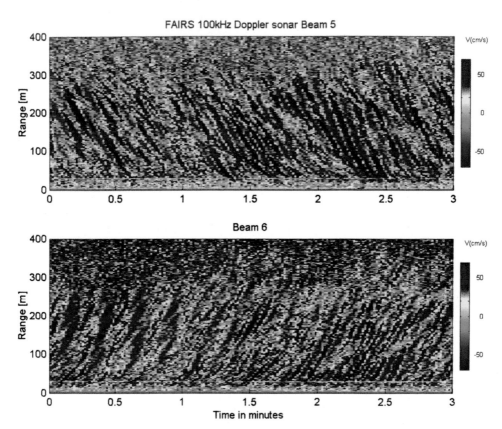

Fig. 10.10(a). (Farmer) Time series example of two Doppler. Sidescan images of the ocean surface acquired with two orthogonally directed 100 kHz narrow beam sonars. The time series shown the orbital velocities of waves propagating away from (upper image) and towards (lower image) the instrument.

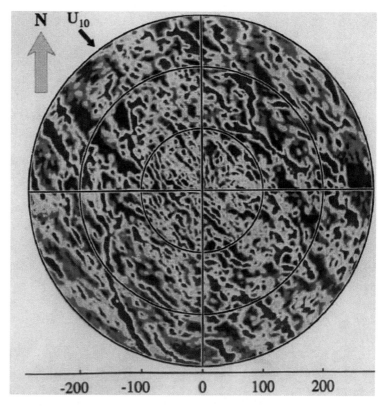

Fig. 10.11. (Farmer) A polar view of backscatter from the sea surface acquired with a combination of four rotating sonars. The image is acquired in about 30 s. Wind direction is from the northwest (U_{10}). Red indicates higher backscatter intensity corresponding is denser bubble distributions.

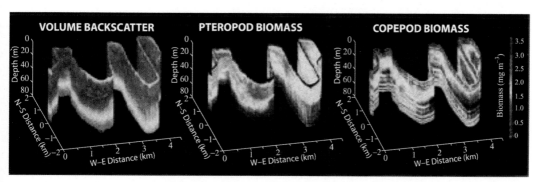

Fig. 12.7. Volume scattering measurements at 420 kHz were made with a predecessor of BIOMAPER II and used to estimate the spatial distributions of pteropod and Copepod biomass along a transect on Georges Bank, off the eastern coast of the State of Massachusetts, USA. Echogram of volume scattering strength (left) used in combination with *in situ* video images of the organisms to estimate distribution of two types of zooplankton (middle and right). From Wiebe *et al.* (1997).

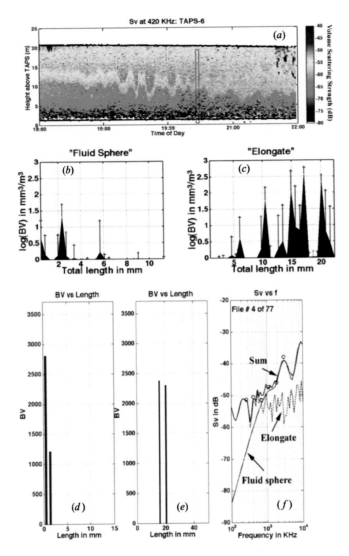

Fig. 12.8. A small gyre often forms in the northeastern part of Monterey Bay, CA during the late summer and fall. As the water at the center of this gyre warms, the water column stratifies, phytoplankton grows within the gyre, and zooplankton responds to an increased availability of food. The 420 kHz data displayed in panel (*a*) were collected at 80-second intervals with an upward-looking, bottom-mounted TAPS. The sensor was positioned within this gyre with the specific intent of examining the zooplankton response to changes in phytoplankton abundance in the upper mixed layer and near the pycnocline. A layer of scatterers was observed during 5 h period near the depth of the thermocline. The pycnocline supported a packet of internal waves at the boundary between the warm water in the upper mixed layer and the underlying water mass. During this particular observation period the thermocline deepened and the layer became more diffuse as the tide receded and evening approached. Panels (*b*)–(*h*) illustrate the results of inverse processing for the time interval and depths included in the red box of panel (*a*). These subplots and the analyses that led to each are discussed in the text.

Fluid Sphere Shapes

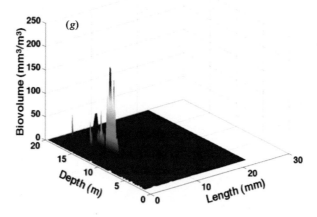

Elongate Shapes

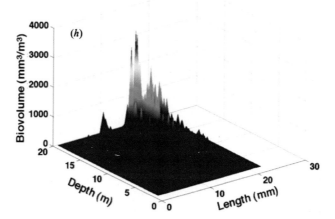

Fig. 12.8. (*Continued*)

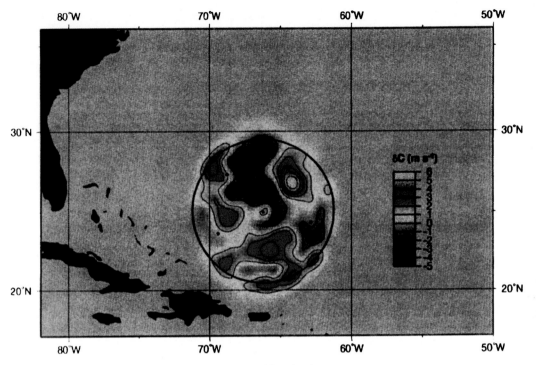

Fig. 18.4. This map of sound speed perturbations at 700 m depth was obtained during the 1991 Acoustic Mid-Ocean Dynamics Experiment. Five transceivers were moored for a year in a pentagonal array East of Florida, and a sixth was placed at the center, yielding a total of 15 paths. Twice during the year a ship circumnavigated the array stopping every 25 km to receive signals from the six moored instruments. During the period of ship circumnavigation, about 750 paths were available, thus providing much increased horizontal resolution over the rest of the year. The map shown was created from data collected during one of the periods when the additional ship data was available (AMODE-MST Group, 1994).

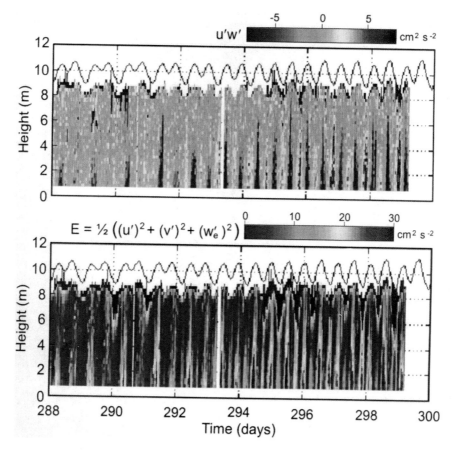

Fig. 20.2. (Di Iorio and Gargett) Turbulent shear stress (upper panel) and turbulent kinetic energy (lower panel) determined over a period of several days by a bottom-mounted ADCP in a shallow tidal channel off San Francisco Bay (data courtesy of M. Stacey and S. Monismith).

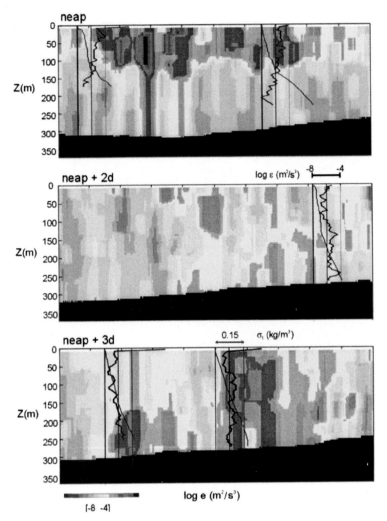

Fig. 20.3. (Di Iorio and Gargett) Color coded fields of the logarithm of turbulent kinetic energy dissipation rate ε, determined from the vertical velocity field measured by the vertical beam of a VADCP crossing a tidal front in the coastal waters of British Columbia, as tides increase from neap (top) towards spring. Superimposed profiles of ε (heavy lines) are used to calibrate the acoustic technique, while those of density (σ_t, light lines) show the effect of turbulence in eroding the stratification of the water column.

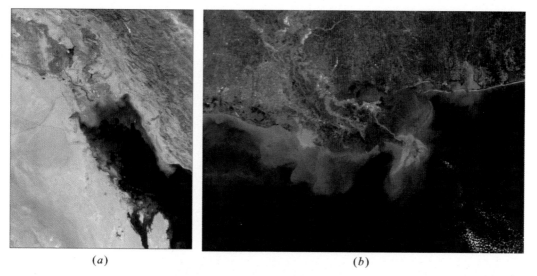

(a)   (b)

Fig. 21.6. (Leighton and Heald) True-color satellite images (from the Moderate Resolution Imaging Spectroradiometer MODIS carried by NASA) of sediment carried by rivers into the sea. (a) As a result of flooding by the confluence of the Tigris and Euphrates Rivers (at center), the sediment-laden waters of the Persian Gulf (November 1, 2001) appear light brown where they enter the northern end of the Persian Gulf and then gradually dissipate into turquoise swirls as they drift southward. (Image courtesy Jacques Descloitres, MODIS Land Rapid Response Team at NASA GSFC.) (b) The Mississippi River carries roughly 550 million metric tonnes of sediment into the Gulf of Mexico each year. Here (March 5, 2001 at 10:55 am local time) the murky brown water of the Mississippi mixes with the dark blue water of the Gulf two days after a rainstorm. The river brings enough sediment from its 3 250 000 square km (1 250 000 square miles) basin to extend the coast of Louisiana 91 m (300 ft) each year. (Image courtesy Liam Gumley, Space Science and Engineering Center, University of Wisconsin-Madison and the MODIS science team.) Source: NMASA, reproduced with the permission of the Lunar and Planetary Institute.

Fig. 21.14. (Leighton and Heald) The percentage error between the actual natural frequency of a bubble and the radiation damping constant, respectively, if the free-field formulations of Minnaert and Devin are used in the 15° wedge shown in Fig. 21.13. The result is shown for three bubble depths, for a bubble situation 1 m from the water's edge (where the water depth is 26 cm). (Figure by T. G. Leighton and P. R. White.)

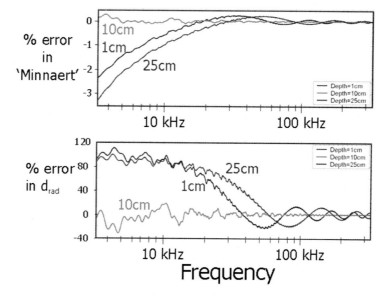

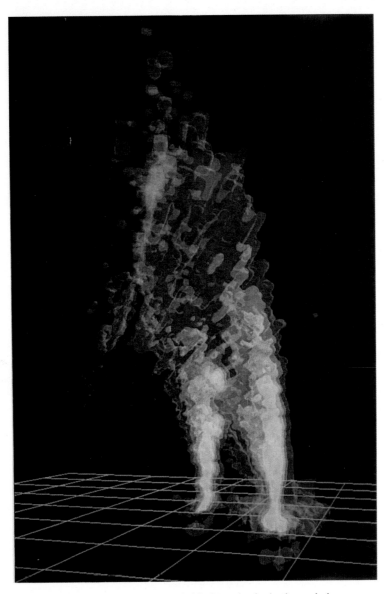

Fig. 22.2. (Palmer and Rona) Buoyant black smoker hydrothermal plumes discharging from adjacent chimneys reconstructed in 3D from our volume backscattering data at latitude 21° N on the East Pacific Rise (Rona *et al.*, 1991). The rectilinear grid on the sea floor is 5 m by 5 m by 5 m (*x, y, z*).

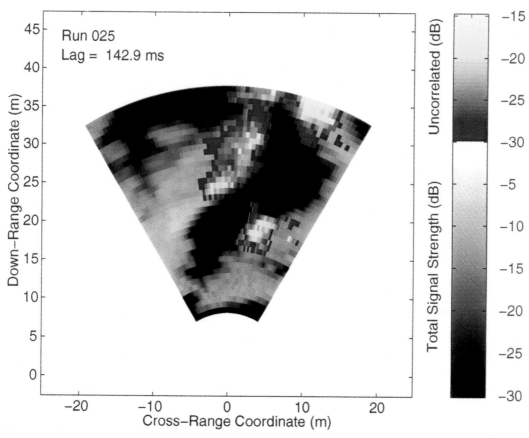

Fig. 22.3. (Palmer and Rona) Mapping of diffuse flow on the Juan de fuca Ridge using acoustical correlation techniques (Rona *et al.*, 1997). The brightly colored patches are areas of decorrelation corresponding to diffuse flow.

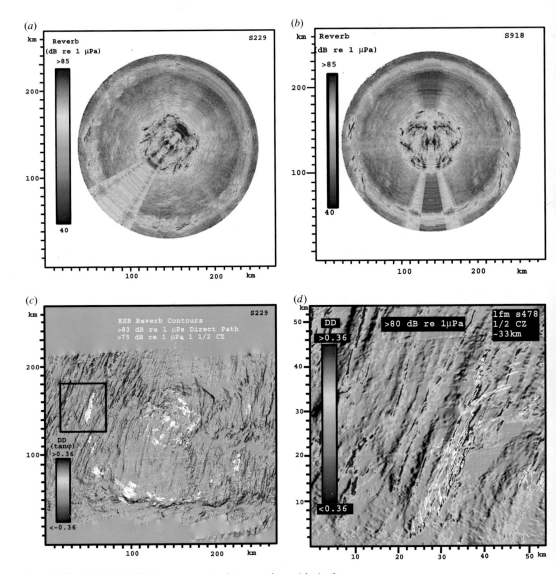

Fig. 23.7. (Makris) (a). Wide area acoustical image taken with single transmission from Cory at center of eastern star. (b) Same as (a) but with a different array orientation to break ambiguity about receiver array axis. (c) Prominent reverberation in white overlain on color directional derivative of bathymetry in the direction of monostatic source-receiver towed from Cory. Red indicates steep slopes facing source-receiver, blue indicates steep slopes facing away from source-receiver. (d) Zoom in of white reverberation overlain on color directional derivative at B' seamount over box shown in (c) but for a different transmission.

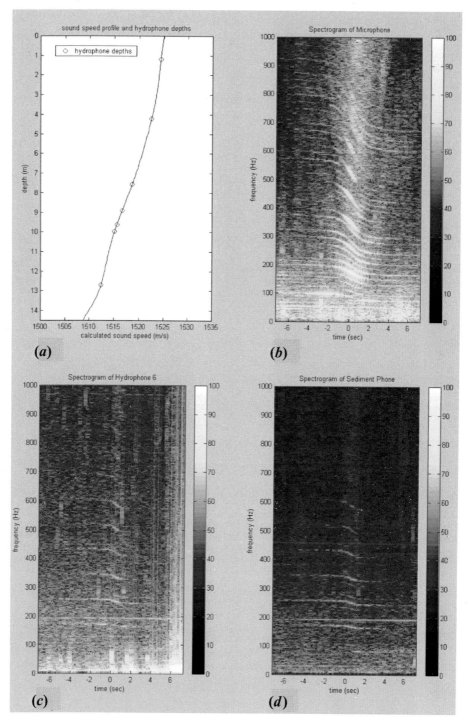

Fig. 24.5. (Buckingham) (*a*) Average sound speed profile from SeaBird, with small circles depicting depths of hydrophones in water column. (*b*)–(*d*) Calibrated spectrograms from flight of Tobago at altitude of 66 m on July 2, 2002 showing Doppler shifted propeller harmonics (all color bars: dB re 1 μPa^2/Hz): (*b*) microphone 1 m above sea surface; (*c*) hydrophone at depth of 10 m in water column; and (*d*) hydrophone buried 75 cm deep in sediment.